# Recent Advances in Mechanisms of Fracture and Fatigue

# Recent Advances in Mechanisms of Fracture and Fatigue

Editors

**Jaroslav Pokluda**
**Reinhard Pippan**

MDPI • Basel • Beijing • Wuhan • Barcelona • Belgrade • Manchester • Tokyo • Cluj • Tianjin

*Editors*
Jaroslav Pokluda
Brno University of Technology
Brno, Czech Republic

Reinhard Pippan
Erich Schmid Institute of Materials Science
Leoben, Austria

*Editorial Office*
MDPI
St. Alban-Anlage 66
4052 Basel, Switzerland

This is a reprint of articles from the Special Issue published online in the open access journal *Materials* (ISSN 1996-1944) (available at: https://www.mdpi.com/journal/materials/special_issues/mechanisms_fracture_fatigue).

For citation purposes, cite each article independently as indicated on the article page online and as indicated below:

LastName, A.A.; LastName, B.B.; LastName, C.C. Article Title. *Journal Name* **Year**, *Volume Number*, Page Range.

**ISBN 978-3-0365-7232-1 (Hbk)**
**ISBN 978-3-0365-7233-8 (PDF)**

# Contents

*materials*

MDPI

*Article*

# Inverse Method to Determine Fatigue Properties of Materials by Combining Cyclic Indentation and Numerical Simulation

**Hafiz Muhammad Sajjad [1,*], Hamad ul Hassan [1], Matthias Kuntz [2], Benjamin J. Schäfer [1,2], Petra Sonnweber-Ribic [2] and Alexander Hartmaier [1]**

1 Interdisciplinary Centre for Advanced Material Simulation (ICAMS), Ruhr-Universität Bochum, Universitätsstr 150, 44801 Bochum, Germany; hamad.ulhassan@rub.de (H.u.H.); BenjaminJosef.Schaefer@de.bosch.com (B.J.S.); alexander.hartmaier@icams.rub.de (A.H.)

2 Robert Bosch GmbH—Corporate Sector Research and Advance Engineering, 71272 Renningen, Germany; Matthias.Kuntz2@de.bosch.com (M.K.); petra.sonnweber-ribic@de.bosch.com (P.S.-R.)

* Correspondence: Hafiz.sajjad@rub.de; Tel.: +49-234-32-22-443

Received: 3 June 2020; Accepted: 8 July 2020; Published: 13 July 2020

**Abstract:** The application of instrumented indentation to assess material properties like Young's modulus and microhardness has become a standard method. In recent developments, indentation experiments and simulations have been combined to inverse methods, from which further material parameters such as yield strength, work hardening rate, and tensile strength can be determined. In this work, an inverse method is introduced by which material parameters for cyclic plasticity, i.e., kinematic hardening parameters, can be determined. To accomplish this, cyclic Vickers indentation experiments are combined with finite element simulations of the indentation with unknown material properties, which are then determined by inverse analysis. To validate the proposed method, these parameters are subsequently applied to predict the uniaxial stress–strain response of a material with success. The method has been validated successfully for a quenched and tempered martensitic steel and for technically pure copper, where an excellent agreement between measured and predicted cyclic stress–strain curves has been achieved. Hence, the proposed inverse method based on cyclic nanoindentation, as a quasi-nondestructive method, could complement or even substitute the resource-intensive conventional fatigue testing in the future for some applications.

**Keywords:** cyclic indentation; Vickers hardness; inverse analysis; numerical simulations; cyclic material properties; fatigue life

## 1. Introduction

Depth-sensing indentations or instrumented indentations are very useful means to characterize and determine mechanical properties (i.e., Young's modulus and hardness) of thin films as well as of bulk materials [1–5]. Hyung [6] and Suresch et al. [7,8] have proposed two novel methods to identify the elastic modulus, yield strength, and the hardening exponent through nano-indentation. In addition, instrumented indentation experiments make it possible to determine further material properties, such as the strain hardening coefficient and yield strength [9]. Schmaling and Hartmaier [10] have introduced a method to identify plastic material properties (i.e., yield strength and work hardening rate) by using an inverse analysis for the remaining imprint after indentation. A comprehensive comparison of the hardness measurement approaches at diverse scales (i.e., nano, micro, and macro) of Brinell, Vickers, Meyer, Rockwell, Shore, IHRD, Knoop, and Buchholz was performed by Broitman [11]. He has not only described each indentation method but has also presented its inadequacies in evaluating results. Furthermore, he has discussed the effects of elasticity, plasticity, pileup, sink-in, grain size, and indentation size on determining hardness by means of depth-sensing indentation techniques at the micro- and nanoscale.

The applications of instrumented nanoindentation are not limited to the identification of conventional material properties, such as elastic modulus and hardness; the ease of performing this method has also attracted the attention of some authors who have employed it to estimate complex material properties like fatigue life or even the famous S-N (Wöhler) curve [12–14]. The determination of these quantities requires tedious and difficult conventional fatigue experiments; consequently, there are some analytical relationships available that allow the determination of the S-N curve and fatigue strength by using indentation hardness. For example, the Strzelecki model [12] has described the relationship between fatigue life and material hardness. Similarly, by utilizing the Murakami [13] formulation, Bandara [14] has presented full range S-N curves for six different medium steel grades. However, this suggested formulation is material-specific and requires two input parameters, i.e., Vickers hardness and ultimate tensile strength. Furthermore, Bandara [15] has used the Brinell hardness value in order to determine material fatigue strength. The three [12,14,15] aforementioned analytical approaches (mainly dependent on the hardness value of the material) have been comprehensively summarized by Strzelecki and Tomaszewski in [16], where they have expounded the merits and demerits of each individual approach and concluded that both models are good for predicting tensile strength and yield strength, while the Strzelecki model [12] is superior in the prediction of fatigue life as compared to the Bandara model [14,15].

In classical indentation experiments, only single loading and unloading on the specimen is performed for determining the desired mechanical properties. However, to study the fatigue life in materials, the cyclic indentation has attracted the attention of many authors [17–20]. For example, Lyamkin et al. [17] have shown the potential of cyclic indentation for studying the fatigue properties. As the cycles of indentations are repeated at the same location, the elastoplastic area of high cyclic indentation demonstrates fatigue under these conditions. For instance, a nanoimpact fatigue test has been studied by Faisal et al. [18] by using cyclic indentation, and they have concluded that the indenter geometrical shape (Berkovich or conical) and the indentation loading history are crucial in inducing film failure. Moreover, Haghshenas et al. [19] have employed cyclic nanoindentation in order to determine the indentation size effects and the strain rate sensitivity for tantalum. Similarly, Prakash [20] has demonstrated fatigue damage in materials with the help of two different nondestructive techniques, of which one is spherical cyclic indentation. In addition, he has investigated fatigue properties with a cyclic small punch test and with cyclic automated ball indentation and has concluded that the stiffness in the weld region drops more quickly in comparison to the base metal [21]. Xu et al. [22] have performed the numerical analyses of flat cylindrical indentation for polycrystalline copper. They have concluded that strain accumulation reached a steady-state indentation depth rate in sinusoidal cyclic loading just like in real fatigue experiments.

From this literature review, it is concluded that using instrumented indentation for determining the fatigue life of materials would reveal significant advantages of indentation testing (quasi-nondestructive technique, less time- and cost-intensive) over conventional fatigue experiments. Some analytical relations are indeed available to predict the S-N curve for fatigue life based on the hardness value, but these methods are mostly limited to steels and not applicable to softer materials, for instance, copper. Furthermore, it is not possible to analyze the microstructural influences. Hence, there is a need for a more general approach that can be applied to a larger range of materials. In addition, a method that uses the indentation experimental data for predicting complete uniaxial stress–strain hysteresis is still missing in the literature according to the authors' knowledge. Yet, this uniaxial stress–strain hysteresis has a key role in fatigue life determination. Although some attempts have already been made, as discussed earlier, the literature still lacks a hybrid method that predicts the material fatigue life and the complete uniaxial stress–strain hysteresis with a combination of the numerical analysis and Vickers indentation.

## 2. Material and Experiment

### 2.1. Materials Specifications

In this study, we used a conventional quenched and tempered martensitic high-strength steel SAE 4150 (German denomination 50CrMo4, Judenburg GmbH, Judenburg, Austria) that exhibits a hardness of 38 HRC. The cyclic properties of this material were analyzed in detail by Schäfer et al. [23]. The second material used in this study, technically pure copper, is a conventional Cu-ETP R290 (Electrolytic Tough-Pitch, Wieland-Werke AG, Ulm, Germany).

### 2.2. Indentation and Fatigue Testing

Indentation testing was performed by using a small Zwick load indenter and was adapted from Kramer et al. [24]. The test was performed in a cyclic fashion; therefore, the indenter tip was repeatedly indented at the same spot. To hold the sample in the testing position, a constant minimal load remained between the cycles. A minimum load of 2 N and a maximum load of 50 N were used for martensitic steel. Further experiments were also performed for higher force amplitudes, i.e., at 75 and 100 N, and for higher hardness, such as 47 HRC at 50 N. In addition, a cyclic indentation test was performed for Cu with a maximum load of 10 N and a minimum load of 1 N. The load increased at a rate of 5 N/s. As an example of indentation testing, only the first complete cycle for 50CrMo4 and Cu is demonstrated in Figure 1a,c. Fatigue experimental data of uniaxial cyclic stress–strain curves were obtained from the study of Schäfer et al. [23]. Similarly, stress–strain hystereses are shown for the aforementioned two materials (50CrMo4 and Cu) in Figure 1b,d.

The complete cycle of the indentation test for martensitic steel (50CrMo4) is demonstrated in Figure 2 with three different colors. The first (blue) part of the curve indicates the loading, the green section displays the unloading, and the red portion demonstrates the reloading for the force–displacement. It can be seen that the unloading and reloading curves make a closed loop, which is named as the force–displacement loop (FD loop), and the intersection point is named ΔF. This loop will be used as a target curve for the parameter identification in the following section.

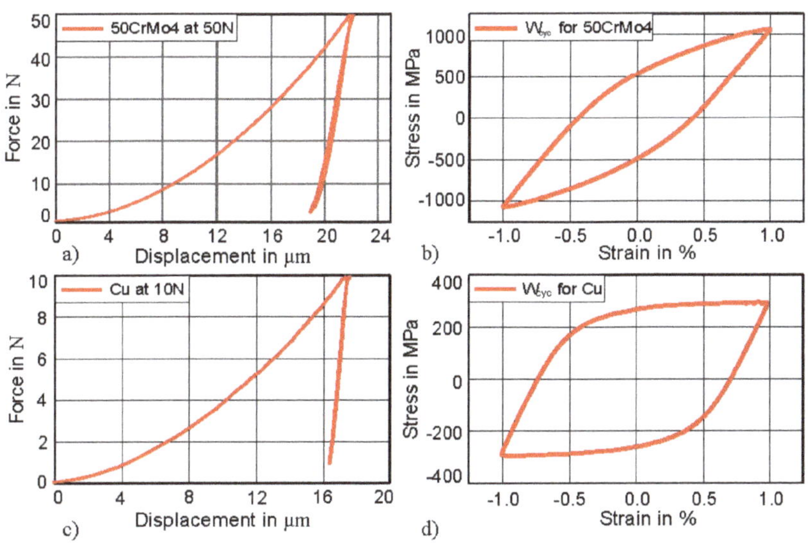


**Figure 1.** Experimental results for Cu and 50CrMo4: (**a**) indentation cycle at 50 N for 50CrMo4; (**b**) stress–strain from fatigue experiments of 50CrMo4; (**c**) indentation cycle at 10 N for Cu (**d**) Stress–strain from fatigue experiments of Cu.

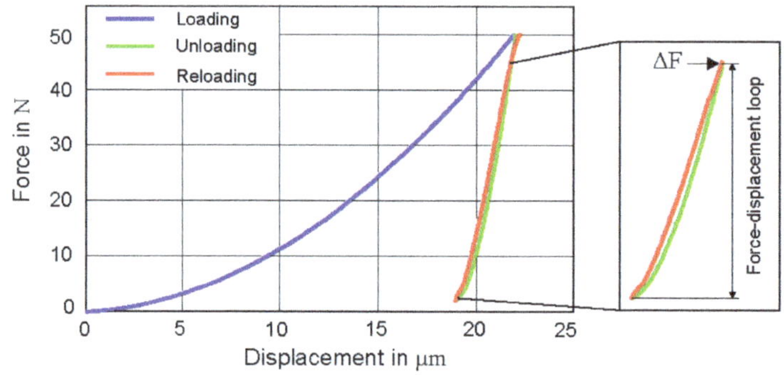


**Figure 2.** Force–displacement curve from the cyclic indentation curve. Components of the cycle are shown in different colors for clarity.

## 2.3. Numerical Models

The finite element model used in the investigation is depicted in Figure 3a. The indenter was modeled as a rigid body. The specimen was fixed from the bottom, and a vertical load was applied at the center by using a Vickers indenter. The square pyramid had an opposite face angle of 136° (DIN EN ISO 6507-2: 2005) [25]. For example, the applied force amplitude of 50 N in the simulation is explained as follows: The specimen was indented with the preselected applied force amplitude (i.e., 50 N), which was unloaded until the force of approximately 2 N and reloaded to the maximum force. Similarly, the simulations were run at other indentation force amplitudes of 25, 75, and 100 N. The force–time history from the experiments was used as an input for the simulations so that the loading in experiments could be fully depicted in the simulations.

It is our goal to identify the material parameters, which requires many simulations to be performed and is computationally very costly. Therefore, the size of the simulation model (2 mm × 2 mm and extruded to 1.5 mm) was optimized with a mesh size of 4 μm (Figure 3b); after the mesh convergence study, C3D8 linear elements with a full integration scheme were chosen for this paper. The friction effect between the indenter and the specimen was also studied, and it did not show a considerable effect on the simulation results in the scope of this work.

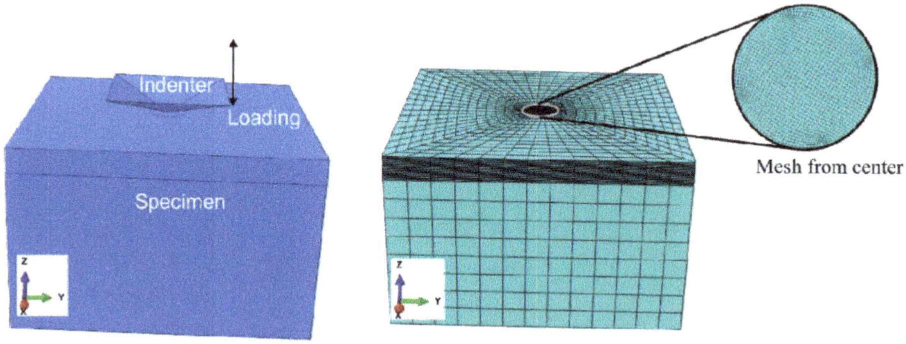


(**a**) Geometry with boundary conditions (**b**) Meshed specimen

**Figure 3.** Details of the numerical model used. (**a**) Specimen is held fixed from the bottom and the indenter is placed in the center of the specimen that moves in and out during loading and unloading, respectively. (**b**) The fine meshing is performed at the center of the specimen.

### 2.4. Material Model

In this investigation, the Chaboche material model was used, because it is one of the most efficient and convenient constitutive models incorporating the cyclic plasticity behavior of materials during cyclic loading. Furthermore, the ratcheting behavior of the material, which rises under cyclic loading, can also be analyzed with it. Details of this model for J2 plasticity are given in the subsequent section.

According to the von Mises yielding criterion [26], the yielding of material starts once the second deviatoric stress invariant J2 reaches a critical value. The common formula of this yielding criterion f is as follows:

$$f = \sqrt{\frac{3}{2}(\mathrm{S}-\alpha) : (\mathrm{S}-\alpha)} - (\sigma_0 + \mathrm{R}) \tag{1}$$

where $\sigma_0$, $\alpha$, and $S$ represent the initial yield stress, backstress, and deviatoric stress, respectively, and R represents the isotropic hardening (i.e., constant growth of yield surface) [27]. The increase in plastic strain with respect to the gradient of the yield surface will lead to the definition of the associative flow rule used in this study.

$$R = Q\left(1 - e^{-b\varepsilon_{eq}}\right) \tag{2}$$

where $\varepsilon_{eq}$ is the equivalent plastic strain, $b$ determines the rate of isotropic hardening, and $Q$ is the maximum change in the size of the yield surface [28,29].

In order to model the cyclic behavior of materials, a nonlinear kinematic hardening model was proposed by Armstrong and Frederick [30]. This kinematic hardening model contains only one backstress term ($\alpha$) and was extended by Chaboche by decomposing the single backstress term into several backstress terms, making the Chaboche material model [28] capable of capturing the complex kinematic hardening behavior.

The decomposed backstress terms of Chaboche kinematic hardening [28] model are described in the following equation:

$$\alpha = \sum_{i}^{n} \alpha_i;\ d\alpha_i = \frac{2}{3}C_i d\varepsilon_p - \gamma_i \alpha_i d\varepsilon_{eq}, \tag{3}$$

where $\gamma_i$ describes the reduction rate of the related modulus with respect to the plastic strain $d\varepsilon_p$, while $C_i$ represents the kinematic hardening moduli. The change in the yield surface of the combined hardening evolution for monotonic tension and in the stress space is graphically presented in Figure 4 [27].

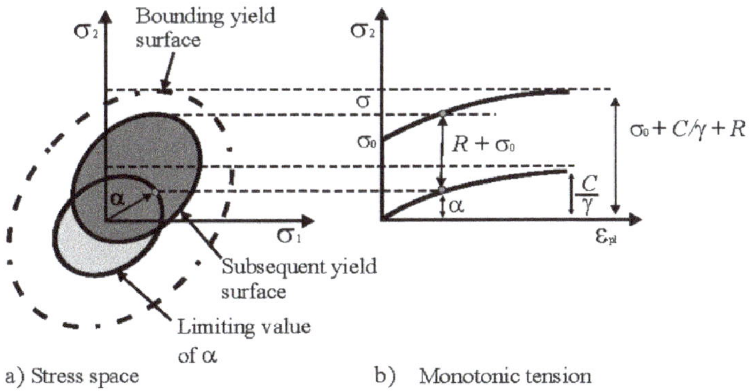


**Figure 4.** Graphical depiction of the combined hardening growth in the (**a**) stress space shown by the yield surface and (**b**) under monotonic tension presented as a stress–strain diagram, redrawn from [27] under the CC-BY license.

In the present study, three backstress terms, which comprise six unknowns, and isotropic softening with two unknown parameters are used initially, leading to a total of eight unknowns which are identified by using an inverse modeling technique. Furthermore, an effort has been made to use only two backstress terms, which reduces the unknown terms to six with the almost same quality of results in our case [31]. Therefore, in this study, only the results of two backstress terms are shown.

### 3. Inverse Parameter Identification

In order to capture the experimental force–displacement loop by simulation, the commercially available LS-Opt optimizer (DYNAmore GmbH, Stuttgart, Germany) [32] was used. The inverse identification technique was applied, where the difference between experimental and simulated values of the force–displacement loop, shown in Figure 2, was minimized by varying the material parameters in an iterative procedure, as shown schematically in Figure 5. The quality of fit between the simulated and the experimental force–displacement loops was evaluated by using the normalized mean square error (NMSE):

$$\text{NMSE} = \frac{1}{N}\sum_{i}\frac{(E_i - S_i)^2}{\overline{E}\overline{S}}, \tag{4}$$

an objective function, where $\overline{E}$ and $\overline{S}$ represent the averages of the experimental values $E_i$ and simulation values $S_i$, respectively, for the displacement at the same force, and $N$ is the total number of data points ($N = 75$).

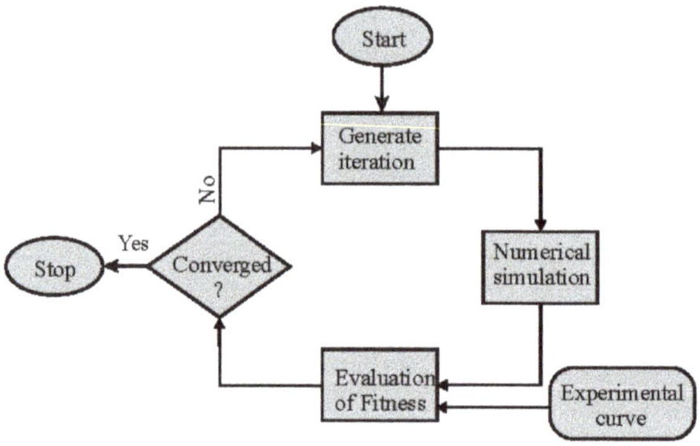


**Figure 5.** Optimization loop used to identify the material parameters by fitting the experimental curve with the simulation curve.

Hence, the force–displacement loop resulting from the partial unloading and reloading during indentation was used as the target for the optimization. For determining the material parameters that meet the given objective (i.e., that minimize the value of NMSE), a genetic algorithm [33] was used because it does not require a good initial guess for the target parameters. This algorithm generates the first iteration, which contains certain material parameter sets (one set contains four parameters for kinematic hardening and two for isotropic hardening) for the identification process. The force–displacement loop from the simulation was taken out using a postprocessing script written in Python 3. Based on the fitness results obtained at the end of the first iteration, the algorithm updates the material parameters in the subsequent iterations as the optimization loop continues. In each iteration, the algorithm calculates the fitness of the obtained force–displacement loop with the experimental force–displacement loop by using NMSE. This optimization loop continues until the convergence criterion (i.e., NMSE = $3 \times 10^{-5}$) is met or the maximum allowed iterations are reached. The yield stress and Young's modulus are kept constant at 1060 MPa and 204 GPa, respectively, based on monotonic stress–strain experimental data. It is known that the yield strength cannot be uniquely determined only based on force–displacement curves with sharp indenters. Hence, this material parameter must be assumed as known and can be determined by other methods, e.g., tensile tests or other inverse methods based on indentation (e.g., see [1–11]).

Two kinds of optimization procedures were studied in this work: "objective function 1" includes a free optimization of the objective function defined in Equation (4) until the convergence criterion is reached; "objective function 2" is also based on the objective function of Equation (4), but the minimization occurs under the side condition that the height of the force–displacement loop is restricted to the experimentally found value, i.e., $\Delta F_{sim} = \Delta F_{exp}$.

## 4. Results and Discussion

### *4.1. Method Development*

The experimental force–displacement loop of the first indentation cycle is used as a target curve along with ΔF. The material parameters (see Table 1) obtained after the optimization with this strategy show a good agreement for the complete indentation cycle: The normalized mean square error (NMSE) between the simulated force–displacement loop and the experimental force–displacement loop is $2.0 \times 10^{-5}$.

To achieve a comparison with experimentally determined hysteresis from fatigue tests, the hysteresis under a tensile–compressive load is predicted in the next step by using the parameters from Table 1. Comparing this prediction to the experimental results reveals, with a plastic work error of 2.5%, a quite good accuracy, as can be seen in Figure 6. In the scope of this study, the identified material parameters of cyclic macroindentation are used to predict the complete uniaxial stress–strain hysteresis for the first time.

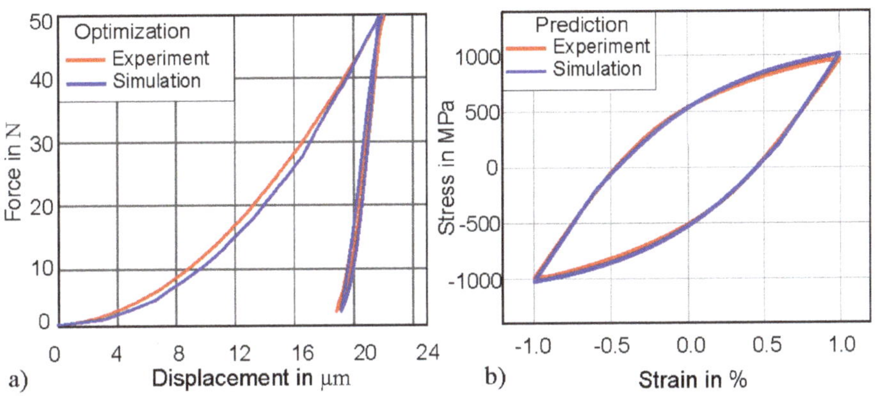


**Figure 6.** (**a**) Complete cycle of the force–displacement curve from indentation, with normalized mean square error (NMSE) = $2.0 \times 10^{-5}$. (**b**) Predicted uniaxial stress–strain hysteresis with a plastic work error of 2.5%.

There is a clear relationship between the ΔF value measured by cyclic indentation and the uniaxial stress–strain hysteresis. This relationship is qualitatively investigated in this study. In Figure 7, the results from the two objective functions can be seen.

By using the "objective function 1", the comparison of the simulated and experimental force–displacement loops seems to be in an acceptable range of agreement (see Figure 7, solid blue FD loop), with NMSE $3.0 \times 10^{-5}$; nevertheless, the value of ΔF from the simulation is lower than the experimental ΔF, which has a direct impact on the uniaxial stress–strain hysteresis prediction. The inclusion of the ΔF into "objective function 2" leads to a better prediction of the uniaxial stress–strain hysteresis, which is shown by the dotted blue line hysteresis in Figure 7.

**Table 1.** Identified material parameters for 50CrMo4 (38 HRC) after fitting of force–displacement at 50 N.

| Symbol | Value |
|---|---|
| $C_1$ (MPa) | 262,197 |
| $\gamma_1$ | 373 |
| $C_2$ (MPa) | 4714 |
| $\gamma_2$ | 0.25 |
| Q (MPa) | −575 |
| b | 262 |

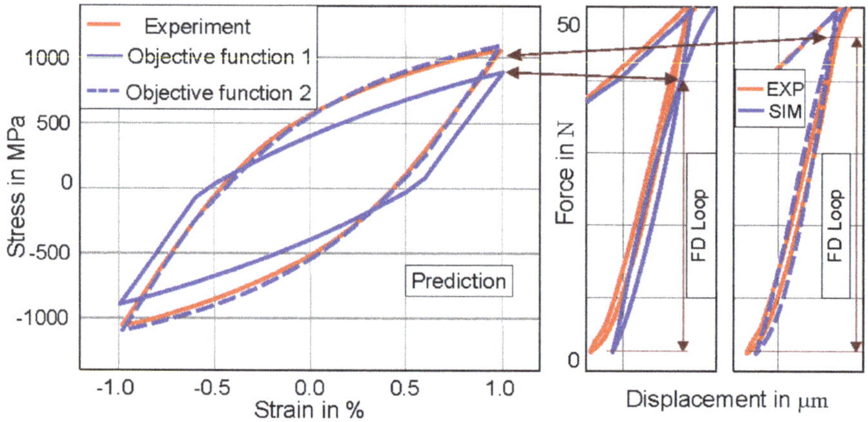


**Figure 7.** Effect of ΔF on the uniaxial stress–strain hysteresis prediction. The solid blue force-displacement (FD) loop displays the fitting of the FD loop to the blue experimental FD loop by using objective function 1, while the dotted solid stress–strain hysteresis is the prediction of stress–strain hysteresis. Similarly, the dotted blue FD loop shows the fitting of the FD loop by using objective function 2, while the dotted blue stress–strain hysteresis represents the prediction of the stress–strain hysteresis.

It can be observed that if the simulated ΔF has a lower value than the experimental ΔF, the prediction of the uniaxial stress–strain hysteresis reveals a larger disagreement (plastic work error = 20%) between the experimental and the simulated stress–strain hysteresis. On the other hand, when the value of ΔF is comparable to the experimental ΔF value, the uniaxial stress–strain hysteresis provides an acceptable prediction (plastic work error = 2.0%), as demonstrated by the dotted solid curve in Figure 7.

From now on, we will only present the results obtained by "objective function 2" after optimization. As the Chaboche material model is also capable to capture the ratcheting behavior in cyclic loading, further simulations are performed with multiple cycles of indentation to compare the experimental ratcheting effect of force–displacement by using the identified parameters from the complete indentation cycle. The ratcheting observed in simulation and experiment is slightly overestimated. In Figure 8b, force–displacement curves for 13 consecutive cycles are compared with the experiment.

### 4.2. Validation

The material parameters, which have been identified at a 50 N force amplitude, are also tested at higher forces of 75 and 100 N to check the validity of the obtained parameters. The comparison of the simulation curve and the experimental curve at higher force amplitudes reveals that they are also in good agreement at 75 N force amplitude (NMSE = $3.3 \times 10^{-4}$; Figure 9a) and 100 N force amplitude (NMSE = $1.6 \times 10^{-4}$; Figure 9b).

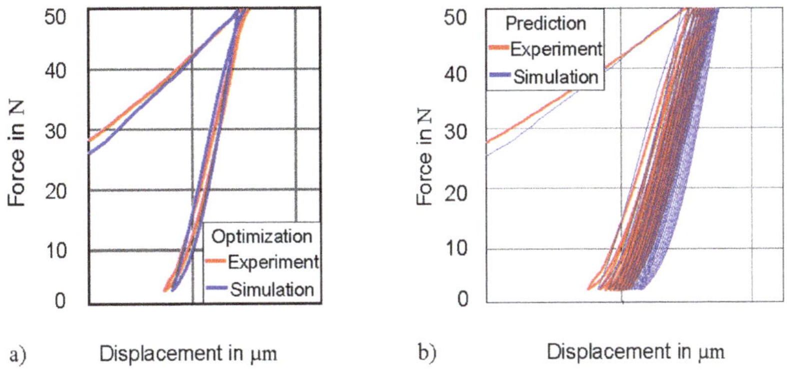


**Figure 8.** (**a**) Simulated force–displacement loop. (**b**) Predicted force–displacement for 13 cycles.

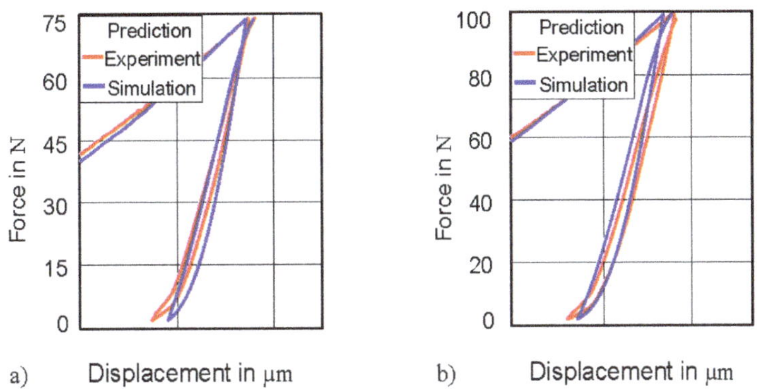


**Figure 9.** Validation of the method at higher force amplitudes: (**a**) predicted force–displacement at 75 N; (**b**) predicted force–displacement at 100 N.

Figure 10a demonstrates the comparison of experimental and simulated uniaxial stress–strain hystereses for the 10th cycle of the same material. The red hysteresis, which is obtained by using the above-obtained fitted material parameters from the cyclic indentation force–displacement curve, shows a quite good agreement (plastic work error = 3.5%) with the experimental hysteresis. Figure 10b demonstrates the maximum and minimum stress on the vertical axis, while the horizontal axis displays the cycle number.

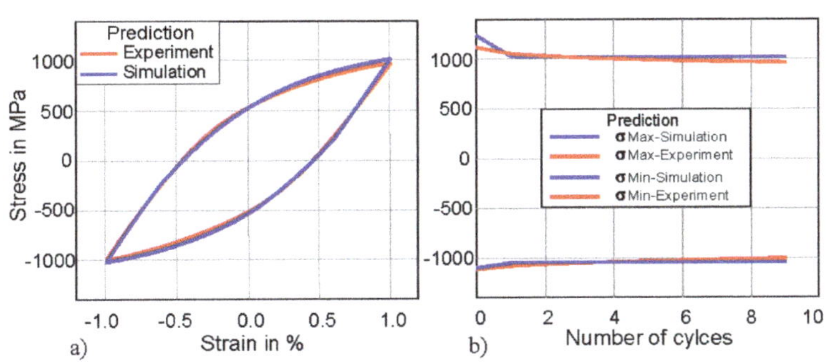


**Figure 10.** (**a**) Predicted uniaxial stress–strain hysteresis for the 10th cycle. (**b**) Stress amplitude over the number of cycles for the first 10 cycles.

It is evident from Figure 10 that the material parameters obtained from the cyclic indentation force–displacement curve can predict higher uniaxial stress–strain hysteresis very accurately, and the difference is less than 4%. This uniaxial stress–strain hysteresis is obtained without any initial input from uniaxial stress–strain hysteresis. The uniaxial stress–strain hysteresis has a key role in determining material fatigue life. As already mentioned, performing fatigue experiments is quite expensive both in terms of cost and time, and performing indentation tests is quite easy and requires fewer resources. By using this technique of identifying kinematic hardening material parameters, the need to perform fatigue experiments will be required only for the validation process.

### 4.3. *Transferability of the Method*

#### 4.3.1. Transferability to Higher Force Amplitude (75 N)

Until now, we have used a 50 N indentation force amplitude for identifying material parameters and then predicted the uniaxial stress–strain hysteresis by using these identified parameters. To check the robustness and transferability of our method, we have decided to also fit the 75 N indentation curve and to try to predict the uniaxial stress–strain hysteresis by using these parameters. The rest of the optimization setting and procedure were kept the same as explained before. Figure 11a displays the comparison of the cyclic force–displacement curve, while Figure 11b demonstrates the comparison of the uniaxial stress–strain between simulation and experiment. The results are in good agreement (NMSE = $4.0 \times 10^{-5}$) between the experimental and the simulated curves for both force–displacement and uniaxial stress–strain hysteresis. The parameters obtained after the simulation are reported in Table 2 and not much different from the parameters obtained for the 50 N force amplitude.

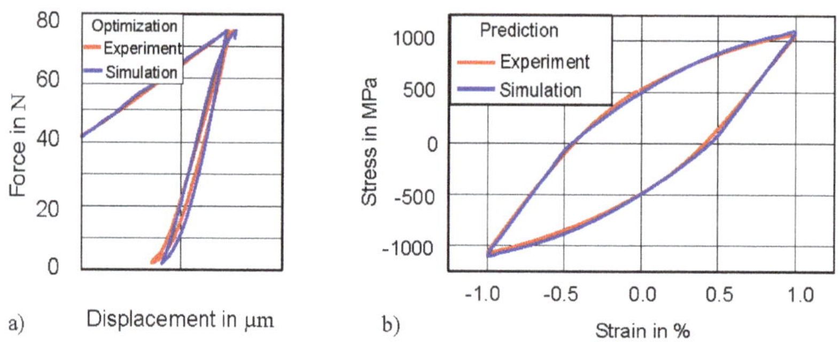


**Figure 11.** Transferability of method: (**a**) indentation of 38 HRC at 75 N; (**b**) prediction of stress–strain hysteresis of 38 HRC.

**Table 2.** Identified material parameters for 50CrMo4 (38 HRC) after fitting of force–displacement at 75 N force amplitude.

| Symbol | Value |
|---|---|
| $C_1$ (MPa) | 257,503 |
| $\gamma_1$ | 354 |
| $C_2$ (MPa) | 3663 |
| $\gamma_2$ | 0.2837 |
| Q (MPa) | −611 |
| b | 163 |

#### 4.3.2. Transferability to Higher Hardness (47 HRC)

Similarly, the force–displacement loop at 50 N force amplitude of 47 HRC hardness is used for material parameter identification by the inverse method. The value of yield strength used (1400 MPa) was obtained from the monotonic loading experiments. The results of the force–displacement curve after the optimization show a good agreement between the experimental and simulated force–displacement loops, as shown in Figure 12. The prediction of the uniaxial stress–strain curve (by using the material parameters from Table 3) also depicts a sufficient agreement between simulation and experiment. The difference in energy dissipation between the experiment and the prediction is 4.5%.

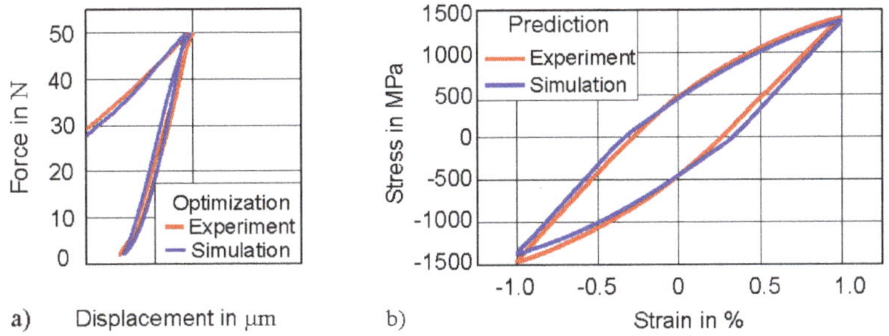


**Figure 12.** Transferability of method: (**a**) simulated indentation force–displacement at 50 N for 47 HRC; (**b**) prediction of uniaxial stress–strain hysteresis of 47 HRC.

**Table 3.** Identified material parameters for 50CrMo4 (47 HRC) of force–displacement at 50 N force amplitude.

| Symbol | Value |
|---|---|
| $C_1$ (MPa) | 337,885 |
| $\gamma_1$ | 374 |
| $C_2$ (MPa) | 6681 |
| $\gamma_2$ | 2.3 |
| Q (MPa) | −724 |
| b | 273 |

### 4.3.3. Transferability for Other Material (Cu)

The extensive study of using different force amplitudes and different hardnesses has been done with our method for martensitic steel in the previous section. In this section, the aim is to test our methodology on a different metallic material. For this purpose, copper (Cu), a relatively softer material is selected, and therefore, instead of using a 50 N force amplitude, a smaller force amplitude of 10 N is applied for cyclic indentation. The optimization is performed by using the same setup with two backstress terms (Table 4), and the results after optimization are shown in Figure 13. Figure 13a demonstrates the comparison of the force–displacement at 10 N between simulation and experiment, while Figure 13b shows the predicted uniaxial stress–strain at 1% total strain amplitude. It can be seen from Figure 13a that the force–displacement has quite a good fit after optimization. The same is true when we predict the uniaxial stress–strain hysteresis by using these identified parameters and compare it with the experimental uniaxial stress–strain hysteresis. The difference in dissipated energy and between predicted and experimental stress–strain hysteresis is only 3.5%.

**Table 4.** Identified material parameters of the force–displacement loop for Cu at 10 N force amplitude.

| Symbol. | Value |
|---|---|
| $C_1$ (MPa) | 154,790 |
| $\gamma_1$ | 2,257 |
| $C_2$ (MPa) | 11,586 |
| $\gamma_2$ | 82 |
| Q (MPa) | −12 |
| b | 47 |

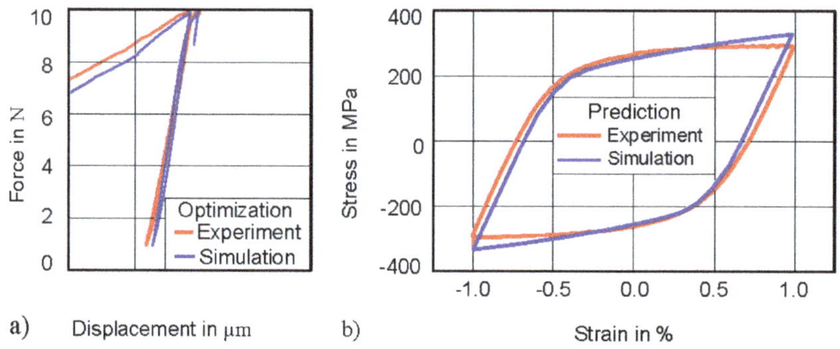


**Figure 13.** Transferability of method: (**a**) simulated indentation force–displacement of Cu at 10 N; (**b**) prediction of uniaxial stress–strain hysteresis.

## 5. Conclusions

A novel hybrid method for the inverse analysis of fatigue properties of metals has been introduced. The method combines cyclic Vickers indentation experiments and finite element simulations in an inverse method, by which the material parameters are determined in an iterative way by an optimization scheme. It has been demonstrated that this method can be used to determine the parameters of the Chaboche model for kinematic hardening. Based on these parameters, the model has been successfully employed to predict the cyclic stress–strain responses of a tempered martensitic steel, SAE 4150 (German denomination 50CrMo4), with different heat treatments and of technically pure copper. The error in the parameters determined with the inverse method has been evaluated as less than 4% on average. It has been observed that the difference between the maximum and minimum force of the force–displacement loop obtained from cyclic indentation has a direct correlation with the stress amplitude of the hysteresis loop measured in strain-controlled uniaxial fatigue tests; thus, it plays a crucial role in predicting the uniaxial stress–strain hysteresis accurately. By applying the method to high-strength martensitic steel, on which it has been validated for different maximum forces, and also to technically pure copper, its validity has been demonstrated for a wide variety of materials and process parameters. The prediction of a complete cyclic stress–strain curve by using data from cyclic indentation has great potential to reduce time- and cost-intensive fatigue experiments and can thus open a new and economic way to predict the fatigue life of materials with a quasi-nondestructive test method.

**Author Contributions:** Conceptualization, H.M.S., H.u.H., M.K., and A.H.; methodology, H.M.S., H.u.H., M.K., and A.H.; investigation, H.M.S., B.J.S., and P.S.-R.; data curation; B.J.S., P.S.-R., and M.K.; writing—original draft, H.M.S.; visualization, H.M.S.; writing—review & editing, M.K., H.u.H., and A.H., supervision, A.H. All authors have read and agreed to the published version of the manuscript.

**Funding:** This research received no external funding.

**Acknowledgments:** The authors gratefully acknowledge the financial support of the Higher Education Commission (HEC) of Pakistan, DFG Open Access Publication Funds of the Ruhr-Universität Bochum, and the contributions of Robert Bosch GmbH in carrying out the experiments.

**Conflicts of Interest:** The authors declare no conflict of interest.

## References

1. Huber, N.; Tsakmakis, C. Determination of constitutive properties from spherical indentation data using neural networks. Part I: The case of pure kinematic hardening in plasticity laws. *J. Mech. Phys. Solids* **1999**, *47*, 1569–1588. [CrossRef]

2. Huber, N.; Tsakmakis, C. Determination of constitutive properties from spherical indentation data using neural networks. Part II: Plasticity with nonlinear isotropic and kinematic hardening. *J. Mech. Phys. Solids* **1999**, *47*, 1589–1607. [CrossRef]
3. Wymysłowski, A.; Dowhań, Ł. Application of nanoindentation technique for investigation of elasto-plastic properties of the selected thin film materials. *Microelectron. Reliab.* **2013**, *53*, 443–451. [CrossRef]
4. Oliver, W.C.; Pharr, G.M. Measurement of hardness and elastic modulus by instrumented indentation: Advances in understanding and refinements to methodology. *J. Mater. Res.* **2004**, *19*, 3–20. [CrossRef]
5. Peirce, D.; Asaro, R.J.; Needleman, A. Material rate dependence and localized deformation in crystalline solids. *Acta Metall.* **1983**, *31*, 1951–1976. [CrossRef]
6. Hyung-Yil, L. Ball Indenter Utilizing Fea Solutions for Property Evaluation. WO2003010515A1, 2002. Available online: https://patents.google.com/patent/WO2003010515A1 (accessed on 10 July 2020).
7. Suresch, A.; Alcala, S.; Giannakopoulos, J. Depth Sensing Indentation and Methodology for Mechanical Property Measurements. WO1997039333A2, 1996. Available online: https://patents.google.com/patent/WO1997039333A2 (accessed on 10 July 2020).
8. Suresh, T.A.; Dao, S.; Chollacoop, M.; Van, N.; Venkatesh, K.V. Systems and Methods for Estimation and Analysis of Mechanical Property Data. WO2002073162A2, 2002. Available online: https://patents.google.com/patent/WO2002073162A2 (accessed on 10 July 2020).
9. Fontanari, V.; Beghini, M.; Bertini, L. Method and Apparatus for Determining Mechanical Features of a Material with Comparison to Reference Database. WO2006013450A2, 2004. Available online: https://patents.google.com/patent/WO2006013450A2/en (accessed on July 10 2020).
10. Schmaling, B.; Hartmaier, A. Method for Testing Material, particularly for Hardness Testing, Involves Producing Impression in to Be Tested Material in Experimental Manner with Test Body with Known Geometry and with Known Test Load. DE102011115519A1, 2011. Available online: https://patents.google.com/patent/DE102011115519A1/de (accessed on 10 July 2020).
11. Broitman, E. Indentation Hardness Measurements at Macro-, Micro-, and Nanoscale: A Critical Overview. *Tribol. Lett.* **2017**, *65*, 23. [CrossRef]
12. Strzelecki, P. *Analytical Method for Determining Fatigue Properties of Materials and Construction Elements in High Cycle Life*; Uniwersytet Technologiczno-Przyrodniczy w Bydgoszczy: Bydgoszcz, Poland, 2014.
13. Murakami, Y. Effects of small defects and nonmetallic inclusions on the fatigue strength of metals. *JMSE Int. J.* **1989**, *32*, 167–180. [CrossRef]
14. Bandara, C.S.; Siriwardane, S.C.; Dissanayake, U.I.; Dissanayake, R. Developing a full range S-N curve and estimating cumulative fatigue damage of steel elements. *Comput. Mater. Sci.* **2015**, *96*, 96–101. [CrossRef]
15. Bandara, C.S.; Siriwardane, S.C.; Dissanayake, U.I.; Dissanayake, R. Full range S-N curves for fatigue life evaluation of steels using hardness measurements. *Int. J. Fatigue* **2016**, *82*, 325–331. [CrossRef]
16. Strzelecki, P.; Tomaszewski, T. Analytical models of the S-N curve based on the hardness of the material. *Procedia Struct. Integr.* **2017**, *5*, 832–839. [CrossRef]
17. Lyamkin, V.; Starke, P.; Boller, C. Cyclic indentation as an alternative to classic fatigue evaluation. In Proceedings of the 7th International Symposium on Aircraft Materialsno, Compiegne, France, 24–26 April 2018.
18. Faisal, N.H.; Prathuru, A.K.; Goel, S.; Ahmed, R.; Droubi, M.G.; Beake, B.D.; Fu, Y.Q. Cyclic Nanoindentation and Nano-Impact Fatigue Mechanisms of Functionally Graded TiN/TiNi Film. *Shape Mem. Superelasticity* **2017**, *3*, 149–167. [CrossRef]
19. Haghshenas, M.; Klassen, R.J.; Liu, S.F. Depth-sensing cyclic nanoindentation of tantalum. *Int. J. Refract. Met. Hard Mater.* **2017**, *66*, 144–149. [CrossRef]
20. Prakash, R.V. Evaluation of fatigue damage in materials using indentation testing and infrared thermography. *Trans. Indian Inst. Met.* **2010**, *63*, 173–179. [CrossRef]
21. Prakash, R.V. Study of Fatigue Properties of Materials through Cyclic Automated Ball Indentation and Cyclic Small Punch Test Methods. *Key Eng. Mater.* **2017**, *734*, 273–284. [CrossRef]
22. Xu, B.X.; Yue, Z.F.; Chen, X. Numerical investigation of indentation fatigue on polycrystalline copper. *J. Mater. Res.* **2009**, *24*, 1007–1015. [CrossRef]
23. Schäfer, B.; Song, X.; Sonnweber-Ribic, P.; Hassan, H.U.; Hartmaier, A. Micromechanical Modelling of the Cyclic Deformation Behavior of Martensitic SAE 4150—A Comparison of Different Kinematic Hardening Models. *Metals (Basel)* **2019**, *9*, 368. [CrossRef]

24. Kramer, H.S.; Starke, P.; Klein, M.; Eifler, D. Cyclic hardness test PHYBALCHT - Short-time procedure to evaluate fatigue properties of metallic materials. *Int. J. Fatigue* **2014**, *63*, 78–84. [CrossRef]
25. DIN EN ISO 6507-2. *Metallic Materials—Vickers Hardness Test—Part 2: Verification and Calibration of Testing Machines*; NSAI: Dublin, Ireland, 2005.
26. Mises, R.V. Mechanik der festen Körper im plastisch- deformablen Zustand. *Nachrichten von der Gesellschaft der Wissenschaften zu Göttingen, Mathematisch-Physikalische Klasse* **1913**, *1913*, 582–592.
27. Srnec Nova, J.; Benasciutti, D.; De Bona, F.; Stanojević, A.; De Luca, A.; Raffaglio, Y. Estimation of Material Parameters in Nonlinear Hardening Plasticity Models and Strain Life Curves for CuAg Alloy. *IOP Conf. Ser. Mater. Sci. Eng.* **2016**, *119*, 12020. [CrossRef]
28. Chaboche, J.L.L. Constitutive equations for cyclic plasticity and cyclic viscoplasticity. *Int. J. Plast.* **1989**, *5*, 247–302. [CrossRef]
29. Lemaitre, J.; Chaboche, J.-L. *Mechanics of Solid Materials*; Cambridge University Press: Cambridge, UK, 1990.
30. Frederick, C.O.; Armstrong, P.J. A mathematical representation of the multiaxial Bauschinger effect. *Mater. High Temp.* **2007**, *24*, 1–26. [CrossRef]
31. Sajjad, H.M.; Hanke, S.; Güler, S.; ul Hassan, H.; Fischer, A.; Hartmaier, A. Modelling cyclic behaviour of martensitic steel with J2 plasticity and crystal plasticity. *Materials* **2019**, *12*, 1767. [CrossRef] [PubMed]
32. About LS-OPT—DYNAmore GmbH. Available online: https://www.dynamore.de/de/produkte/opt/ls-opt (accessed on 26 April 2020).
33. Chaparro, B.M.; Thuillier, S.; Menezes, L.F.; Manach, P.Y.; Fernandes, J.V. Material parameters identification: Gradient-based, genetic and hybrid optimization algorithms. *Comput. Mater. Sci.* **2008**, *44*, 339–346. [CrossRef]

*materials*

MDPI

*Article*

# Addressing H-Material Interaction in Fast Diffusion Materials—A Feasibility Study on a Complex Phase Steel

**Agustina Massone [1,2,*], Armin Manhard [3], Andreas Drexler [4], Christian Posch [1], Werner Ecker [1], Verena Maier-Kiener [5] and Daniel Kiener [2]**

1 Materials Center Leoben, Forschungs GmbH, Roseggerstrasse 12, 8700 Leoben, Austria; Christian.Posch@mcl.at (C.P.); werner.ecker@mcl.at (W.E.)
2 Department Materials Science, Chair of Materials Physics, Montanuniversität Leoben, Jahnstrasse 12, 8700 Leoben, Austria; daniel.kiener@unileoben.ac.at
3 Max-Planck-Institut für Plasmaphysik, Boltzmannstr. 2, D-85748 Garching, Germany; manharar@ipp.mpg.de
4 Institut für Werkstoffkunde, Fügetechnik und Umformtechnik, Technische Universität Graz, Rechbauerstrasse 12, 8010 Graz, Austria; andreas.drexler@tugraz.at
5 Department Materials Science, Chair of Physically Metallurgy and Metallic Materials, Montanuniversität Leoben, Roseggerstrasse 12/Max-Tendler-Strasse 9, 8700 Leoben, Austria; verena.maier-kiener@unileoben.ac.at
* Correspondence: agustina.massone@mcl.at; Tel.: +43-(0)3842-804-410

Received: 26 August 2020; Accepted: 16 October 2020; Published: 20 October 2020

**Abstract:** Hydrogen embrittlement (HE) is one of the main limitations in the use of advanced high-strength steels in the automotive industry. To have a better understanding of the interaction between hydrogen (H) and a complex phase steel, an in-situ method with plasma charging was applied in order to provide continuous H supply during mechanical testing in order to avoid H outgassing. For such fast-H diffusion materials, only direct observation during in-situ charging allows for addressing H effects on materials. Different plasma charging conditions were analysed, yet there was not a pronounced effect on the mechanical properties. The H concentration was calculated while using a simple analytical model as well as a simulation approach, resulting in consistent low H values, below the critical concentration to produce embrittlement. However, the dimple size decreased in the presence of H and, with increasing charging time, the crack propagation rate increased. The rate dependence of flow properties of the material was also investigated, proving that the material has no strain rate sensitivity, which confirmed that the crack propagation rate increased due to H effects. Even though the H concentration was low in the experiments that are presented here, different technological alternatives can be implemented in order to increase the maximum solute concentration.

**Keywords:** advanced high-strength steels; hydrogen embrittlement; in-situ testing; scanning electron microscopy; plasma charging

## 1. Introduction

One of the main goals in the automotive industry is the reduction of weight while maintaining an adequate strength and toughness, at low cost, and enhancing both safety and fuel economy. In this scenario, advanced high-strength steels (AHSS) are excellent candidates for this application, since they combine both high strength and low weight [1,2]. The most important ones are dual phase (DP), ferritic-bainitic (FB), martensitic, transformation-induced plasticity (TRIP), and complex phase (CP) steels. DP steels consist of martensitic islands in a ferritic matrix and they combine low yield strength with high ultimate tensile strength. FB steels have a microstructure of fine ferrite and bainite and strengthening is obtained by both grain refinement and second phase hardening with bainite. Martensitic steels show the highest tensile strength level and they are often subjected to post-quench tempering in order to improve ductility. TRIP steels exhibit superior strength and good formability as a result of the transformation of metastable retained austenite to martensite during deformation. The microstructure of CP steels contains small amounts of martensite, retained austenite, and pearlite within a ferrite/bainite matrix. An extreme grain refinement is created in this material by retarded recrystallization or precipitation of microalloying elements, such as Ti or Nb [3].

Despite their remarkably good mechanical properties, AHSS are susceptible to hydrogen embrittlement (HE), and this can lead to a loss of ductility. When AHSS are electroplated with a sacrificial metal, typically Zn, H can be absorbed during the coating deposition, as the process is not 100% efficient. Moreover, if the sacrificial coating corrodes in service, the exposed areas of the steel substrate will act as cathodic sites and H can be absorbed into the material [4]. H can also be introduced into the material during the painting process of a body in white structure. Lovicu et al. [5] measured the H content absorbed during the production process of autobody components, in which cathodic reactions in water solution take place in the phosphatizing and electrophoresis stages of the painting process. During these reactions, atomic H can form on the steel surface and diffuse into the material. The absorbed H during the painting process was lower than about 0.4 wppm. In this context, H is one of the main limitations in the use of AHSS, since it can reduce the ultimate tensile strength, ductility, fatigue strength, and/or fracture toughness of the steels [6]. This degradation becomes apparent when the material is under residual or applied tensile stresses. Absorbed H diffuses through the metal facilitating crack propagation and, the higher the stress level of the material, the more susceptible it is to undergo detrimental HE effects [7].

The mechanism of HE has been under discussion for many years, leading to different interpretations and controversial findings [8–16]. Nevertheless, it is accepted that HE can only occur for a critical combination of local stress and local H concentration [6,17–19]. This critical value for H concentration may depend on applied stress, microstructure, trapping state, and tensile strength level, among others. Moreover, it is believed that there is a saturation level above which there is a minimal change in HE susceptibility [17]. Drexler et al. confirmed that this is the case for AHSS [18,19].

High-strength materials are more prone to HE due to an increased number of potential fracture initiation sites, with martensite usually being the most susceptible steel phase [7]. The morphology of H-assisted fracture in AHSS depends on the steel microstructure and H concentration and it can be either unaffected or changed from ductile microvoid coalescence to quasi-cleavage, cleavage, or intergranular failure [17–20].

The effect of H on high-strength steels has been widely studied [5,6,18–35], but there are few studies regarding the interaction between H and CP steels. Malitckii et al. [32] investigated the role of retained austenite in CP steel and proposed that fatigue intergranular areas might be formed due to H accumulation at the austenite/martensite interfaces, followed by H-induced decohesion. Loidl et al. studied the effect of H on a CP1200, among other different AHSS. They showed that the degree of embrittlement was similar to martensitic steels and that TRIP steels presented the highest tendency to HE [33]. Lovicu et al. [5] concluded that martensitic microstructures exhibit great susceptibility to HE and, the higher the tensile strength, the lower the critical H concentration to produce damage in the material. Duprez et al. [29] studied the effect of H on a DP steel, a TRIP steel, a FB steel, and a ferrite-cementite grade. They demonstrated that the ductility of all the samples was reduced after electrochemical charging, with TRIP and DP steels the most susceptible ones. Nevertheless, after discharging the samples for one week, a large part of the ductility was recovered. This proved that the damage was caused by the intrinsic presence of H and not by an irreversible damage mechanism. Rehrl et al. [34] investigated the effect of different loading rates in four grades of AHSS and showed that, at high strain rates, there was no effect of H on the mechanical properties. Only with slow strain rate testing, the elongation at failure was reduced. This was explained when considering that at a high strain rate, the H diffusion is too slow to reach highly stressed microstructure regions. Drexler et al. studied the local H accumulation and its effect on HE for cold formed, punched, and heat treated CP1200 and DP1200 [18–20].

In all of the studies mentioned above, the investigations were made either with ex-situ H charging or without in-situ observation. These two approaches can lead to a misinterpretation of the results. When the materials are ex-situ charged, there is a risk of H outgassing before the test is performed, especially in some steels where H diffusion is very fast [35]. Without in-situ observation, the H effect can only be analysed post-mortem and not during the test. Even though the effect on mechanical properties can be determined, details on crack dynamics cannot be investigated without an in-situ approach. In order to overcome these limitations, several studies have already been made with in-situ charging [36–40] and, in the present work, the interaction between H and a CP steel was investigated by implementing an in-situ method, which allows for in-situ H charging during a tensile test inside a scanning electron microscope (SEM).

## 2. Materials and Methods

### *2.1. Material*

A CP1200 steel was investigated in this work. The main microstructural phases are martensite, tempered martensite, and bainite, with a small content (<2%) of retained austenite. Table 1 shows the chemical composition of the material.

**Table 1.** Chemical composition of the investigated industrial steel grade.

| | C | Mn | Si | Cr | S | Nb | Ti | Al |
|---|---|---|---|---|---|---|---|---|
| wt% | <0.20 | <2.6 | <0.8 | <1.00 | <0.010 | <0.05 | <0.15 | 0.015–1.0 |

The microstructure analysis was conducted on a field emission SEM Zeiss LEO 1525 (Carl Zeiss GmbH, Oberkochen, Germany) while using an acceleration voltage of 20 kV and by EBSD using a pixel size of 59 nm and a working distance of 15.7 mm. Figure 1 depicts the microstructure of the material. Figure 1a exhibits the secondary electron image and Figure 1b shows the inverse pole figure map in the normal direction of the same region. The average prior austenite grain size was measured to be ~3 µm.

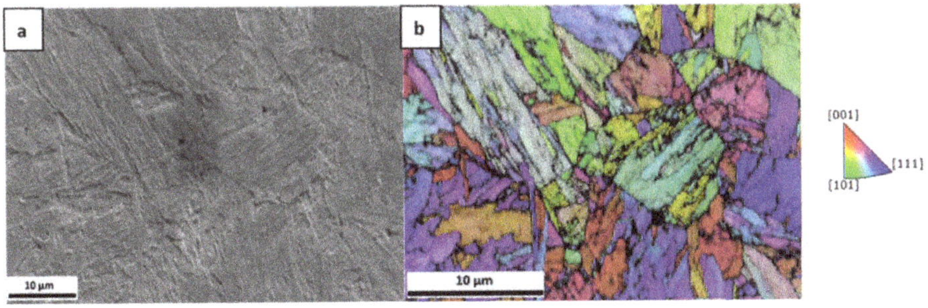


**Figure 1.** (**a**) Secondary electron image; and (**b**) inverse pole figure in the normal direction of the investigated CP1200 microstructure.

The as-delivered material had a thickness of 1.2 mm and tensile samples with the geometry of 32 mm length, 5 mm width, and 0.5–1.10 mm thickness were fabricated by electrical discharge machining parallel to the rolling direction. The samples were then ground and polished with 9 µm, 3 µm and 1 µm diamond paste to remove surface damage from machining. In the centre of the sample, in a rounded area of 4 mm diameter, the thickness was reduced to ~120–200 µm to reach steady-state permeation through the thickness of the charged samples more quickly. Table 2 displays the thicknesses of the samples.

**Table 2.** Case studies and sample thicknesses.

| Case Study | Overall Thickness (mm) | Center-Reduced Thickness (mm) |
|---|---|---|
| Rate Dependence of Flow Properties (Tensile Samples) | 1.06 ± 0.01 | – |
| Effect of H Pre-Charging Time | 0.5 ± 0.01 | 0.12 ± 0.01 |
| Effect of Plasma Parameters | 1.10 ± 0.01 | 0.18 ± 0.01 |
| Effect of H Charging-Discharging | 1.10 ± 0.01 | 0.20 ± 0.01 |

In-situ mechanical tests were performed in an SEM Zeiss Stereoscan 440 (Carl Zeiss GmbH, Oberkochen, Germany) operating at an acceleration voltage of 10 kV for imaging. A Kammrath & Weiss tensile stage (Kammrath & Weiss GmbH, Dortmund, Germany) equipped with a 10 kN load cell and an inductive displacement sensor was used for the tensile tests. The accuracy of the load cell is in the order of ~1 N, and the accuracy of the displacement sensor is ~0.5 μm.

The effect of strain rate was studied by tensile testing uncharged specimens with nominal strain rates of $3 \times 10^{-5}$ $s^{-1}$, $1.5 \times 10^{-4}$ $s^{-1}$, and $3 \times 10^{-4}$ $s^{-1}$. For these tests, samples with uniform thickness and no thinned area were used. To further study the strain rate sensitivity and measure depth-dependent properties, nanoindentation testing was performed on a KLA G200 platform nanoindenter (KLA, Milpitas, CA, USA) that was equipped with a three-sided diamond Berkovich tip (Synton-MDP). Six constant strain rate indentations with an applied indentation strain rate of 0.05 $s^{-1}$ and additionally five strain-rate jump tests with a strain-rate profile of 0.05 $s^{-1}$, 0.005 $s^{-1}$, 0.05 $s^{-1}$, 0.001 $s^{-1}$, and 0.05 $s^{-1}$ (applied changes in the strain rate every 500 nm) were executed [41]. The continuous stiffness measurement technique was utilized to continuously measure the contact stiffness and, thereby, the hardness and Young's modulus over indentation depth. This technique involves applying a dynamic load that is then used to measure the stiffness, which is further processed in order to calculate the modulus and hardness of the material. This method allows for the measurement of the depth-dependent properties of materials [41].

### 2.2. Plasma H Charging

The effect of H-material interaction was studied on the CP steel while using a dedicated in-situ design. H was charged into the material via localised loading by a deuterium plasma cell, allowing for the combination of in-situ charging, mechanical testing, and observation in an SEM. The applied method consists of a miniaturized radio frequency-plasma cell, in which two electrodes are confined in a vacuum vessel and deuterium gas is supplied. The tensile sample, operating as the grounded electrode, can, in this way, be charged from the bottom side with the H isotope, providing a contamination free top surface for observation. Only the thinned centre of the samples was charged and the observation was localized in this region. The main advantage of this method, when compared to conventional electrochemical charging, is that it allows having a high-resolution observation during H charging and deformation. Furthermore, there is no risk for the SEM, since the plasma turns off automatically when the sample fractures. For more details of the method, see ref. [40].

The influence of different H charging conditions was examined by applying different power settings, obtaining plasmas with different bias voltages. A higher bias voltage can be correlated with a higher ion flux and implantation range and, consequently, higher ion concentrations. Power levels of 5 W, 6 W, 8 W, and 11 W were applied, resulting in DC bias voltages of 110 V, 127 V, 173 V, and 174 V, respectively. All of the samples were consistently only charged during the tensile test, i.e., no pre-charging was performed. It was demonstrated [42] that, in the presence of H, the reduction of fracture area increases with decreasing deformation rate and that only with slow strain rate testing the elongation to failure of a material can be decreased [34]. With a slow deformation rate, diffusible H has more time to migrate towards the crack tip, which results in an embrittlement of the material. For this reason, the tensile tests were performed with a slower strain rate than the previous cases, using a displacement rate of 0.1 μm/s. Afterwards, the fracture surfaces were observed in the FEG-SEM LEO 1525.

A sample was charged for 4 h and then discharged for 12 h before starting the test in order to address the potential reversibility of HE and the possibility of plasma damage. The results were subsequently compared with an uncharged specimen.

Moreover, maintaining plasma parameters fixed, the effect of pre-charging time was investigated. Three samples were tested: an uncharged sample for reference and two charged samples with 3 h and 6 h of H pre-charge, respectively. After the pre-charging time, the charging was also maintained during the tensile test (~1 h duration). Afterwards, the surfaces were analysed using EBSD and fractography was performed with the FEG-SEM using the previously mentioned parameters.

### 2.3. H Uptake and Diffusion Simulation

The model described in [43] was utilised in order to assess the present H concentration. This is a model for ion-driven permeation of H in plasma-facing materials at steady state. In steady-state, the incident flux is balanced with the reflected and permeated fluxes. Depending on the relative rate of recombination and diffusion on front and back sides of the membrane, there are three possible ion-driven permeation processes: diffusion-limited on both sides (DD regime), recombination-limited on both sides (RR regime), and recombination-limited on one side and diffusion-limited on the other side (RD regime). Equation (1) shows the estimation of the maximum concentration for a RR and RD regimes. In these regimes, the surface recombination is the rate-limiting step and is valid for fast-diffusion materials. Because the parameters on the back side do not affect the maximum concentration at steady state [43], the equation is the same for both cases.

$$C_{max} = \sqrt{\frac{\eta}{k_f}} \tag{1}$$

where $C_{max}$ is the maximum lattice H concentration, $\eta$ the ion flux, and $k_f$ the recombination coefficient. The ion flux was calculated from Langmuir Probe measurements and it is described in detail in [40]. A Monte-Carlo program, SDTrim.SP 6.0 [44] was used for estimating the implanted fraction/particle reflection yield and the mean implantation range. The value of $k_f$ was calculated according to three different sources [43,45,46].

For a more detailed analysis, the H concentration was calculated using a more sophisticated diffusion model [47,48], developed and implemented as subroutine (UMATHT) in the finite element simulation software package Abaqus (version 2019) [49]. Newton–Raphson scheme is used to solve the system of equations, whereas Crank–Nicholson procedure is used for the time integration. This is a sequentially coupled diffusion-mechanical model, which considers concentration driven diffusion, stress driven diffusion, as well as physically meaningful boundary conditions. The interplay between trapped and lattice H is considered by the following relationship written in its multiple trap formulation [50]:

$$\frac{\gamma_{lattice}\left(1-\gamma_{trap,k}\right)}{\gamma_{trap,k}\left(1-\gamma_{lattice}\right)} = \exp\left(-\frac{\Delta E_k}{R_g T}\right), \quad (k = 1,\ldots,m) \tag{2}$$

With $\gamma_{lattice}$ as site fraction of lattice H, $\gamma_{trap,k}$ as trap site fraction and $\Delta E_k$ as trapping energy for the k$^{th}$ sort of trap. The parameters of the model can be either determined from permeation experiments [51,52] or from thermal desorption spectroscopy measurements [53,54]. The applied model for the CP1200 steel only considers one effective trap with a trap energy of 30 kJ/mol and a trap density of $5.7 \times 10^{-7}$ [18]. A trapping energy of about 30 kJ/mol is representative for dislocations and some kinds of grain boundaries and it is related to rather shallow traps. The three-dimensional (3D) sample geometry was modelled making use of a quarter symmetry and it was discretised by finite elements with linear shape functions (DC3D8) and an element size of interest of $2 \times 10^{-3}$–$2 \times 10^{-2}$ mm, resulting in a number of 13 elements alongside the flux direction. A mesh convergence study was done in order to avoid mesh dependency of the results. The applied flux boundary condition on the plasma-oriented surface in the 3D model results from the incident ion intake flux $\phi_i$ of $10^{20}$ $m^{-2}s^{-1}$ in 1.15 nm depth and on the recombination flux $\phi_r = k_f c_{lattice}^2$, with $c_{lattice}$ being the locally corresponding lattice hydrogen concentration. Due to narrow distance between the flux due to plasma loading and the recombination flux, the one-dimensional (1D) hydrogen permeation FE model calculated net influx is used to prescribe the hydrogen intake in the 3D model. Figure 2 depicts the boundary conditions applied to the 1D hydrogen permeation model. In fact, the boundary conditions on the SEM-oriented face of the sample was chosen corresponding to, both, the RR ($flux = k_f c_{lattice}^2$) and the RD ($c_{lattice} = 0$) regimes, and the differences were negligible. Therefore, only the model and results of the RD case are shown in the present paper.

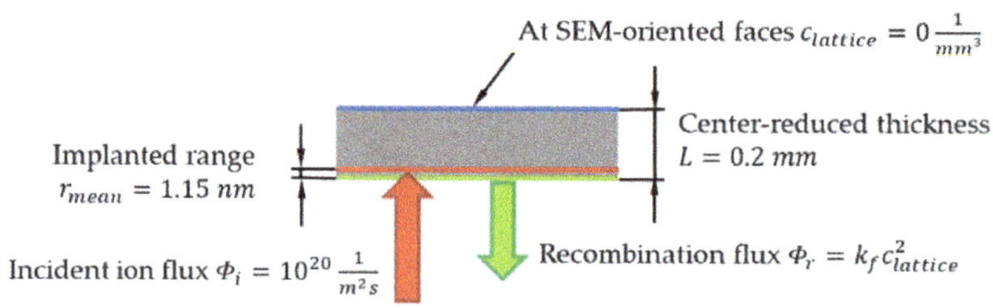


**Figure 2.** RD boundary conditions of the one-dimensional (1D) hydrogen permeation model. The time history of the sum of the incident ion flux and the recycling flux is applied as boundary condition at the H charging area in the three-dimensional (3D) model.

## 3. Results

### *3.1. Rate Dependance of Flow Properties*

Figure 3a shows the stress–strain curves of three specimens that were tested with $3 \times 10^{-5}$ $s^{-1}$, $1.5 \times 10^{-4}$ $s^{-1}$, and $3 \times 10^{-4}$ $s^{-1}$. The initial part of the curve that corresponds to $3 \times 10^{-5}$ $s^{-1}$ strain rate does not start with zero stress, due to possible friction effects, but after approximately 1% strain, the curve exhibits a normal behaviour. There is almost no difference between the three curves; the yield stress is ~1060 MPa, the tensile strength ~1200 MPa, and strain to failure ~9%. The strain rate sensitivity of a material can be verified by the value of the strain rate sensitivity index, m, from a simple power-law relationship [55]:

$$\sigma = \varepsilon^m \tag{3}$$

where σ is the flow stress and ε the strain rate. The value of m can be then determined as the slope of the plot of lnσ vs. lnε. Figure 3b depicts the flow stress vs. strain rate for the strains that are indicated in the box in Figure 3a. The calculated *m* for three curves was almost zero, proving that the material exhibits no significant strain rate sensitivity in the investigated strain rate regime.

At a macroscopic level, the three samples exhibited a high degree of ductility (necking). Figure 4 shows the fracture surfaces of the three specimens, where the same ductile failure behaviour was observed for all specimens, i.e., microvoid coalescence, leading to the presence of dimples on the fracture surfaces. The size of approximately 50 dimples in each of the three samples were estimated from the SEM images, giving a bimodal distribution with sizes of 14.5 ± 0.8 μm and 6.6 ± 0.6 μm. No effect of strain rate was observed in the tested range.

From the constant strain rate indentation tests, the average hardness and Young's Modulus were calculated to be 5.13 GPa and 233 GPa, respectively. Similar hardness values were reported for tempered martensite and martensitic steels with similar C content [56,57]. Figure 5a depicts the load-indentation depth plot that corresponds to the strain rate jump tests and Figure 5b shows the exemplarily resulting hardness and Young's modulus over indentation depth. The Young's modulus is independent of the applied strain rate and the differences in data points is due to the differences in testing times with different strain rates. The hardness levels clearly shift with strain rate, even though by only a small amount. The decreasing hardness at very shallow indentation depths is related to the indentation size effect [58], but it does not affect the analysis of strain rate sensitivity. The calculation of the strain rate sensitivity m in nanoindentation experiments is made with the hardness, which is directly related with the stress through Equation (4):

$$\sigma = H/C^{*} \tag{4}$$

where $\sigma$ is the flow stress, $H$ the hardness and $C^{*}$ a constraint factor. The calculated strain rate sensitivity $m$ was on average 0.006 ± 0.0005. This result confirms that the material exhibit almost no strain rate sensitivity within the tested parameter range.

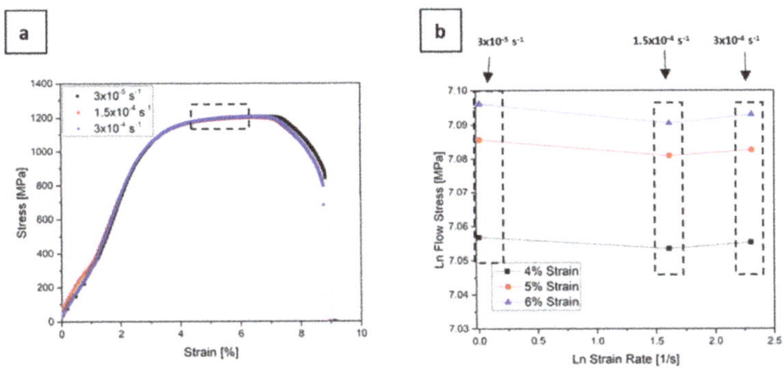


**Figure 3.** (**a**) Stress-strain curves of CP 1200 steel tested strain rates of $3 \times 10^{-5}$ s$^{-1}$, $1.5 \times 10^{-4}$ s$^{-1}$, and $3 \times 10^{-4}$ s$^{-1}$; (**b**) ln flow stress-ln strain rate plot for the strains indicated in the box in (**a**), showing no significant strain rate sensitivity.

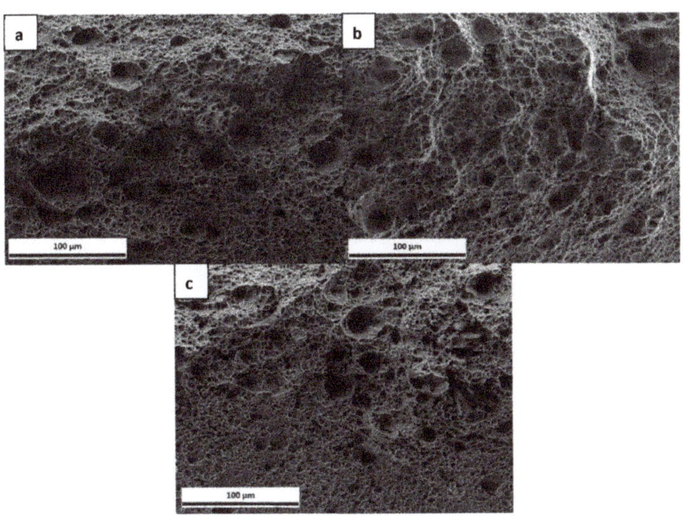


**Figure 4.** Fracture surfaces of complex phase (CP) steel tested with strain rates of (**a**) $3 \times 10^{-5}$ s$^{-1}$; (**b**) $1.5 \times 10^{-4}$ s$^{-1}$; (**c**) $3 \times 10^{-4}$ s$^{-1}$.

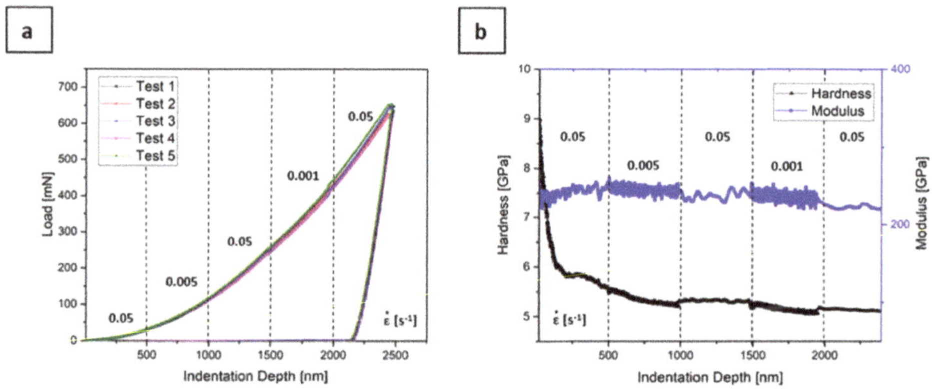


**Figure 5.** (**a**) Load-indentation depth plots corresponding to jump tests with 0.05 $s^{-1}$, 0.005 $s^{-1}$, 0.05 $s^{-1}$, 0.001 $s^{-1}$, and 0.05 $s^{-1}$, with a change in the strain rate every 500 nm; (**b**) the exemplarily resulting hardness and Young's modulus.

## 3.2. *Effect of H Charging*

### 3.2.1. Effect of Pre-Charging Time

Figure 6 depicts the load-elongation curves of an uncharged sample and two charged samples with different pre-charging times, both also continuously charged during the tensile test. The first linear part of the curve is not affected by the presence of H. However, a trend can be observed in the plastic part, after the crack onset, which is indicated by an arrow. With increasing H-charging time, the elongation to failure decreased slightly, and there was a more pronounced drop in the load at ~200 μm, which is when cracking was initiated.

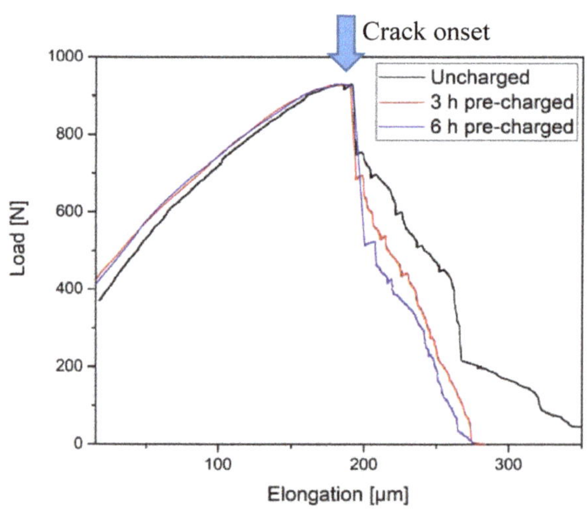


**Figure 6.** Load-elongation curves of three CP steel samples: uncharged, 3 h, and 6 h pre-charged.

In Figure 7, the fracture surfaces of the three samples are presented. The uncharged sample, Figure 7a, exhibits a ductile failure. When comparing with the uncharged samples from Figure 4, they do not look very similar, since, in this set of samples, the thickness was reduced in the centre of the sample, as described above, which gives rise to a somewhat different appearance of the fracture surface. The charged samples presented in Figure 7b,c, on the other hand, exhibit some differences when compared to the uncharged specimen. In Figure 7b, in addition to voids, some flat regions can be seen. In the sample with the longer charging time, Figure 7c, there is a marked difference between the upper and lower part of the fracture surface, as indicated with a red line. Only in the upper part of the surface microvoids can be observed, while the lower part, where the H was supplied, shows another failure mechanism with more localized damage. Even though the lower part of Figure 7a also exhibits some surface damage, this is less pronounced than for the charged sample.

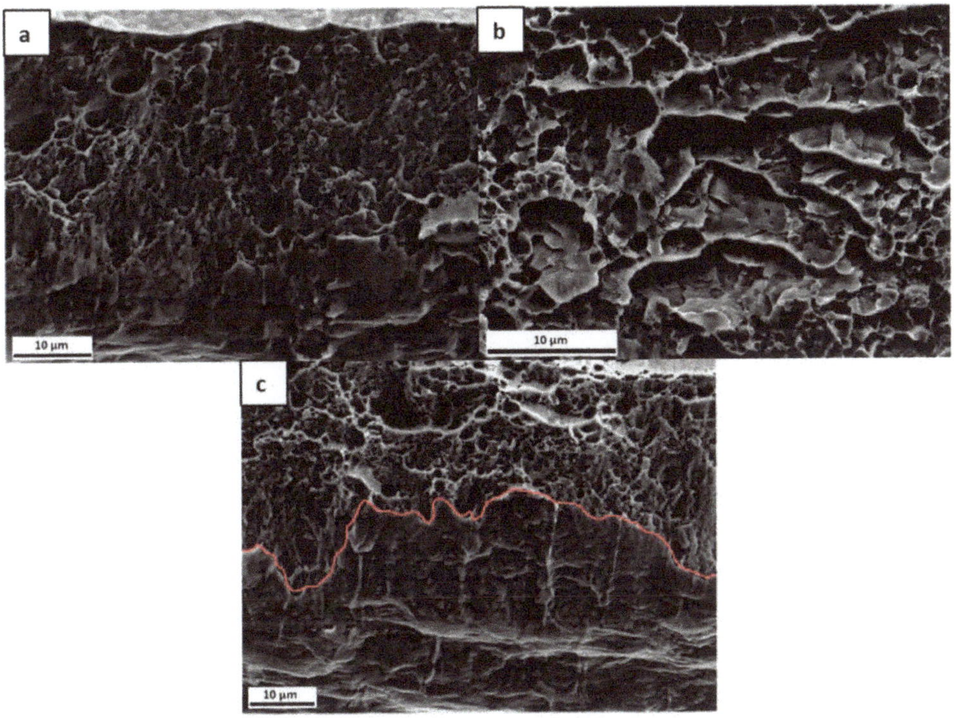


**Figure 7.** Fracture surfaces of the (**a**) uncharged; (**b**) 3 h pre-charged; and, (**c**) 6 h pre-charged CP steel samples.

From in-situ SEM images that were recorded during the tests, the linear intercept of the crack length was estimated at different loading steps for each sample, as shown in Figure 8a. For comparison purposes, the crack initiation times were normalized to 0 s. When comparing the slopes of the curves for each sample, it can be seen that the crack propagation rate increased with charging time. Figure 8b shows a linear estimate of the crack growth rate for the three samples. While the crack growth rate for the uncharged sample was ~1.5 µm/s, the 3 h pre-charged sample exhibited a rate of 2.5 µm/s, while, for the 6 h pre-charged one, it further increased to 3 µm/s.

The surface of samples in regions near the fracture site was analysed using EBSD. The Inverse Pole Figure maps in the normal direction are shown in Figure 9. Secondary cracks, highlighted with arrows, can only be seen in the sample with the longer charging time, Figure 9c,d. Nevertheless, while the secondary cracks seem to follow interfaces in Figure 9c, it was not possible to determine whether the cracks started at grain boundaries or inside the grains. No secondary cracks were observed in the uncharged and 3 h pre-charged samples.

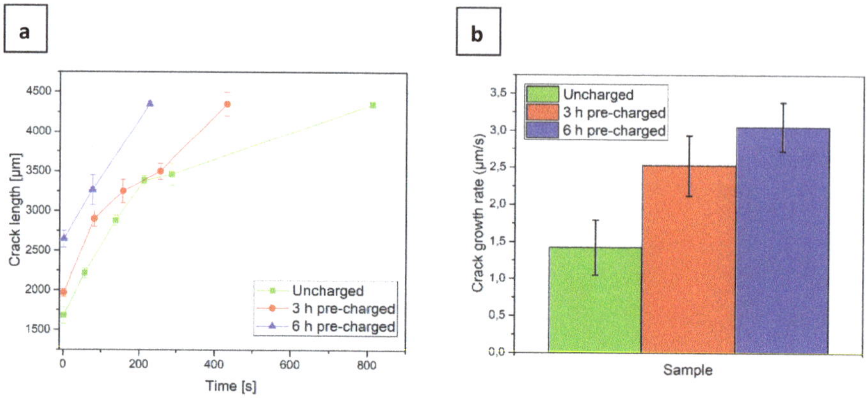


**Figure 8.** (**a**) Normalized crack length evolution; (**b**) crack growth rate of the uncharged, 3 h pre-charged and 6 h pre-charged CP steel samples.

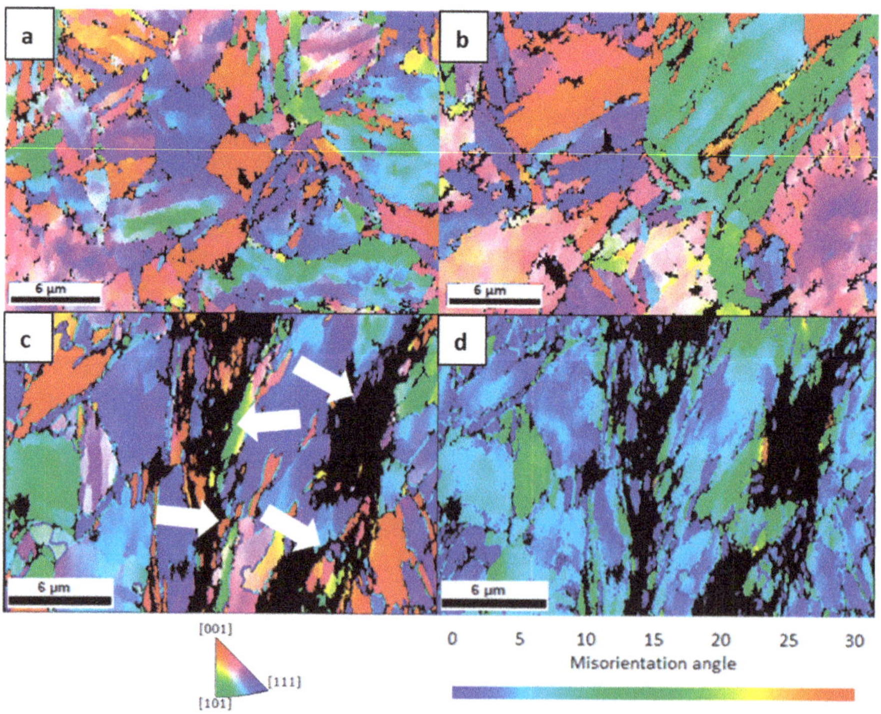


**Figure 9.** Inverse Pole Figure maps in the normal direction of the (**a**) uncharged, (**b**) 3 h pre-charged, (**c**) 6 h pre-charged CP steel samples near the fracture and (**d**) misorientation map of 6 h pre-charged CP steel sample.

### 3.2.2. Effect of Plasma Parameters

Figure 10 shows the load-elongation curves of samples tested under different plasma conditions. The noise that was observed in some of the curves was due to interference between the radio frequency power supply and the electronics in the tensile module, despite the shielding of the cables.

While inconvenient, this does not affect the data. It is important to mention that, since the samples used for each case study correspond to a different machining set, the difference in thickness gives rise to different maximum loads.

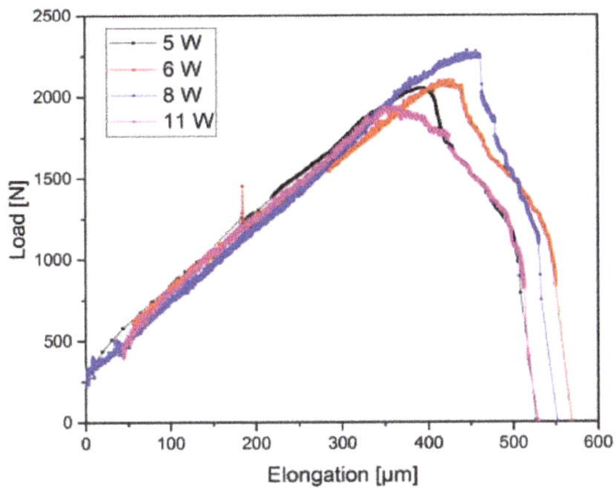


**Figure 10.** Load-elongation curves of CP 1200 steel tested under different plasma charging conditions. Power levels of 5 W, 6 W, 8 W, and 11 W were applied for running the plasma.

The linear loading regimes of the curves are approximately the same for all tested samples. Nevertheless, there are some differences in the maximum load (tensile strength) and elongation to failure. It was expected that the samples with higher power and, therefore, higher H concentration, would be the first ones to fail, but no clear trend was observed. Because load is plotted instead of stress due to the locally thinned sample geometry, the differences in tensile strength could be attributed to small differences in the thickness of the samples due to fabrication. Even a slight difference of ~0.02 mm in the thicknesses of the thinned area of the samples tested could lead to an error of around 5.5% and 3.5% in the maximum load and elongation to failure, respectively, which is consistent with the differences observed in Figure 10. Considering that the H concentration increases with the power (a higher power results in a higher plasma bias voltage), the sample that was tested with 11 W should have the smallest elongation to failure, while the sample with 5 W, the largest. In Figure 10 it can be seen that, even though the sample with 11 W has lower elongation to failure than the samples with 6 W and 8 W, the one with 5 W has approximately the same.

Figure 11 displays the fracture surfaces of the tested samples within the reduced thickness area. Although all of them failed in a ductile manner, they exhibited less degree of necking than the uncharged specimens and there were some differences at the microscopic level. Figure 11a,b shows a very similar morphology. While the upper part of the surfaces exhibit the presence of dimples, the lower parts show a rather smooth surface with less dimples. Figure 11c, on the other hand, shows a mixture of dimples with areas (marked with a box) that could be shear fracture or grains specially oriented. In Figure 11d, the fracture surface morphology is more uniform than the previous cases, being dominated by the presence of small dimples.

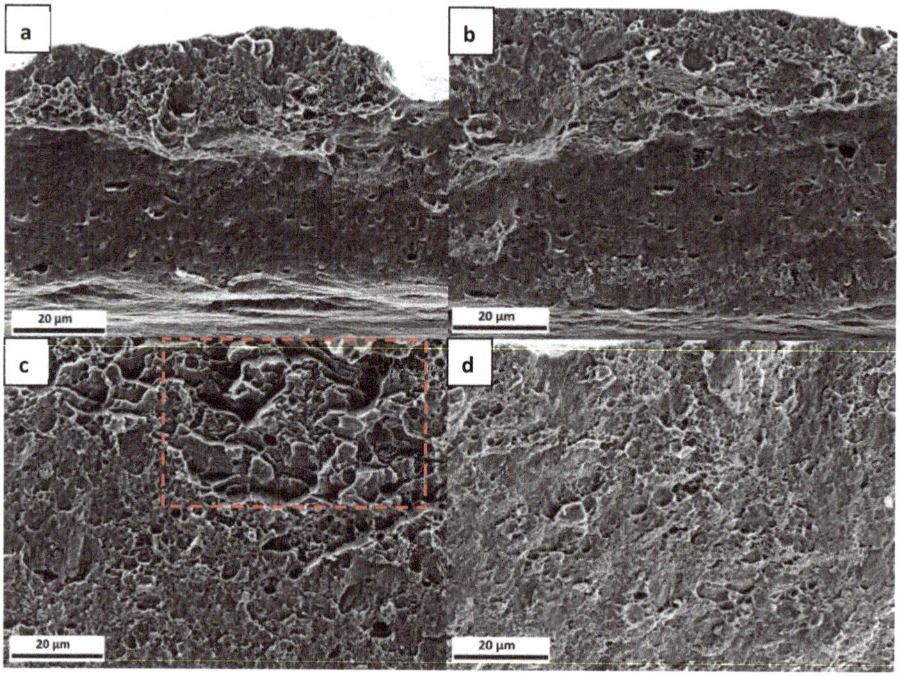


**Figure 11.** Fracture surfaces of the CP steel specimens tested with (**a**) 5 W; (**b**) 6 W; (**c**) 8 W; and, (**d**) 11 W. The red box in (**c**) indicates an area with shear fracture or specially oriented grains.

3.2.3. Effect of H Charging-Discharging

In Figure 12, the stress-strain characteristic of an uncharged sample is compared to a sample that was charged and tested after a discharging time of 24 h. There are some differences in the data, but these are related to small dimensional differences. This would support the notion that HE is a reversible effect. Fractography was also performed to fully support this statement.

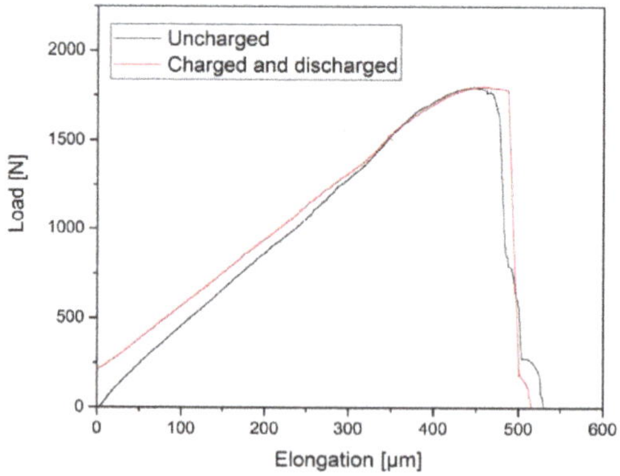


**Figure 12.** Load-elongation curves of an uncharged CP 1200 steel sample and one initially charged for 4 h and tested after discharging for additional 12 h in vacuum.

Figure 13 shows fracture surfaces corresponding to the uncharged and charged and discharged samples. It can be seen that, while the uncharged sample exhibits a typical ductile microvoid coalescence failure, see Figure 4, the charged and discharged sample display some regions where the morphology is similar to the one presented in Figure 11c, with a mixture of dimples and more flat features.

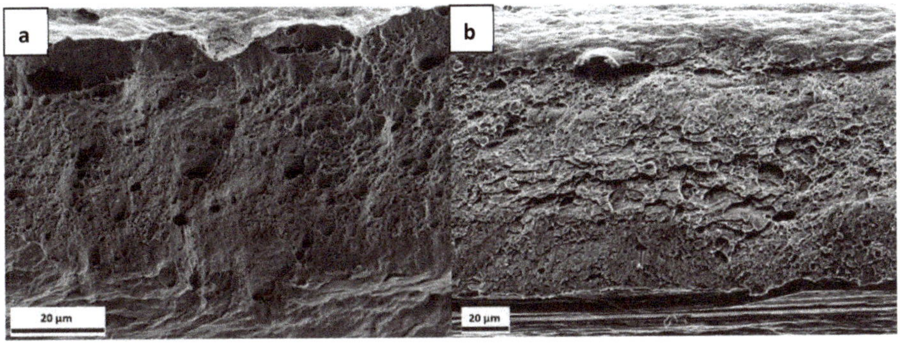


**Figure 13.** Fracture surfaces of the (**a**) uncharged and (**b**) charged and discharged CP steel specimens.

### 3.3. H Concentration

The Monte-Carlo program STrim.SP 6.0D gave a $r_{mean}$=1.15 nm and that 41% of the particles are implanted. From Langmuir Probe measurements [40], the ion flux was calculated to be $10^{20}$ $m^{-2}s^{-1}$. Table 3 shows the recombination coefficients that were calculated according to refs. [43,45,46] and the resulting H concentration. Because the different sources resulted in different concentration values, it is considered that they give a range for H concentration. It is important to mention that these values correspond to lattice H, i.e., the analytical calculation does not consider the trapping effect and, therefore, a higher concentration is expected in the material.

**Table 3.** Recombination coefficient and lattice H concentration calculated with different sources.

| Source | $k_f$ ($m^4$ $s^{-1}$) | $C_{lattice}$ (wppm) |
|---|---|---|
| Shu et al. [43] | $1.75 \times 10^{-21}$ | $1.31 \times 10^{-4}$ |
| Zhou et al. [45] | $5.56 \times 10^{-21}$ | $7.34 \times 10^{-5}$ |
| Baskes [46] | $5.00 \times 10^{-24}$ | $2.45 \times 10^{-3}$ |

Figure 14 shows more elaborated simulation results. Here, Figure 14a depicts the cross section with two different views of the tensile samples, where the arrows indicate the direction of H charging. The concentration decreases through the thickness of the sample, resulting in 0 wppm at the top surface. The lattice concentration shows very good agreement with the analytical calculation made with ref. [46]. However, the total H concentration summing over the H stored in the interstitial lattice positions and in H traps reaches 0.82 wppm, and it is around a factor of 350 higher than the corresponding lattice H. Figure 14b shows the corresponding flux for the three sources in the evaluation node. Even though the values differ, the three cases show that a steady state is reached after approximately 500 s.

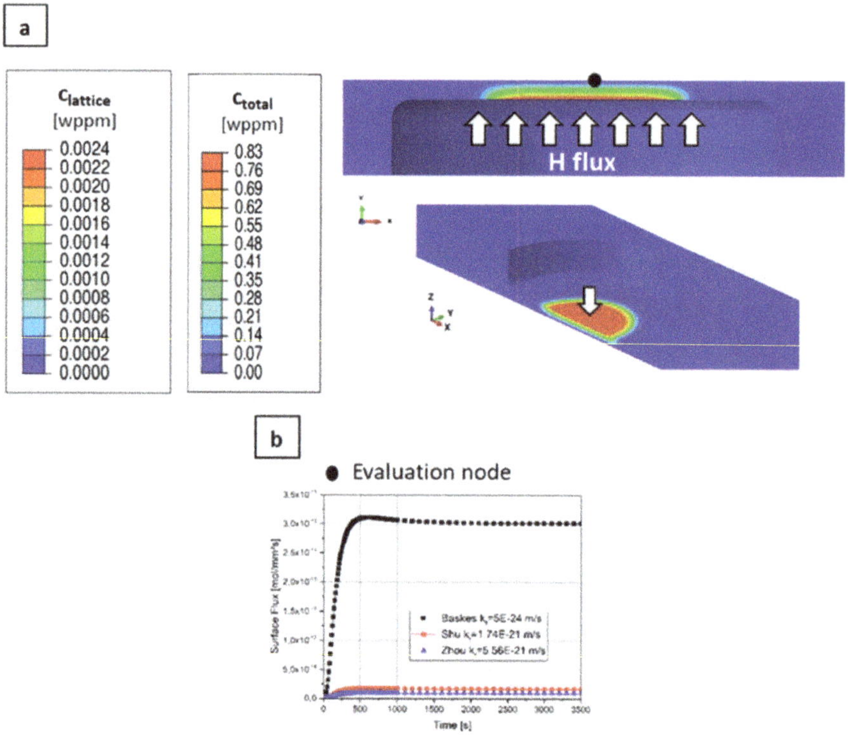


**Figure 14.** (**a**) H concentration distribution through the thickness of the CP steel sample, the arrows indicate the direction of H charging from the bottom of the samples; and, (**b**) H flux over time in an evaluation node on the top surface of the sample.

## 4. Discussion

The H degradation behaviour is known to be dependent on the concentration of the absorbed H [58]. Zackroczmski et al. studied the effect of H concentration on a duplex stainless steel and concluded that the intensity of HE was strongly dependent on the concentration of the absorbed H [59]. Furthermore, HE does not occur below a critical H concentration value [6,17]. Generally speaking, a higher DC self-bias voltage leads to a higher ion concentration [60]. This means that, increasing the power supplied to the RF discharge, an increase in the H concentration should be obtained. Taking this into account, it was expected that, with increasing power in Figure 10, the total elongation (related to the ductility of the specimens) should decrease. As mentioned before, this behaviour was not observed in the samples that were tested with 5 W, 6 W, 8 W, and 11 W, and considering that the accuracy of the displacement sensor is ~0.5 µm, the small differences in elongation to failure could be attributed to the 3.5% error arising from the slight differences in thickness.

Although there were no marked differences in the load-elongation curves, the fracture surfaces exhibited different characteristics. When comparing with the fracture surfaces of the uncharged samples shown in Figure 4, there are some differences with the charged samples in Figure 11. The uncharged samples exhibit a typical ductile failure, formed by the coalescence of large voids involving the nucleation and growth of small voids leading to the presence of large and small dimples on the fracture surface with a bimodal distribution [61]. On the other hand, although, in the charged samples, dimples can also be observed, these have smaller size than in the uncharged case. The dimple size of the samples tested with 5 W and 6 W, as estimated from the SEM images, was 1.4 ± 0.2 µm. The sample tested with 8 W exhibited a dimple size of 2.4 ± 0.5 µm and the 11 W sample, 1.8 ± 0.4 µm. While these values do not follow a specific trend, they are smaller than the uncharged specimens, which presented a bimodal distribution with sizes of 14.5 ± 0.8 µm and 6.6 ± 0.6 µm, as before mentioned. This is in agreement with many previous studies, which showed that the presence of H reduces the dimple size on the fracture surface [58–66].

Moreover, the samples depicted in Figure 11a,b show similar features: the lower part of the fracture surface is rather flat with less dimple features, while the upper part looks more similar to ductile failure. This flat part could be the result of the plastic deformation of the sample. Nevertheless, the dimples in this case are uniform in size and there is not a bimodal distribution present as compared to the uncharged specimens. The similarities between these two samples, with 5 W and 6 W, could be attributed to similar H concentration values. When considering that the H charging is conducted from the bottom of the sample, it is reasonable that with a gradient in H concentration, a gradient in the fracture surface morphology is obtained. In contrast to these two specimens, the ones that were tested with 8 W and 11 W exhibit a more uniform fracture surface, less plastically deformed, and also different to the ones seen in the uncharged specimens. Both of the samples have smaller dimples than the uncharged ones and the main difference between the two charged samples is that in the 8 W condition there are some with flat features. It has been reported [67] that a reduction in dimple size can be a consequence of H effect and it represents a large increase in dimple nucleation.

The load-elongation curves presented in Figure 12 demonstrate that there is either no effect on the mechanical properties caused by the presence of H, or that this effect is reversible, and the properties of the material are restored once the H is eliminated from the samples. These facts would be in line with the premise that diffusible H, which can allegedly diffuse out of the specimens at room temperature, is responsible for HE [68], while trapped H has little to no effect. It has been reported [16] that, in this material, H is mainly trapped at dislocations and martensitic lath boundaries, which are not deep traps. Therefore, it is believed that all of the H effuses when discharging the material. Anyways, the observation of the fracture surfaces in Figure 13 demonstrates that the failure characteristic is somewhat affected.

In Figure 6, the load-elongation curves depict a different trend than in the previous cases. Notably, the three specimens tested converged to approximately the same maximum load. This is mostly set by the unaffected uncharged specimen parts. However, once the failure in the thinned charged parts of the samples started, there was a larger drop in the load with increasing charging time. Furthermore, the crack propagation rate, as shown in Figure 8, also increased in the presence of H and the ultimate strain was reduced by the presence of H. Even though the crack propagation rate was accelerated with increasing charging time, the final elongation was approximately the same for the charged samples. This is in agreement with the result presented in Figure 14b, where it is shown that a steady-state is reached after a very short charging time and the H concentration does not further increase. It is also worth mentioning that the material exhibits diminishing strain rate sensitivity (see Section 3.1.), which means that the changes in crack propagation rate does not relate to different loading situations, but due to H effects on crack propagation.

The fracture surfaces in Figure 7 depict that, in the presence of H, the fracture morphology changed. In the 3 h pre-charged sample, as in Figure 7b, there are "flat" regions, while, in the sample with 6 h of H pre-charge, there is damage in the bottom part of the surface. Although none of the samples exhibited brittle fracture characteristics, the fracture morphology was affected by the presence of H for both pre-charging times, (Figure 7b,c). It is known that the morphologies of H-enhanced fracture surfaces can be changed from microvoid coalescence to quasi-cleavage, cleavage or intergranular fracture, or they can also be unaffected [69]. Furthermore, dimple size was also calculated from the SEM images for these samples, again confirming again that, in the presence of H, the dimple size was reduced. The uncharged specimen exhibited a dimple size of 3.2 ± 0.9 µm, while the sizes of the 3 h pre-charged and 6 h pre-charged samples were 1.3 ± 0.1 and 1.4 ± 0.2 µm, respectively.

To gain a better understanding on the H effect on these samples, the surfaces of the samples were also investigated through EBSD. Analysing the EBSD scans next to the fracture site, it was evident that in the 6 h pre-charged sample secondary cracks formed, and a larger amount of local misorientation in their surroundings was present, as shown in Figure 9d. Misorientation maps can be used as an approach for visualizing plastic deformation. They are a measure of the geometric dislocations in the crystalline lattice and, therefore, measures of plastic deformation at the microstructural level [70]. Although there is a lack of quantitative measures of deformation (strain, strain gradient, dislocation density, etc.), it can be locally related to the density of geometrically necessary dislocations. This indicates that in the sample with 6 h pre-charging, there is a localized plastic deformation in the areas next to the secondary cracks, which means that H could promote a greater dislocation activity in some regions of the microstructure that eventually lead to the formation of many secondary cracks.

The calculation of the present H concentration using two different approaches, a simple analytical method and a more sophisticated simulation approach, generally results in a low value. As expected, there is a gradient through the thickness of the sample, giving the lowest value at the top surface subjected to observation. On the plus side, a steady state concentration profile is reached after only 500 s. This is almost immediately after starting the charging, meaning that varying the pre-charging times would not have any effect in the results, as confirmed by the final elongation of the charged samples presented in Figure 6. Nevertheless, this does not explain why the crack propagation rate increased with charging time and why secondary cracks were only observed in the 6 h pre-charged sample. A possible explanation is that with pre-charging deeper traps are filled up causing more damage.

The low H concentrations could explain why there were no major differences in the load-elongation curves presented in Figure 10, i.e., the mechanical properties were not strongly affected. According to the simulation results that are shown in Figure 14, the maximum H concentration in the plasma-exposed surface is 0.83 wppm and Drexler et al. [18] found that for the same CP1200 steel there is a minor effect on the fracture strain at this H exposure. When considering that during body in white painting process, steel absorbs around 0.4 wppm of H content [5], it would be safe to assume that the CP steel will maintain its mechanical properties during this production process. Drexler et al. [18] showed that a pronounced HE starts at a H concentration of around 2 wppm. Nevertheless, as mentioned previously, some differences in the fracture surfaces were observed, especially when compared with the uncharged specimens, already indicating an onset on material behaviour modification at quite low H concentration, but without losing their strength and ductility.

Low H concentrations will always be a challenge for this high-diffusivity material, but different technological alternatives can be implemented in order to solve this issue. In the future, for example, impermeable nm thick coatings, transparent to the electron beam, could be used to limit the outgassing on the top surface of the sample. For instance, Baskes et al. [46] proposed that permeation barriers made of, for example, Ti or Zr could be used to reduce the recombination coefficient by more than 10 orders of magnitude giving a significant increase in the total H concentration in the material. Another alternative could be to reduce the sample temperature. A higher temperature enhances diffusion and, thereby, reduces the maximum concentration [71].

Even though it is not possible to fully eliminate the gradient in the H concentration, the maximum solute concentration would be higher. Furthermore, the swift diffusion of H defines that only by using an in-situ charging approach any modification can be observed. Ex-situ charged samples would immediately return to their original condition, as shown in Figure 12.

## 5. Conclusions

The presented methodology allowed for having a direct observation of H effect on a CP steel during tensile tests, enabling monitoring the deformation and crack propagation with their correlation to the load-elongation curves. With more conventional charging methods, such as H gas or electrochemical charging, it can be very challenging to have in-situ charging and observation and, for such fast diffusing materials, only direct observation during in-situ charging, as established here, will allow addressing H effects on materials.

Different approaches were performed in order to investigate the behaviour of the steel under the effect of H. The combination of the applied charging conditions and a material exhibiting high H diffusivity resulted in a low H concentration, slightly below the critical value in order to produce embrittlement in the samples. Even though there was a constant supply of H, which led to a dynamic equilibrium of H concentration, this value was not sufficient to have a large impact on the mechanical properties of the material. When comparing these results with the H concentration absorbed during the body in white painting process, it can be concluded that the CP steel is a suitable material to build body in white components.

As aforementioned, there are already slight modifications in fracture characteristics and crack propagation rates, even noticeable at these concentrations. Furthermore, a number of possible future alternatives to generally increase the total H concentration in the material microstructure exist.

**Author Contributions:** Conceptualization, A.M. (Agustina Massone) and D.K.; methodology, A.M. (Agustina Massone), A.M. (Armin Manhard) and A.D.; software, A.D., C.P. and W.E.; formal analysis, A.M. (Agustina Massone) and D.K.; investigation, A.M. (Agustina Massone), V.M.-K.; writing—original draft preparation, A.M. (Agustina Massone); writing—review and editing, A.M. (Agustina Massone), A.M. (Armin Manhard), A.D., C.P., W.E., V.M.-K. and D.K.; supervision, D.K. All authors have read and agreed to the published version of the manuscript.

**Funding:** The authors gratefully acknowledge the financial support under the scope of the COMET program within the K2 Center "Integrated Computational Material, Process and Product Engineering (IC-MPPE)" (Project No 859480). This program is supported by the Austrian Federal Ministries for Climate Action, Environment, Energy, Mobility, Innovation and Technology (BMK) and for Digital and Economic Affairs (BMDW), represented by the Austrian research funding association (FFG), and the federal states of Styria, Upper Austria and Tyrol.

**Acknowledgments:** The authors also acknowledge Voestalpine Stahl GmbH for the donation of the CP1200 material.

**Conflicts of Interest:** The authors declare no conflict of interest. The funders had no role in the design of the study; in the collection, analyses, or interpretation of data; in the writing of the manuscript, or in the decision to publish the results.

## References

1. Matlock, D.K.; Speer, J.G.; De Moor, E.; Gibbs, P.J. Recent developments in advanced high strength sheet steels for automotive applications: An overview. *Jestech.* **2012**, *15*, 1–12.
2. Bergmann, C.; Mraczek, K.; Kröger, B.; Sturel, T.; Jürgensen, J.; Yagodzinskyy, Y.; Guo, X.; Vucko, F.; Kuhlmann, M.; Veith, S.; et al. Hydrogen embrittlement resistance evaluation of advanced high strength steel for automotive applications. In Proceedings of the Steely Hydrogen, Ghent, Belgium, 29–31 May 2018; pp. 1–15.
3. Eurostat. Available online: https://ec.europa.eu/eurostat/statistics-explained/index.php/Construction_production_(volume)_index_overview (accessed on 15 July 2020).
4. Figueroa, D.; Robinson, M. The effects of sacrificial coatings on hydrogen embrittlement and re-embrittlement of ultra high strength steels. *Corros. Sci.* **2008**, *50*, 1066–1079. [CrossRef]
5. Lovicu, G.; Bottazzi, M.; D'Aiuto, F.; De Sanctis, M.; DiMatteo, A.; Santus, C.; Valentini, R. Hydrogen Embrittlement of Automotive Advanced High-Strength Steels. *Met. Mater. Trans. A* **2012**, *43*, 4075–4087. [CrossRef]
6. Louthan, M.R. Hydrogen Embrittlement of Metals: A Primer for the Failure Analyst. *J. Fail. Anal. Prev.* **2008**, *8*, 289–307. [CrossRef]
7. Hilditch, T.B.; Lee, S.-B.; Speer, J.G.; Matlock, D.K. Response to Hydrogen Charging in High Strength Automotive Sheet Steel Products. *Sae Tech. Pap. Ser.* **2003**, 47–56. [CrossRef]
8. Gerberich, W.W.; Oriani, R.A.; Lji, M.-J.; Chen, X.; Foecke, T. The necessity of both plasticity and brittleness in the fracture thresholds of iron. *Philos. Mag. A* **1991**, *63*, 363–376. [CrossRef]
9. Oriani, R.A.; Josephic, P.H. Equilibrium aspects of hydrogen-induced cracking in steels. *Acta Metall.* **1974**, *22*, 1065–1074. [CrossRef]

10. Lynch, S. Environmentally assisted cracking: Overview of evidence for an adsorption-induced localised-slip process. *Acta Met.* **1988**, *36*, 2639–2661. [CrossRef]
11. Lynch, S. Metallographic contributions to understanding mechanisms of environmentally assisted cracking. *Metallography* **1989**, *23*, 147–171. [CrossRef]
12. Lynch, S.P. Comments on "A unified model of environment-assisted cracking. *Scr. Mater.* **2009**, *61*, 331–334. [CrossRef]
13. Birnbaum, H.; Sofronis, P. Hydrogen-enhanced localized plasticity—A mechanism for hydrogen-related fracture. *Mater. Sci. Eng. A* **1994**, *176*, 191–202. [CrossRef]
14. Robertson, I.; Birnbaum, H.; Sofronis, P. Chapter 91 Hydrogen Effects on Plasticity. *Dislocat. Solids* **2009**, *15*, 249–293.
15. Birnbaum, H.K. Mechanisms of Hydrogen Related Fracture of Metals. In *Mechanisms of Hydrogen Related Fracture of Metals*; Defense Technical Information Center (DTIC): Urbana, IL, USA, 1989; pp. 639–658.
16. Birnbaum, H.K.; Robertston, I.M.; Sofronis, P.; Teter, D. Mechanisms of hydrogen related fracture—A review. In *Corrosion-Deformation Interactions*; Trans Tech Publications: Zurich, Switzerland, 1996; pp. 172–195.
17. Lynch, S. Hydrogen embrittlement phenomena and mechanisms. *Corros. Rev.* **2012**, *30*, 105–123. [CrossRef]
18. Drexler, A.; Bergmann, C.; Manke, G.; Kokotin, V.; Mraczek, K.; Pohl, M.; Ecker, W. On the local evaluation of the hydrogen susceptibility of cold-formed and heat treated advanced high strength steel (AHSS) sheets. *Mater. Sci. Eng. A.* **2020**, *800*, 140276. [CrossRef]
19. Drexler, A.; Ecker, W.; Winzer, N.; Mraczek, K.; Kokotin, V.; Manke, G.; Bergmann, C. A step towards numerical evaluation of the local hydrogen susceptibility of punched and cold-formed advanced high strength steel (AHSS) sheets. In Proceedings of the Steely Hydrogen, Gent, Belgium, 29–31 May 2018.
20. Drexler, A.; Bergmann, C.; Manke, G.; Kokotin, V.; Mraczek, K.; Pohl, M.; Ecker, W. Local hydrogen accumulation after cold forming and heat treatment in punched advanced high strength steel sheets. *J. Alloys Compd.* **2020**, in press.
21. Murakami, Y.; Yokoyama, N.N.; Takai, K. Fatigue. Effect of Hydrogen Trapped by Inclusions on Ultra-Long Life Fatigue Failure of Bearing Steel. *J. Soc. Mater. Sci. Jpn.* **2001**, *50*, 1068–1073. [CrossRef]
22. Todoshchenko, O.; Yagodzinskyy, Y.; Yagodzinska, V.; Saukkonen, T.; Hänninen, H. Hydrogen effects on fracture of high-strength steels with different micro-alloying. *Corros. Rev.* **2015**, *33*, 515–527. [CrossRef]
23. Ravichandran, K.; Dwarakadasa, E. Kishore Some considerations on the occurrence of intergranular fracture during fatigue crack growth in steels. *Mater. Sci. Eng.* **1986**, *83*, L11–L16. [CrossRef]
24. Venezuela, J.; Liu, Q.; Zhang, M.; Zhou, Q.; Atrens, A. A review of hydrogen embrittlement of martensitic advanced high-strength steels. *Corros. Rev.* **2016**, *34*, 153–186. [CrossRef]
25. DePover, T.; Escobar, D.M.P.; Wallaert, E.; Zermout, Z.; Verbeken, K. Effect of hydrogen charging on the mechanical properties of advanced high strength steels. *Int. J. Hydrogen Energy* **2014**, *39*, 4647–4656. [CrossRef]
26. DePover, T.; Wallaert, E.; Verbeken, K. Fractographic analysis of the role of hydrogen diffusion on the hydrogen embrittlement susceptibility of DP steel. *Mater. Sci. Eng. A* **2016**, *649*, 201–208. [CrossRef]
27. McCoy, R.A.; Gerberich, W.W.; Zackay, V.F. On the resistance of TRIP steel to hydrogen embrittlement. *Met. Mater. Trans. A* **1970**, *1*, 2031–2034. [CrossRef]
28. Davies, R.G. Hydrogen embrittlement of dual-phase steels. *Met. Mater. Trans. A* **1981**, *12*, 1667–1672. [CrossRef]
29. Duprez, L.; Verbeken, K.; Verhaege, M. Effect of hydrogen on the mechanical properties of multiphase high strength steels. In Proceedings of the 2008 International Hydrogen Conference: Effects of hydrogen on materials, Jackson, WY, USA, 7–10 September 2008; pp. 62–69.
30. Ronevich, J.A.; Speer, J.G.; Matlock, D.K. Hydrogen Embrittlement of Commercially Produced Advanced High Strength Sheet Steels. *Sae Int. J. Mater. Manuf.* **2010**, *3*, 255–267. [CrossRef]
31. DePover, T.; Laureys, A.; Escobar, D.P.; Eeckhout, E.V.D.; Wallaert, E.; Verbeken, K. Understanding the Interaction between a Steel Microstructure and Hydrogen. *Materials* **2018**, *11*, 698. [CrossRef] [PubMed]
32. Malitckii, E.; Yagodzinskyy, Y.; Vilaça, P. Role of retained austenite in hydrogen trapping and hydrogen-assisted fatigue fracture of high-strength steels. *Mater. Sci. Eng. A* **2019**, *760*, 68–75. [CrossRef]
33. Loidl, M.; Kolk, O.; Veith, S.; Gobel, T. Characterization of hydrogen embrittlement in automotive advanced high strength steels. *Mater. Werkst.* **2011**, *42*, 1105–1110. [CrossRef]

34. Rehrl, J.; Mraczek, K.; Pichler, A.; Werner, E. Mechanical properties and fracture behavior of hydrogen charged AHSS/UHSS grades at high- and low strain rate tests. *Mater. Sci. Eng. A* **2014**, *590*, 360–367. [CrossRef]
35. Bhadeshia, H.K.D.H. Prevention of Hydrogen Embrittlement in Steels. *Isij Int.* **2016**, *56*, 24–36. [CrossRef]
36. Ebner, A.S.; Brinckmann, S.; Plesiutschnig, E.; Clemens, H.; Pippan, R.; Maier-Kiener, V. A Modified Electrochemical Nanoindentation Setup for Probing Hydrogen-Material Interaction Demonstrated on a Nickel-Based Alloy. *JOM* **2020**, *72*, 1–10. [CrossRef]
37. Wan, D.; Deng, Y.; Meling, J.I.H.; Alvaro, A.; Barnoush, A. Hydrogen-enhanced fatigue crack growth in a single-edge notched tensile specimen under in-situ hydrogen charging inside an environmental scanning electron microscope. *Acta Mater.* **2019**, *170*, 87–99. [CrossRef]
38. DePover, T.; Hajilou, T.; Wan, D.; Wang, D.; Barnoush, A.; Verbeken, K. Assessment of the potential of hydrogen plasma charging as compared to conventional electrochemical hydrogen charging on dual phase steel. *Mater. Sci. Eng. A* **2019**, *754*, 613–621. [CrossRef]
39. Kim, J.; Tasan, C.C. Microstructural and micro-mechanical characterization during hydrogen charging: An in situ scanning electron microscopy study. *Int. J. Hydrogen Energy* **2019**, *44*, 6333–6343. [CrossRef]
40. Massone, A.; Manhard, A.; Jacob, W.; Drexler, A.; Ecker, W.; Hohenwarter, A.; Wurster, S.; Kiener, D. An SEM compatible plasma cell for in situ studies of hydrogen-material interaction. *Rev. Sci. Instrum.* **2020**, *91*, 043705. [CrossRef] [PubMed]
41. Maier, V.; Durst, K.; Mueller, J.; Backes, B.; Höppel, H.W.; Göken, M. Nanoindentation strain-rate jump tests for determining the local strain-rate sensitivity in nanocrystalline Ni and ultrafine-grained Al. *J. Mater. Res.* **2011**, *26*, 1421–1430. [CrossRef]
42. Toh, T.; Baldwin, W.M. Ductility of steel with varying concentrations of hydrogen. In *Stress Corrosion Cracking and Embrittlement*; Robertson, W.D., Ed.; Wiley: New York, NY, USA, 1956; pp. 176–186.
43. Shu, W.; Okuno, K.; Hayashi, Y. Ion-driven permeation of deuterium in metals. *Jaeri-M.* **1993**, *43*, 1–62.
44. Mutzke, A.; Schneider, R.; Eckstein, W.; Dohmen, R.; Schmid, K.; von Toussaint, U.; Badelow, G. *SDTrimSP Version 6.00 (IPP 2019-2)*; Max-Planck-Institut für Plasmaphysik: Garching, Germany, 2019; pp. 1–92.
45. Zhou, H.; Hirooka, Y.; Ashikawa, N.; Muroga, T.; Sagara, A. Gas- and plasma-driven hydrogen permeation through a reduced activation ferritic steel alloy F82H. *J. Nucl. Mater.* **2014**, *455*, 470–474. [CrossRef]
46. Baskes, M. A calculation of the surface recombination rate constant for hydrogen isotopes on metals. *J. Nucl. Mater.* **1980**, *92*, 318–324. [CrossRef]
47. Barrera, O.; Tarleton, E.; Tang, H.; Cocks, A. Modelling the coupling between hydrogen diffusion and the mechanical behaviour of metals. *Comput. Mater. Sci.* **2016**, *122*, 219–228. [CrossRef]
48. Fischer, F.; Mori, G.; Svoboda, J. Modelling the influence of trapping on hydrogen permeation in metals. *Corros. Sci.* **2013**, *76*, 382–389. [CrossRef]
49. Dassault Systemes. *ABAQUS User's Manual*; Dassault Systemes: Velizy-Villacoublay, France, 2016.
50. Svoboda, J.; Fischer, F. Modelling for hydrogen diffusion in metals with traps revisited. *Acta Mater.* **2012**, *60*, 1211–1220. [CrossRef]
51. Drexler, A.; Siegl, W.; Tkadletz, M.; Ecker, W.; Mori, G.; Svoboda, J.; Fischer, F.D. Cycled hydrogen permeation through Armco iron—A joint experimental and modeling approach. *Corros. Sci.* **2020**, *176*, 109017. [CrossRef]
52. Svoboda, J.; Mori, G.; Prethaler, A.; Fischer, F. Determination of trapping parameters and the chemical diffusion coefficient from hydrogen permeation experiments. *Corros. Sci.* **2014**, *82*, 93–100. [CrossRef]
53. Drexler, A.; DePover, T.; Verbeken, K.; Ecker, W. Model-based interpretation of thermal desorption spectra of Fe-C-Ti alloys. *J. Alloys Compd.* **2019**, *789*, 647–657. [CrossRef]
54. Drexler, A.; DePover, T.; Leitner, S.; Verbeken, K.; Ecker, W. Microstructural based hydrogen diffusion and trapping models applied to Fe–C X alloys. *J. Alloys Compd.* **2020**, *826*, 154057. [CrossRef]
55. Alturk, R.; Mates, S.; Xu, Z.; Abu-Farha, F. Effects of Microstructure on the Strain Rate Sensitivity of Advanced Steels. In Proceedings of the 3rd Pan American Materials Congress, San Diego, CA, USA, 26 February–2 March 2017; Springer Science and Business Media LLC: San Diego, CA, USA, 2017; pp. 243–254.
56. Canale, L.C.; Yao, X.; Gu, J.; Totten, G. A historical overview of steel tempering parameters. *Int. J. Microstruct. Mater. Prop.* **2008**, *3*, 474. [CrossRef]
57. Grange, R.A.; Hribal, C.R.; Porter, L.F. Hardness of tempered martensite in carbon and low-alloy steels. *Met. Mater. Trans. A* **1977**, *8*, 1775–1785. [CrossRef]

58. Pharr, G.M.; Herbert, E.G.; Gao, Y. The Indentation Size Effect: A Critical Examination of Experimental Observations and Mechanistic Interpretations. *Annu. Rev. Mater. Res.* **2010**, *40*, 271–292. [CrossRef]
59. Zakroczymski, T.; Glowacka, A.; Swiatnicki, W. Effect of hydrogen concentration on the embrittlement of a duplex stainless steel. *Corros. Sci.* **2005**, *47*, 1403–1414. [CrossRef]
60. Selwyn, G.S. Optical diagnostic techniques for rie. In Proceedings of the Sixth Symposium on Plasma Processing, San Diego, CA, USA, 29–31 October 1986; Mathad, G.S., Schwartz, G.C., Gottscho, R.A., Eds.; Electrochemical Society: Pennington, NJ, USA, 1987; pp. 220–253.
61. Lynch, S. Mechanistic and fractographic aspects of stress corrosion cracking. *Corros. Rev.* **2012**, *30*, 105–123. [CrossRef]
62. Thompson, A.W. The behavior of sensitized 309S stainless steel in hydrogen. *Mater. Sci. Eng.* **1974**, *14*, 253–264. [CrossRef]
63. Stenerud, G.; Wenner, S.; Olsen, J.S.; Johnsen, R. Effect of different microstructural features on the hydrogen embrittlement susceptibility of alloy 718. *Int. J. Hydrogen Energy* **2018**, *43*, 6765–6776. [CrossRef]
64. Tiwari, G.; Bose, A.; Chakravartty, J.; Wadekar, S.; Totlani, M.; Arya, R.; Fotedar, R. A study of internal hydrogen embrittlement of steels. *Mater. Sci. Eng. A* **2000**, *286*, 269–281. [CrossRef]
65. Marchi, C.S.; Somerday, B.; Tang, X.; Schiroky, G. Effects of alloy composition and strain hardening on tensile fracture of hydrogen-precharged type 316 stainless steels. *Int. J. Hydrogen Energy* **2008**, *33*, 889–904. [CrossRef]
66. Matsuo, T.; Yamabe, J.; Matsuoka, S.; Murakami, Y. Influence of Hydrogen and Prestrain on Tensile Properties of Type 316L Austenitic Stainless Steel. In Proceedings of the 2008 International Hydrogen Conference: Effects of Hydrogen on Materials, Jackson, WY, USA, 7–10 September 2008; pp. 105–112.
67. Thompson, A.W. Ductile fracture topography: Geometrical contributions and effects of hydrogen. *Met. Mater. Trans. A* **1979**, *10*, 727–731. [CrossRef]
68. Takai, K.; Watanuki, R. Hydrogen in Trapping States Innocuous to Environmental Degradation of High-strength Steels. *Isij Int.* **2003**, *43*, 520–526. [CrossRef]
69. Robertson, I.M.; Sofronis, P.; Nagao, A.; Martin, M.L.; Wang, S.; Gross, D.W.; Nygren, K.E. Hydrogen Embrittlement Understood. *Met. Mater. Trans. A* **2015**, *46*, 1085–1103. [CrossRef]
70. Brewer, L.; Othon, M.; Young, L.; Angeliu, T. Misorientation Mapping for Visualization of Plastic Deformation via Electron Back-Scattered Diffraction. *Microsc. Microanal.* **2006**, *12*, 85–91. [CrossRef]
71. Gao, L.; Manhard, A.; Jacob, W.; Von Toussaint, U.; Balden, M.; Schmid, K. High-flux hydrogen irradiation-induced cracking of tungsten reproduced by low-flux plasma exposure. *Nucl. Fusion* **2019**, *59*, 056023. [CrossRef]

*materials*

MDPI

*Article*

# Strain Localizations in Notches for a Coarse-Grained Ni-Based Superalloy: Simulations and Experiments

**Francesco Sausto [1], Luca Patriarca [1,*], Stefano Foletti [1], Stefano Beretta [1] and Erica Vacchieri [2]**

1 Department of Mechanical Engineering, School of Industrial and Information Engineering, Politecnico di Milano, Via La Masa 1, 20156 Milano, Italy; francesco.sausto@polimi.it (F.S.); stefano.foletti@polimi.it (S.F.); stefano.beretta@polimi.it (S.B.)

2 Ansaldo Energia, SPA, Via N. Lorenzi 8, 16152 Genova, Italy; erica.vacchieri@ansaldoenergia.com

* Correspondence: luca.patriarca@polimi.it

**Abstract:** Alloys used for turbine blades have to safely sustain severe thermomechanical loadings during service such as, for example, centrifugal loadings, creep and high temperature gradients. For these applications, cast Ni-based superalloys characterized by a coarse-grained microstructure are widely adopted. This microstructure dictates a strong anisotropic mechanical behaviour and, concurrently, a large scatter in the fatigue properties is observed. In this work, Crystal Plasticity Finite Element (CPFE) simulations and strain measurements performed by means of Digital Image Correlations (DIC) were adopted to study the variability introduced by the coarse-grained microstructure. In particular, the CPFE simulations were calibrated and used to simulate the effect of the grain cluster orientations in proximity to notches, which reproduce the cooling air ducts of the turbine blades. The numerical simulations were experimentally validated by the DIC measurements. This study aims to predict the statistical variability of the strain concentration factors and support component design.

**Keywords:** coarse-grained Ni-based superalloys; crystal plasticity; notch effect; strain concentration factor

**Citation:** Sausto, F.; Patriarca, L.; Foletti, S.; Beretta, S.; Vacchieri, E. Strain Localizations in Notches for a Coarse-Grained Ni-Based Superalloy: Simulations and Experiments. *Materials* **2021**, *14*, 564. https://doi.org/10.3390/ma14030564

Academic Editor: Jaroslav Pokluda
Received: 10 December 2020
Accepted: 11 January 2021
Published: 25 January 2021



## 1. Introduction

Nowadays, renewable energy plants are increasingly used for energy production, even if they cannot guarantee a continuous supply of electrical energy. Thus, turbo-gas engines are still used as an alternative source of energy able to maintain the continuous production of energy. For this reason, industries are still working on improving the design of gas turbines. In particular, frequent start-ups and shut-downs of these power plants lead to fatigue problems in the engine components, which are particularly burdensome for blades and disks in the high pressure and high temperature turbine stages. One of the materials available on the market which can sustain the high mechanical and thermal loads is the Nickel-based superalloy René80 [1]. To guarantee high creep and fatigue performance at high temperatures, René80 features a coarse-grained microstructure. This microstructure directly influences the mechanical behaviour of this material, leading to high anisotropy in its mechanical properties, which promotes a large scatter in the fatigue life. The usual practice for taking into account the great fatigue variability of this material is the employment of a large safety factor. Due to the pressing requirement to improve the performance of these turbo-gas engines, new design methodologies are required, which properly consider the variability introduced by the coarse microstructure.

The possibility to predict the scatter of the fatigue life of René80 correlated with the coarse-grained microstructure and to adopt a proper probabilistic approach can be achieved by combining advanced simulation tools and a dedicated experimental campaign. In [2], Gottschalk and co-workers computed the Schmid factors grain by grain from electron back-scattered diffraction (EBSD) data of the René80 specimens tested. The Schmid factor depends on the orientation of the grain considered in relation to the principal loading axis, and it is hence a parameter that depends on the material's microstructure. Gottschalk et al. proposed correlating the material fatigue life with the Schmid factor, by means of a modified version of the Coffin–Manson curve. The experimental points correlate quite well with the proposed curve, and the authors suggested using the statistical distribution of the Schmid factors in correlation with the proposed curve in the design phase in order to estimate the failure probability of the components.

One of the drawbacks of this approach is that the Schmid factor depends not only upon the grain orientation in relation to the loading axis, but also on the local stress tensor generated between the neighbouring grains. To overcome this limitation, Engel et al. [3] performed several Monte Carlo simulations of the material microstructure of René80; the polycrystalline models of the specimens obtained were then simulated considering a local elastic anisotropic model by means of the finite elements (FE) technique. From the FE simulations of the microstructure, the Schmid factor was then calculated for each node of the elements. The statistical distribution of numerical values was used in a probabilistic framework and showed a better correlation with experimental results than the work by Gottschalk et al. [2]. This model neglects the plastic strain that a grain can accumulate, even if the material behaves elastically at a macroscopic level.

A refined microstructural model, that takes into account both plastic and elastic strain accumulation at a grain level, is the Crystal Plasticity (CP). The backbone of the model's CP is based on the theory of deformation of the crystals proposed by Asaro [4,5]. CP can be implemented in finite element (FE) software to simulate the local behaviour of the material's microstructure, which was proved to correctly match the experimental observations [6–9]. This numerical technique can be used to estimate the impact of geometrical discontinuities, such as notches or defects, on the local stress and strain behaviour of the material. CPFE simulations were used in the work by Battaile et al. [10] to estimate the scatter of the plastic stress inside the microstructure for several defect dimensions. The main result of this work was that a pore influences the mechanical behaviour of the microstructure when its dimensions are comparable with the grains' diameter, leading to a high level of scatter of the plastic strain accumulation. Following the results of Battaile et al. [10], Prithivirajan and co-workers [11] studied the critical pore size that interacts with the microstructure of Inconel718 produced by additive manufacturing (AM). The authors showed that the microstructure of this material is influenced by pores or defects that are bigger than 20 μm; pores or defects smaller than this dimension produced a stress gradient that is almost the same as if one considers the material as isotropic and homogeneous.

In this paper, we analyse the notch effect on the mechanical behaviour of the Ni-based superalloy René80. The performance of a gas turbine blade can be increased by air cooling by means of air ducts inside its body. These ducts can be seen as notches, and hence preferential locations for fatigue crack nucleation. The dimensions of these features are comparable with the material's microstructure, producing an even more accentuated life scatter compared to smooth specimens tested in the laboratory. To estimate the scatter due to the interaction between the notches and the microstructure, the distribution of the strain concentration factor $K_\varepsilon$ was estimated both experimentally and numerically by means of CPFE simulations. The model's CP parameters were estimated from tensile tests on micro-tensile specimens integrated with local, grain-scale, strain measurements using the Digital Image Correlations (DIC) technique. The scatter of the strain concentration factor represents the uncertainty due to the effect of the microstructure, and this has to be accounted for along with the material's fatigue life variability in the design phase to make a safe assessment of the component considered.

This paper is organized as follows. In Section 2, the material and the experimental set-up are presented and discussed. In Section 3, the experimental results are shown. Section 4 describes the procedure adopted to fit the model's CP parameters. In Section 5, both the numerical and the experimental results are discussed.

## 2. Material and Experiments

### 2.1. Microstructure of René80

René80 is a coarse-grain material and the grain size is observed to be quite inhomogeneous, as depicted in Figure 1a for test bar metallographic sections and for real thin wall airfoil components. The starting microstructure for René80 is composed of a bimodal distribution of $\gamma'$ phase; the primary particle size is about 400 nm and secondary particles are smaller than 20 nm, with a total volume fraction that reaches about 50%, as shown in Figure 1b.

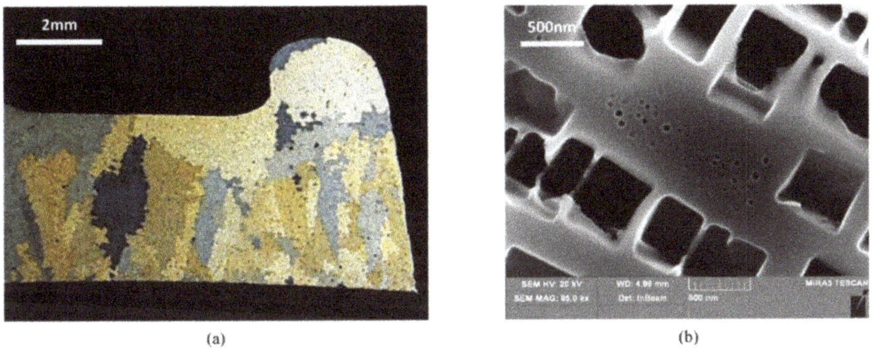


**Figure 1.** Stereo-microscope micrographs of René80: (**a**) cross section of an actual component; (**b**) $\gamma/\gamma'$ phase microstructure at high magnification in FEG-SEM in the as-delivered condition after metallographic etching to highlight $\gamma'$ phase.

### 2.2. Digital Image Correlation Set-Up

Calibration of the CP parameters was performed based on two dedicated tensile tests. Four specimens were cut by electro-discharge machining (EDM), and their geometry is depicted in Figure 2a. The specimens were initially polished to a surface quality for electron back-scattered diffraction (EBSD) analysis by means of emery papers with a grit size of P800 to P2500, 1 µm diamond paste, and final polishing performed using colloidal silica. Three EBSD maps (approximate dimension 3 mm × 2.5 mm) for each specimen side were acquired by means of an FEG-SEM Mira3 manufactured by Tescan (Brno, Czech Republic) equipped with a Hikari EBSD camera manufactured by EDAX-TSL (Mahwah, NJ 07430, USA). The maps were successively stitched to cover the entire gauge length, and a total of two EBSD maps for each tensile specimen were then obtained. Two tensile specimens were tested, and the stress-strain curves were used to calibrate the material's CP parameters. The additional EBSD maps measured on the other two specimens were only used to consolidate the CP simulations and define the scatter in the material behaviour induced by the coarse-grained microstructure. The EBSD maps of the two specimens specifically used for CP calibration are shown in Figure 2b,c. The orientation maps are viewed from the two sample sides to properly observe the grain morphology's development throughout the specimen's thickness. Despite most of the grain sizes being larger than the specimen's thickness, the majority of the grains detected on one side of the specimen differ from the grains observed on the opposite side. Calibration of the material's CP parameters was then performed by analysing both of the EBSD maps for each specimen.

The DIC measurements were performed according to two strategies which, in general, depend on the resolution and dimension of the region of interest (ROI) required. The ex-situ DIC measurements refer to strain measurements performed before and after specimen deformation. Providing a fixed image resolution, the ROI can also be enlarged since multiple images can be acquired and then stitched. The strain maps obtained refer to the un-loaded configuration (residual strains). The second acquisition strategy is labelled as in-situ, and it enables us to measure real time strain fields. The ROI is fixed and the images are continuously acquired during the specimen deformation.

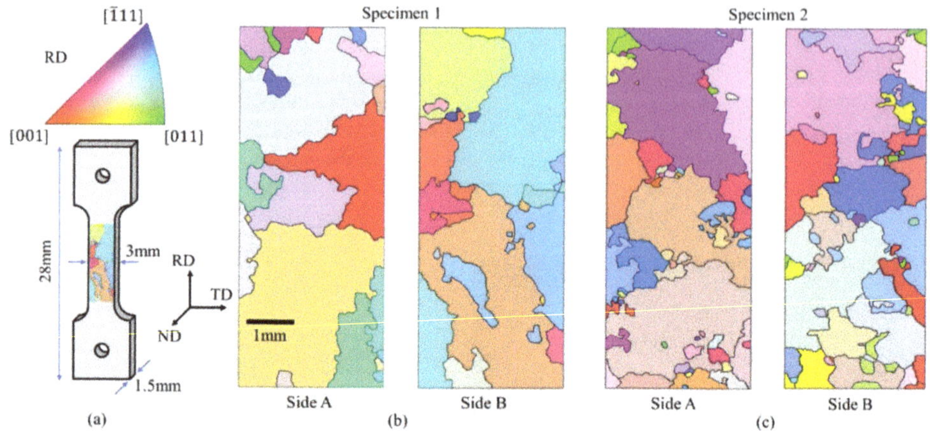


**Figure 2.** Experimental set-up adopted to calibrate the material's Crystal Plasticity (CP) parameters: (**a**) the dog-bone tensile specimens; (**b**,**c**) the electron back-scattered diffraction (EBSD) orientation maps obtained on both sides of specimen 1 and specimen 2.

Following the EBSD measurements, the tensile specimens were successively prepared firstly for DIC measurements. A central area with dimensions of 7 mm × 3 mm was marked with a tape and successively etched with a solution of 95 vol.% of hydrochloric acid and 5 vol.% of hydrogen peroxide for 12 s. This region corresponds with the EBSD map acquired previously. A set of optical images of the grains was then captured and stitched. The reconstructed grain geometry was used to properly correlate the DIC measurements with the microstructure. In fact, the EBSD measurements introduce some degree of distortion when the scanned area is large, as in this analysis. To avoid this problem, the geometry was reconstructed using an optical microscope, while the associated grain orientation was extracted from the EBSD data file.

The specimen surfaces were painted with a black paint using an IWATA airbrush with a characteristic nozzle diameter of 0.18 mm. The images for the DIC measurements were captured with a final resolution of approximately 2 µm/px, and the ROI covered by the single image was found to be 3.2 mm × 2.4 mm as indicated in Figure 3. The DIC set-up consists of: (i) a 2 megapixel Allied vision Manta G201B CDD digital camera; (ii) a set of lenses manufactured by Optem; (iii) a set of linear micro-stages; (iv) a Schott ACE I EKE rim light fibre optic illuminator. DIC measurements were then performed ex-situ and in-situ. With ex-situ, the area of investigation can be larger than the single ROI as multiple images can be captured to cover the entire speckled surface. Three images were captured along the loading direction before specimen loading, and we captured another three images of the same regions after specimen deformation, at zero load. The correlation of the images made it possible to calculate the strain maps of the entire gauge length of the specimen and correlate this with the EBSD data acquired previously. The strain maps acquired with the ex-situ DIC strategy show the residual strains on the tensile test. The DIC was then used in the in-situ configuration. This set-up requires that the position of the microscope be fixed during the experiment, meanwhile a video is recorded during deformation. The ROI is fixed and the strain measurements are performed continuously. This set-up makes it possible to perform real-time acquisition, and the data can be used to calculate either the averaged stress-strain curve or the strain maps.

The tests were performed under a Deben load machine with a load capacity of ±5 kN. The tests were conducted in displacement control with a displacement rate of 1.5 mm/min at room temperature, and the nominal (bulk) axial strain was measured from the in-situ DIC strain fields.

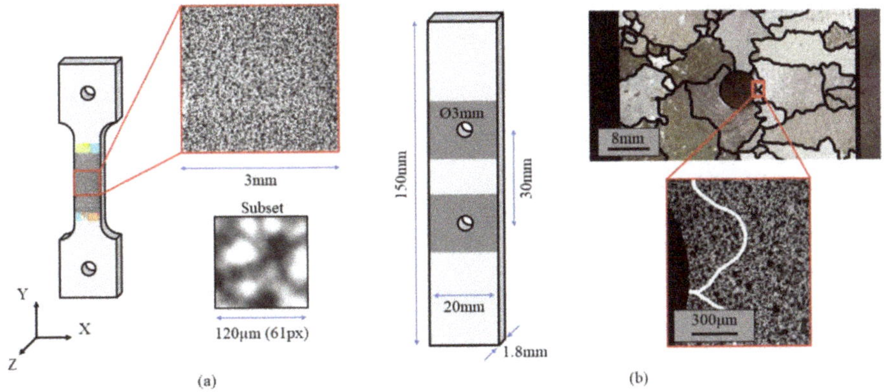


**Figure 3.** Details of the speckle pattern adopted for the Digital Image Correlations (DIC) measurements performed on the specimens tested: (**a**) speckle pattern of the tensile micro specimen with details of the monitored DIC in-situ region; (**b**) details of the notched specimen with details of the microstructure surrounding the manufactured hole and the speckle pattern adopted to perform the DIC.

The notched specimens were machined by means of electro discharge machining (EDM). This process leaves a layer of oxides on the machined surface that are not suitable for testing. To eliminate the material layer affected, specimens were mechanically polished by means of emery papers with meshing of the abrasive particles ranging between 800 and 2500. Two holes were machined in the centre of the plate to act as stress-raisers (Figure 3b).

Before testing, the specimens were airbrushed with black paint to obtain a random speckle on the surfaces, see the inset positioned at the bottom right of Figure 3b. The strain gradient due to the effect of the notch is highly influenced by the microstructure due to the comparable dimensions between the radius of the notch and the nominal size of the grains.

A total of four notched specimens were prepared and tested at room temperature. The tests were conducted in force control by employing an MTS 810 servo-hydraulic testing machine with a maximum load cell capacity of 100 kN. The applied axial force was set to obtain a nominal stress of 350 MPa. The test consisted of two main parts. The initial loading was applied incrementally; at each increment the image of the deformed speckle was recorded in order to evaluate the strain map's evolution around the notches. Then, the specimens were cyclically loaded at stress ratio $R_\sigma = 0$ and a frequency of 15 Hz, with a maximum stress of 350 MPa. Fatigue tests were performed discontinuously with interruptions every $N_i$ cycles—this represents a fatigue block. The DIC acquisition was performed at the end of each fatigue block, making it possible to both inspect the presence of a crack and to compute the evolution of the plastic strain accumulation during the test. The material behaves elastically at 350 MPa, reaching a local yielding at the root of the notches.

## 3. Experimental Results

### 3.1. Tensile Tests

Two specimens of the type depicted in Figure 3a were tested in order to calibrate the CP parameters. The tensile curves, obtained from the correlation between the signal from the testing machine's load cell and the mean strain measured with DIC in the ROI, are reported for specimens 1 and 2 in Figure 4a,b, respectively, by means of the solid black line.

For each test, the strain maps are shown for the ROI related to three points, A B and C, on the tensile curves; all the shown strain maps were computed using a Green-Lagrange tensor. Point A corresponds to the initial yielding phase; at this point most of the grains are subjected to elastic deformation, with sporadic zones in which the plastic strain starts to accumulate. In point B, the unloading phase starts. This point corresponds to the maximum level of macroscopic strain. From the microstructural point of view, the softer grains are subjected to local high yielding as shown in Figure 4. As can be noted, a few numbers of the observed grains, for both tests, show high yielding level, while the rest are subjected to a lower level of deformation. Point C corresponds to the end of the test. At a microstructural level, the residual strain accumulated in the most deformed grains can be observed.

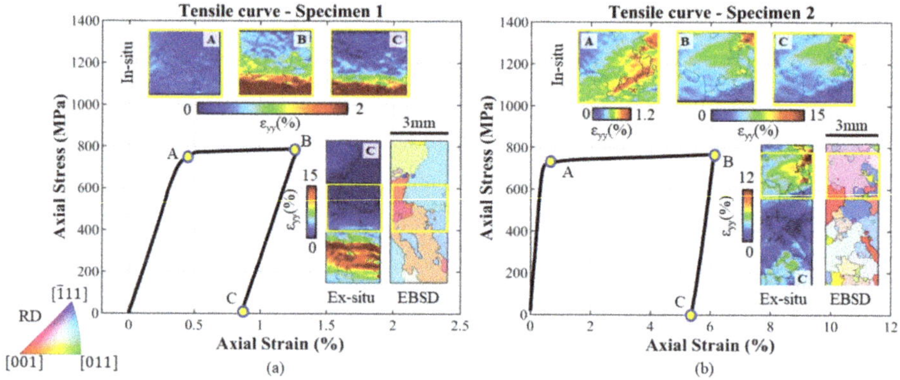


**Figure 4.** Tensile behaviour and DIC strain measurements for (**a**) Specimen 1 and (**b**) Specimen 2. The strain fields reported in the top were captured in-situ at points A, B and C. The ex-situ DIC strain maps were acquired for the entire specimen region and are reported along with the EBSD data. The yellow box in the ex-situ and in the EBSD map represents the position of the region of interest (ROI) acquired in the in-situ configuration.

The strain evolution of the ROI was monitored at each time instant of the test. The whole specimen surface was investigated ex-situ for both tests, and this makes it possible to pinpoint the grains with the highest level of accumulated plastic strain. The resulting strain maps corresponding to the ex-situ analyses are reported in Figure 4, together with the experimental tensile curves. On analysing the ex-situ strain map for the first specimen, it can be noted that the position of the ROI is far from the zone of maximum accumulated strains. At the beginning of the first test, only the EBSD map was available, hence it was difficult to estimate a priori the zone of the surface which is more prone to deform. For this reason, the central part of the specimen was monitored in-situ. Differently from the first test, the microstructure of the second specimen was simulated by CPFE with the parameters found in the first test. Thanks to this preliminary simulation, it was possible to monitor the zone with the higher local plastic strain evaluated via FE. As can be noted, the ROI of the second specimen corresponds to the zone of maximum accumulated plastic strain highlighted in the ex-situ map.

The EBSD maps obtained for the two specimens' observed surfaces are reported together with the strain maps from the ex-situ analysis. In the first test, the most deformed grain featured an orange colour on the EBSD map. In the stereographic triangle (see Figure 2), this colour corresponds to a grain oriented with the [1] crystallographic direction close to the main loading axis. The DIC analysis in Figure 4a confirms that this grain is prone to plastic deformation. In the second test, the violet bigger grain is the one which features the higher level of deformation. Looking at the stereographic triangle of Figure 2, this grain is oriented with the [111] crystallographic direction close to the main loading axis.

### 3.2. Fatigue Tests on Notched Specimens

The first load reversal was applied incrementally; after each load increment the image of the deformed speckle was acquired. The correlation between these images makes it possible to calculate the evolution of the strain map at different load levels. The load was applied considering an increment of 50 MPa from the initial unloaded state to the maximum deformed state. The resulting strain evolution around the notch of one of the specimens tested is reported in Figure 5a.

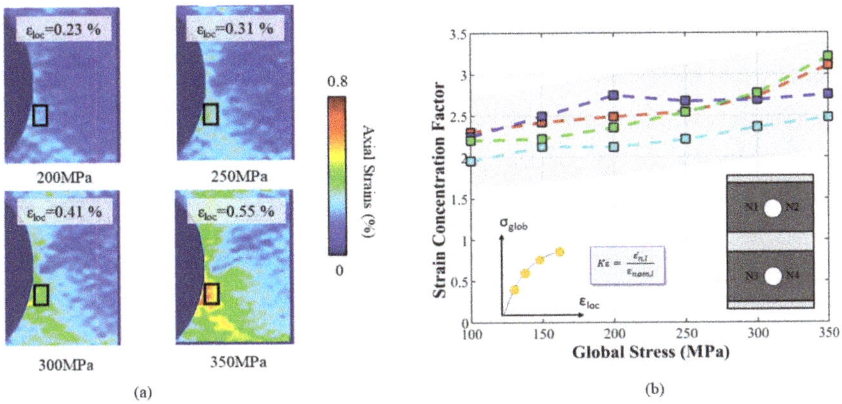


**Figure 5.** Local strain fields captured during the first loading cycle for the notched specimen. (**a**) The local strains are calculated according the point of local maximum and are averaged in a small rectangular area. (**b**) The local strain data were used to calculate the strain concentration factor $K_\varepsilon$ as a function of the applied remote stress $\sigma_{glob}$ and the local microstructure, which determines the final experimental scatter. Each root of the hole represents a notch, hence for each specimen four notches can be evaluated as shown in the scheme.

The maximum deformation was localized below the centreline of the hole. On increasing the load, one observes that the strain continues to accumulate. Due to the microstructure, the strain concentration is accumulated in the softer grain along the notch's circumferences, resulting in deformed regions different from the notch root.

At each loading increment, it was then possible to evaluate the evolution of the strain concentration factor from the DIC analysis. The strain concentration factor $K_\varepsilon$ at each *i*-th loading increment was computed according to Equation (1):

$$K_{\varepsilon,i} = \frac{\varepsilon_{i,loc}}{\varepsilon_{i,nom}} \tag{1}$$

where $\varepsilon_{i,nom}$ is the specimen's nominal strain computed as the ratio of the applied stress and the material's Young's modulus, and $\varepsilon_{i,loc}$ is the local maximum strain. The local maximum strain $\varepsilon_{i,loc}$ was computed from DIC strain maps as the average strain inside a control area as shown in Figure 5a. This procedure was necessary due to the intrinsic experimental noise that could affect precise strain calculations from DIC. The size of the control area was set constant for the whole set of experimental measures, with main dimensions of 210 × 310 μm$^2$. The position of the control area was chosen manually for each test in order to precisely localize the maximum zone of deformation.

The local strains obtained from DIC analysis at different applied loads are reported in Figure 5a. In Figure 5b, the experimental strain concentration factor $K_\varepsilon$, evaluated at four notch locations, i.e., the four notch roots of the manufactured holes, of one test is reported for different levels of applied load. All the locations feature the same incremental trend, with slight fluctuations due to the unavoidable experimental uncertainty. As can be noted, the values of $K_\varepsilon$ show great variability at each load increment from the same specimen. This variability is influenced by the microstructure, which rules the concentration of strain in zones randomly distributed along the notch radius. To appreciate the statistical variability of the experimental findings, a scatter band is depicted in Figure 5b and is shaded in grey, together with the $K_\varepsilon$ curves obtained.

After the first load reversal, the specimens were cyclically tested considering a stress ratio of $R_\sigma = 0$ with a maximum stress level of 350 MPa in the specimen's net section. The tests were interrupted after a certain number of loading cycles. At each test interruption, a DIC acquisition was performed at the maximum and at the minimum applied force. After correlation, the plastic strain accumulations at the highly local deformed regions of each notch were measured. An example of the evolution of the accumulated plastic strain taken from one experimental test is reported in Figure 6. The red solid curve represents the evolution of the local maximum strain in the most deformed zones after each interruption, while the blue solid curve is the minimum strain (i.e., the residual plastic strain). The local strain features an initial stabilization after the first load reversal, followed by a rapid increment which indicates the crack nucleation and then the crack propagation. It was noted that all the final cracks nucleated in the most deformed zones highlighted by DIC analysis of the first reversal cycle.

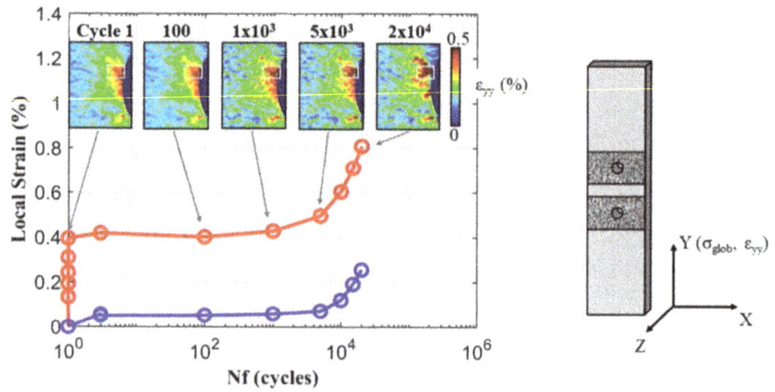


**Figure 6.** The evolution of the local strains according the number of cycles for a notch showing a fatigue crack. At first, a localization of strains can be observed, which stabilizes in the first cycles. Successively, the rapid increment of local strains indicates the nucleation and propagation of a fatigue crack.

## 3.3. Experimental Distribution of Strain Concentration Factors

As stated in the previous sections, the position of the localization of the maximum deformation is randomly distributed along the notch's radius. This is due to the microstructure, which features randomly oriented grains that may or may not accommodate the dislocation motion. The strain maps resulting from the DIC analysis of four notches of one specimen tested at the highest applied force are reported in Figure 7. The position of the notches in relation to the local axis is schematically reported in Figure 7a. The local strains were computed as the averaged values inside the control areas as shown in Figure 7a, and the positions and values vary from one notch to another, even if the specimen is the same.

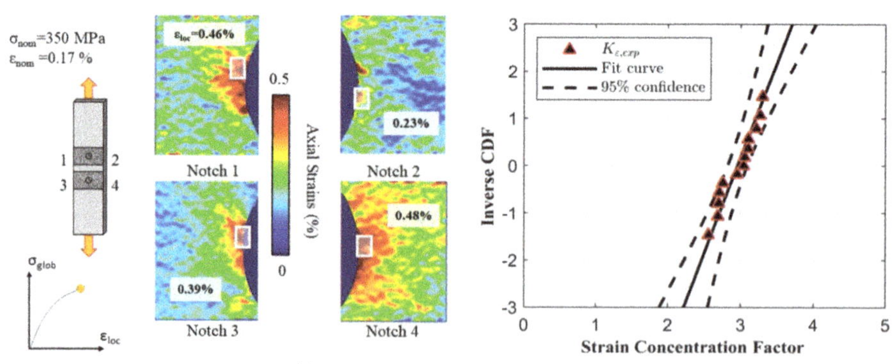


**Figure 7.** Strain heterogeneities as a function of the local microstructure. The maximum local strain was computed from DIC observation (**a**), and the obtained data were found to follow a Gaussian statistical distribution as shown in the probability plot (**b**).

The computed strain concentration factors $K_\varepsilon$ were plotted into a Gaussian probability plot as shown in Figure 7b. Together with the experimental data set, Figure 7b also shows the fitted statistical distribution by a solid black line, and the mean μ and standard deviation σ parameters are reported in Table 1. As can be seen, all the experimental points fall within the 95% confidence bands plotted in dashed black lines, meaning that the chosen distribution is a good approximation of the experimental results.

**Table 1.** Parameters of the Gaussian statistical distribution describing the experimental strain concentration factors $K_\varepsilon$.

| $\mu$ | $\sigma$ |
|---|---|
| 2.96 | 0.25 |

## 4. Crystal Plasticity Simulations

### 4.1. Crystal Plasticity Model

Crystal Plasticity (CP) finite element simulations were performed with Warp3D [12]. This software employs an *implicit* framework to solve the global non-linear equations of nodal equilibrium, using an incremental-iterative approach. The code is based on the deformation model proposed by Asaro et al. [4,5], consisting of a multiplicative decomposition of the deformation gradient:

$$F = F_e \cdot F_p \tag{2}$$

where $F_e$ and $F_p$ are the elastic and the plastic part of the deformation gradient, respectively. To define the current deformation state, it is necessary to define the velocity gradient, which can be broken down into its elastic and plastic parts:

$$L = \dot{F} \cdot F^{-1} = \dot{F}_e \cdot F_e{}^{-1} + F_e \cdot \dot{F}_p \cdot F_p{}^{-1} \cdot F_e{}^{-1} = L_e + L_p \tag{3}$$

According to [5], the plastic deformation of a crystal is supposed to occur by dislocation slip, hence the plastic deformation gradient $L_p$ can be written as:

$$L_p = \sum_{\alpha=1}^{n} \dot{\gamma}^{(\alpha)} \cdot \left( s^{(\alpha)} \times m^{(\alpha)} \right) \tag{4}$$

where $\dot{\gamma}^{(\alpha)}$ is the slipping rate associated with the α-slip plane, while $s^{(\alpha)}$ and $m^{(\alpha)}$ are the unit vectors describing the slip direction and the normal to the slip plane on the α-slip system, respectively. In Warp3D, the slip rate in the α-th slip system is represented as a power law function of the resolved shear stress $\tau^{(\alpha)}$ and the slip resistance $\tilde{\tau}$:

$$\dot{\gamma}^{(\alpha)} = \frac{\dot{\gamma}_0}{\tilde{\tau}} \left| \frac{\tau^{(\alpha)}}{\tilde{\tau}} \right|^{n-1} \tau^{(\alpha)} \tag{5}$$

where $\dot{\gamma}_0$ is the reference slip rate and n is the hardening exponent (typically n is equal to 20 to force the material to approximate an independent flow rate).

Warp3D allows two different numerical methods for the computation of the slip system's strength, which are: the phenomenological constitutive models, in which a critical resolved shear stress, $\tau_c{}^{(\alpha)}$, is associated with each α-slip system; and the physics-based constitutive models, in which the behaviour of the material is linked with the energy required to overcome lattice defects. In this paper, a physics-based constitutive model was used, and in particular from those available in the Warp3D routine, the Mechanical Threshold Stress model (MTS) was chosen. During the deformation, the material experiences the hardening mechanism due to the presence of obstacles that limit the mobility of the dislocations. A metal experiences four different stages of work hardening; in particular, the first hardening stage is appreciable only with material in the form of single crystals [13,14].

The MTS model provides temperature and rate dependent hardening in Stage III, while Stage IV is described by a local strain gradient of the elastic distortion. The total strength can be written in the following general form [12]:

$$\widetilde{\tau} = \tau_a + \tau_y(T,\dot{\varepsilon}) \cdot \frac{\mu(T)}{\mu_0} + \overline{\tau}(\varepsilon_p, T, \dot{\varepsilon}) \cdot \frac{\mu(T)}{\mu_0} \tag{6}$$

where $\tau_a$ is the yielding stress and $\tau_y$ and $\overline{\tau}$ are the scale factor function of the temperature and the strain rate. The term $\overline{\tau}$ depends on the accumulated plastic strain and can be expressed as:

$$\frac{d\overline{\tau}}{dt} = \theta_0 \cdot \left(1 - \frac{\overline{\tau}}{\tau_v(T,\dot{\varepsilon})}\right)^m \sum_{\alpha=1}^{n} \left|\dot{\gamma}^{(\alpha)}\right| \tag{7}$$

where $\tau_v(T,\dot{\varepsilon})$ is the saturation strength of work hardening for a given temperature and strain rate without dependence on accumulated plastic strain $\varepsilon^p$. The parameter $\theta_0$ determines the initial Stage II hardening slope. The parameters $\tau_y(T,\dot{\varepsilon})$ and $\tau_v(T,\dot{\varepsilon})$ have the following expressions:

$$\tau_y(T,\dot{\varepsilon}) = \hat{\tau}_y \left[1 - \left(\frac{k \cdot T}{\mu(T) \cdot b^3 \cdot g_{0,y}} \cdot \ln \frac{\dot{\varepsilon}_{0,y}}{\dot{e}}\right)^{1/q_y}\right]^{1/p_y} \tag{8}$$

$$\tau_v(T,\dot{\varepsilon}) = \hat{\tau}_v \left[1 - \left(\frac{k \cdot T}{\mu(T) \cdot b^3 \cdot g_{0,v}} \cdot \ln \frac{\dot{\varepsilon}_{0,v}}{\dot{e}}\right)^{1/q_v}\right]^{1/p_v} \tag{9}$$

where $\dot{e}$ has the following expression:

$$\dot{e} = \sqrt{\frac{2}{3} D : D} \tag{10}$$

In Equations (7) and (8), $k$ is the Boltzmann constant, $T$ is the absolute temperature, $b$ is the magnitude of the Burger's vector, $g_{y,v}$ are the normalized activation energies, $\dot{\varepsilon}_{y,v}$ are the reference strain rates and $g_{y,v}$ and $p_{y,v}$ are the constants related to the shape of the activation energy barrier.

### 4.2. Crystal Plasticity Model

The CP model's parameters were fitted from the tensile curves presented in Section 3.1. The strategy adopted consists of a trial and error procedure, in which the tensile local curve of the ROI obtained from DIC was compared with that obtained from FE simulation.

To obtain the numerical tensile curves, a precise simulation of the specimen's microstructure was required. As highlighted in Section 2.2, a grain observed on a certain face could result in a totally different shape or even disappear on the second face, resulting in a 3D texture. A simulation in which the microstructure is fully reconstructed even in the thickness direction could be prohibitive from the point of view of both numerical effort and the difficulty of obtaining an experimental 3D map of the grains. To overcome these limitations, the microstructure was assumed to remain the same as that observed on the flat surface until the mid-plane of the specimen (i.e., columnar grains). An example of the procedure adopted to reconstruct one half of the parallel section of the tensile specimen is reported in Figure 8a,b.

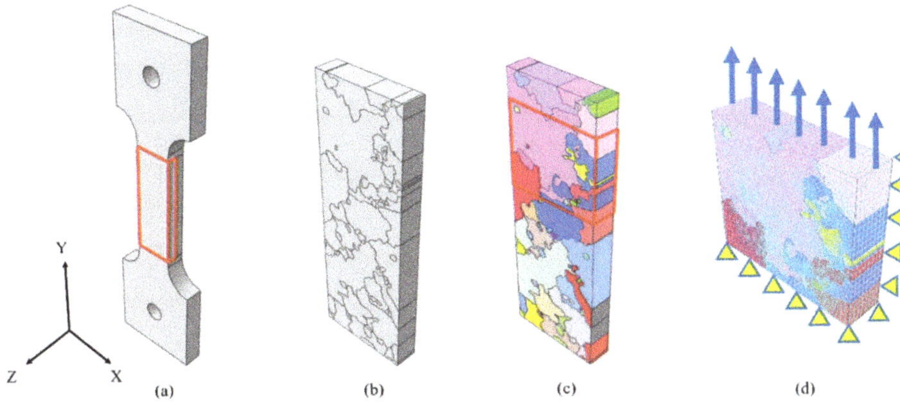


**Figure 8.** CP simulations of tensile specimens' scheme: (**a**) whole specimen geometry with the half parallel section numerically simulated highlighted in red; (**b**) model of the half parallel section simulated considering the microstructure; (**c**) EBSD representation of the simulated model; (**d**) reduced model aimed to simulate the ROI investigated experimentally via DIC.

To reduce the computational effort, only the part of the "half-parallel-section" which corresponds to the observed ROI was numerically simulated with the DIC in in-situ configuration. An example of one simulated ROI is provided in Figure 8c. The surface microstructure was obtained by optical observations after chemical etching. This was performed in order to avoid possible distortions of the grains due to EBSD analysis, which is particularly evident when scanning large areas. From the microscope images, the grains map was mapped manually. Then, using a Matlab code, the grains map was converted into a cloud of points, which was then imported into Abaqus/CAE 6.14 via a Python script. The points were then connected in order to form closed curves, which were then used to create partitions that can be individually meshed. The models were meshed with the dedicated module available in Abaqus/CAE 6.14, and one of the meshed models obtained is shown in Figure 8d. The models were meshed with linear hexahedral FEs, with a mean element size of 0.15 mm.

The ROI models were simulated in Warp3D [12] by considering the following loading and boundary conditions:

- Fixed displacement along the Y direction for all the nodes lying on the XZ plane.
- Fixed displacement along the Z direction for all the nodes lying on the XY plane.
- Fixed displacement along the X direction for the nodes positioned at the origin to avoid numerical errors.
- An imposed displacement along the Y direction for all the nodes that belong to the upper model surface.

The experimental tensile curves were obtained considering the average strain computed by DIC analysis, that is hence representative of the observed region taking into account the specimen's whole thickness. To take into account the effect of the microstructure on the specimen's thickness, the back face was also simulated using CP. The specimen's Side A was meshed with 4962 nodes and 3925 elements, while Side B was meshed with 15,936 nodes and 13,070 elements. The two numerical curves obtained were compared with the experimental one, and the CP parameters were changed until satisfactory fitting was obtained. The final curves obtained in the last iteration were plotted against the experimental ones in Figure 9a, where Side B corresponds to the specimen's face observed using DIC. The macroscopic tensile behaviour of the material is well reproduced by the FE simulation, featuring a maximum error of 3.29%. The microscopic behaviour is reported in Figure 9b for three relevant points, A, B and C, by comparing the experimental and the numerical strain maps. The strain maps simulated using FE are also satisfactory in relation to the experimental findings for both the points considered. The parameters of the model obtained are reported in Table 2.

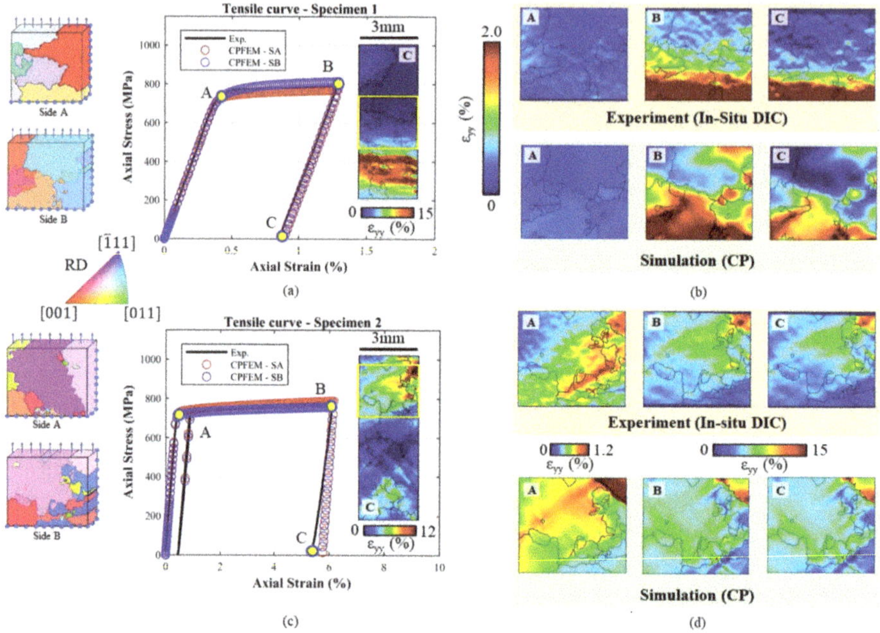


**Figure 9.** Comparison between CP and finite element (FE) simulations of the tensile specimens and the experimental results obtained: (**a**) macroscopic tensile behaviour of the first tensile specimen with the ex-situ DIC strain map; (**b**) comparison between the microscopic results at specific locations on the tensile curve from FE and experimental testing of the first specimen; (**c**) macroscopic tensile behaviour of the second tensile specimen with the ex-situ DIC strain map; (**d**) comparison between the microscopic results at specific locations on the tensile curve from FE and experimental testing of the second specimen.

**Table 2.** Fitted parameters of the CP model for describing the mechanical behaviour of René80.

| Property | Description | Type | Value |
|---|---|---|---|
| E | Young's modulus | Fitted | 185.0 MPa |
| ν | Poisson's ratio | Fitted | 0.33 |
| $\mu_0$ | Shear modulus | Fitted | 75.5 MPa |
| b | Burgers' vector | Literature [15] | $3.5 \times 10^{-7}$ mm |
| $\hat{\tau}_a$ | Athermal slip resistance | Literature [12] | 0 MPa |
| $\hat{\tau}_y$ | MTS strength for intrinsic barrier (yield) | Fitted | 330.0 MPa |
| $g_{0,y}$ | Normalized activation energy for intrinsic barriers | Literature [15] | 0.37 |
| $q_y$ | Shape coefficient for intrinsic barriers | Literature [16] | 1.5 |
| $p_y$ | Shape coefficient for intrinsic barriers | Literature [16] | 1.5 |
| $\varepsilon_{0,y}$ | Strain rate sensitivity for intrinsic barriers | Literature [17] | $1.0 \times 10^{-9}$ s$^{-1}$ |
| $\hat{\tau}_v$ | MTS strength for work hardening | Fitted | 13.0 MPa |
| $g_{0,v}$ | Normalized activation energy for work hardening | Literature [15] | 1.6 |
| $q_v$ | Shape coefficient for work hardening | Literature [16] | 0.667 |
| $p_v$ | Shape coefficient for work hardening | Literature [16] | 1.2 |
| $\varepsilon_{0,v}$ | Strain rate sensitivity for work hardening | Literature [17] | $1 \times 10^{-7}$ s$^{-1}$ |
| $\vartheta_0$ | Initial hardening slope | Fitted | 60.0 MPa |
| $k_0$ | Geometric hardening parameter | Fitted | 1.0 |

The second tensile specimen was simulated using CP before testing, employing the material parameters found in the first test. This makes it possible to have a general idea of the local strain accumulation. The analysis showed that the big upper grain on Side B was expected to undergo high local deformation, and for this reason, the ROI by DIC was positioned in this region. The specimen's Side A was meshed with 33,759 nodes and 29,536 elements, while Side B was meshed with 41,650 nodes and 36,819 elements. The results of the DIC in an in-situ configuration of the selected region confirmed the FE estimations. The results of the macroscopic tensile curves of the simulated ROIs for the two faces are compared with the experimental curve in Figure 9c; the comparison between the experimental and numerical strain maps is then shown in Figure 9d. From both the macroscopic and microscopic point of view, the FE simulations of the second tensile specimen agree well with the experimental findings.

### 4.3. Simulation of the Notched Specimens

The variability of the strain concentration factor (SCF) obtained from the experimental results in Section 3.3 has to be attributed to the random microstructure. To reproduce the experimental scatter, a series of FE simulations is required, taking into account the variability due to the microstructure. The notched specimens tested were then simulated using the CP model fitted in Section 4.1.

To reduce the numerical efforts, only one half of the specimen was simulated. To further simplify the model, only the portion of material surrounding the notch was modelled in terms of the microstructure, while the remaining material was considered to have isotropic behaviour. The schematic of the specimen considered and the different material models adopted in the simulations are reported in Figure 10a,b respectively. The position and shape of each grain were obtained by using the Voronoi tessellation method, which has been shown to reproduce the microstructure of metallic materials well [18–20]. This methodology consists of dividing the space into sub-regions, which represent the locus of the points with the minimum distance from the one specific seed. The random aggregation of grains was obtained using a Matlab code. This code makes it possible to obtain a microstructure that features: (i) random grain orientation and position of the corresponding seeds, and (ii) random growing direction of the grains. A random growing direction of the grains was chosen to approximate the real grain geometry in the thickness direction.

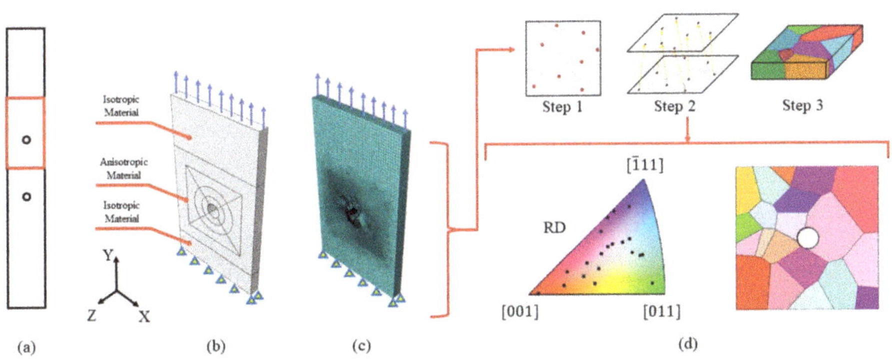


**Figure 10.** Scheme of the simulated model to reproduce the experimental material scatter: (**a**) whole scheme of the specimen considered with the simulated geometry highlighted in red; (**b**) adopted numerical models to describe the behaviour of sections of the specimen; (**c**) final meshed model with boundary and loading conditions. The model was meshed with linear hexahedron FEs, resulting in 117,887 nodes and 104,960 elements; (**d**) EBSD of the obtained random microstructure with the orientation of the grains inside the stereographic triangle.

The algorithm starts by randomly distributing the grains' seeds. This was obtained by drawing a number of coordinates from a uniform random distribution, limited to the feasible domain represented by the specimen's external surfaces. A growing direction was associated with each seed by randomly drawing two angles from a uniform distribution, limited between −20° and 20°. The number of drawings was calculated as the average grain density times the surface area considered. The grain orientations were assigned by changing the value of the Euler's angles in a random way by extractions from uniform distributions obtained from the values of the EBSD analyses. The first layer of FE elements was associated with belonging to a certain grain, according to the partitions obtained from the Voronoi tessellation of the surface. Once the first layer of FE elements was associated with a certain grain index, the analysis continued until the last layer. The positions of the initial seeds were updated for each layer according to the growing direction. The three main steps performed by the developed algorithm are schematically reported in Figure 10d, where the finite elements, shown in Figure 10c, were mapped into a random microstructure. The final microstructure features coarse boundaries, represented by the elements' faces. The orientations of the grains for the obtained EBSD map are reported in black dots inside the stereographic triangle in Figure 10d, and as it can be noted, they are well distributed inside the domain. In reality, the grains are smoother, but it was demonstrated that the approximation of considering the boundary surfaces to be rough did not influence the results [20].

Once the geometry and the microstructure of the specimens were randomly generated, a total of ten CP simulations were run using the free finite elements code Warp3D [12]. From each simulation, a set of four strain concentration factors can be computed along the edge of the notch: two from the frontal face and two from the rear face. Each simulation was obtained by imposing a nominal stress of 350 MPa on the top face in order to reproduce the experimental conditions described in Section 3.3. To be consistent with the experimental findings presented in Section 3.3, the numerical values of the strain concentration factors were computed by averaging the FE nodes inside a control area of 210 × 310 μm$^2$. Due to the relatively large number of notched surfaces, the computation of the $K_\varepsilon$s was performed using Matlab scripts that are able to automatically locate the most critical surface control area. The script considers a moving window that scans all the nodes inside a circular crown with a thickness of 0.21 mm. The lower and the upper bounds of the windows have a relative angle, which is computed in order to make the area of the moving window the same as that used in the experimental characterization of $K_\varepsilon$s from DIC analyses. At each iteration, the nodal stresses inside the window were averaged; the location of the window with the highest averaged stress was considered to be critical, and hence it was the one for which the strain intensity factor was to be evaluated.

In Figure 11, the results in terms of total axial strain of one of the ten FE simulations are reported. Comparing these results with the strain maps obtained via DIC analysis, it is possible to note how the different strain concentrations are well reproduced via numerical simulations.

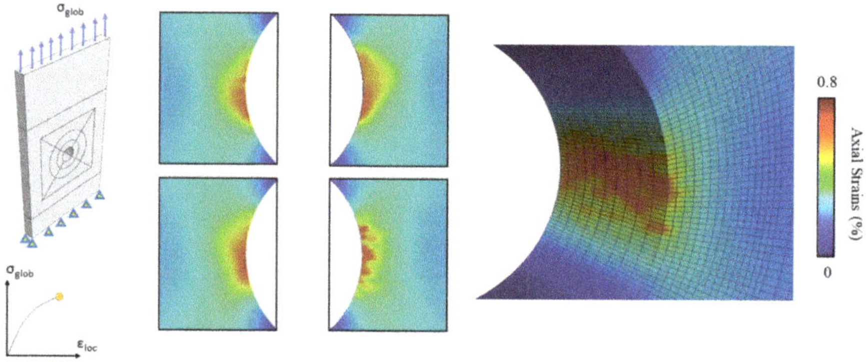


**Figure 11.** Results of CPFE simulation of the notched specimen. The axial strain concentrations observed experimentally via DIC analysis are well estimated by the numerical simulations, considering the effect of random microstructure generated using a Matlab code.

## 5. Discussion

The strain concentration factors $K_\varepsilon$ computed from the numerical simulations of the notched specimens are compared with the experimental values in Figure 12, and the comparison is provided in terms of cumulative density function. The fitted distributions are shown by a solid black line and a solid blue line for the experimental and the numerical data, respectively.

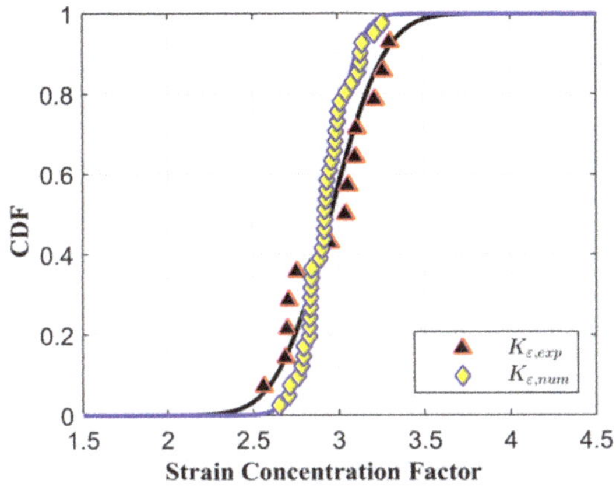


**Figure 12.** Comparison between the distribution of the strain concentration factor estimated experimentally using DIC and that calculated numerically using CPFE.

The fitted statistical distribution is a Gaussian distribution, and the slopes of the simulated and numerical distributions are very similar. The fitted parameters of the experimental data set and the numerical ones are reported in Table 3; the mean value of the two distributions is very similar, while the experimental standard deviation is almost double the numerical one. From the numerical CP simulations, the local tensile curves for the most stressed region of the notches are available; knowing the maximum level of stress and strain and the value of the stress and strain concentration factors, the local hysteresis cycle can be estimated using Neuber's rule supposing an elastic shake down. The local stress and strain concentration factors will be:

$$K_\sigma = \frac{\sigma_{FEM,loc}}{\sigma_n}, \quad K_\varepsilon = \frac{\varepsilon_{FEM,loc}}{\varepsilon_n} \tag{11}$$

where $\sigma_{FEM,loc}$ and $\varepsilon_{FEM,loc}$, respectively, are the local stress and strain in the most stressed region of the notch computed from the FE simulation, and $\sigma_n$ and $\varepsilon_n$, respectively, are the remote stress and strain applied to the specimen. The stress and strain range can be computed simply using Equation (12):

$$\begin{cases} \Delta\sigma = \sqrt{K_\sigma \cdot K_\varepsilon \cdot (\Delta\sigma_n)^2} \\ \Delta\varepsilon = \Delta\sigma / E \end{cases} \tag{12}$$

**Table 3.** Parameters of the Gaussian statistical distribution describing the experimental and numerical strain concentration factors $K_\varepsilon$.

| | $\mu$ | $\sigma$ |
|---|---|---|
| **Experimental** | 2.96 | 0.25 |
| **Numerical** | 2.93 | 0.14 |

Knowing the stress and strain ranges of each numerical simulation, the number of cycles to failure can be computed. In this work, the Fatemi–Socie (FS) multiaxial criterion was adopted. This criterion was chosen due to the non-zero mean stress level of the hysteresis loops. The parameters of the Coffin–Manson curve adopted for estimating the fatigue lives are not reported due to intellectual property rights. The experimental results of the fatigue lives of the notched specimens are reported in Table 4; the number of cycles to failure corresponds to that after which a fatigue crack can be visible, detected using DIC analysis.

**Table 4.** Number of cycles required to detect a visible crack using DIC analysis. All the specimens were tested with a nominal alternate stress of 350 MPa and a stress ratio $R\sigma = 0$.

| Notch Index | Crack | Cycles |
|---|---|---|
| R1_N1 | Yes | 50,000 |
| R1_N2 | Yes | 50,000 |
| R1_N3 | Yes | 50,000 |
| R1_N4 | Yes | 50,000 |
| R2_N1 | Yes | 20,000 |
| R2_N2 | Yes | 5000 |
| R2_N3 | Yes | 30,000 |
| R2_N4 | Yes | 30,000 |
| R3_N1 | Yes | 10,000 |
| R3_N2 | No | NA |
| R3_N3 | No | NA |
| R3_N4 | No | NA |
| R4_N1 | No | NA |
| R4_N2 | No | NA |
| R4_N3 | Yes | 10,000 |
| R4_N4 | Yes | 10,000 |

The fatigue data reported in Table 4 can be fitted using a log-normal distribution. The standard deviation of the log-life is shown to be ${\sigma^{exp}}_{logN} = 0.3623$. The dispersion of the fatigue life of the specimens tested can be divided into two main contributions: (i) the first is due to the coarse microstructure that induced a variability of strain accumulation which is accentuated by the notch; (ii) the second contribution is due to the intrinsic fatigue scatter of the material which comes from tests of a uniformly stressed material. This observation can be expressed in mathematical terms as:

$$ {\sigma^{exp}}_{logN} = \sqrt{{\sigma_{logN,K\varepsilon}}^2 + {\sigma_{logN,mat}}^2} \tag{13} $$

where $\sigma_{logN,K\varepsilon}$ is the fatigue scatter due to the combined effect of the microstructure and the notch effect, and $\sigma_{logN,mat}$ is the intrinsic scatter of the material. From the fatigue life estimations considering the hysteresis cycles approximated using Neuber's rule and the variability of the $K_\varepsilon$, the fatigue scatter is found to be $\sigma_{logN,K\varepsilon} = 0.1809$. Considering the expression in Equation (13) and the experimental data for the notched samples, the fatigue scatter associated with the material will be $\sigma_{logN,mat} = 0.3139$. This value is in line with the fatigue scatter from Low Cycle Fatigue (LCF) data of smooth specimens. For a typical engineering material, such as 9CrMo steel, for which data are analysed by Beretta et al. in [21] and by Zhu et al. in [22], the scatter on the log life is about 0.1325; this value is much lower than the one found for René80, meaning that a design based only on the Coffin–Manson curve is not enough for critical fatigue components.

In light of the previous observations, during the design phase of a critical mechanical component featuring one or more notches (such as the air ducts of a gas turbine blade, which are of the same order of dimension as the considered microstructure), one has to take into account both the variability of fatigue life of the uniformly stressed material and the scatter of the stress and strain accumulation due to the interaction of the microstructure and geometrical features.

## 6. Conclusions

In this work, the statistical variability of the strain concentration factor in notched specimens of René80 superalloy is analysed and discussed. This superalloy features a coarse-grained microstructure that is mainly responsible for the high scatter in both strain localizations and fatigue life. To correctly model the interaction of the notches with the microstructure, a Crystal Plasticity model was fitted with the tensile testing of micro specimens. The local microstructural deformation was evaluated using DIC, and the strain maps obtained were compared with numerical simulations, showing a good correlation. The model was then used to estimate the scatter of the strain concentration due to the interaction between the notches and the microstructure. From these results, it was then possible to estimate the fatigue scatter associated with the notch effect. The following conclusions hold:

- Crystal Plasticity finite element simulations showed high accuracy in the estimation of strain concentrations inside the material's microstructure.
- The strain maps computed using CP showed good accuracy in relation to the experimental ones evaluated using the DIC technique.
- The numerical estimations of the strain concentration factors from finite element simulations of the notched specimens considering random microstructures can be well described by means of a Gaussian distribution. The numerical distribution is in accordance with the experimental distribution computed from DIC acquisitions.
- In the design phase of a critical component which features a coarse-grained microstructure of the same dimensional order of geometrical features, such as René80, the failure probability has to be computed, taking into account not only the fatigue scatter distribution of the flat specimens, but also the material scatter due to the interaction between the notches and the microstructure.

**Author Contributions:** F.S. numerical simulation, data analysis and writing of the original draft; L.P. experimental campaign, analysis of DIC data and contributing to the original draft preparation; S.F. conceptualization and reviewing of the manuscript; S.B. supervision and project administrator; E.V. microstructure and EBSD analysis of the material. All authors have read and agreed to the published version of the manuscript.

**Funding:** The authors did not receive any external funding.

**Institutional Review Board Statement:** Not applicable.

**Informed Consent Statement:** Not applicable.

**Data Availability Statement:** The data presented in this study are available on request from the corresponding author. The data are not publicly due to disclosure agreement.

**Acknowledgments:** The authors acknowledge A. Riva from Ansaldo Energia (AEN) for the revision of the paper and for supplying the material data. AEN is acknowledged for supplying the testing materials and for their permission to publish this research activity.

**Conflicts of Interest:** The authors declare no conflict of interest.

## Nomenclature

| | |
|---|---|
| $\hat{\tau}_a$ | Athermal slip resistance |
| $\hat{\tau}_y$ | MTS strength for intrinsic barrier (yield) |
| $\hat{\tau}_v$ | MTS strength for work hardening |
| $K_\varepsilon$ | Strain concentration factor |
| $\varepsilon_{glob}$ | Global applied strain |
| $\varepsilon_{loc}$ | Local strain |

$g_{0,y}$ Normalized activation energy for intrinsic barriers
$g_{0,v}$ Normalized activation energy for work hardening
$k_0$ Geometric hardening parameter
$p_y$ Shape coefficient for intrinsic barriers
$p_v$ Shape coefficient for work hardening
$q_y$ Shape coefficient for intrinsic barriers
$q_v$ Shape coefficient for work hardening
$\varepsilon_{0,y}$ Strain rate sensitivity for intrinsic barriers
$\varepsilon_{0,v}$ Strain rate sensitivity for work hardening
$\mu_0$ Shear modulus
$\vartheta_0$ Initial hardening slope
μ Mean of the Gaussian statistical distribution
b Burgers' vector
E Young's modulus
ν Poisson's ratio
σ Standard deviation of the Gaussian statistical distribution

**Acronyms**

CP Crystal plasticity
FE Finite element
SCF Strain concentration factor

## References


1. Antolovich, B.F. Fatigue and Fracture of Nickel-Base Superalloys. *Fatigue Fract.* **2018**, *19*, 854–868. [CrossRef]
2. Gottschalk, H.; Schmitz, S.; Seibel, T.; Rollmann, G.; Krause, R.; Beck, T. Probabilistic Schmid factors and scatter of low cycle fatigue (LCF) life. *Materwiss. Werksttech.* **2015**, *46*, 156–164. [CrossRef]
3. Engel, B.; Mäde, L.; Lion, P.; Moch, N.; Gottschalk, H.; Beck, T. Probabilistic modeling of slip system-based shear stresses and fatigue behavior of coarse-grained ni-base superalloy considering local grain anisotropy and grain orientation. *Metals* **2019**, *9*, 813. [CrossRef]
4. Asaro, R.J. Crystal plasticity. *J. Appl. Mech.* **1983**, *50*, 921–934. [CrossRef]
5. Asaro, R.J. Micromechanics of Crystals and Polycrystals. *Adv. Appl. Mech.* **1983**, *23*, 115.
6. Dunne, F.P.E.; Wilkinson, A.J.; Allen, R. Experimental and computational studies of low cycle fatigue crack nucleation in a polycrystal. *Int. J. Plast.* **2007**, *23*, 273–295. [CrossRef]
7. Lim, H.; Carroll, J.D.; Battaile, C.C.; Buchheit, T.E.; Boyce, B.L.; Weinberger, C.R. Grain-scale experimental validation of crystal plasticity finite element simulations of tantalum oligocrystals. *Int. J. Plast.* **2014**, *60*, 1–18. [CrossRef]
8. Jiang, J.; Dunne, F.P.E.; Britton, T.B. Toward Predictive Understanding of Fatigue Crack Nucleation in Ni-Based Superalloys. *JOM* **2017**, *69*, 863–871. [CrossRef]
9. Delaire, F.; Raphanel, J.L.; Rey, C. Plastic heterogeneities of a copper multicrystal deformed in uniaxial tension: Experimental study and finite element simulations. *Acta Mater.* **2000**, *48*, 1075–1087. [CrossRef]
10. Battaile, C.C.; Emery, J.M.; Brewer, L.N.; Boyce, B.L. Crystal plasticity simulations of microstructure-induced uncertainty in strain concentration near voids in brass. *Philos. Mag.* **2015**, *95*, 1069–1079. [CrossRef]
11. Prithivirajan, V.; Sangid, M.D. The role of defects and critical pore size analysis in the fatigue response of additively manufactured IN718 via crystal plasticity. *Mater. Des.* **2018**, *150*, 139–153. [CrossRef]
12. Healy, B. WARP3D. Available online: http://www.warp3d.net/ (accessed on 21 January 2021).
13. Rollett, A.D.; Kocks, U.F. A Review of the Stages of Work Hardening. *Solid State Phenom.* **1993**, *35–36*, 1–18. [CrossRef]
14. Mecking, H.; Kocks, U.F. Kinetics of flow and strain-hardening. *Acta Metall.* **1981**, *29*, 1865–1875. [CrossRef]
15. Gray, G.T.; Chen, S.R.; Vecchio, K.S. Influence of grain size on the constitutive response and substructure evolution of MONEL 400. *Metall. Mater. Trans. A* **1999**, *30*, 1235–1247. [CrossRef]
16. Kok, S.; Beaudoin, A.J.; Tortorelli, D.A. A polycrystal plasticity model based on the mechanical threshold. *Int. J. Plast.* **2002**, *18*, 715–741. [CrossRef]
17. Banerjee, B. The Mechanical Threshold Stress model for various tempers of AISI 4340 steel. *Int. J. Solids Struct.* **2007**, *44*, 834–859. [CrossRef]
18. Zhang, P.; Balint, D.; Lin, J. An integrated scheme for crystal plasticity analysis: Virtual grain structure generation. *Comput. Mater. Sci.* **2011**, *50*, 2854–2864. [CrossRef]
19. Zhang, P.; Karimpour, M.; Balint, D.; Lin, J.; Farrugia, D. A controlled Poisson Voronoi tessellation for grain and cohesive boundary generation applied to crystal plasticity analysis. *Comput. Mater. Sci.* **2012**, *64*, 84–89. [CrossRef]
20. Zhang, K.S.; Ju, J.W.; Li, Z.; Bai, Y.L.; Brocks, W. Micromechanics based fatigue life prediction of a polycrystalline metal applying crystal plasticity. *Mech. Mater.* **2015**, *85*, 16–37. [CrossRef]
21. Beretta, S.; Foletti, S.; Rusconi, E.; Riva, A.; Socie, D. A log-normal format for failure probability under LCF: Concept, validation and definition of design curve. *Int. J. Fatigue* **2016**, *82*, 2–11. [CrossRef]
22. Zhu, S.P.; Foletti, S.; Beretta, S. Probabilistic framework for multiaxial LCF assessment under material variability. *Int. J. Fatigue* **2017**, *103*, 371–385. [CrossRef]

*materials*

*Article*

# Analysis of the Root Causes of Damage to the Edges of Tank Manholes on the Main Deck of Handy-Size Bulk Carriers

**Leszek Chybowski * and Katarzyna Gawdzińska**

Faculty of Marine Engineering, Maritime University of Szczecin, Wały Chrobrego 1-2, 70-500 Szczecin, Poland; k.gawdzinska@am.szczecin.pl

* Correspondence: l.chybowski@am.szczecin.pl; Tel.: +48-91-480-9412

**Abstract:** This study analyzes the root causes of cracks in the deck plating around tank manholes. Four handy-size bulk carriers built in one shipyard were analyzed. In all cases, deck cracks were found near manholes, and the average time from the commencement of operation until the occurrence of cracks was 1356 days. Due to this short wear-life of the vessel's structural material, the authors believed that it was unlikely to be caused by corrosion fatigue. The authors hypothesized that main decks cracked around manholes because of very poor-quality welded joints and poor-quality steel (large amounts of non-metallic impurities) used to make the manholes. In order to verify this hypothesis, on each of the vessels, material samples were collected from near the cracks and then examined thoroughly. Each sample was subjected to the macroscopic examination of the natural surfaces of cracks and their vicinity, microscopic examination of the material, mechanical property tests, and scanning electron microscope fractography for samples obtained after impact tests. The examination and test results were used to draw detailed conclusions for each case study. The general conclusions based on examination of the whole damage population validated the authors' hypothesis that main decks cracked around manholes because of very poor-quality welded joints and poor-quality steel used to make the manholes.

**Keywords:** hull fracture; inspection manholes; bulk carrier; cracks; corrosion fatigue; steel quality

**Citation:** Chybowski, L.; Gawdzińska, K. Analysis of the Root Causes of Damage to the Edges of Tank Manholes on the Main Deck of Handy-Size Bulk Carriers. *Materials* **2021**, *14*, 632. https://doi.org/10.3390/ma14030632

Academic Editor: Jaroslav Pokluda
Received: 21 December 2020
Accepted: 25 January 2021
Published: 29 January 2021



## 1. Introduction

Components of ship hulls are subject to variable loads, which result in fatigue and cracking of their structural materials. This problem concerns ships of various types and has existed for many years [1,2]. Since marine vessels are subject to cyclical and time-varying loads and operate in a very corrosive environment (seawater, salty air), it is commonly agreed that ship hull components crack primarily because of corrosion fatigue [3–5]. This phenomenon is the result of galvanic corrosion and the variable stresses caused by severe pitting, followed by a transition into cracks filled with corrosion products.

Reliability analysis involves the analysis of physical phenomena that cause the fatigue deterioration of a material and a probabilistic assessment of how the damage may have occurred. The physical aspects of the fatigue destruction of metals have so far been explained using many theories and hypotheses [6], but no comprehensive and coherent theory has exhaustively explained the processes underlying fatigue destruction. This is because the process is very complex, and it is impossible to comprehensively account for all factors. The most important hypotheses of fatigue-related destruction of metals include the following [6–8]:

(1) Theories of strengthening and weakening based on experiments conducted by Dehlinger [9] and Gough [10], who showed that strengthening occurs near the glide planes of a crystal lattice under the action of variable loads [11];
(2) Dislocation and vacancy theories are based on the impact of crystal lattice dislocations in metals and interactions [12,13]. These theories were developed by Fujita [14] and Mott [15];

(3) Statistical hypotheses are based on the assumption that the causes of fatigue are in the random (stochastic) material properties. This assumption is based on the fact that every polycrystalline metal is a conglomerate of linked grains containing pores and nonmetallic inclusions [16]. This field was pioneered by Afanasjev [17] and Freudenthal [18];

(4) Energy hypotheses explain how crystal lattice glide bands form slits due to local temperature fluctuations caused by varying loads [8,19]. This hypothesis was proposed by Freudenthal and Weiner [19], and their work was continued by Zakrzewski, Gołaski as well as Oding, [20] and Chowdhury [21];

(5) Decohesion hypotheses assume that a polycrystalline metal contains stress concentrators, such as vacancies, surface irregularities, nonmetallic inclusions, etc. [7]. Fatigue decohesion occurs when the maximum normal stress around a concentrator is equal to the metal cohesive strength. This area was studied by Gołaski, Zakrzewski, Lardner, Bergsmo [22] and others.

There are many corrosion development models for the probabilistic analysis of damage, including probabilistic models invented by Paik et al. [23,24], Akpan et al. [25], Mansour and Wirsching [26], and Downes and Pu [27]. Crack propagation models were developed for a similar approach [28–30]. Probabilistic models were approved by the International Maritime Organisation (IMO), allowing the use of the Formal Safety Assessment (FSA) to evaluate a ship's operational risk [31–33]. These models have been progressively developed since 1997. During the reliability analysis of ship hulls, probabilistic methods are commonly used by classification societies [34–36].

The bulk carrier structural elements most prone to cracks are shown in Figure 1. According to the classification of the US Ship Structure Committee, these elements are located in area IX (ii)—"Fractures in deck plating around access manholes." These cracks are the subject of the analysis in this article. Detailed information about all areas presented in the figure is provided in the US Coast Guard document [37].

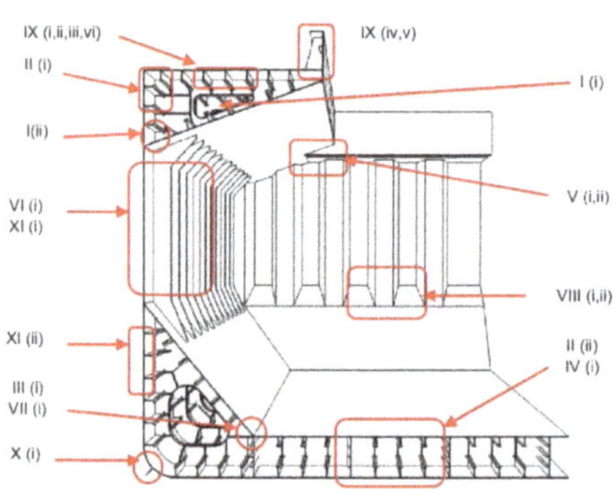


**Figure 1.** Structural areas prone to fractures in bulk carriers [37].

Because the average time from the start of ship operation until the damage was found was about 1356 days for the four analyzed bulk carriers (it is at least six times faster degradation than in highly loaded energetic machines [38]), it does not appear that this damage was caused by corrosion fatigue.

Because ships were built under the supervision of a renowned classification society, it was assumed that they were designed in accordance with the engineering principles and the society's requirements. Moreover, it was assumed that the damages were not caused by mistakes in ship operation or maintenance [39], as the observed incidents occurred simultaneously on four different ships operating in different locations around the world.

Nonmetallic inclusions in the material or poor-quality material joints can also reduce the service life of lids [40–42]. To determine if this is the cause of the damage, studies must be carried out and the results compared [43]. The authors hypothesized that the main decks cracked around manholes because of very poor-quality welded joints and poor-quality steel used to make manholes, which contain large amounts of non-metallic impurities. These factors were accompanied by excessive mechanical loads associated with hull reactions to waves and operation in a corrosive and low-temperature environment [44–46]. Another important factor is the influence of residual stresses resulting from welding on the appearance of cracks inside the welds. This issue was deliberated on in the literature [47,48]. The combination of these factors led to the cracking of the hulls on the examined ships. To analyze the influence of the combination of different factors on a given object failure, decision support tools [49,50] and finite element methods [51–53] might be applied.

## 2. Materials and Methods

During operation, cracks were found in orifice edges in the deck plating around the access manholes of the ballast tanks. These cracks were examined and tested on-board four handy-size bulk carriers of a major shipowner. Examples of the lids on the main deck of a ship are shown in Figure 2.

**Figure 2.** The main deck and tank inspection manhole covers for ballast tanks.

All ships were built by the same shipyard from 2011 to 2020, and the basic data of the ships are shown in Table 1.

**Table 1.** Basic data of ships whose manholes were analyzed.

| No. | DWT | Year of Manufacturing | Length (m) | Beam (m) | Working Time (Days) |
|---|---|---|---|---|---|
| 1 | 30,182 | 2012 | 189 | 24 | 1206 |
| 2 | 30,206 | 2012 | 190 | 23 | 1200 |
| 3 | 30,210 | 2011 | 190 | 24 | 1412 |
| 4 | 30,185 | 2011 | 189 | 24 | 1606 |

On each of the analyzed ships in operation, the edges of the holes were found to be cracking in deck plating around access manholes. The estimated times between launch and the examination of causes of the damage to individual ships were 1206 days, 1200 days, 1412 days, and 1606 days, respectively. The time to damage $t$ can be described by a two-parameter Weibull distribution, which is universal for various data populations, and its probability density function is

$$f(t) = \frac{\beta}{\eta}\left(\frac{t}{\eta}\right)^{\beta-1} e^{-\left(\frac{t}{\eta}\right)^{\beta}} \quad (1)$$

where: $\eta$ is the scale parameter, and $\beta$ is the shape parameter.

The analysis carried out with Weibull++ 9.0 (ReliaSoft Synthesis, HBM Prenscia, Tucson, AZ, USA) enabled us to plot the most important reliability characteristics [54,55], shown in Figure 3. For the analyzed population data, the scale parameter was 1429.74 days, and the shape parameter was 7.91.

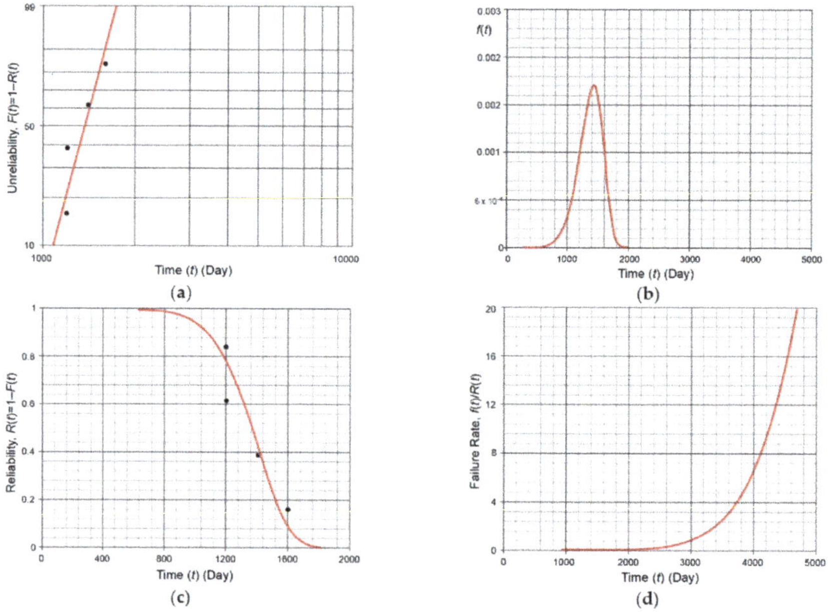

**Figure 3.** Reliability analysis of observed failures: (**a**) Weibull diagram of the unreliability of the analyzed damage population; (**b**) probability density function of the analyzed damage population; (**c**) the graph of the function of the reliability of the analyzed manholes; and (**d**) the graph of the function of the failure rate for the analyzed manholes.

Preliminary reliability analysis showed that for the analyzed manholes, a crack formed 2000 days after the commencement of the ship operation. The expected lifetime of a ship is usually between about a dozen years and a few decades [56], so if cracks occurred so early into a ship's operation, the cause must be investigated.

In order to diagnose the root causes of the cracks at the edges of manholes, detailed tests were conducted on each of the ships, including the collection of material samples from around the cracks and carrying out thorough laboratory tests. The chemistry was analyzed using an X-ray microanalyzer. In each of the cases analyzed, the following items were conducted:

(1) A macroscopic visual examination of the natural surfaces of cracks and their surrounding area using a photo camera;
(2) Microscopic examination of materials was performed using optical (including stereoscopic) microscopy. The number of inclusions was evaluated by comparing to patterns in PN-64/H-04510 [57] and PN-66/H-04505 [58];
(3) Mechanical properties examination included static tensile tests and impact tests. Static tensile tests were conducted in accordance with the standard PN-EN 10002 [59]. Impact tests were conducted in accordance with PN-EN 10045-1 [60], PN-EN 10045-2 [61] and PN-79/H-04371 [62]. The standards PN-EN 10045-1 and PN-79/H-04371 are commonly used for this type of testing, and the values given therein do not differ significantly from those required by classification societies;
(4) The fractography analyses were conducted on samples obtained from the impact tests. The morphology of samples with cracked surfaces was examined using a MIRA 3

scanning electron microscope (TESCAN, Brno, Czech Republic) with high-resolution imaging. Components were mapped on the scanned area based on energy-dispersive spectroscopy (EDS, Oxford Instruments, Concord, MA, USA).

For each of the analyzed case studies, specific conclusions were drawn, then presented in the following sections, and used to draw and compile general and final conclusions on the entire ship population.

## 3. Results and Discussion

### 3.1. Case Study 1—Handy-Size Bulk Carrier 30182 DWT

#### 3.1.1. Macroscopic Examination

A preliminary macroscopic examination was performed to determine the condition of the cracked material. Figure 4a shows two crack origins in the form of the edges of holes and the edges of the manhole stiffening plate perpendicular to the deck. Examples of both origins of cracking seen from the inside of the tank are shown in Figure 4b.

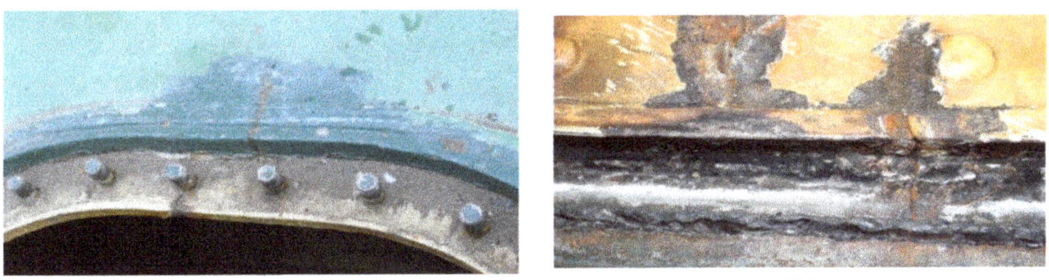

(a) (b)

**Figure 4.** Pictures of origins of cracks: (**a**) view from the outside and (**b**) view from the inside of the tank (handy-size bulk carrier 30182 DWT). Macroscopic examination.

The crack on the right side of Figure 4b developed across a visible weld and then penetrated the insert between the manhole structure and the hole cut out in the deck. The crack does not reach the deck plate (Figure 5), but it is clear that further operation would lead to reaching it.

**Figure 5.** The insert and the welds that join the manhole structure and the deck (handy-size bulk carrier 30182 DWT). Macroscopic examination.

What contributes to the formation and development of cracks is the very poor quality of weld seams, their number, and the presence of the insert compensating for a hole's dimensional differences in the deck. There were no cracks in the deck plates, and they were only found in the manhole structure (material examination was limited to the manhole structure).

3.1.2. Microscopic Examination of Materials

Sections of the manhole structure with cracks were selected and cut out for examination. These locations are shown in Figure 6.

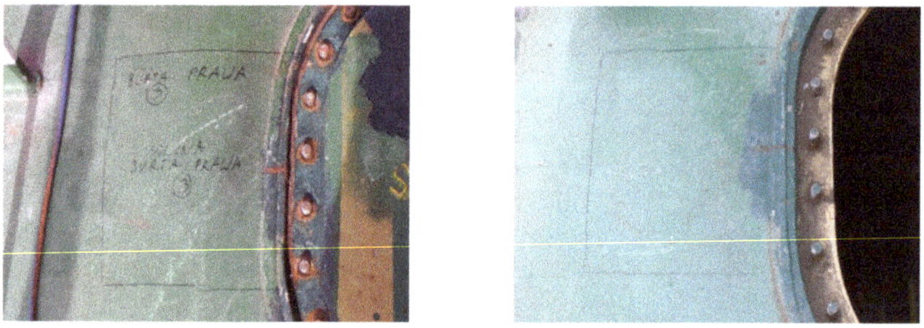

**Figure 6.** The sections of the manhole structure cut out for examination (handy-size bulk carrier 30182 DWT). Macroscopic examination.

Microscopic examination was performed for the front of the crack that originated at the edge of the manhole. A metallographic specimen was made by merely polishing the piece using diamond pastes with 9 μm and 3 μm grain. Figure 7a shows a crack filled with corrosion products.

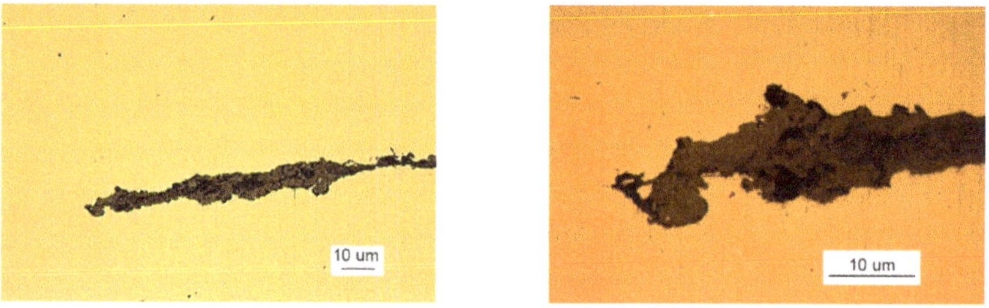


(**a**) (**b**)

**Figure 7.** A crack view: (**a**) a crack filled with corrosion products and (**b**) branching crack face (handy-size bulk carrier 30182 DWT). Optical microscopy.

Despite the severe corrosion, observation of the crack face makes it possible to conclude that the crack developed along grain boundaries and was branched (Figure 7b).

After etching the metallographic specimen with nitric acid and alcohol solution, traces of plastic strain followed by the crack front are seen (Figure 8a).

(**a**) (**b**)

**Figure 8.** Optical microscopy views: (**a**) the face of the crack after etching with Nital and (**b**) the ferritic and pearlitic structure of the examined material (handy-size bulk carrier 30182 DWT).

Figure 8b shows that the examined material has a ferrite structure with a small amount of severely-degraded pearlite. At the boundaries of ferrite grains, an intermittent cementite network is visible.

The cracking process can be defined as stress corrosion caused by varying mechanical loads and an aggressive marine environment. The effect of stress within deck plates can be very important under conditions of extreme load in the hull in port or offshore conditions.

#### 3.1.3. Mechanical Property Tests

For the analyzed case study, material samples were tested for their impact strength. The observed material structure containing cementite network precipitates led to testing the steel's impact strength. Four samples were cut perpendicularly (marking 1) and four parallel (marking 2) to the edge of the manhole opening. The temperature was chosen experimentally to determine the testing range, which were 0 °C, −20 °C, and −40 °C. The results are shown in Table 2.

**Table 2.** The results of the impact strength tests of the samples taken from handy-size bulk carrier 30182 DWT.

| Test Temperature (°C) | Sample No./Test No. | Impact Energy KV 100 (J) |
|---|---|---|
| 0 | 2/1 | Above 100 |
| −20 | 2/1 | 44 |
| −20 | 2/2 | 91 |
| −20 | 1/1 | 21 |
| −20 | 1/2 | 76 |
| −40 | 2/1 | 11 |
| −40 | 1/1 | 10 |
| −40 | 1/2 | 13 |

The material has a significant impact strength at 0 °C. At −20 °C, a large scatter of results is observed, indicating that there is a threshold of brittleness at this very temperature. At −40 °C, the material's impact strength is low.

#### 3.1.4. Fractography Analysis

The severe corrosion of cracks prevented the fractography analysis of these surfaces. Fractography was performed on the surface of fractures in samples from the impact strength testing at −20 °C and −40 °C. Fractures formed at −20 °C exhibit varied structures. Apart from brittle cracks, along the cleavage planes, there are areas where a crack was preceded by plastic deformation (Figure 9a).

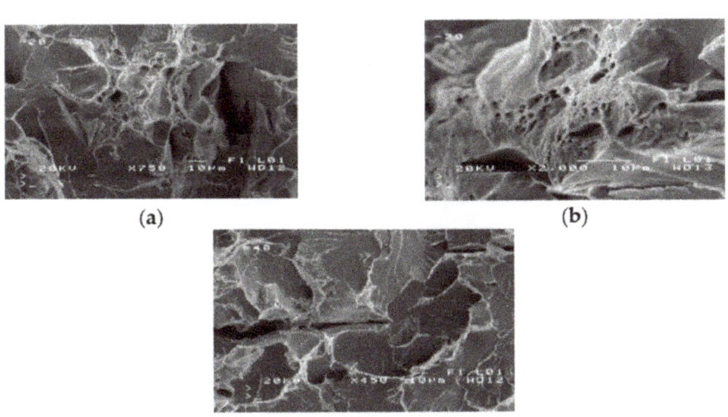

**Figure 9.** Scanning electron microscope (SEM) views: (**a**) a fracture with a mixed structure.; (**b**) plastic fracture with a crater-like structure; and (**c**) brittle fracture obtained at −40 °C (handy-size bulk carrier 30182 DWT).

The area of a ductile crack with a crater-like structure is shown in Figure 9b. Figure 9c shows the fractures obtained in tests conducted at −40 °C.

Fractures obtained in tests performed at −40 °C have a brittle structure. Running along cleavage planes, the cracks have a typical "river basin" appearance. Large fracture-based cracks are also observed.

#### 3.1.5. Summary

The analysis of test results for samples taken from handy-size bulk carrier 30182 DWT makes it possible to draw the following conclusions:

(1) Cracks are found in the material of the manhole;
(2) The cracks originated at the edges of manholes and the edges of stiffening plates positioned perpendicular to the deck surface;
(3) The developing cracks crossed the welds and reached the inserts between the hole cut out in the deck and the manhole structure;
(4) The examined cracks occurred because the manhole structure strain under the ship operating conditions exceeded the strength of the steel used.

### *3.2. Case Study 2—Handy-Size Bulk Carrier 30206 DWT*

#### 3.2.1. Macroscopic Examination

The next tests were performed to establish the causes of cracking in the studied manhole structure using a section of a marine manhole from the handy-size bulk carrier 30206 DWT (Figure 10).

**Figure 10.** The section of a marine manhole from the handy-size bulk carrier 30206 DWT. Macroscopic examination.

The manhole structure with the marking of components and their surfaces are shown in the enclosed sketches in Figures 11 and 12. Part I is a component with bolts installed in it, as shown in Figure 1. Part II is a combination of part I with element III, which is the deck named in Figure 10. Part IV is not visible in Figure 10.

Figure 10 shows two cracks, also numbered 1 and 2 in Figure 12a,b. Crack 1 probably originates on edge A of element I and runs next to the bolt toward element II. Crack 2 originates on edge C of element II (Figure 11). To analyze crack 1, a bar of the plate (12 mm in size, perpendicular to parts I and II) that contained crack 1 was cut out. Figure 13a shows the sample obtained after cutting off a large part of the cracked element I.

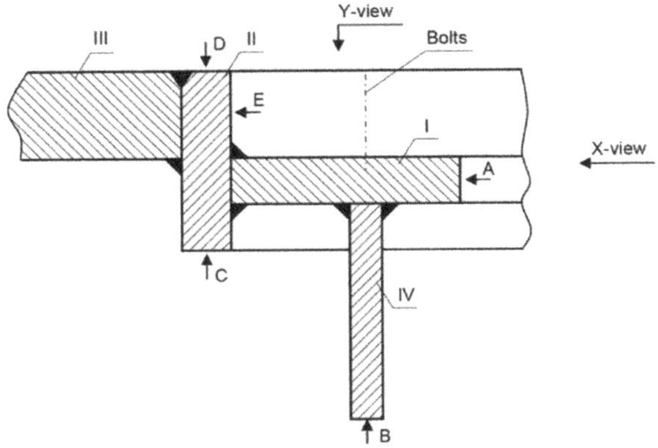


**Figure 11.** The indication of individual manhole parts from the handy-size bulk carrier 30206 DWT.

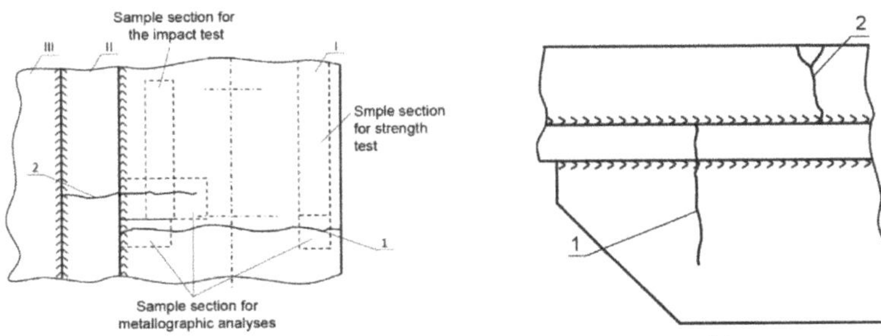


(a) (b)

**Figure 12.** The picture of *y* cracks in a manhole from a handy-size bulk carrier 30206 DWT: (**a**) Y-view and (**b**) X-view.

(a) (b) (c)

**Figure 13.** A macroscopic examination of crack 1 (handy-size bulk carrier 30206 DWT): (**a**) the bar cut from elements I and II, containing crack 1; (**b**) the joint between elements I and II and (**c**) the exit of crack 1.

A cross section of the joint between the two elements is shown in Figure 13b. The right part in Figure is element II, and the left part is the remainder of the cut-off element I. In the cross-section shown, both elements are virtually not welded together. The joints concerned are similar to two other cross-sections made at a distance of about 100 mm and 120 mm from the first section. In the condition of the area being examined, element I acts independently of the entire manhole structure—they do not form a whole. Because element

I is not reinforced by other structural components in this section, it undergoes significant local strain due to the work load. This probably initiates cracking on edge A, although massive corrosion on the fracture surfaces makes it impossible to pinpoint the origin of the crack. The crack extends towards element II, but it does not create another crack since there was no connection between them. It then passes through the weld and exits (Figure 13c).

Crack I ends in element I, stays in it, and does not affect the manhole structure, with one exception. The crack is initiated in part IV (Figure 12b), which is welded to part I, and the former reinforces the latter in the plane perpendicular to the deck (Figure 14).

**Figure 14.** The crack in element IV (handy-size bulk carrier 30206 DWT). Macroscopic examination.

It is very important that element IV is not destroyed in a manner typical for stiffening ribs, in which a crack initiates on its free outer edge as a result of cyclic strain in the plane perpendicular to the deck. Thus, it can be concluded that deformations and the related stresses did not exceed the structure's ultimate strength in this plane. Element IV was destroyed due to the crack in element I.

Crack 2 was tested on a 26 mm thick bar containing the analyzed crack, which was cut out perpendicular to edge A of part I (Figure 15a). The right part is element III, and the left part is element II. The figure shows the bottom view of the deck. The crack initiated near surface C of element II. The end view of the bar is shown in Figure 15b. It shows a subsequent section perpendicular to the length of part II and passing through part I parallel to its length. The image of the surfaces obtained from this cut is shown in Figure 15c.

(a) (b) (c)

**Figure 15.** A macroscopic examination of crack 2 (handy-size bulk carrier 30206 DWT): (**a**) the crack running across parts I, II, and III; (**b**) the cross-cut of part II in the area of the origin of crack 2; and (**c**) the cross-cut of part II in the area of the origin of crack 2.

There is a very large metallurgical inclusion, approximately 10 mm in size, in the cross-sectional plane, situated near the origin of the crack. The inclusion is located in the material of part II, near surface C, and caused the initiation of crack 2. The formed crack developed across the entire height of element II up to surface D and reached part III. Next, it ran inside the element over a significant length (Figure 15a) crossing successive welds and spacer pads between the hole excised in the deck and the manhole structure. The crack also developed in element I (Figure 12a).

Crack 2 originated in part II next to surface X (Figure 12a) and extended toward part III. To analyze this crack, from part II, a 55 mm wide bar containing this crack was cut out perpendicular to the edge of the manhole. The cut-out bar seen from beneath the deck is shown in Figure 16a, and the cutting surface is shown in Figure 16b. In addition to defects of the welds joining the structural parts of the manhole, another cut near the crack origin was made 12 mm off surface X (Figure 12a). That cut created the surfaces depicted in Figure 15a. At the origin of crack II, a large metallurgical inclusion was found, located on the surface of part II adhered to part 1 and is very close to surface X (Figure 12a).

(**a**) (**b**)

**Figure 16.** Macroscopic examination of the surface X surroundings (handy-size bulk carrier 30206 DWT): (**a**) the examined bar cut out from parts II and III with a visible crack and (**b**) the lateral surface of the bar cut out (the location of the next cut is shown).

To show the extensiveness of the inclusion (Figure 17b), another cut was made perpendicular to the surface (Figure 17a), 5 mm off the line between parts I and II. The structure of the inclusion is shown in Figure 18.

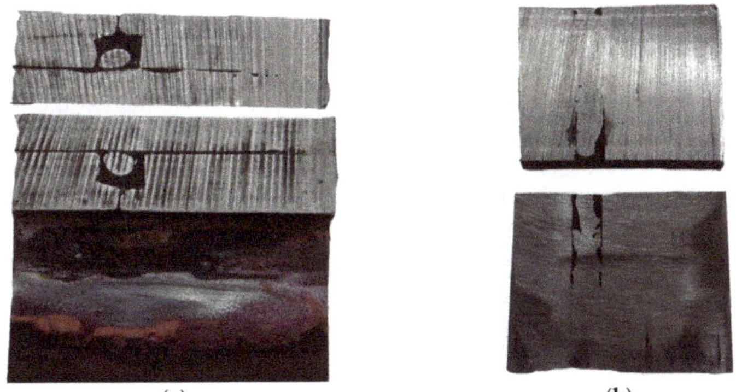

(**a**) (**b**)

**Figure 17.** The extensiveness of the inclusion analyzed with the use of macroscopic examination (handy-size bulk carrier 30206 DWT): (**a**) surfaces created by the cut shown in Figure 11 and (**b**) metallurgical inclusion running parallel to the surface of part II.

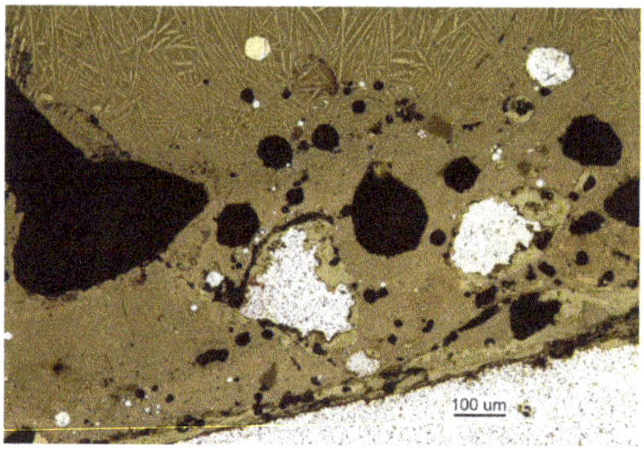


**Figure 18.** Fragment of the metallurgical inclusion that initiated crack 2 (handy-size bulk carrier 30206 DWT). Stereoscopic optical microscopy.

X-ray microanalysis of the item was performed using a Hitachi scanning electron microscope (Hitachi, Ltd., Chiyoda, Tokyo, Japan), and the average chemical composition is given in Table 3.

**Table 3.** The chemical composition of the sample taken from the handy-size bulk carrier 30206 DWT. X-ray microanalysis.

| Element | Content (% m/m) |
|---|---|
| C (carbon) | 0.65 |
| O (oxygen) | 31.14 |
| Na (sodium) | 1.43 |
| Mg (magnesium) | 1.31 |
| Al (aluminum) | 3.65 |
| Si (silicon) | 12.4 |
| K (potassium) | 3.27 |
| Ca (calcium) | 4.55 |
| Ti (titanium) | 21.74 |
| Cr (chromium) | 0.1 |
| Mn (manganese) | 8.29 |
| Fe (iron) | 10.40 |

The analyzed inclusion is a portion of the scoured brick lining of a furnace mixed with charge. The analysis shows that the inclusion did not produce a corrosive but rather a mechanical effect that concentrated the stress and thereby initiated the crack and its development. When the crack reached the surface and contacted water, it started to rapidly expand due to stress corrosion that covered a large part of element II but did not reach element III; however, the crack passed through the weld and reached element I.

3.2.2. Microscopic Examination of Materials

The metallographic examination was conducted on samples cut from elements I and III. A sample of element I was cut from the front of crack 2. The analysis of a non-etched specimen showed a very large amount of nonmetallic inclusions as spots of aluminum oxides and carbonitrides (Figure 19a). There were more oxides than in the TP5a pattern, and the amount of carbonitrides was higher than in the AA5b reference. The examined specimen was etched with an alcoholic solution of nitric acid to reveal the ferritic-pearlitic structure of the steel (Figure 19b). The pearlite has a varied structure with the largest interplate distance corresponding to pattern 7 on scale 1 according to PN-66/H-04505 [58].

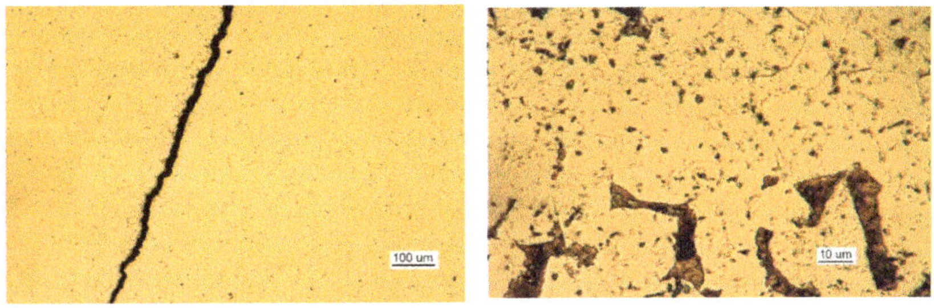


(**a**) (**b**)

**Figure 19.** Optical microscopy views (handy-size bulk carrier 30206 DWT): (**a**) the crack and a very large number of spot-wise oxides and carbonitrides. Non-etched specimen; and (**b**) the ferritic and pearlitic structure of the material of element I.

Over its entire length, including the front (Figure 20), the crack was filled with corrosion products, which indicates that cracking was based on stress and corrosion.

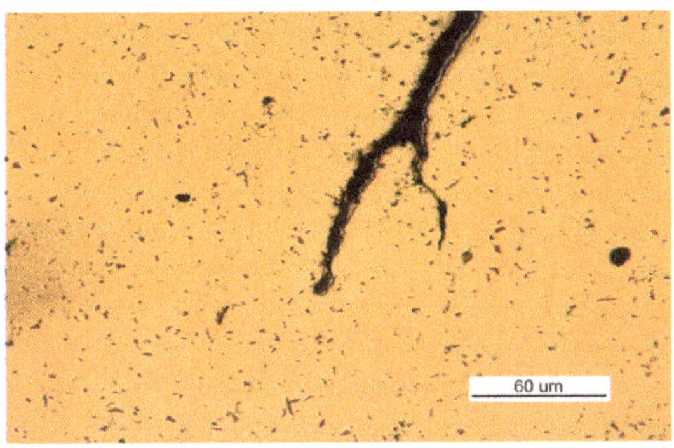


**Figure 20.** The crack front. Non-etched metallographic specimen (handy-size bulk carrier 30206 DWT). Optical microscopy.

A sample of element III was cut from the front of crack 2. The examination of the non-etched specimen (Figure 21a) shows a large number of nonmetallic inclusions as spot-wise oxides, and this number corresponds to TP5a pattern and globular silicates corresponding to KN5b pattern according to PN-64/H-04510 [57].

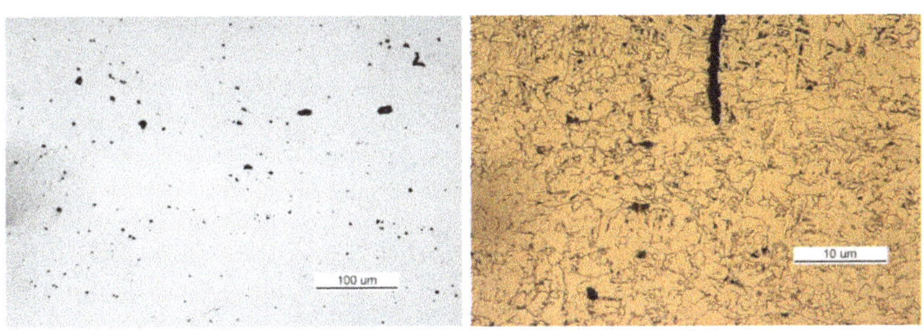


(**a**) (**b**)

**Figure 21.** Optical microscopy views (handy-size bulk carrier 30206 DWT): (**a**) nonmetallic inclusions in the structure of element III (non-etched metallographic specimen) and (**b**) structure of the steel near the front of crack 2.

The structure of steel in this area is typical for a heat-affected zone, and it contained ferrite supersaturated with carbon and traces of pearlite (Figure 21b).

Over its entire length (including the front), the crack was filled with corrosion products, which indicates that cracking was based on stress and corrosion. At the origin of the crack, a very large metallurgic inclusion (Figure 22) was found that contained sulfides and other substances. Because the inclusion was located near the surface, water more easily entered the inclusion during vessel operation. In the presence of water, sulfides dissociate to form sulfuric acid; thus, a strong electrolyte is present, creating conditions for intense crack development, indicating that the crack is stress-corrosive.

**Figure 22.** The passage of the crack from part II into part I through the weld that joins the two parts (handy-size bulk carrier 30206 DWT). Macroscopic examination.

The structure of the steel used to make part II is ferritic and pearlitic (Figure 23a), with an intermittent cementite network at the ferrite grain boundaries (Figure 23b). The metallographic specimen was etched with an alcoholic solution of nitric acid, and it was observed on a Nikon MM-40 optical microscope (Nikon Instech Co., Ltd., Kawasaki, Kanagawa, Japan).

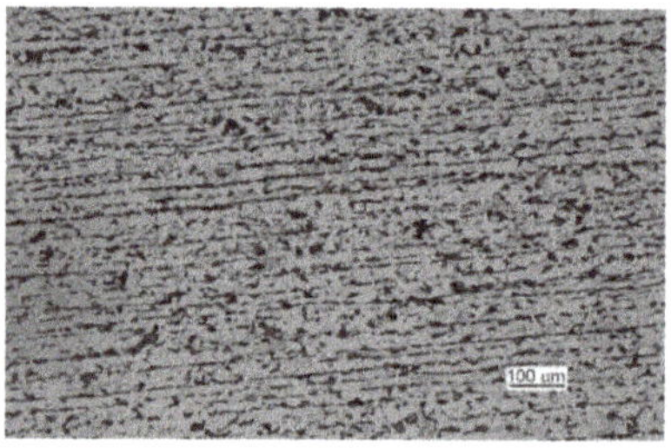



(**a**)

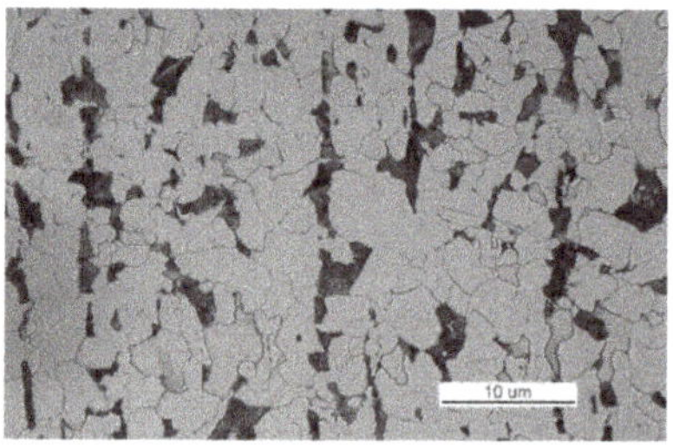



(**b**)

**Figure 23.** Optical microscopy views (handy-size bulk carrier 30206 DWT): (**a**) the ferritic and pearlitic structure of the steel of part II and (**b**) structure of the steel of part II (the cementite network is intermittent at the grain boundaries).

The non-etched specimen exhibited rolled manganese and iron sulfates corresponding to the S3b/4a pattern [57] with adjacent silicates (Figure 24a) and numerous oxide spots (Figure 24b) corresponding to pattern TP3a [57].

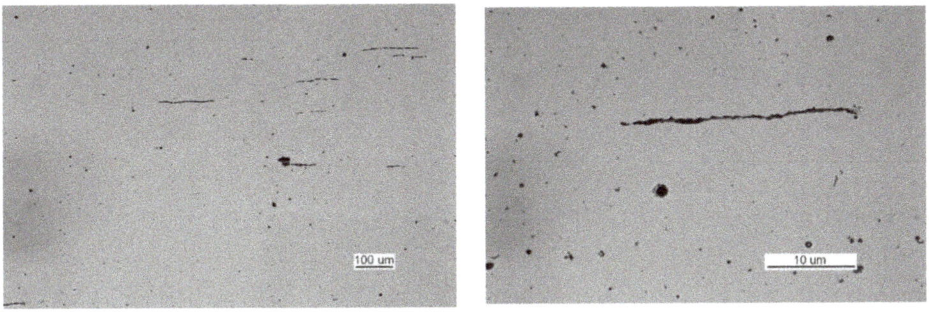


**(a)** **(b)**

**Figure 24.** Optical microscopy views (handy-size bulk carrier 30206 DWT): (**a**) numerous rolled-out sulfides and oxide spots and (**b**) rolled-out manganese and iron sulfides with adjacent silicates and numerous oxide spots.

### 3.2.3. Mechanical Property Tests

Static tensile tests were conducted for the analyzed samples. Samples for the strength tests were cut from element I from the location shown in Figure 12a. The results are shown in Table 4.

**Table 4.** Results of the static tensile tests of samples from the handy-size bulk carrier 30206 DWT.

| Sample No. | The Diameter of the Measured Part (mm) | Ultimate Tensile Strength $R_m$ (MPa) | Elongation at Break $A$ (%) |
|---|---|---|---|
| 1 | 6 | 465 | 32 |
| 2 | 6 | 458 | 35 |

An impact test was also conducted using samples cut from element I from the location shown in Figure 12a. The results are shown in Table 5.

**Table 5.** Results of the impact strength test of samples from the handy-size bulk carrier 30206 DWT.

| Sample No. | Impact Energy $KV^0100$ (J) |
|---|---|
| 1 | >98 |
| 2 | 54 |
| 3 | 83 |

The test was only conducted at 0 °C, because the impact strength is very low.

### 3.2.4. Fractography Analysis

The observed fracture has a mixed structure (Figure 25b,c) that includes ductile, crater-shaped areas (Figure 25a) adjacent to areas of the brittle structure (Figure 25b). There are numerous non-metallic inclusions located inside the craters, as shown in Figure 25c.

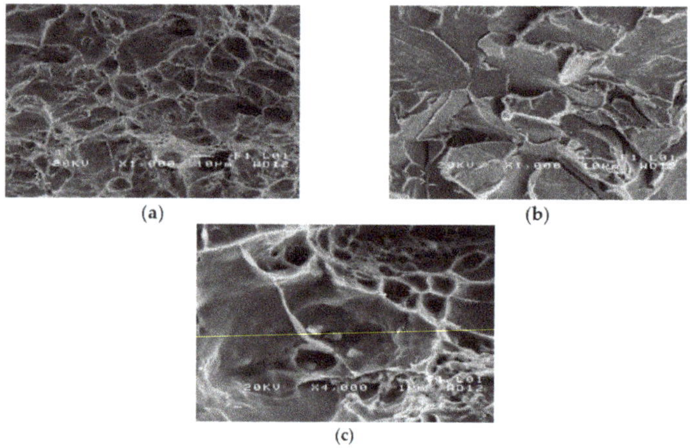

**Figure 25.** Scanning electron microscope (SEM) views (handy-size bulk carrier 30206 DWT): (**a**) the ductile structure of the fracture; (**b**) the brittle structure of the fracture; and (**c**) nonmetallic inclusions inside craters.

#### 3.2.5. Summary

Cracks formed in the examined manhole due to the very poor quality of the welded joints and because poor-quality steel with a large number of non-metallic impurities was used to produce the manhole.

From the condition of element IV, it can be concluded that the loads in the plane perpendicular to the deck did not exceed the strength of the manhole structure. It cannot be concluded from crack 1 that the loads in the plane of the deck exceeded the strength of the manhole structure because only element I was cracked due to the poor quality of its joining with the whole structure. Crack 2 was caused by a metallurgical defect in the material used for element II.

### *3.3. Case Study 3—Handy-Size Bulk Carrier 30210 DWT*

#### 3.3.1. Macroscopic Examination

Testing was performed to establish the causes of cracking in the examined manhole structure. The testing focused on a section of a manhole supplied from the handy-size bulk carrier 30210 DWT. The section seen from the lid mounting is shown in Figure 26a and from the opposite side in Figure 26b.

**Figure 26.** Macroscopic examination results (handy-size bulk carrier 30210 DWT): (**a**) view of a manhole section from the top of the deck and (**b**) view of a manhole section from the bottom of the deck.

Two cracks were observed in the examined manhole section. Crack I (visible in Figure 26) is also shown in the close-up photo in Figure 27a, and crack II (visible in Figure 26b) is shown in the close-up photo in Figure 27b.

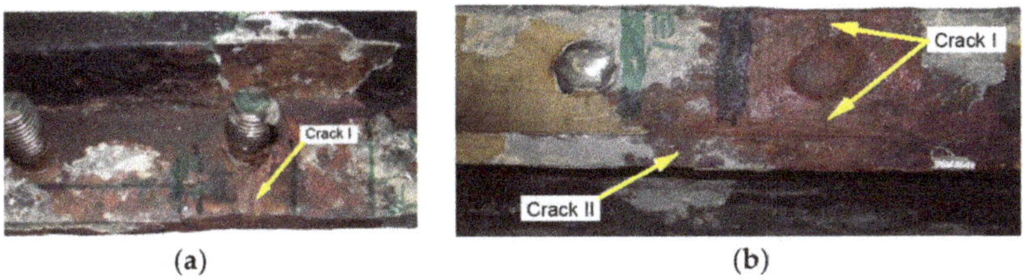


**Figure 27.** Macroscopic examination (handy-size bulk carrier 30210 DWT): (**a**) view of crack I, (**b**) view of crack II (crack I is visible on the right-hand side).

Crack I in part 1 originates on the edge of the manhole and runs to the bolt and then toward part 2 (Figure 26a). The area near the origin of the crack is corroded, which prevents examination of its structure and the detection of a possible material defect there.

3.3.2. Microscopic Examination of Materials

To analyze crack I in part 1, a 32 mm-wide bar containing this crack was cut out perpendicular to the edge of the manhole. The cross-cut surfaces are shown in Figure 28. The sample contains part 1, to which part 2 is perpendicular and the residues of part 3 remain after blasting the manhole structure off the deck. The cross-cut shows the weld joining parts 1 and 2 and the weld joining parts 2 and 3 together with their defects. The excised bar was cut across in plane 1-1 marked in Figure 28, and the resulting cutting plane is shown in Figure 29a.

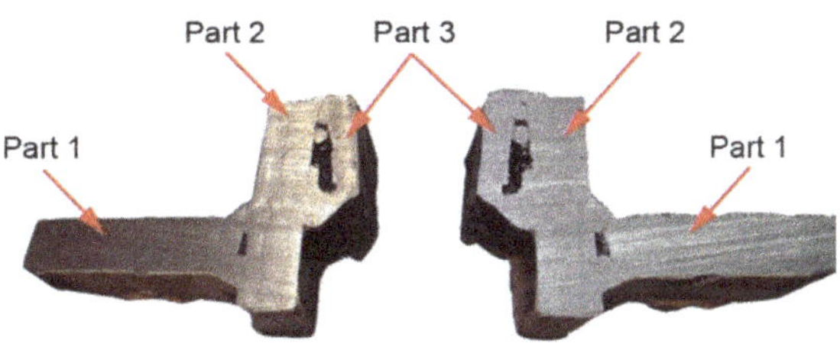


**Figure 28.** The cross-cut surface. Weld seam defects are shown (handy-size bulk carrier 30210 DWT). Macroscopic examination.

**Figure 29.** Macroscopic examination (handy-size bulk carrier 30210 DWT): (**a**) crack I and the cutting plane 2-2 and (**b**) the branching end of crack I.

This last cut created the surfaces depicted in Figure 29b. The branching crack I reaches part 2 and is stopped there.

The analysis of the images obtained from successive cross-cuts indicates that the crack formed on the edge of the manhole in part 1. It then ran towards part 2, and once it reached this part, it branched but did not initiate another crack because the two parts did not contact each other in the cutting plane. The welds joining the two parts are, in fact, tack welds with a relatively small sectional area located on surfaces without full penetration (Figure 28).

Crack I only exists in part 1 and does not cover the entire manhole structure. The crack is corroded along its entire length, and the degree of corrosion here it reaches part 2 is shown in Figure 30.

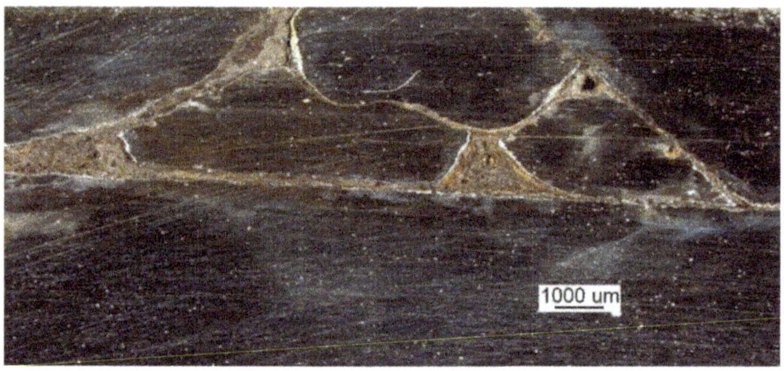


**Figure 30.** Degree of corrosion of crack I where it branches (handy-size bulk carrier 30210 DWT). Stereoscopic optical microscopy.

The microscopic analysis of the metallographic specimen of the sample taken from part 1 (Figure 31) revealed large amounts of spot oxides with an intensity exceeding the 5a pattern according to PN-64/H-04510 [57]—Determination of the degree of steel contamination by non-metallic inclusions. After etching with an alcoholic solution of nitric acid, the ferritic-pearlitic structure shown in Figure 32a was obtained.

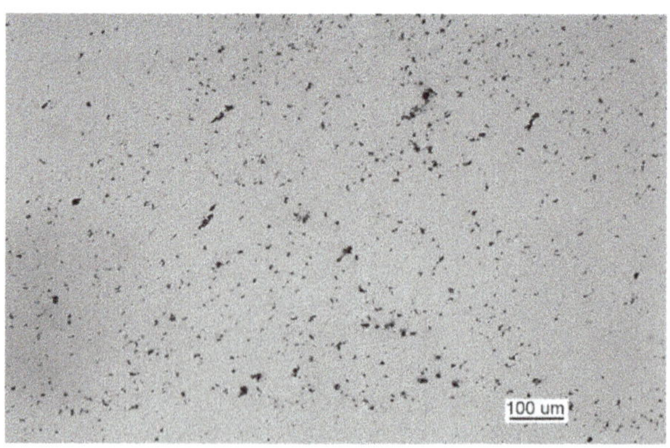


**Figure 31.** Non-etched metallographic specimen of part 1 showing the presence of spot-wise oxides (handy-size bulk carrier 30210 DWT).

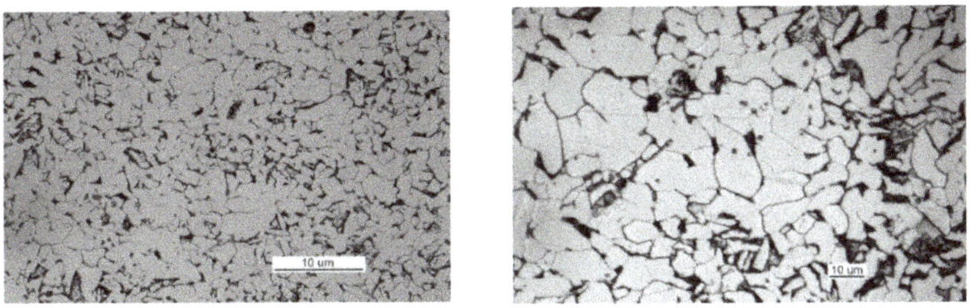


(a) (b)

**Figure 32.** Optical microscopy views (handy-size bulk carrier 30210 DWT): (**a**) the ferritic and pearlitic structure of the material in part 1 and (**b**) the ferritic and pearlitic structure of the material, with visible non-continuous cementite network at the ferrite grain boundaries.

At the grain boundaries of ferrite, cement was precipitated as a non-continuous network, as illustrated in Figure 32b.

#### 3.3.3. Mechanical Property Tests

Static tensile tests were conducted on two samples with a diameter of ϕ6 mm cut parallel to the edge of part 1 [59], and the results are shown in Table 6.

**Table 6.** Results of the static tensile tests of samples from the handy-size bulk carrier 30210 DWT.

| Sample No. | Yield Point $R_e$ (MPa) | Ultimate Tensile Strength $R_m$ (MPa) | Elongation after a Break $A$ (%) | Waist Formation $Z$ (%) |
|---|---|---|---|---|
| 1 | 277 | 416 | 23 | 62 |
| 2 | 330 | 458 | 34 | 58 |

For condition J2 of this steel, the required impact energy was 27 J, and the result obtained from the impact strength test was only 25% of the required value.

The Charpy impact test was conducted at −20 °C on samples taken from part 1 perpendicular to its edge, and the results are shown in Table 7.

**Table 7.** Results of the impact strength test of samples from the handy-size bulk carrier 30210 DWT.

| Sample No. | Impact Energy $KV^{-20}$ (kGm) | Impact Energy $KV^{-20}$ (J) |
|---|---|---|
| 1 | 0.5 | 4.9 |
| 2 | 0.9 | 8.8 |
| 3 | 0.7 | 6.9 |

The impact strength test was performed no lower than –20 °C because the obtained impact energy values were very low.

#### 3.3.4. Fractography Analysis

The results of the fractographic analysis of samples from handy-size bulk carrier 30210 DWT are shown in Figure 33.

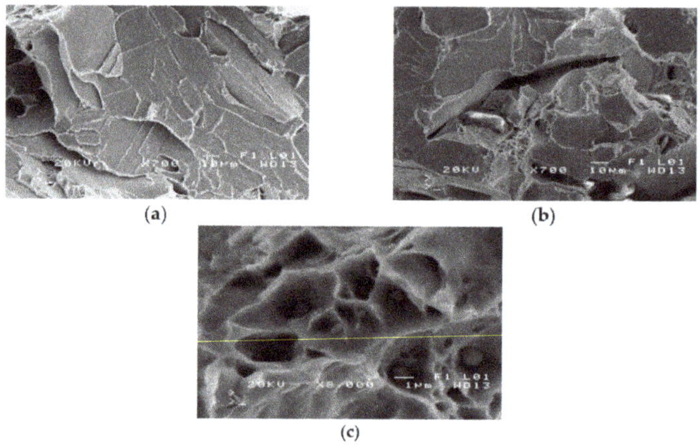

**Figure 33.** Scanning electron microscope (SEM) views (handy-size bulk carrier 30210 DWT): (**a**) the brittle fracture running along the cleavage planes; (**b**) fracture-based crack and small crater-like areas; and (**c**) spherical precipitates of silicon and aluminum oxides at the bottom of craters.

The fractographic analyses revealed brittle fracture (Figure 33a) with numerous fracture-based cracks (Figure 33b) and negligible traces of permanent strain as small breaks. There were also a few areas with small craters, the bottom of which contained spherical precipitates of silicon or aluminum oxides (Figure 33c).

#### 3.3.5. Summary

For crack I, deformations of the deck and the associated stresses were transferred to part 1 across the welds joining parts 1 and 2. They are located only on the surface, have no weld penetration, and have relatively small cross sections. Despite this, no cracks were found in these welds except insignificant, negligible cases; thus, it is unclear why part 1 cracked. It had a relatively large cross section. Analysis of crack I indicates that the crack was caused by stress corrosion, which means that the crack was caused by the synergistic action of two factors—mechanical loads and the corrosive environment.

Crack II was initiated by stress concentration caused by a massive metallurgic inclusion located very close to the surface. Its chemical composition indicates that it is a scoured furnace lining element that is not corrosive but only has a mechanical effect. However, after the crack reaches the surface and contacts water, it propagates by stress corrosion in the manner described for crack I. It is highly probable that cracks of this type are mainly caused by excessive tensile loads affecting deck plates as the hull reacts to waves. In contrast, during continued operation, the corrosive environment becomes active after the crack has been formed.

The root cause of cracking in the elements of the examined manhole structure is the use of low-quality steel heavily contaminated by nonmetallic inclusions. The latter include large notches that concentrate working stresses and initiate cracks. All of these inclusions significantly reduce the impact strength (ductility) of the steel to an unacceptably low level. The second cause of cracking in manhole structural elements are poor-quality welded joints and welds without penetration, which create notches that initiate cracks. The effect of stress in the deck plates may be the third cause under the conditions of extreme hull loads in port and offshore conditions.

### 3.4. *Case Study 4—Handy-Size Bulk Carrier 30185 DWT*

#### 3.4.1. Macroscopic Examination

This examination was aimed to determine the causes and circumstances of cracks in the manhole section structure. The delivered test piece is a ship's manhole section marked

5WBTS. The section seen from the lid mounting is shown in Figure 34a and from the opposite side in Figure 34b.

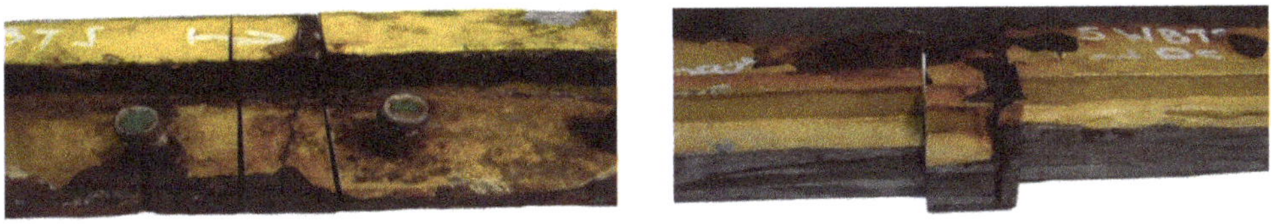

(a) (b)

**Figure 34.** Macroscopic examination (handy-size bulk carrier 30185 DWT): (**a**) lid view of the manhole section marked 5WBTS and (**b**) bottom view of the manhole section marked 5WBTS.

Figure 35 shows a cross-section of the manhole structure with the individual parts marked. Part 2 is a combination of part 1 (the screws fitted in it are visible in Figure 34, and part 3 is part of the deck.

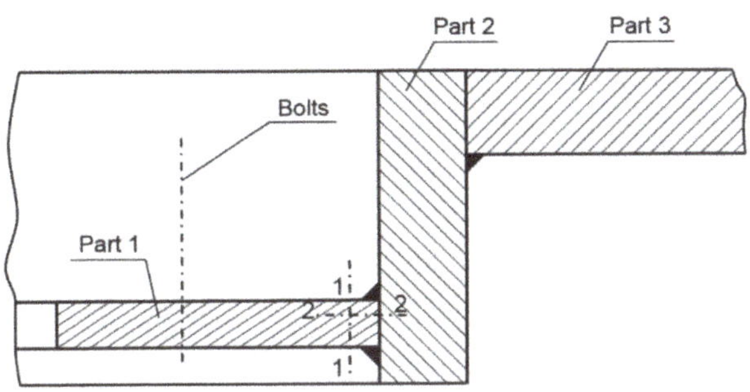


**Figure 35.** Cross-section diagram of the manhole structure (handy-size bulk carrier 30185 DWT).

Two cracks were identified in the section shown in Figure 36. The right-hand crack, visible in the cut-out bar, is hereafter referred to as crack I, and the left-hand crack passing through the bolt hole is referred to as crack II. The lateral surface of the bar containing crack I is shown in Figure 36a.

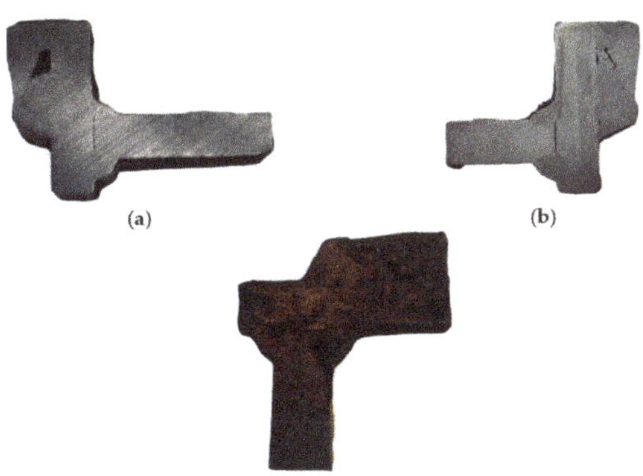

**Figure 36.** Macroscopic examination (handy-size bulk carrier 30185 DWT): (**a**) the lateral surface of the bar containing crack I; (**b**) view of the second side surface of the bar after etching; and (**c**) fracture surface.

There is a lack of penetration in the joint bonding parts 1 and 2 and a defective joint between parts 2 and 3. The second side surface was sanded and etched (Figure 36b). Defects in the weld seams and corrosion pitting running from the crack surface are visible.

Part 1 was cut 25 mm off the part's edge (Figure 35, cut 1-1) so that the cracked part of the bar was separated, and the fracture surface was exposed (Figure 36c).

The crack surface was very corroded, making fractographic analysis impossible. Part 2 exhibited only a single large pitting, which probably remained due to local material heterogeneity and, presumably, is the origin of the crack. The view of the pitting under a stereoscopic microscope is shown in Figure 37.

**Figure 37.** The deep corrosive pit is the presumed origin of the crack (handy-size bulk carrier 30185 DWT). Stereoscopic optical microscopy.

The crack covered all of part 2 and developed in part 3 (to an unknown depth) and part 1 through both welds that join parts 1 and 3 to part 2. The front of the crack in part 1 was further examined. The sample covering the crack front area was cut parallel to the surface of part 1 at a depth of about 6 mm. The front of the crack is shown in Figure 38a.

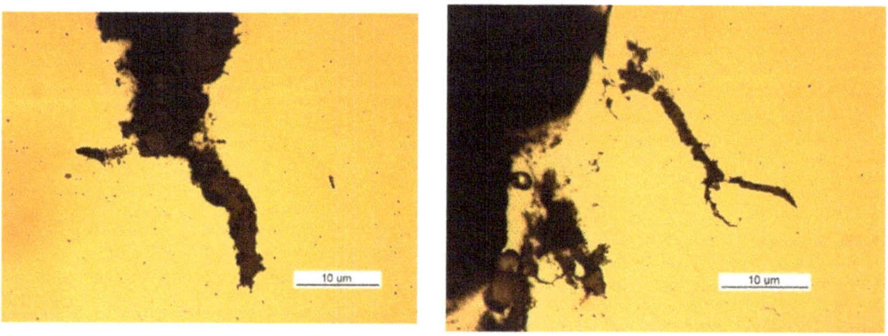


(**a**) (**b**)

**Figure 38.** Optical microscopy views (handy-size bulk carrier 30185 DWT): (**a**) the front of the crack developing in part 1 and (**b**) secondary cracks filled with corrosion products.

Corrosion products were observed within the entire crack, including its front. No part contains a crack that would be caused only by a mechanical load because the branches of the crack, including its face, are filled with corrosion products (Figure 38b).

The results show the important role of corrosion processes in the cracking of materials. The corrosion processes intensify due to corrosive conditions, which can be regarded as being non-variant in the first approximation, and due to the material's purity and homogeneity. It should be noted that cracks near the edge of the deck manhole can also occur if the permissible stress (i.e., mechanical load) is significantly exceeded in this area.

Figure 39 shows a close-up photo of a 5WBTS section with crack II running from the edge of part 1 to the hole in which the bolt is placed and then to part 2.

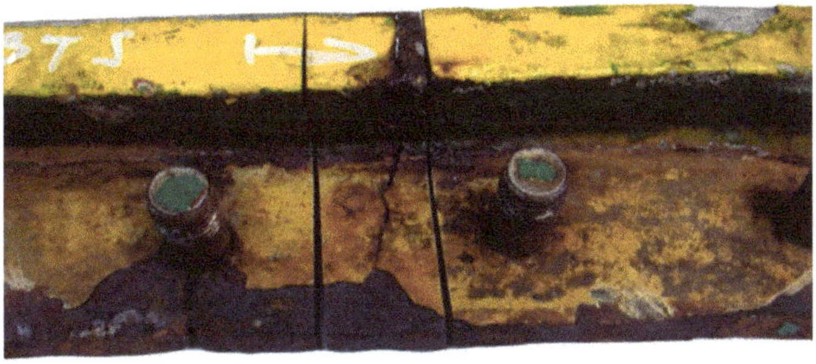

**Figure 39.** Crack II running through a bolt hole (handy-size bulk carrier 30185 DWT). Macroscopic examination.

The origin of the crack is located near the edge of part 1. The large corrosion cavities near the origin (visible in the figure) make a fractographic analysis of the fracture surface impossible; thus, it is not possible to determine the cause of the crack at this location. It can only be concluded from the numerous and deep corrosion pits located near the edge of part 1 that the origin of the crack was located here because of material heterogeneity due to non-metallic inclusions (Figure 40).

**Figure 40.** Corrosion pits at the edge of part 1 (handy-size bulk carrier 30185 DWT). Stereoscopic optical microscopy.

The camera's relatively small depth of field does not allow the full depiction of the plasticity of the stereoscopic microscope image.

The crack expanding across the bolt hole was split at this point into two parallel cracks, which then merged into one crack, what is visible in the cross section made 50 mm from the edge of part 1 (Figure 41).

Figure 42 shows where the cracks merged, which is a special point because it can have one of two natures. It can be either a rolled, large, non-metallic inclusion (as its shape indicates), or a very large number of products of intensive corrosion caused by stress at the crack face.

**Figure 41.** The merging of cracks previously branched at the bolt thread (handy-size bulk carrier 30185 DWT). Macroscopic examination.

**Figure 42.** (**a**,**b**) Part of the location where the cracks merged (handy-size bulk carrier 30185 DWT). Stereoscopic optical microscopy.

The first case confirms the poor quality of the steel used, and the second case proves the effect of stress on the expansion of cracks and the synergistic effect of stress and corrosion. Cut 2-2 (Figure 35) revealed the route of the crack until it reached part 2, where it branched and severe corrosion occurred (Figure 43a). At this point, the crack did not enter part 2 because there is no connection between the parts, but it expanded and penetrated the joint on the surface of both parts (Figure 43b).

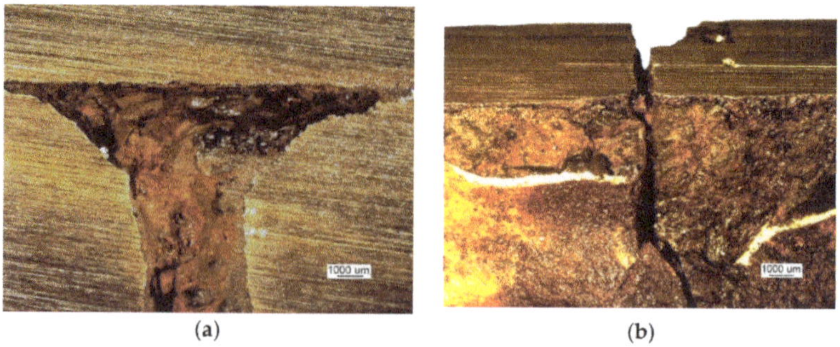


**Figure 43.** Stereoscopic optical microscopy views (handy-size bulk carrier 30185 DWT): (**a**) crack reaches part 2 and (**b**) crack transition from part 1 (upper part of the figure) to the joint.

3.4.2. Microscopic Examination of Materials

The examination of non-etched specimens made from samples taken from parts 1 and 2 showed very large amounts of nonmetallic inclusions, mainly oxide spots, which exceed the TP5a pattern [57] (Figure 44).

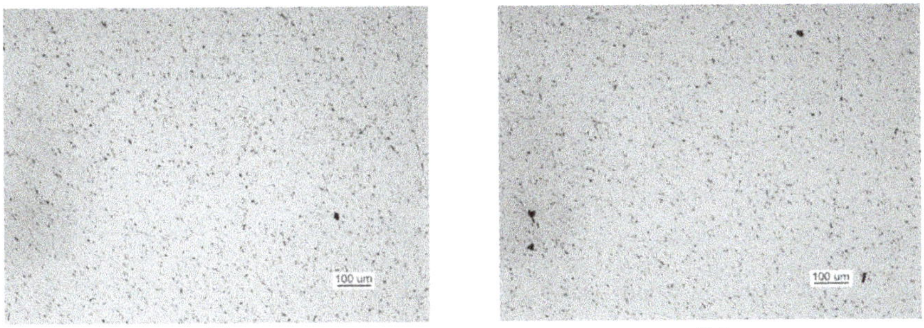

(a) (b)

**Figure 44.** Optical microscopy views (handy-size bulk carrier 30185 DWT): (**a**) oxide spots in the material of part 1 (non-etched metallographic specimen) and (**b**) oxide spots in the material of part 2 (non-etched metallographic specimen).

Such large amounts of impurities significantly reduced the ductile properties of the material, particularly its impact strength. After being etched with an alcoholic solution of nitric acid, the metallographic specimens revealed the ferritic-pearlitic structure of the material from part 1 with an intermittent mesh of precipitated cementite at the ferrite grain boundaries (Figure 45).

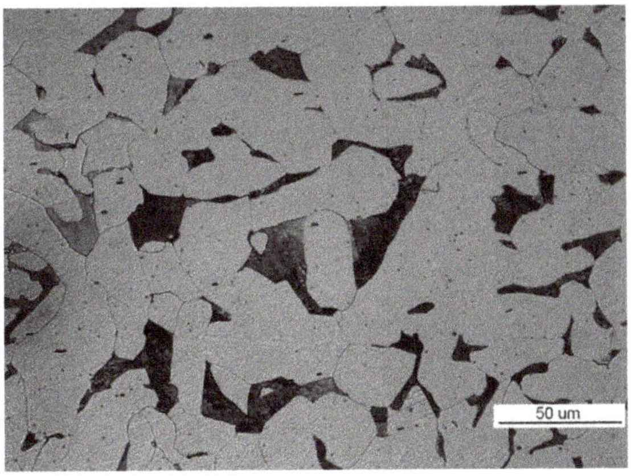


**Figure 45.** The structure of the steel in part 1 (handy-size bulk carrier 30185 DWT). Optical microscopy.

The structure of the material in part 2 is shown in Figure 46. The material of part 2 has a ferritic-pearlitic, band-like structure (Figure 46a) with cementite precipitates at grain boundaries, numerous silicates, and rolled sulfides (Figure 46b).

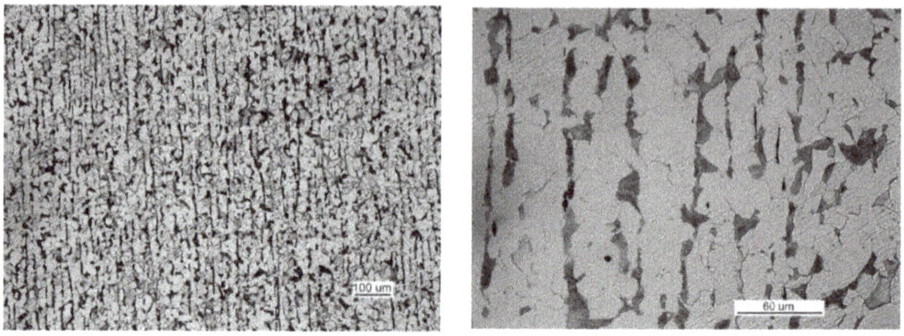

(**a**) (**b**)

**Figure 46.** Optical microscopy views (handy-size bulk carrier 30185 DWT): (**a**) the band-like structure of part 2 and (**b**) material structure of part 2 with cementite, silicate, and sulfide precipitates.

### 3.4.3. Mechanical Property Tests

A static tensile test was conducted according to PN-EN 10002-1 [59] on φ6 mm samples taken from part 1. The results are shown in Table 8, which corresponds to S275 steel according to PN-EN 10,025 [63].

**Table 8.** Results of the static tensile test of samples from the handy-size bulk carrier 30185 DWT.

| Sample No. | Yield Point $R_e$ (MPa) | Ultimate Tensile Strength $R_m$ (MPa) | Elongation after a Break $A$ (%) | Waist Formation $Z$ (%) |
|---|---|---|---|---|
| 1 | 263 | 416 | 34 | 64 |
| 2 | 256 | 416 | 40 | 58 |
| 3 | 284 | 430 | 28 | 58 |

Charpy impact tests were conducted at −20 °C on samples taken from part 1 perpendicular to its edge, and the results are shown in Table 9. The obtained impact energy values are very low.

**Table 9.** Results of the Charpy impact strength tests of samples from the handy-size bulk carrier 30185 DWT.

| Sample No. | Impact Energy $KV^{-20}$ (kGm) | Impact Energy $KV^{-20}$ (J) |
|---|---|---|
| 1 | 1.2 | 11.8 |
| 2 | 0.9 | 8.8 |
| 3 | 0.7 | 6.9 |

### 3.4.4. Fractography Analysis

Fractographic analyses were conducted using an SEM with samples obtained from the impact tests. The fractures are shown in Figure 47.

The fractures displayed brittle structures, with cracks running along the cleavage planes to form a distinct river-basin-like structure (Figure 47a). Fracture-based cracks were found (Figure 47b). Small bands revealing plastic strain (Figure 47c) with a crater-like structure are observed locally, with spherical precipitates of iron and silicon oxides visible at the bottom of the craters.

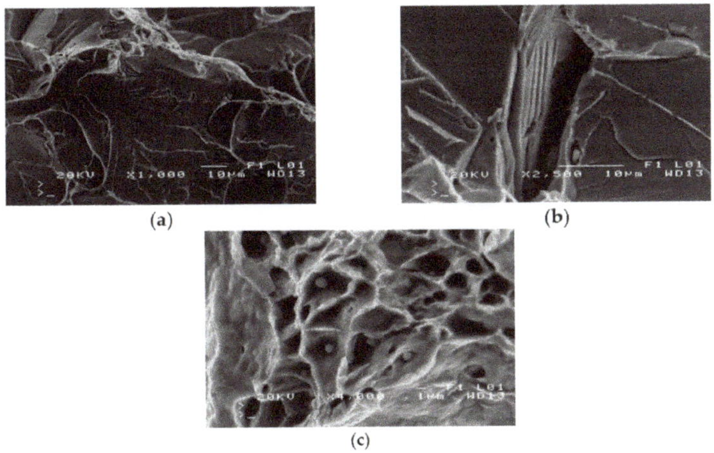

**Figure 47.** Scanning electron microscope (SEM) views (handy-size bulk carrier 30185 DWT): (**a**) brittle fracture produced after impact testing at −20 °C; (**b**) fracture-based crack that runs through a pearlite grain; and (**c**) a fragment of crater-like ductile fracture (spherical precipitates of Fe and Si oxides are visible at the bottom of craters).

#### 3.4.5. Summary

In the section of the manhole structure, cracking analysis was largely limited by the severely advanced corrosion processes that damaged the fracture surface, making it impossible to directly determine the causes of cracking, the origin, and the course of development.

Crack I, which was initiated in part 2 and expanded across the welds in part 1, was caused by stress in the material and corrosion. The corrosion factor was very important, and the intensity of the corrosion processes was greatly influenced by the purity and homogeneity of the steel used. In the investigated case, both of these factors are sources of concern. The results imply that the applied paint coatings do not sufficiently protect manhole elements because they are easily damaged at the edges, especially with the highly-contaminated steel used in this case. The test pieces are operated under bending load. In contrast to the uniformity of the cross-section subject to a tensile load, where the crack originates in the axis of the item affected by a load, the area most stressed during bending is the outermost layer of the piece, and this is where the crack was initiated. The crack then propagates most intensively in the outermost layer.

Being highly contaminated, the tested steel has a very low impact strength, which results in easy crack initiation due to dynamic loads, especially at low temperatures.

### *3.5. Root Causes Synthesis*

In summary, the research results presented in the article and referring to the conducted literature analysis are connected with a number of root causes contributing to the cracking in the structural elements of manholes. The root causes can be categorized according to one of the concepts of Ishikawa diagrams [64] into the following groups: material, method, environment (external factors), operating conditions, and human factor.

Table 10 presents a classification of these groups of causes and their involvement in basic fatigue destruction factors, i.e., material properties, residual stresses, and cyclic load characteristics. The causes contributing to appropriate factors are marked by an X sign. Particular items and their description (cause symbols) are presented in the text following the table.

**Table 10.** List of the factors affecting cracks of manholes.

| Causes Group | Cause Symbol | Fatigue Damage Factors | | |
|---|---|---|---|---|
| | | **Material Properties** | **Residual Stresses** | **Cyclic Load Characteristics** |
| Material | MA1 | X | | |
| | MA2 | X | | |
| | MA3 | X | X | |
| Method | ME1 | X | X | |
| | ME2 | | X | |
| | ME3 | X | X | |
| | ME4 | X | X | |
| | ME5 | X | X | |
| Environment | EX1 | X | X | X |
| | EX2 | X | X | X |
| | EX3 | | | X |
| Operating conditions | OC1 | | | X |
| | OC2 | | | X |
| Human factor | HF1 | X | X | |
| | HF2 | X | | X |
| | HF3 | X | X | X |
| | HF4 | X | X | |
| | HF5 | X | X | X |
| | HF6 | X | X | X |
| | HF7 | X | X | X |

The material factors related to the material from which the described damaged main deck components were made include the following:

(1) MA1—Technological contamination of the material (uncleaned, unprepared material);
(2) MA2—Contamination resulting from the influence of external factors, e.g., corrosion;
(3) MA3—Susceptibility to residual stresses.

The factors related to the method understood as joining structural elements of the deck by welding, riveting, bolting, etc. include the following:

(1) ME1—Failure to meet technological parameters, including the processes of joining sheets (e.g., welding);
(2) ME2—Occurrence of welding stresses, including residual stresses, in shaping and joining of materials;
(3) ME3—Inadequate calibration of technological devices (machine tools, welders, ultrasonic detectors for tests, etc.);
(4) ME4—Inadequate working conditions in the production plant;
(5) ME5—Inadequate quality control discipline (lack of an appropriate non-destructive testing approach such as ultrasonic testing to access the quality and integrity of welds and a material quality assurance program).

The influence of the environment (external factors, including the atmospheric ones) on the operation object, i.e., the influence of the environment on the ship operation at a specific time and geographical location, includes the following factors:

(1) EX1—Electrochemical interaction of the environment (seawater, water mist, salt, etc.);
(2) EX2—Thermodynamic influence of the environment (water temperature, air temperature, humidity, atmospheric pressure, etc.);
(3) EX3—The influence of hydrometeorological factors (wind force and direction, variable currents, eddies, waves, tides, etc.).

The ship operating conditions associated with cyclic mechanical deformations of the hull during the operation of the ship at sea includes the following factors:

(1) OC1—Effect of shipload (cargo distribution, trim, heel, etc.);

(2) OC2—The influence of the sailing area (sailing in limited and shallow waters).

The human factor, including human interaction during the construction and operation of a ship, concerns the following:

(1) HF1—Lack of appropriate qualifications of the yard's employees;
(2) HF2—Lack of proper professional preparation of the ship's service;
(3) HF3—Lack of decision and negligence by the shipyard's management and ship's personnel;
(4) HF4—Negligence during quality control in the shipyard;
(5) HF5—Insufficient technological discipline among the shipyard's employees and ship's crew;
(6) HF6—Psychophysical disability of the shipyard's employees and ship's crew (illness, exhaustion, influence of psychoactive substances);
(7) HF7—Time pressure on the shipyard and shipowner's employees.

The synthesis of different root causes presented in Table 10 proves that the most significant influence is connected with material properties (80% of all fatigue damage contribution factors). Next, are residual stresses and cyclic load characteristics, which account for 70% and 50%, respectively.

## 4. Conclusions

In all examined cases, the cracking in the structural elements of manholes was caused by the low quality of the steel used. Local strain concentration—caused by various kinds of notches—was not relieved by local plastic strain. This property and the low ductility of the steel, especially when kept at low temperatures, caused strain to be relieved by cracking. The effect of stress within deck plates under extreme loads in the hull in port or at sea can also cause cracking. The examined cracks occurred due to the manhole structure strain under the ship operating conditions, which exceeded the strength of the steel used, indicating structural defects in the vessel.

This confirms the hypothesis proposed in the introduction that the main decks crack around manholes because of the very low-quality welded joints and because the manholes were made from poor-quality steel that contained large amounts of non-metallic impurities. The prevention of similar cracks in the future requires:

(1) For the structural parts of the manhole, using steel with suitable mechanical properties, including an impact energy corresponding to at least marking J2; use steel tested for defects in materials that can initiate cracks;
(2) Developing the right technology for welding work and closely monitoring compliance with respective regulations;
(3) Applying non-destructive testing (e.g., eddy-current, magnetic-particle, liquid penetrant, radiographic, ultrasonic, or acoustic emission) to access the quality and integrity of welds and material quality assurance program;
(4) Modifying the ship's design in the context of variable mechanical loads carried by the hull.

Further research on the presented topic of ship machinery destruction in terms of corrosion fatigue and complex load situations might include a combination of the material analysis as presented in this paper with the issues of residual stresses resulting from welding. Moreover, the finite element method may be involved to receive a holistic approach to the ship's operational safety and reliability.

**Author Contributions:** Conceptualization, L.C. and K.G.; methodology, L.C. and K.G.; software, L.C. and K.G.; validation, L.C. and K.G.; formal analysis, L.C. and K.G.; investigation, L.C. and K.G.; resources, L.C. and K.G.; data curation, L.C. and K.G.; writing—original draft preparation, L.C. and K.G.; writing—review and editing, L.C. and K.G.; visualization, L.C. and K.G.; supervision, L.C. and K.G.; project administration, L.C. and K.G.; funding acquisition, L.C. and K.G. All authors have read and agreed to the published version of the manuscript.

**Funding:** The research was prepared under Grant 1/S/KPBMIM/20 "Improving the effectiveness of technical systems by changing their structure and using modern materials" financed by the Ministry of Science and Higher Education of Poland. The APC was funded from authors' own sources.

**Institutional Review Board Statement:** Not applicable.

**Informed Consent Statement:** Not applicable.

**Data Availability Statement:** All results are provided in the paper.

**Conflicts of Interest:** The authors declare no conflict of interest.

## Abbreviations

| | |
|---|---|
| $A$ | Elongation after a break |
| AA5b | Material structure pattern according to standard [57]. |
| DWT | Deadweight tonnage |
| FSA | Formal Safety Assessment |
| IMO | International Maritime Organisation |
| KN5b | Material structure pattern according to standard [57]. |
| $KV^{XX}$ | V-notched Charpy impact strength test absorbed energy at the temperature $XX$. |
| PN | Polish Standard |
| SEM | Scanning Electron Microscope |
| TP5a | Material structure pattern according to standard [57]. |
| $R_e$ | Yield point |
| $R_m$ | Ultimate tensile strength |
| $Z$ | Waist formation |

## References

1. Özgüç, Ö. Typical cracks in deck of ship-shaped structures and ways to modify and improve the design. *Sak. Univ. J. Sci.* **2017**, *21*, 1. [CrossRef]
2. Kozak, J.; Górski, Z. Fatigue strength determination of ship structural joints. *Pol. Marit. Res.* **2011**, *18*, 28–36. [CrossRef]
3. Ship Structure Committee. *Long-Term Corrosion Fatigue of Welded Marine Steels—SSC-32*; US Coast Guard: Washington, DC, USA, 1984.
4. Fricke, W. Fatigue and fracture of ship structures. In *Encyclopedia of Maritime and Offshore Engineering*; John Wiley & Sons Ltd.: Chichester, UK, 2017; pp. 1–12.
5. Madsen, H.O. Fatigue reliability of marine structures. In *Stochastic Approach to Fatigue*; Springer: Vienna, Austria, 1993; pp. 243–301.
6. Bathias, C.; Pineau, A. *Fatigue of Materials and Structures*; Bathias, C., Pineau, A., Eds.; John Wiley & Sons Inc.: Hoboken, NJ, USA, 2013; ISBN 9781118616789.
7. Ziemba, S. *Fizyczne Aspekty Trwałości i Niezawodności Obiektów Technicznych*; Polska Akademia Nauk IPPT: Warsaw, Poland, 1976.
8. Suresh, S. *Fatigue of Materials*, 2nd ed.; Cambridge University Press: Cambridge, MA, USA, 2012; ISBN 9780521570466.
9. Dehlinger, U. *Theoretische Metallkunde*; Springer: Berlin/Heidelberg, Germany, 1968; ISBN 978-3-642-49085-9.
10. Gough, H.J. Engineering steel under combined cyclic and static stress. *Proc. Inst. Mech. Eng.* **1949**, *60*, 417–440. [CrossRef]
11. Mroziński, S. The influence of loading program on the course of fatigue damage cumulation. *J. Theor. Appl. Mech.* **2011**, *49*, 83–95.
12. Botvina, L.R.; Limar', L.V. Relationship of the spacing of the fatigue striations to the range in the stress intensity factor. *Sov. Mater. Sci.* **1985**, *21*, 144–152. [CrossRef]
13. Tamura, M. Relation between sub-grain size and dislocation density during steady-state dislocation creep of polycrystalline cubic metals. *J. Mater. Sci. Res.* **2018**, *7*, 26. [CrossRef]
14. Fujita, F. Dislocation theory of fracture of crystals. *Acta Metall.* **1958**, *6*, 543–551. [CrossRef]
15. Mott, N.F. Dislocations and the theory of solids. *Nature* **1953**, *171*, 234–237. [CrossRef]
16. Jimenez-Martinez, M. Fatigue of offshore structures: A review of statistical fatigue damage assessment for stochastic loadings. *Int. J. Fatigue* **2020**, *132*, 105327. [CrossRef]
17. Afanasyev, N.N. *Statistical Theory of Fatigue Strength of Metals*; Izd. Akad. Nauk. USSR: Kiev, Ukraine, 1953. (In Russian)
18. Freudenthal, A.M. The statistical aspect of fatigue of materials. *Proc. R. Soc. Lond. Ser. A* **1946**, *187*, 416–429. [CrossRef]
19. Freudenthal, A.M.; Weiner, J.H. On the thermal aspect of fatigue. *J. Appl. Phys.* **1956**, *27*, 44–50. [CrossRef]
20. Kostochkin, Y.V.; Oding, I.A. Failure of metals due to thermal fatigue. *ARS J.* **1960**, *30*, 1062–1064. [CrossRef]
21. Chowdhury, P.; Sehitoglu, H. Mechanisms of fatigue crack growth—A critical digest of theoretical developments. *Fatigue Fract. Eng. Mater. Struct.* **2016**, *39*, 652–674. [CrossRef]

22. Bergsmo, A.; Dunne, F.P.E. Competing mechanisms of particle fracture, decohesion and slip-driven fatigue crack nucleation in a PM nickel superalloy. *Int. J. Fatigue* **2020**, *135*, 105573. [CrossRef]
23. Kee Paik, J.; Kyu Kim, S.; Kon Lee, S. Probabilistic corrosion rate estimation model for longitudinal strength members of bulk carriers. *Ocean Eng.* **1998**, *25*, 837–860. [CrossRef]
24. Paik, J.K.; Frieze, P.A. Ship structural safety and reliability. *Prog. Struct. Eng. Mater.* **2001**, *3*, 198–210. [CrossRef]
25. Akpan, U.O.; Koko, T.S.; Ayyub, B.; Dunbar, T.E. Reliability assessment of corroding ship hull structure. *Nav. Eng. J.* **2003**, *115*, 37–48. [CrossRef]
26. Mansour, A.E.; Wirsching, P.H. Sensitivity factors and their application to marine structures. *Mar. Struct.* **1995**, *8*, 229–255. [CrossRef]
27. Downes, J.; Pu, Y. Reliability-based sensitivity analysis of ships. *Proc. Inst. Mech. Eng. Part M* **2005**, *219*, 11–23. [CrossRef]
28. Liu, Y.; Yi, H.; Chen, L. Submarine pressure hull butt weld fatigue life reliability prediction method. *Mar. Struct.* **2014**, *36*, 51–64. [CrossRef]
29. Tsurui, A.; Ishikawa, H. Application of the Fokker-Planck equation to a stochastic fatigue crack growth model. *Struct. Saf.* **1986**, *4*, 15–29. [CrossRef]
30. Zhu, W.Q.; Lin, Y.K.; Lei, Y. On fatigue crack growth under random loading. *Eng. Fract. Mech.* **1992**, *43*, 1–12. [CrossRef]
31. International Maritime Organization. *Revised Guidelines for Formal Safety Assessment (FSA) for Use in the IMO Rule-Making Process—MSC-MEPC.2/Circ.12/Rev.2*; IMO: London, UK, 2017.
32. Gucma, S.; Ślączka, W. Comprehensive method of formal safety assessment of ship manoeuvring in waterways. *Sci. J. Marit. Univ. Szczecin/Zesz. Nauk. Akad. Morskiej Szczecinie* **2018**, *54*, 110–119. [CrossRef]
33. Bai, Y.; Jin, W.-L. *Marine Structural Design*; Elsevier: Oxford, UK, 2016; ISBN 9780080999975.
34. DNV. *Classification Notes No. 30.7: Fatigue Assessment of Ship Structures*; Norwegian University of Science and Technology: Høvik, Norway, 2014.
35. DNV. *DNVGL-RP-C210: Probabilistic Methods for Planning of Inspection for Fatigue Cracks in Offshore Structures*; Norwegian University of Science and Technology: Høvik, Norway, 2015.
36. DNV. *DNVGL-RP-C203: Fatigue Design of Offshore Steel Structures.*; Norwegian University of Science and Technology: Høvik, Norway, 2016.
37. Ship Structure Committee. *Review of Current Practices of Fracture Repair Procedures for Ship Structures—SSC-462*; US Coast Guard: Washington, DC, USA, 2012.
38. Dobosiewicz, J.; Zbroińska-Szczechura, E. Warunki powstawania uszkodzeń zmęczeniowo-korozyjnych w walczakach kotłów wysokoprężnych. *Energetyka* **2012**, *6*, 310–314.
39. Kerdabadi, M.; Sakaki, A.; Izadi, A. Evaluation of ship structure reliability during design, maintenance, and repair phases. *Zesz. Nauk. Akad. Mor. Gdyni* **2018**, *53*, 19–27. [CrossRef]
40. Gawdzińska, K.; Chybowski, L.; Przetakiewicz, W.; Laskowski, R. Application of FMEA in the quality estimation of metal matrix composite castings produced by squeeze infiltration. *Arch. Metall. Mater.* **2017**, *62*, 2171–2182. [CrossRef]
41. Gawdzińska, K.; Chybowski, L.; Przetakiewicz, W. Proper matrix-reinforcement bonding in cast metal matrix composites as a factor of their good quality. *Arch. Civ. Mech. Eng.* **2016**, *16*, 553–563. [CrossRef]
42. Gawdzińska, K. Methods of the detection and identification of structural defects in saturated metallic composite castings. *Arch. Foundry Eng.* **2017**, *17*, 37–44. [CrossRef]
43. Chwalczuk, T.; Przestacki, D.; Szablewski, P.; Felusiak, A. Microstructure characterization of Inconel 718 after laser assisted turning. *MATEC Web Conf.* **2018**, *188*, 02004. [CrossRef]
44. Li, Z.; Jiang, X.; Hopman, H. Surface crack growth in offshore metallic pipes under cyclic loads: A literature review. *J. Mar. Sci. Eng.* **2020**, *8*, 339. [CrossRef]
45. Colombi, P.; Fava, G.; Sonzogni, L. Fatigue crack growth in CFRP-strengthened steel plates. *Compos. Part B* **2015**, *72*, 87–96. [CrossRef]
46. Dong, Y.; Guedes Soares, C. Stress distribution and fatigue crack propagation analyses in welded joints. *Fatigue Fract. Eng. Mater. Struct.* **2019**, *42*, 69–83. [CrossRef]
47. Lee, J.-H.; Jang, B.-S.; Kim, H.-J.; Shim, S.H.; Im, S.W. The effect of weld residual stress on fracture toughness at the intersection of two welding lines of offshore tubular structure. *Mar. Struct.* **2020**, *71*, 102708. [CrossRef]
48. Perić, M.; Nižetić, S.; Garašić, I.; Gubeljak, N.; Vuherer, T.; Tonković, Z. Numerical calculation and experimental measurement of temperatures and welding residual stresses in a thick-walled T-joint structure. *J. Therm. Anal. Calorim.* **2020**, *141*, 313–322. [CrossRef]
49. Sika, R.; Rogalewicz, M.; Popielarski, P.; Czarnecka-Komorowska, D.; Przestacki, D.; Gawdzińska, K.; Szymański, P. Decision support system in the field of defects assessment in the metal matrix composites castings. *Materials* **2020**, *13*, 3552. [CrossRef] [PubMed]
50. Jahan, A.; Edwards, K.L.; Bahraminasab, M. The importance of decision support in materials selection. In *Multi-criteria Decision Analysis for Supporting the Selection of Engineering Materials in Product Design*; Elsevier: Amsterdam, The Netherlands, 2016; pp. 1–23.
51. Dunaj, P.; Berczyński, S.; Chodźko, M.; Niesterowicz, B. Finite element modeling of the dynamic properties of composite steel–polymer concrete beams. *Materials* **2020**, *13*, 1630. [CrossRef] [PubMed]

52. Dounar, S.; Iakimovitch, A.; Mishchanka, K.; Jakubowski, A.; Chybowski, L. FEA simulation of the biomechanical structure overload in the university campus planting. *Appl. Bionics Biomech.* **2020**, *2020*, 1–13. [CrossRef] [PubMed]
53. Karliński, J.; Ptak, M.; Chybowski, L. A Numerical analysis of the working machine tyre inflation process to ensure operator safety. *Energies* **2019**, *12*, 2971. [CrossRef]
54. ReliaSoft Corporation. *System Analysis Reference. Reliability, Availability and Optimization*; ReliaSoft Corporation: Tucson, AZ, USA, 2007.
55. ReliaSoft Corporation. *Accelerated Life Testing Reference*; ReliaSoft Corporation: Tucson, AZ, USA, 2015.
56. Chybowski, L.; Gawdzińska, K.; Laskowski, R. Assessing the unreliability of systems during the early operation period of a ship—A case study. *J. Mar. Sci. Eng.* **2019**, *7*, 213. [CrossRef]
57. Polski Komitet Normalizacyjny. *PN-64/H-04510:1980—Determination of content of non-metallic inclusions in steel*; Polish Committee for Standardization: Warsaw, Poland, 1980.
58. Polski Komitet Normalizacyjny. *PN-66/H-04505:1966—Microstructure of Steel Products—Templates and Designations*; Polish Committee for Standardization: Warsaw, Poland, 1966.
59. Polski Komitet Normalizacyjny. *PN-EN 10002-1:2004—Metallic Materials—Tensile Testing—Part. 1: Method of Test at Ambient Temperature*; Polish Committee for Standardization: Warsaw, Poland, 2004.
60. Polski Komitet Normalizacyjny. *PN EN 10045-1:1994—Metallic Materials—Charpy Impact Test—Part 1: Test Method*; Polish Committee for Standardization: Warsaw, Poland, 1994.
61. Polski Komitet Normalizacyjny. *PN-EN 10045-2:1996—Metallic materials—Charpy Impact Test—Testing Pendulum Hammers)*; Polish Committee for Standardization: Warsaw, Poland, 1996.
62. Polski Komitet Normalizacyjny. *PN-H-04371:1979—Metallic materials—Low Temperature Impact Test*; Polish Committee for Standardization: Warsaw, Poland, 1979.
63. Polski Komitet Normalizacyjny. *PN EN 10025-1:2007—Hot Rolled Products of Structural Steels—Part 1: General Technical Delivery Conditions*; Polish Committee for Standardization: Warsaw, Poland, 2007.
64. Ishikawa, K. *Guide to Quality Control*; JUSE: Tokyo, Japan, 1968.

materials

MDPI

*Article*

# Incorporation of Obstacle Hardening into Local Approach to Cleavage Fracture to Predict Temperature Effects in the Ductile to Brittle Transition Regime

**Maria S. Yankova [1,*], Andrey P. Jivkov [1] and Rajesh Patel [2]**

1 Department of Mechanical, Aerospace and Civil Engineering, University of Manchester, Oxford Rd, Manchester M13 9PL, UK; andrey.jivkov@manchester.ac.uk
2 National Nuclear Laboratory, Sperry Way, Stonehouse, GL10 3UT, UK; rajesh.patel@uknnl.com
* Correspondence: maria.yankova@manchester.ac.uk

**Abstract:** Ductile-to-brittle-transition refers to observable change in fracture mode with decreasing temperature—from slow ductile crack growth to rapid cleavage. It is exhibited by body-centred cubic metals and presents a challenge for integrity assessment of structural components made of such metals. Local approaches to cleavage fracture, based on Weibull stress as a cleavage crack-driving force, have been shown to predict fracture toughness at very low temperatures. However, they are ineffective in the transition regime without the recalibration of Weibull stress parameters, which requires further testing and thus diminishes their predictive capability. We propose new Weibull stress formulation with thinning function based on obstacle hardening model, which modifies the number of cleavage-initiating features with temperature. Our model is implemented as a post-processor of finite element analysis results. It is applied to analyses of standard compact tension specimens of typical reactor pressure vessel steel, for which deformation and fracture toughness properties in the transition regime are available. It is shown that the new Weibull stress is independent of temperature, and of Weibull shape parameter, within the experimental error. It accurately predicts the fracture toughness at any temperature in the transition regime without relying upon empirical fits for the first time.

**Keywords:** cleavage fracture; finite element analysis; local approach; Weibull stress

**Citation:** Yankova, M.S.; Jivkov, A.P.; Patel, R. Incorporation of Obstacle Hardening into Local Approach to Cleavage Fracture to Predict Temperature Effects in the Ductile to Brittle Transition Regime. *Materials* **2021**, *14*, 1224. https://doi.org/10.3390/ma14051224

Academic Editors: Jaroslav Pokluda and Reinhard Pippan

Received: 17 February 2021
Accepted: 2 March 2021
Published: 5 March 2021



## 1. Introduction

Metallic materials with body-centred cubic (bcc) lattices exhibit a unique behaviour with decreasing temperature—their fracture toughness decreases rapidly as the mode of fracture shifts from slow ductile crack growth, typically by void growth and coalescence, to fast brittle fracture, typically by transgranular cleavage. This behaviour is referred to as the ductile-to-brittle transition (DBT). A distinguishing feature of DBT is the scatter in the measured fracture toughness values, which arises from the random spatial distribution of microstructural features controlling the fracture processes. DBT presents a challenge for a number of industries, where structural components are made of bcc metals, with a very important example being ferritic steels. Stress-critical applications of ferritic steels include reactor pressure vessels (RPV) in light water reactors and pressurized equipment in hydrocarbon processing. These applications require reliable assessments of the fracture toughness in the DBT regime. Such assessments are challenging because the fracture mechanisms encompass multiple length scales—from atomic to component—and remain an active area of research.

The microstructure processes leading to ductile and cleavage failure modes are well known from experimental observations [1–3]. In ferritic steels, cleavage initiates at sharp microcracks formed in brittle carbides, sulphides, or other second phase particles. Alternatively, second phase particles could either decohere from the matrix or break and then blunt and form voids (prerequisites for ductile crack growth). Microstructural parameters of the particles such as their size, shape and orientation with respect to the applied load, have been shown to affect their tendency to break or decohere [1–4]. Additional factors such as irradiation and welds in the structural component need to be considered as they are known to modify the DBT behaviour. The main factor that determines the outcome in the competition between cleavage fracture and ductile tearing is the ability of the plastic deformation to relieve the applied stresses ahead of a crack. On a microstructural scale, plastic flow in bcc metals is controlled by the thermally activated movement of screw dislocations, which is dependent on a combination of an internal lattice resistance and an interaction between the dislocations and obstacles in the microstructure [5–7]. A recently published dislocation-obstacle model showed that the activation energy for plastic flow is mainly determined by the kink (which represents a step of atomic dimension in a dislocation line) formation energy [7]. Swinburne and Dudarev [7] were able to predict the ductile to brittle transition temperature in both unirradiated and irradiated ferritic-martensitic steels, where the irradiated temperature increased up to twice that of the unirradiated material in agreement with multiple experimental data. Their dislocation-obstacle interaction model is used as a key ingredient in our proposal.

From an engineering perspective, integrity assessments can be done by two different approaches. In the traditional global approach, a single mechanics parameter, such as the $J$ integral, representing a crack-driving force, is evaluated against a material parameter—an experimentally obtained value representing the resistance to fracture, or the fracture toughness. This can be highly conservative, since the fracture toughness is typically obtained with a high-constraint (deep) crack, whereas the mechanics parameter could be obtained from analysis of a component with a low-constraint (shallow) crack. Apart from the inherent conservatism, the global approach cannot predict DBT behaviour—its use requires measurements of fracture toughness values at any temperature of interest and the decision which of the scattered values to evaluate against the crack driving force.

Local approaches to cleavage fracture (LAF) were developed as alternatives to the global approach with the intention to capture the mechanistic understanding of the cleavage process described above. These are probabilistic models, which incorporate weakest link statistics of microcracks, controlled by local stresses, to model the observed scatter of fracture toughness values. The majority of local approaches to cleavage fracture are based on the Beremin model [8], which defines a microstructural crack-driving force, the so-called Weibull stress, and the probability of fracture as a Weibull distribution. The Weibull stress depends on two material-dependent parameters—a shape parameter $m$, linked to the shape of the probability density of the micro-crack sizes, and a scale parameter $\sigma_u$, linked to the elastic properties and surface energy of the material. Typically, the Weibull stress parameters are fitted to experimental data with shape parameters varying between 17 and 20 for different materials in the lower-shelf [9]. Considering that measured size distributions of second phase particles would provide significantly smaller Weibull shape parameters (around 4–5, see, e.g., [10]), such calibrations suggest large differences between the shapes of particle and micro-crack size distributions.

While the Beremin model predicts well the cleavage fracture toughness in the lower-shelf of the DBT curve, the accurate prediction of cleavage fracture toughness in the DBT temperature regime has proven a challenge. Petti and Dodds [11] and Wasiluk et al. [12] proposed that the shape parameter should be independent of temperature, while the scale parameter should increase with temperature and should be calibrated to achieve agreement with the experimental data for the material. This is equivalent to the assumption that the shape of the micro-crack size distribution does not change with temperature, while the variation of cleavage fracture toughness is controlled by the variation of the material's elastic properties and surface energy with temperature. That assumption is problematic because the changes of material's elastic properties and surface energy are very small in the DBT temperature regime. Another option is to consider that the shape parameter changes with temperature, while the scale parameter is constant due to the argument in the previous sentence. For the case of small-scale yielding (SSY), the Hutchinson-Rice-Rosengren [13,14] fields have been used to calibrate the shape parameter in the DBT regime [15]. While the approach is reasonable, it still requires fracture toughness data at multiple temperatures, and assessments will be conservative, similarly to the global approach, due to the SSY conditions being used. The third option is to consider both Weibull stress parameters to be dependent on temperature. Wiesner and Goldthorpe [16] reported variation of $m$ between 13 and 23 and of $\sigma_u$ between 1700 and 3700 MPa. As in the previous two cases, this option relies on extensive experimental data at a number of temperatures. A model for cleavage fracture toughness predictions in the DBT regime without relying on empirical fittings to a large number of toughness measurements is yet to be achieved.

In the present study, we extend our previous approach [17] of incorporating a thinning function in the Weibull stress by presenting a theoretical formulation as opposed to an empirical one. The newly developed thinning function determines the number of cleavage initiators at a given temperature in the DBT regime. It is based on the kink-formation free energy from [7], and thus entirely theoretically based without any reliance on empirical fit to experimental data. We calibrated the model using a typical RPV steel with data from the Euro fracture dataset [18]. We were able to predict the characteristic fracture toughness at a few temperatures in the DBT regime based on experimental toughness data of a single temperature, which has not been done before. The paper is structured as follows: after a brief introduction to the Beremin model and some of the major previous modifications of the Weibull stress, we outline our new model and the required parameters, followed by a review of the finite element analysis in Section 2. Section 3 presents our results of applying the model and predicting the fracture toughness at three temperatures in the DBT region. Section 4 discusses the accuracy of the predictions and the importance of the findings, followed by a summary of the main conclusions in Section 5.

## 2. Methods

### 2.1. The Beremin Model

The Beremin model [8] was the first local approach to cleavage fracture, that is based on the weakest-link assumption, according to which the macroscopic cleavage event depends on the failure of a single cleavage initiating microcrack. This approximation is considered reasonable for the physical cleavage process, which consists of three main stages—firstly, microcracks initiate at randomly distributed second phase particles, assisted by the plastic deformation of the surrounding matrix; next, some microcracks propagate across the particle-matrix interface (alternatively, they can form voids, determined by the local mechanical conditions); and finally, cleavage occurs when a single microcrack propagates across the matrix without being arrested at a grain boundary [19,20]. From a statistical point of view, the cleavage process is represented as an inhomogeneous spatial Poisson point process [21] with points corresponding to the cleavage initiators (CI). There are three conditions to be met for a Poisson process including: failures of non-overlapping volumes are independent of each other; the probability of failure of a unit volume is proportional to its volume and the probability of more than one failure in the unit volume is zero. If $\mu_c$ is the probability density of sharp microcracks eligible for cleavage, the probability of failure $\delta P_f$ of a unit volume $\delta V$ and that of the whole solid $P_f$ are:

$$\delta P_f = \mu_c \delta V = \left(\frac{1}{V_0}\int_{a_c}^{\infty} f_c(a)da\right)\delta V \text{ and} \tag{1}$$

$$P_f = 1 - \exp\left(-\int_V \mu_c dV\right), \tag{2}$$

where $f_c(a)$ is the microcrack size distribution, $a_c$ is the critical size of a penny-shaped crack of size $a$ according to Griffith's criterion, and $V_0$ is a reference volume. The critical microcrack size can therefore be expressed as:

$$a_c = \frac{\pi E \gamma}{2(1-\nu^2)\sigma_1^2}, \tag{3}$$

where $E$ and $\nu$ are the elastic modulus and Poisson's ratio of the steel, respectively, $\sigma_1$ is the maximum principal stress, and $\gamma$ is a measure of fracture energy.

In the original Beremin model, plastic deformation is assumed to be a prerequisite for cleavage initiation, and the microcrack size distribution is assumed to follow a power law with scale and shape parameters, $\beta$ and $\alpha$, as follows:

$$f_c(a) = \theta\left(\frac{\beta}{a}\right)^{\alpha}, \tag{4}$$

where $\theta$, the fraction of particles converted into eligible micro-cracks, is equal to the Heaviside step function, $\theta = H(\varepsilon_p)$, or in other words all cleavage initiators nucleate at the onset of plasticity. In the terminology of the statistical Poisson process, $\theta$ is a thinning function, which creates a new Poisson process by only including some of the points of the original Poisson process. Integrating the microcrack size distribution and rearranging leads to the expression for the total failure probability in terms of the Weibull stress $\sigma_w$:

$$P_f(\sigma_w) = 1 - \exp\left(\frac{\sigma_w}{\sigma_u}\right)^{m}, \text{ and} \tag{5}$$

$$\sigma_w = \left(\frac{1}{V_0}\int_V \theta\sigma_1^m dV\right)^{1/m}, \tag{6}$$

where the Weibull shape and scale parameters are: $m = 2\alpha - 2$, and $\sigma_u = \left(\frac{\pi E \gamma}{2(1-\nu^2)\beta}\right)^{1/2}$, respectively. We note that the shape parameter $m$ is linked to the shape parameter of the microcrack size distribution, where it is assumed that the distribution follows the same shape as that of the particle size distribution, whereas the scale parameter $\sigma_u$ incorporates the elastic properties, the surface energy, and the scale of the size distribution.

## 2.2. Proposed Model

In bcc metals, the plastic flow is controlled by the thermally activated motion of screw dislocations and their interaction with obstacles. Dislocations move through the lattice via kinks, minimizing the Peierls potential of the dislocation. Swinburne and Dudarev [7] developed a model of obstacle hardening of bcc materials that accounts for the kink mechanism, which is temperature and shear stress dependent, in the dislocation-obstacle interaction. Their model demonstrated that the kink activation energy halves when a critical length becomes smaller than the average dislocation segment.

Equations (7) and (8) show the critical length $L^*$ and the kink formation free energy $F_k$, analogous to [7], where we substitute the maximum shear stress using Tresca criterion as an approximation to the Peierls stress, $b$ is the dislocation Burger's vector, $T$ is the temperature, $k$ is the Boltzmann's constant, $\sigma_p$ is the critical Peierls stress, $T_{ath}$ is the athermal temperature and we set the free energy to zero in elements with negative energy due to numerical reasons:

$$L^*(\tau_{max}, T) = b\,\exp[F_k(\tau_{max}, T)/kT], \tag{7}$$

$$F_k(\tau_{max}, T) = \begin{cases} 0.0, & \tau_{max} > \sigma_p(1 - T/T_{ath})^2 \\ U_k\left(1 - \frac{T}{T_{ath}} - \frac{\tau_{max}/\sigma_p}{1 - T/T_{ath}}\right), & \tau_{max} < \sigma_p(1 - T/T_{ath})^2 \end{cases}. \tag{8}$$

Setting the average dislocation segment to the average distance between carbides $\langle d \rangle$, we calculate a normalized free energy as:

$$f_k(\tau_{\max}, T) = \begin{cases} 2F_k(\tau_{\max}, T)/kT, & L^* \geq \langle d \rangle \\ F_k(\tau_{\max}, T)/kT, & L^* \leq \langle d \rangle \end{cases}. \quad (9)$$

Next, we obtain a thinning function $\theta$ for the Poisson process, dependent on the normalized free energy as:

$$\theta(\tau_{\max}, T) = \frac{f_k(\tau_{\max}, T)}{f_k(0\ \text{MPa}, -200\ ^\circ\text{C})}. \quad (10)$$

The as-constructed thinning function's behaviour is shown in Figure 1 as a function of the shear stress and temperature. At very low temperatures and stresses, the thinning tends to 1, which corresponds to all particles being converted to cleavage initiators, whereas increasing the temperature and the maximum shear stress reduces the value of the thinning as physically expected by the reduction in particles that form sharp micro-cracks.

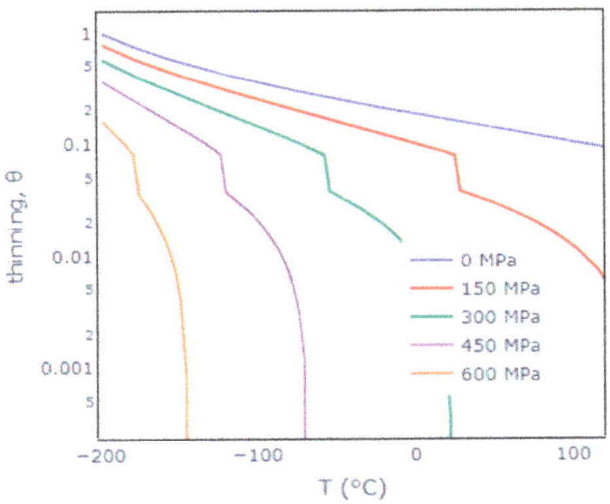


**Figure 1.** Poisson thinning vs. temperature for a range of shear stresses.

Figure 2 presents a flow chart of the developed model—for a given temperature $T_{\text{given}}$, at which at least 8 to 10 fracture toughness measurements are available, we perform a finite element (FE) analysis and for each load increment, we find the elements within fracture process zone (FPZ), based on two conditions for the maximum principal stress and the equivalent plastic strain, typically applied in the local approaches to fracture, as follows: $\sigma_I \geq \lambda\sigma_0$ and $\varepsilon^p_{\text{eq}} \geq \varepsilon^p_{\text{offset}}$, where $\lambda$ is a scalar equal to between 1 and 2.5, and $\varepsilon^p_{\text{offset}}$ is equal to 0.2%. The maximum $\sigma_I$ and minimum $\sigma_{I\ II}$ principal stresses and the volumes $V$ for all elements in the FPZ are stored as arrays. Then, the local approach is applied as a post-processor to the FE data—we loop over all load increments and for each increment calculate the maximum shear stress based on Tresca criterion, the kink-formation free energy and characteristic length scale [7], the normalized free energy (Equation (9)) and the thinning function (Equation (10)), the maximum principal stress within the load history up to the current increment. Next, we compute the Weibull stress using *any* shape parameter $m$ and unit volume $V_0$ (here, set to the volume of a spherical grain with a 10 μm radius, which corresponds to the approximate grain size in RPV steels [22]). The procedure is repeated for any temperature of interest $T_i$, at which we would like to predict the fracture toughness. Finally, we plot the Weibull stress curves as a function of $J$ as shown in the schematic in Figure 2 and compute the predicted $J$ for a given percentile, for instance the characteristic $J_0$ at 63.2%, at each temperature $T_i$ corresponding to the equivalent Weibull stress for the given temperature. There is no need to calibrate a scale parameter $\sigma_u$, since the probability curve at $T_{\text{given}}$ can be used to find the corresponding Weibull stress at each required $J$ percentile, and therefore, find this fracture toughness percentile for each temperature $T_i$.

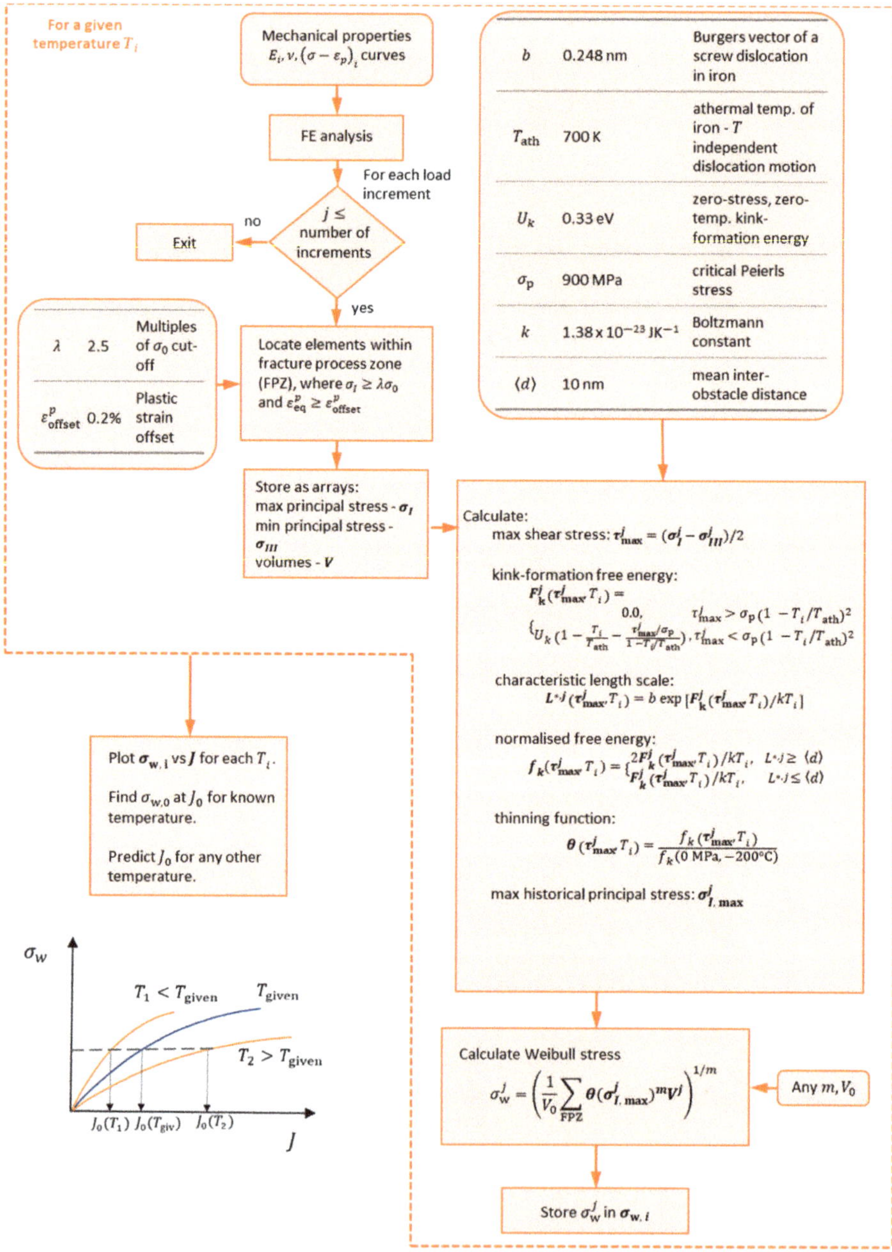


**Figure 2.** Flow chart of the developed model.

### 2.3. Finite Element Analysis

The model was developed using experimental fracture toughness and mechanical properties for the reactor pressure vessel steel 22NiMoCr37, which are available in the literature within the lower shelf and in DBT. Fracture toughness data [18] for a standard 1T compact tension specimen with thickness $B = 25$ mm, width $W = 50$ mm and crack length $a$ to width $W$ ratio of 0.5 at three temperatures, −91 °C, −60 °C and −40 °C, were used. The toughness data were ranked in ascending order with a rank probability $F(J_c^i) = (i - 0.3)/(N + 0.4)$, where $i = 1, \ldots, N$ and $N$ is the total number of data points. The temperature dependence of the Young's modulus $E$, the yield stress $\sigma_Y$ and the ultimate tensile stress $\sigma_{UTS}$ were approximated by [23]:

$$E = -90T + 206{,}000, \tag{11}$$

$$\sigma_Y = 421.2 + 63.9\exp(-T/91) \text{ and} \tag{12}$$

$$\sigma_{UTS} = 564.1 + 70.2\exp(-T/108). \tag{13}$$

Abaqus 2017 [24] was used to model two-dimensional models of a half standard 1T compact tension specimen, 1T-C(T) with the above material properties in the large strain finite element analyses. A mesh with finite crack tip radius was designed, where the corresponding radius for each temperature was chosen approximately five times smaller than the crack tip opening displacement at the characteristic fracture toughness $J_0$. The radii equal to 10, 15 and 20 μm for −91 °C, −60 °C and −40 °C, respectively, were verified against boundary layer models in a previous study for each of the corresponding temperatures [25]. Figure 3a shows the half-C(T) model with a 10-μm radius. The local approaches have been found to be more sensitive to the mesh design compared to the global approaches. Thus, we designed a mesh with a uniform element size within the area of interest, as shown in Figure 3b. There is a small fan section of the mesh near the crack tip with a side of no more than five times the initial radius, which is excluded from the Weibull stress calculation based on the yield stress cut-off condition specified above. This uniform mesh design was shown to lead to a more consistent Weibull stress calculations across temperatures compared to a traditional fan mesh due to the change of the stress and strain fields ahead of the crack with increasing temperatures. The experimental high-strength maraging steel loading pin [26] was replaced by a purely elastic wedge, which was found as a good approximation and provided shorter computational times. The material was modelled as elastic-plastic with flow properties computed using Equations (11)–(13) and specified in a tabular form in the FE software.

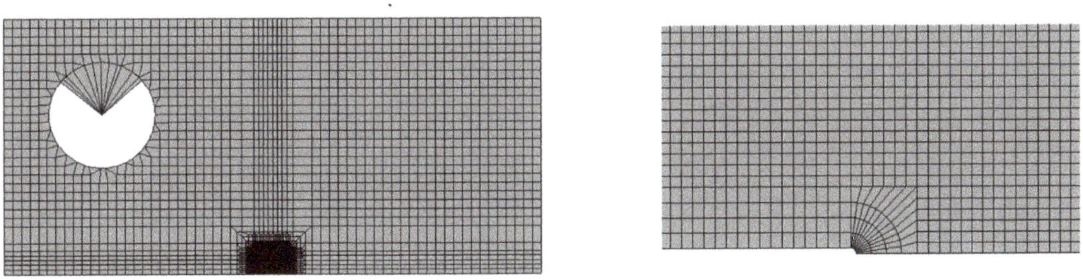

(a) (b)

**Figure 3.** (**a**) 2D half 1T compact tension specimen model; (**b**) Uniform mesh design with a finite radius crack tip.

## 3. Results

Figure 4a,b show the normalised maximum principal stresses and the plastic strain fields ahead of the crack tip, where the distance from the crack tip has been normalized as well. These mechanical parameters behave as expected with the increase of temperature.

The critical length $L^*$ as calculated per Equation (7) decreases with increasing temperature, as seen in Figure 4c. It represents the effect of easier dislocation motion through the lattice at higher temperatures due to a reduced lattice resistance, and accordingly an increase in the plastic flow. The average dislocation segment $\langle d \rangle$ was assumed to be equal to 10 nm and is indicated with a grey line. The thinning ahead of the crack is also plotted in Figure 4d, where we note the step of doubling the value at the critical length $L^*$ crossing $\langle d \rangle$.

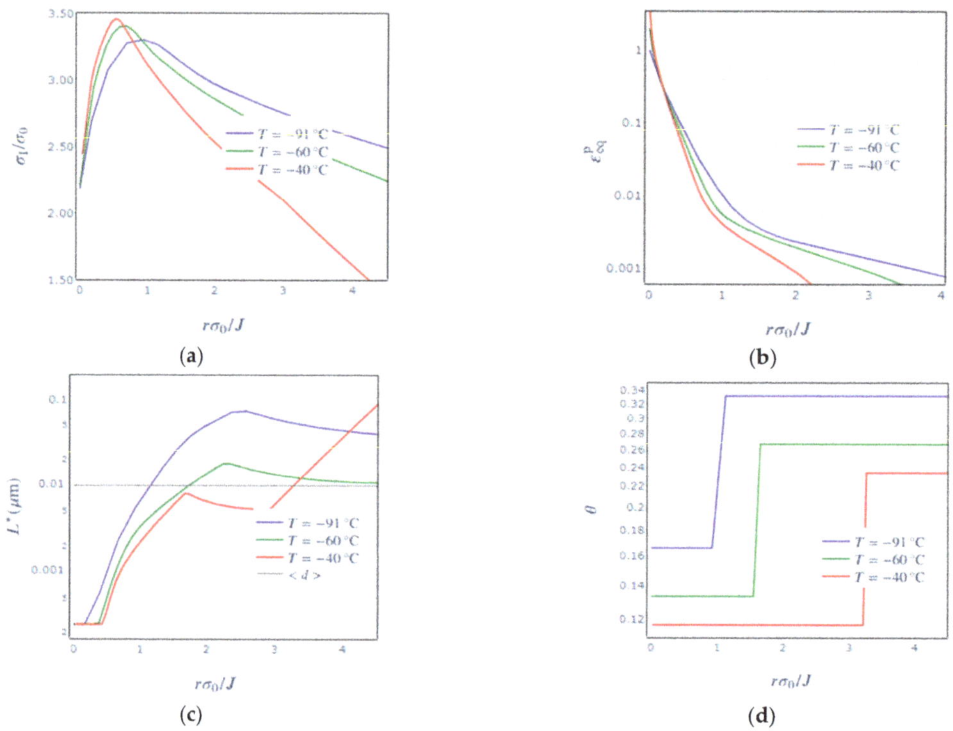


**Figure 4.** (**a**) Normalised maximum principal stress, (**b**) equivalent plastic strain, (**c**) critical length and mean interparticle distance; and (**d**) thinning as a function of the normalized distance from the crack tip for the 1T C(T) models at all temperatures.

Figure 5 presents the Weibull stress calculated for each of the three temperatures considered as a function of the shape parameter $m$ using the original Beremin model in (a) and using the developed model in (c). Furthermore, the Weibull stress differences between the two higher temperatures and the lowest temperature are obtained, as shown in Figure 5b,d, for the original and the developed models, respectively. Qualitatively, we observe the Weibull stress curves to diverge, specifically at low $m$ values, when the original Beremin model is used. On the contrary, the developed model brings the curves to coincide. The importance of this coincidence is that the proposed model works with any shape parameter within reasonable limits, or inversely, that the shape parameter to be used in cleavage fracture toughness predictions is independent of temperature. Quantitatively, we observe a reduction in the difference between the Weibull stress curves from up to about 10% in the case of the original Beremin model, to up to about 4.5% in the case of the developed model. This is within the uncertainty limits for the deformation properties of the material measured at different temperatures and used for the analyses in the work.

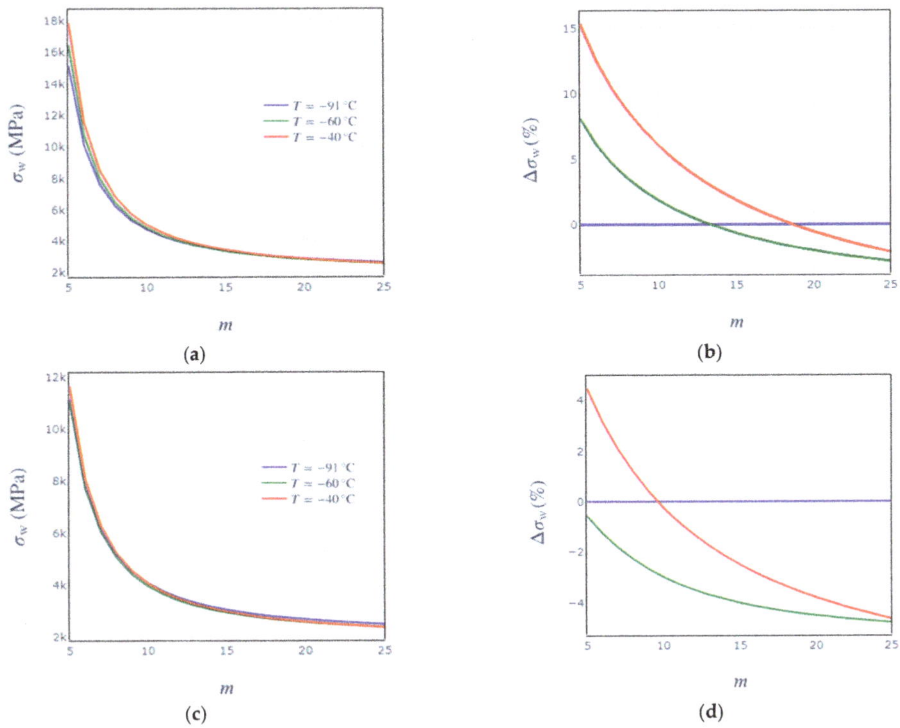


**Figure 5.** (**a**) Weibull stress and (**b**) Weibull stress difference in percentage with respect to the Weibull stress at −91 °C, using the original Beremin model as a function of the shape parameter $m$; (**c**) Weibull stress and (**d**) Weibull stress difference in percentage with respect to the Weibull stress at −91 °C, using the developed model as a function of the shape parameter $m$.

Next, we consider the characteristic fracture toughness predictions, as listed in Table 1. There is a good agreement between the measured and the predicted values using simulations. Figure 6 presents the cumulative probability distribution fits using the Maximum Likelihood (ML) method to the experimentally measured $J_c$ as well as the simulated fits based on a calibration of the developed model to the toughness data at T = −91 °C and a Weibull shape parameter $m$ = 7. At the fitted temperature of −91 °C, there is a complete coincidence with the experimentally fitted curve. At temperatures of −60° and −40 °C, there is better agreement at the lower toughness values, with some under- and overestimation at the higher toughness values, respectively. This is due to the larger number of experimental data points at the lower toughness values at −91 °C, and so there is a better agreement with the predicted values.

**Table 1.** Characteristic fracture toughness $J_0$ calculated based on a Weibull distribution fit to the experimental data using the maximum likelihood method and predicted using the developed method based on the fracture toughness experimental data at T = −91 °C, −60 °C and −40 °C, and a shape parameter $m$ = 10.

| $T$ (°C) | $J_0^{EXP}$ (Nmm$^{-1}$) | $J_0^{SIM,\,T=-91\,°C}$ (Nmm$^{-1}$) | $J_0^{SIM,\,T=-60\,°C}$ (Nmm$^{-1}$) | $J_0^{SIM,\,T=-40\,°C}$ (Nmm$^{-1}$) |
|---|---|---|---|---|
| −91 | 57.6 | - | 41.0 | 57.5 |
| −60 | 112.3 | 153.8 | - | 153.8 |
| −40 | 234.9 | 237.6 | 177.2 | - |

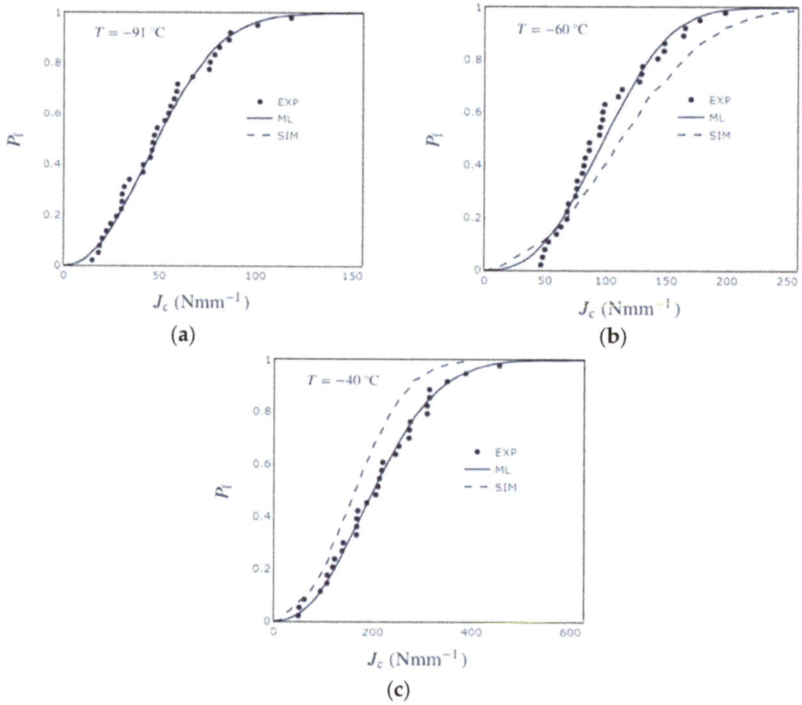


**Figure 6.** Cumulative probability function of experimentally measured $J_c$ using the Maximum Likelihood (ML) method and Weibull fit as predicted by the proposed method for (**a**) T = −91 °C, (**b**) T = −60 °C and (**c**) T = −40 °C, where the toughness data at T = −91 °C is used for calibration of the model at $m$ = 7.

## 4. Discussion

We showed previously [17,24] that the mechanical fields alone are not enough to capture the large exponential difference between the Weibull stress curves, which in general do not cross if no correction is applied. This means the use of a single shape parameter $m$ would be impossible to predict the toughness across temperatures. Additionally, under the assumption of cleavage fracture as a Poisson point process, it was found that for the same percentile $J_p$, the number of cleavage initiators are equal across different temperatures, but they are spread across a different plastic zone volume, i.e., the volume density of CI changes. The proposed thinning function mathematically captures this effect—it represents the decrease of microcracks that convert to CI with the increase of temperature. As shown in Figure 1, the thinning is close to one at low temperatures and shear stresses, corresponding to the case of all microcracks to convert into cleavage initiators. As the temperature and the shear stress (corresponding to the increase of plasticity due to loss of constraint) increase the thinning function reduces, and so a larger fraction of the microcracks will convert into voids rather than cleavage initiators. The biggest advantage of the current form of the thinning function and the developed model is the theoretical basis and so lack of empirical fittings required. The model includes microstructural parameters of the crystal structure of iron and the Boltzmann constant, as shown in Figure 2. All constant parameters have been computed for the screw dislocation in iron in the literature using atomistic simulations [4]. The only parameter calibrated as part of the model is the mean inter-obstacle distance—the value was chosen to describe an assumed mean inter-obstacle distance for the microstructure of a typical RPV steel and it did not have significant effects on the results. The model also includes two engineering parameters, which have values typically chosen in the local approaches to cleavage models as listed in Figure 2, namely the multiples of $\sigma_0$ cut-off and the plastic strain offset.

The major benefit of the developed local approach to cleavage fracture is the ability to predict the probability of cleavage at any temperature in the ductile-to-brittle regime by calibration through fracture toughness experiments at a single temperature. Previous approaches [9,15] require fracture toughness experiments at multiple temperatures, which is costly and time consuming. The developed model is of great industrial relevance, as it reduces the cost and time of performing these assessments.

Furthermore, the developed local approach to cleavage has the potential to predict the changes of fracture toughness due to irradiation. It is known that the macroscopic effects of irradiation are increased yield stress and reduced hardening and ductility. In fact, the kink formation energy model [7] used here has been developed to predict precisely these changes with irradiation. The application of the proposed model promises to capture the change of fracture toughness solely due to these changes of deformation properties, because at a given temperature the changes will result in reduced shear stress and from there in increased density of eligible micro-cracks in the plastic zone, leading eventually to higher probability of cleavage compared to unirradiated material at the same temperature. This needs to be explored in future work, although the validation would be even more challenging due to the very limited fracture toughness data of irradiated materials.

## 5. Conclusions

The main findings of this study are summarised below:

- An exponential correction to the density of micro-cracks eligible for cleavage, which is dependent on a material property, that changes with temperature, was proposed. It represents a thinning function of the Poisson process, which creates a new Poisson process with a subset of the points of the original Poisson process.
- The developed procedure for probability of cleavage estimation would only require experimental fracture toughness tests at a given temperature, and deformation properties at any other temperature of interest.
- Very good agreement was shown between the predicted and experimentally measured characteristic fracture toughness as well as the predicted and experimentally fitted probability distributions.
- The value of the work stems from the possibility of significant reduction in the necessary experimental fracture toughness testing.
- The validation of the model requires four to six deformation tests, and at least 30 fracture toughness tests at three constraint conditions with a ratio, $a/w$ of 0.5, 0.2 and 0.1.
- These would allow the prediction of cleavage toughness in the ductile to brittle transition regime using a decoupled model, which makes it advantageous for engineering purposes.

**Author Contributions:** Conceptualization, M.S.Y. and A.P.J.; methodology, M.S.Y., A.P.J. and R.P.; software, M.S.Y.; validation, M.S.Y.; formal analysis, M.S.Y.; investigation, M.S.Y.; resources, M.S.Y., A.P.J. and R.P.; data curation, M.S.Y. and A.P.J.; writing—original draft preparation, M.S.Y.; writing—review and editing, M.S.Y., A.P.J. and R.P.; visualization, M.S.Y.; supervision, A.P.J. and R.P.; funding acquisition, A.P.J. and R.P. All authors have read and agreed to the published version of the manuscript.

**Funding:** This research was funded by the National Nuclear Laboratory and Électricité de France. Furthermore, A. P. Jivkov acknowledges the financial support of the Engineering and Physical Sciences Research Council UK (EPSRC) via grant EP/N026136/1.

**Institutional Review Board Statement:** Not applicable.

**Informed Consent Statement:** Not applicable.

**Data Availability Statement:** Data supporting the findings of this study are available from the corresponding author upon request.

**Acknowledgments:** The authors would like to acknowledge the assistance given by Research IT and the use of the Computational Shared Facility at The University of Manchester. The authors thank Andrew Sherry for fruitful discussions.

**Conflicts of Interest:** The authors declare no conflict of interest.

## References

1. Curry, D.A.; Knott, J.F. Effects of microstructure on cleavage fracture stress in steel. *Met. Sci.* **1978**, *12*, 511–514. [CrossRef]
2. Mcmahon, C.J.; Ferriten, V. Initiation of cleavage in polycrystalline iron. *Acta Metall.* **1965**, *13*, 591–604. [CrossRef]
3. Gurland, J. Observations on the fracture of cementite particles in a spheroidized 1.05 % C steel deformed at room temperature. *Acta Metall.* **1972**, *20*, 735–741. [CrossRef]
4. Armstrong, R.W. Material grain size and crack size influences on cleavage fracturing. *Philos. Trans. R. Soc. A Math. Phys. Eng. Sci.* **2015**, *373*. [CrossRef]
5. Roberts, S.; Noronha, S.; Wilkinson, A.; Hirsch, P. Modelling the initiation of cleavage fracture of ferritic steels. *Acta Mater.* **2002**, *50*, 1229–1244. [CrossRef]
6. Tang, M.; Marian, J. Temperature and high strain rate dependence of tensile deformation behavior in single-crystal iron from dislocation dynamics simulations. *Acta Mater.* **2014**, *70*, 123–129. [CrossRef]
7. Swinburne, T.D.; Dudarev, S.L. Kink-limited Orowan strengthening explains the brittle to ductile transition of irradiated and unirradiated bcc metals. *Phys. Rev. Mater.* **2018**, *2*, 73608. [CrossRef]
8. Beremin, F.M. A local criterion for cleavage fracture of a nuclear pressure vessel steel. *Metall. Trans. A* **1983**, *14*, 2277–2287. [CrossRef]
9. Tanguy, B.; Besson, J.; Piques, R.; Pineau, A. Ductile to brittle transition of an A508 steel characterized by Charpy impact test. Part II: Modeling of the Charpy transition curve. *Eng. Fract. Mech.* **2005**, *72*, 413–434. [CrossRef]
10. Ruggieri, C.; Jivkov, A. A Local Approach to Cleavage Fracture Incorporating the Measured Statistics of Microcracks 2018. Available online: https://www.research.manchester.ac.uk/portal/files/70129677/CP2018_153_Ruggieri_Jivkov_Final.pdf (accessed on 1 February 2021).
11. Petti, J.P.; Dodds, R.H. Calibration of the Weibull stress scale parameter, σu, using the Master Curve. *Eng. Fract. Mech.* **2005**, *72*, 91–120. [CrossRef]
12. Wasiluk, B.; Petti, J.P.; Dodds, R.H. Temperature dependence of Weibull stress parameters: Studies using the Euro-material. *Eng. Fract. Mech.* **2006**, 1046–1069. [CrossRef]
13. Hutchinson, J.W.; Hutchinson, J.W. Generalizing J 2 flow theory: Fundamental issues in strain gradient plasticity. *Acta Mech. Sin.* **2012**, *28*, 1078–1086. [CrossRef]
14. Rice, J.R.; Rosengren, G.F. Plane strain deformation near a crack tip in a power-law hardening material. *J. Mech. Phys. Solids* **1968**, *16*, 1–12. [CrossRef]
15. Andrieu, A.; Pineau, A.; Besson, J.; Ryckelynck, D.; Bouaziz, O. Beremin model: Methodology and application to the prediction of the Euro toughness data set. *Eng. Fract. Mech.* **2012**, *95*, 102–117. [CrossRef]
16. Wiesner, C.S.; Goldthorpe, M.R. The effect of temperature and specimen geometry on the parameters of the "local approach" to cleavage fracture. *J. Phys. IV JP* **1996**, *6*. [CrossRef]
17. Yankova, M.S.; Jivkov, A.P.; Patel, R.; Sherry, A.H. Using local approaches to fracture to quantify the local conditions during the ductile-to-brittle transition in ferritic steels. In Proceedings of the D2-Fracture Mechanics and Structural Integrity, SMiRT-25, IASMiRT, Charlotte, NC, USA, 4–9 August 2019.
18. Heerens, J.; Hellmann, D. Development of the Euro fracture toughness dataset. *Eng. Fract. Mech.* **2002**, *69*, 421–449. [CrossRef]
19. Wallin, K.; Saario, T.; Törrönen, K. Statistical model for carbide induced brittle fracture in steel. *Met. Sci.* **1984**, *18*, 13–16. [CrossRef]
20. Ruggieri, C.; Dodds, R.H. A local approach to cleavage fracture modeling: An overview of progress and challenges for engineering applications. *Eng. Fract. Mech.* **2018**, *187*, 381–403. [CrossRef]
21. Kingman, J.F.C. *Poisson Processes*; Oxford University Press: New York, NY, USA, 2002.
22. Zhao, X.; Lidbury, D.; da Fonseca, J.Q.; Sherry, A.; Neu, R.; Wallin, K.; Thompson, S.R.; Dean, S.W. Introducing Heterogeneity into Brittle Fracture Modeling of a 22NiMoCr37 Ferritic Steel Ring Forging. *J. ASTM Int.* **2008**, *5*, 101562. [CrossRef]
23. James, P.M.M.; Ford, M.; Jivkov, A.P.P. A novel particle failure criterion for cleavage fracture modelling allowing measured brittle particle distributions. *Eng. Fract. Mech.* **2014**, *121–122*, 98–115. [CrossRef]
24. *Abaqus/Standard 2017*; Dassault Systemes Simulia Corp, 2017. Available online: https://www.4realsim.com/documentos/pdf/abaqus-standard-2017-datasheet.pdf (accessed on 1 February 2021).
25. Jivkov, A.P.; Ford, M.; Yankova, M.; Sarzosa, D.; Ruggieri, C. Progress and challenges with local approaches to cleavage fracture. *Procedia Struct. Integr.* **2019**, *23*, 39–44. [CrossRef]
26. *ASTM E399-20a, Standard Test Method for Linear-Elastic Plane-Strain Fracture Toughness KIc of Metallic Materials*; ASTM International: West Conshohocken, PA, USA, 2020. Available online: www.astm.org (accessed on 20 December 2020).

materials

MDPI

*Article*

# Cyclic Deformation Induced Residual Stress Evolution and 3D Short Fatigue Crack Growth Investigated by Advanced Synchrotron Tomography Techniques

**Benjamin Dönges [1], Melanie Syha [2,3], Anne K. Hüsecken [4], Ullrich Pietsch [4], Wolfgang Ludwig [3], Ulrich Krupp [5] and Hans-Jürgen Christ [1,*]**

1 Institut für Werkstofftechnik, Universität Siegen, D-57068 Siegen, Germany; benjamin.doenges@gmx.de
2 Institut für Angewandte Materialien, Karlsruher Institut für Technologie, D-76131 Karlsruhe, Germany; melanie.syha@ds-ing.net
3 European Synchrotron Radiation Facility, F-38043 Grenoble, France; wolfgang.ludwig@esrf.fr
4 Festkörperphysik, Department Physik, Universität Siegen, D-57068 Siegen, Germany; anne.huesecken@gmx.de (A.K.H.); pietsch@physik.uni-siegen.de (U.P.)
5 Institut für Eisenhüttenkunde, RWTH Aachen, D-52072 Aachen, Germany; krupp@iehk.rwth-aachen.de
* Correspondence: hans-juergen.christ@uni-siegen.de; Tel.: +49-271-740-4658

**Abstract:** Diffraction and phase contrast tomography techniques were successfully applied to an austenitic–ferritic duplex stainless steel representing exemplarily a metallic material containing two phases with different crystal structures. The reconstructed volumes of both phases were discretized by finite elements. A crystal plasticity finite-element analysis was executed in order to simulate the development of the experimentally determined first and second order residual stresses, which built up due to the manufacturing process of the material. Cyclic deformation simulations showed the single-grain-resolved evolution of initial residual stresses in both phases and were found to be in good agreement with the experimental results. Solely in ferritic grains, residual stresses built up due to cyclic deformation, which promoted crack nucleation in this phase. Furthermore, phase contrast tomography was applied in order to analyze the mechanisms of fatigue crack nucleation and short fatigue crack propagation three-dimensionally and nondestructively. The results clearly showed the significance of microstructural barriers for short fatigue crack growth at the surface, as well as into the material. The investigation presented aims for a better understanding of the three-dimensional mechanisms governing short fatigue crack propagation and, in particular, the effect of residual stresses on these mechanisms. The final goal was to generate tailored microstructures for improved fatigue resistance and enhanced fatigue life.

**Keywords:** synchrotron tomography; short fatigue crack growth; residual stress; crystal plasticity; very high cycle fatigue

**Citation:** Dönges, B.; Syha, M.; Hüsecken, A.K.; Pietsch, U.; Ludwig, W.; Krupp, U.; Christ, H.-J. Cyclic Deformation Induced Residual Stress Evolution and 3D Short Fatigue Crack Growth Investigated by Advanced Synchrotron Tomography Techniques. *Materials* **2021**, *14*, 1562. https://doi.org/10.3390/ma14061562

Academic Editors: Jaroslav Pokluda and Reinhard Pippan

Received: 15 February 2021
Accepted: 19 March 2021
Published: 22 March 2021



## 1. Introduction

The failure of safety-relevant components by fatigue of materials in form of crack initiation and propagation caused by periodic loading far below the static strength of a material may cause unexpected damage events and, thereby, is a thread not just for components and engineering structures but also for human life. It is known that the phases of fatigue crack initiation and propagation can cover more than 90% of the total fatigue life of a material at loading situations close to the conventional fatigue limit [1], and therefore, this fatigue stage can be considered as determining time to failure of a component. Recent investigations have shown the significance of microstructural aspects, such as (i) anisotropic elasticity of grains, (ii) three-dimensional geometry of grain and phase boundaries, and (iii) microscopic residual stresses, for fatigue crack initiation and propagation [2,3].

Three-dimensional characterization of dual phase materials has already been realized by means of serial sectioning electron back scatter diffraction (EBSD) experiments [4–6]. However, being a destructive technique, serial sectioning does not provide the possibility to follow the microstructure evolution quasi in situ by means of interrupted fatigue experiments. The combined use of X-ray diffraction contrast tomography and multi-distance phase contrast tomography [7] allows for the unique possibility to reconstruct the three-dimensional microstructure of a two-phase material while simultaneously determining the crystallographic orientation of each grain. Due to the very high resolution achieved in holotomography experiments, a high certainty of the location of inter-phase grain boundaries is achieved. The non-destructive nature of these techniques allows for cyclic deformation with intermittent hold times or slow cycles for grain orientation and grain shape determination.

In a previous study [8], the cyclic deformation-induced change of manufacturing-caused residual stress in the investigated austenitic–ferritic duplex stainless steel was experimentally characterized by means of high energy synchrotron radiation for up to one hundred million load cycles. Due to a higher thermal expansion coefficient of the austenite phase in comparison to the ferrite phase, tensile residual stress is generated in austenite grains and compressive residual stress is generated in ferrite grains as a result of prior quenching from 1050 °C to room temperature as the final part of the manufacturing process of the material. The change of residual stress was investigated by measuring the lattice spacing $d$ by means of synchrotron radiation diffraction experiments after predefined numbers of loading cycles (intermittent fatigue test) and determining the relative change of the lattice spacing $\Delta d$ with respect to the initial lattice spacing before cyclic loading $d_{N=0}$.

Only small changes of the relative lattice spacing as a function of the number of load cycles $N$ were found in most of the austenite grains and also in some of the ferrite grains (shown exemplarily for one grain in Figure 1), indicating that no change of residual stress occurred in these grains. In about half of all ferrite grains, the lattice spacing was found to increase continuously during the first $5 \cdot 10^5$ load cycles and, subsequently, to decrease again slightly (Figure 2). Hence, initial compressive residual stress was reduced in these ferrite grains during cyclic deformation. This is a comprehensible effect since it is well-known that plastic deformation usually causes a reduction of initial residual stress. This increase of lattice spacing was not observed in any austenite grain. A third trend showed that the lattice spacing in a few austenite grains appeared to first stay unchanged until $10^5$ load cycles, but then, during subsequent load cycles, the lattice spacing in these grains continuously became smaller (Figure 3), indicating that initial tensile residual stress was reduced in these austenite grains during cyclic deformation, being a comprehensible effect as already stated above. The most interesting observation in this study was that an increase of lattice spacing was found for many ferrite grains, indicating a reduction of the initial compressive residual stress This effect of partial annihilation of compressive residual stress in some ferrite grains is remarkable and seems to promote the observed predominant fatigue crack initiation process in the ferrite phase as described in detail in [9,10]. It should be noted that the saturation or slight reduction of initial tensile residual stress in the austenite grains is in accordance with the observed true fatigue limit found for the material investigated.

The present study shows experimental techniques that enable the nondestructive characterization of three-dimensional polycrystalline microstructures in order to obtain information about the shape and crystallographic orientation of the single grains. Moreover, these techniques enable the investigation of the three-dimensional interaction between short fatigue cracks and grain and phase boundaries. It is shown how this experimental data can be used in the framework of three-dimensional crystal plasticity finite-element simulations to calculate realistically the development of residual stress as a function of the number of load cycles. Such simulations may serve as a basis for a realistic fatigue life assessment model, which may enable the development of tailored microstructures with improved fatigue resistance and enhanced fatigue life in the future.

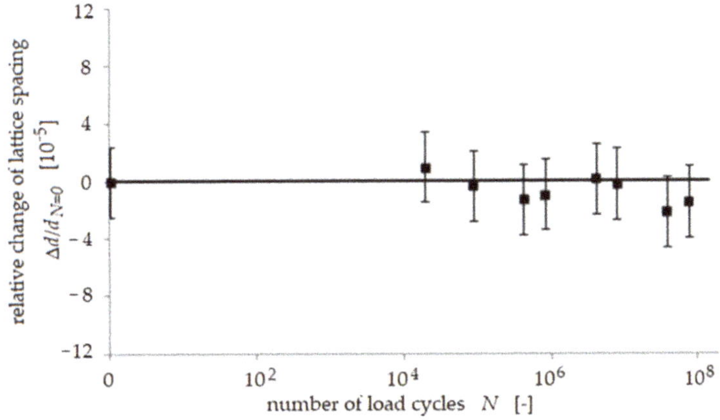


**Figure 1.** Relative change of lattice spacing $d$ in a single grain during cyclic loading at a stress am-plitude of 380 MPa, using the example of an austenite [040] reflection according to [8].

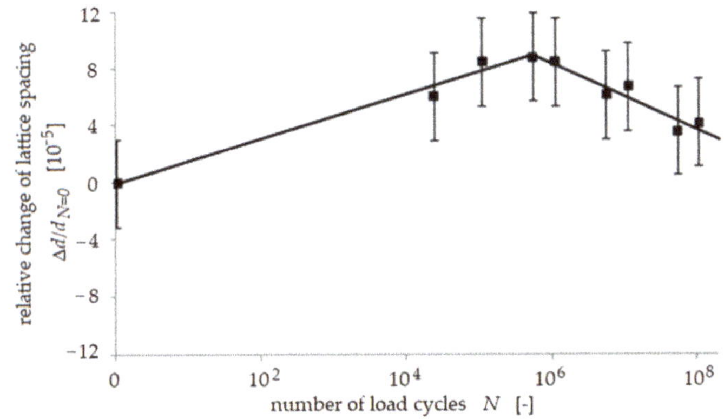


**Figure 2.** Relative change of lattice spacing $d$ in a single grain during cyclic loading at a stress am-plitude of 380 MPa, using the example of a ferrite [031] reflection according to [8].

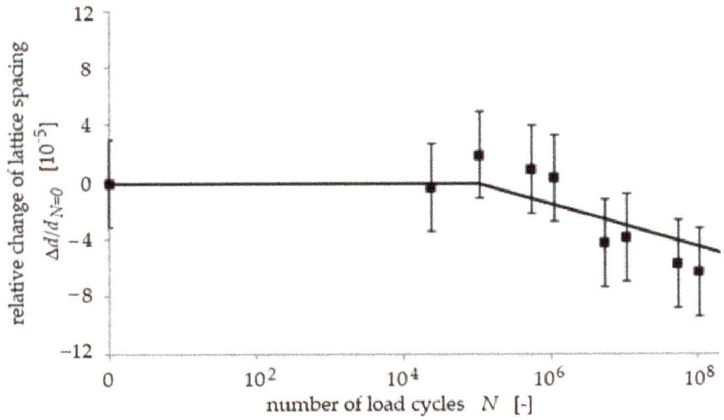


**Figure 3.** Relative change of lattice spacing $d$ in a single grain during cyclic loading at a stress am-plitude of 380 MPa, using the example of another ferrite [031] reflection according to [8].

## 2. Experimental Details

Duplex stainless steels have a high corrosion resistance and relatively high strength, which makes them very attractive for use, e.g., in offshore systems or systems for the chemical and petrochemical industry. The investigated austenitic-ferritic duplex stainless steel 318LN (German designation X2CrNiMoN22-5-3) was delivered as hot rolled and solution annealed bars with a diameter of 25 mm. The material consisted of a fine lamellar microstructure with about 50 vol% austenite and 50 vol% ferrite. The chemical composition of the material is shown in Table 1. In order to ease the experimental investigations, a grain coarsening was carried out by annealing the material at 1250 °C for 4 h followed by a cooling down to 1050 °C within 3 h at constant cooling rate. Subsequently, the material was quenched in water. By means of this heat treatment, the initial volume fraction of both phases was maintained, and the mean grain diameter of the austenite phase and the ferrite phase was increased to 33 µm and 46 µm, respectively. Figure 4 shows the microstructure of the material studied after the heat treatment in a plane parallel to the rolling direction (vertical). The austenite phase is presented in Figure 4a, while the ferrite phase is depicted in Figure 4b. The colors represent the crystallographic orientation of the grains according to the color code inserts, which show the [001] standard triangle. The images were obtained by means of an automated electron backscatter diffraction (EBSD) analysis. In order to execute the EBSD analysis, a material sample was prepared by grinding and electrolytic etching. Mechanical properties of the heat-treated condition (HTC) and the as-received condition (ARC) of the duplex stainless steel were investigated by means of tensile tests in rolling direction. The mechanical properties are presented in Table 2.

**Table 1.** Chemical composition of the investigated duplex stainless steel (mass concentration in %) [11].

| C | Cr | Ni | Mo | N | Fe |
|---|---|---|---|---|---|
| 0.03 | 21.0–23.0 | 4.5–6.5 | 2.5–3.5 | 0.1–0.22 | rest |

**Table 2.** Mechanical properties of the heat-treated condition (HTC) and the as-received condition (ARC) of the investigated duplex stainless steel in rolling direction [11].

| Condition | Young's Modulus [GPa] | 0.2% Yield Strength [MPa] | Tensile Strength [MPa] | Elongation at Fracture [%] |
|---|---|---|---|---|
| HTC | 197 | 535 | 770 | 59 |
| ARC | 197 | 720 | 870 | 33 |

For the application of the diffraction contrast tomography (DCT) technique, a polycrystalline sample was mounted on a rotating sample holder and irradiated by means of a monochromatic X-ray beam (Figure 5). The energy of the X-ray beam had to be sufficiently high or the irradiated sample had to be sufficiently thin that enough X-ray radiation could transmit and that multiple diffraction was prohibited. Crystal planes of single grains diffracted the X-ray beam when fulfilling the Bragg condition. Hereby, single diffraction spots were generated on a semiconductor plate detector, which was positioned behind the sample in the direction of the incident beam. Because the diffracted X-ray beams did not reach the detector in the direction of the incident beam, an extinction spot was generated at the projected area of the diffracting grain in the direction of the incident beam at the detector. Furthermore, a diffraction spot was generated at the position where the diffracted beam reached the detector. The sample was rotated in small angle increments (0.1°) by, in total, 360° around the cylinder axis, and after each rotation step, a diffraction pattern was recorded by the detector. By means of the positions of the diffraction spots in the diffraction patterns and the corresponding rotation angles of the diffraction pattern images, the three-dimensional geometry and crystallographic orientation of each grain could be reconstructed. For this purpose, a sophisticated mathematical algorithm applied. Unfortunately, positioning all reconstructed grains at their correct position inside the sample volume could lead to a grain arrangement that does not represent the irradiated volume perfectly. Rather, occasionally, the reconstruction process gave rise to overlapping of several grains, or voids might be generated, affecting the accuracy of the determination of the local grain boundary orientation [12]. This drawback was partially mitigated by using the 3D phase reconstructions resulting from holotomography, a technique explained in detail in [7]. Due to different absorption coefficients of different phases (here: austenite and ferrite), contrast differences were generated on the semiconductor plate detector.

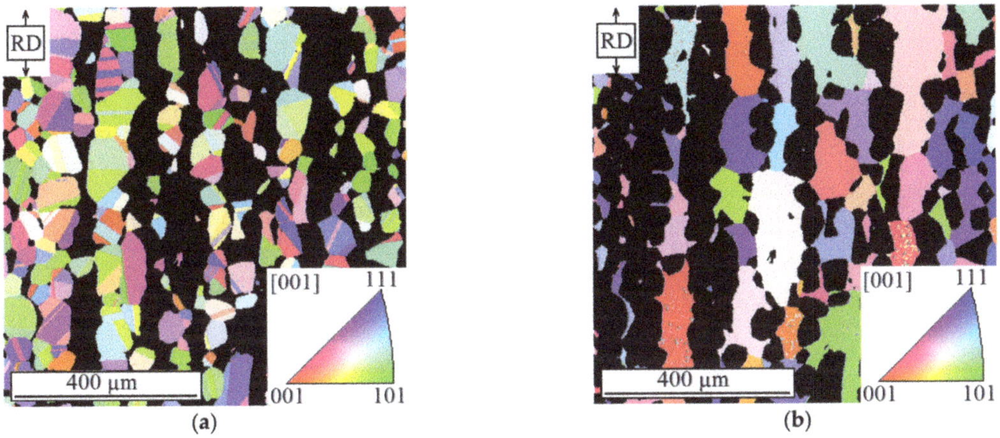


**Figure 4.** Inverse pole figure maps of (**a**) the austenite phase and (**b**) the ferrite phase of the investigated duplex stainless steel, taken parallel to the rolling direction (vertical) according to [3].

In principle, the setup for the phase contrast tomography (PCT) technique was very similar to that of DCT. The sample was rotated in small angle increments by, in total, 360° with respect to its axis, and at each angle, a contrast image was recorded. By overlapping the single contrast images according to the corresponding rotation angles, a three-dimensional volume reconstruction of the individual phases could be generated. In contrast to the microstructure reconstruction, which was generated by means of the DCT technique, as has been described before, the information of the crystallographic orientation of grains and the information about intra-phase grain boundaries were missing in the data, which was obtained by means of the PCT technique. However, the reconstructions that were generated by PCT were more precise regarding the positions of phase boundaries. For a more detailed description of the PCT technique the reader is referred to [14].

The cyclic deformation experiments were executed by means of ultrasonic fatigue testing [15] at a testing frequency of about 20 kHz (Figure 6). The tests were carried out at room temperature and in laboratory atmosphere. A sinusoidal mechanical stress wave was generated by means of a piezoelectric crystal, amplified by means of a horn in its amplitude, and introduced into a fatigue sample holder, on which a miniature fatigue sample was mounted. Resonance occurred if the eigenfrequency of the assembly of miniature fatigue sample and sample holder was identical to the excitation frequency of the piezoelectric crystal. Then, the required stress amplitude in the area of minimum cross section of the miniature fatigue sample was generated. A control system kept this amplitude constant. Due to heat generation during the fatigue tests, a pulse-pause mode (100 ms/1200 ms) and an air-cooling system were required. Hereby, the temperature rise in the fatigue sample was restricted to maximum 5 °C. The chosen pulse-pause mode led to an effective testing frequency of about 1.5 kHz. Hereby, a testing of very high numbers of load cycles was possible in a reasonable testing time (e.g., one billion load cycles in about 7.5 days).

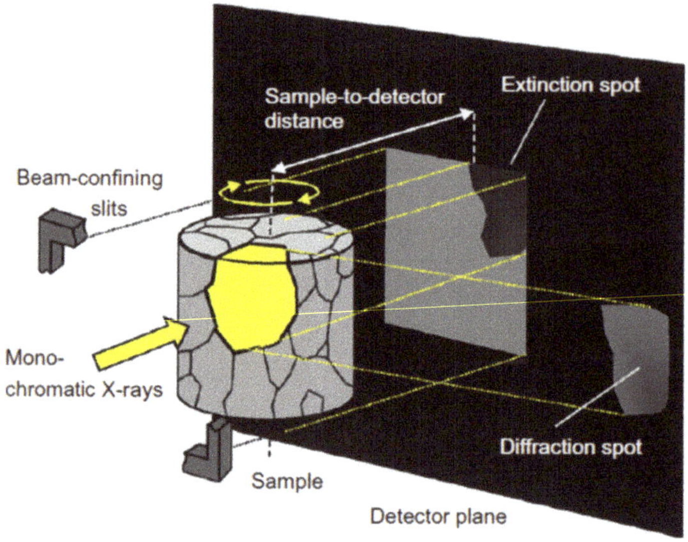


**Figure 5.** Principle of diffraction contrast tomography [13].

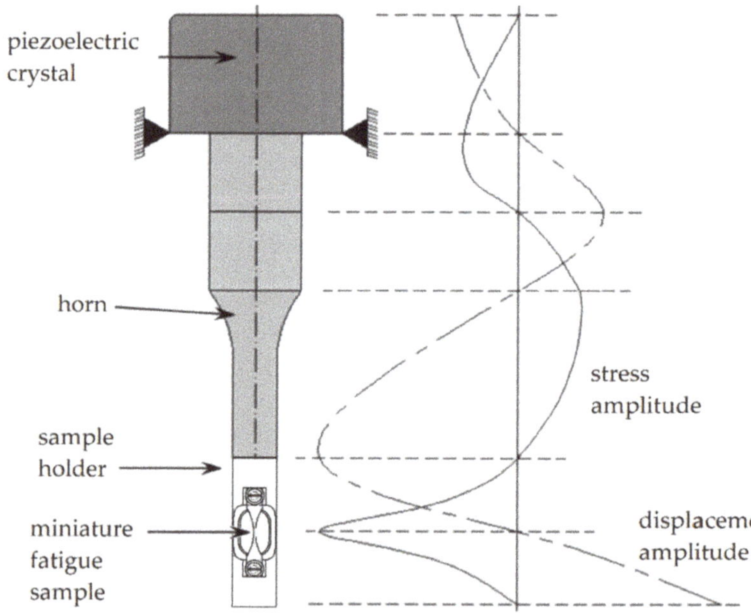


**Figure 6.** Ultrasonic fatigue testing equipment according to [16], in combination with a newly developed sample holder for miniature fatigue samples suitable for DCT and PCT.

## 3. Some Numerical Details

The user-defined material subroutine UMAT, which was developed by Huang [17] for the commercial finite-element program ABAQUS, was applied in this study. This subroutine assumed that the plastic shear on single slip systems was determined by the elastic shear stress acting on the corresponding slip system. The plastic shear rates $\dot{\gamma}_{pl}$ on a single slip system were calculated by means of the following rate dependent flow law [18]:

$$\dot{\gamma}_{pl} = \dot{a}\left(\frac{\tau}{\tau_f}\right)^m \tag{1}$$

where $\dot{a}$ is the reference value of the plastic shear rate, which is reached when the shear stress $\tau$ equals the frictional shear stress $\tau_f$. By means of the power law exponent $m$, the course of the function $\dot{\gamma}_{pl} = f(\tau)$ is determined. An exponent value $m \rightarrow \infty$ corresponds to rate-independent material behavior [18]. The plastic shear was determined by means of numerical integration of the plastic shear rate over the time increment $\Delta t$ as follows:

$$\Delta\gamma_{pl} = \Delta t\Big((1-\theta)\,\dot{\gamma}_{pl,t} + \theta\,\dot{\gamma}_{pl,t+\Delta t}\Big) \tag{2}$$

The implicit integration parameter $\theta$ determines which fraction of the plastic shear rate at the end of a time increment is considered. The parameter can have a value between 0 and 1, whereas a value between 0.5 and 1 is recommended [19]. The plastic shear is iteratively calculated during a time increment. The iteration ends when the calculated plastic shear for a time increment is lower than a user-defined value $\Delta\gamma_{\text{pl,error}}$. In the framework of the simulations, which led to the results presented in the next section, the parameter values shown in Table 3 were used. For a more detailed description of the UMAT subroutine the reader is kindly referred to [17,20].

**Table 3.** Parameters used in this work for the UMAT subroutine [11,17].

| Parameter | Symbol | Value | Dimension |
|---|---|---|---|
| reference shear rate | $\dot{a}$ | $10^{-3}$ [17] | 1/s |
| power law exponent | $m$ | 20 [17] | – |
| frictional shear stress of austenite phase | $\tau_{f\gamma}$ | 68 [11] | MPa |
| frictional shear stress of ferrite phase | $\tau_{f\alpha}$ | 99 [11] | MPa |
| implicit integration parameter | $\Theta$ | 0.5 [17] | – |
| iteration stop criterion | $\Delta\gamma_{pl,error}$ | $10^{-5}$ [17] | – |

## 4. Results and Discussion

In order to reconstruct the microstructure of the investigated duplex stainless steel, the advantages of both X-ray tomography techniques described in Section 2 were combined. The reconstructed 3D-microstructure of the austenite phase (Figure 7a) and of the ferrite phase (Figure 7b) were generated by combining the DCT volume data with the PCT volume data. Before doing this, the overlapping grain volume of the DCT reconstruction data was deleted.

A fatigue crack at the surface of a miniature sample, to which ultrasonic fatigue testing at 20 kHz was applied, is shown in Figure 8a. The fatigue sample was fatigued at a stress amplitude of 400 MPa. The test was stepwise interrupted in order to monitor the short fatigue crack propagation by means of optical microscope images, which were taken after different loading cycles. The fatigue crack presented in Figure 8 initiated at two nucleation sites, i.e., (i) at the phase boundary between the austenite grain $\gamma_3$ and the ferrite grain $\alpha_2$ and (ii) at the phase boundary between the austenite grain $\gamma_4$ and the ferrite grain $\alpha_3$. The transcrystalline crack nuclei grew together and subsequently propagated in both directions as a consequence of further cyclic deformation. The crack growth was decelerated, when a crack tip approached a phase boundary, such as in the case of the phase boundary between the austenite grain $\gamma_7$ and the ferrite grain $\alpha_4$. An acceleration of crack propagation was always detected when the crack overcame the boundary. At the phase boundary between the austenite grain $\gamma_1$ and the ferrite grain $\alpha_1$, the crack was even temporarily stopped.

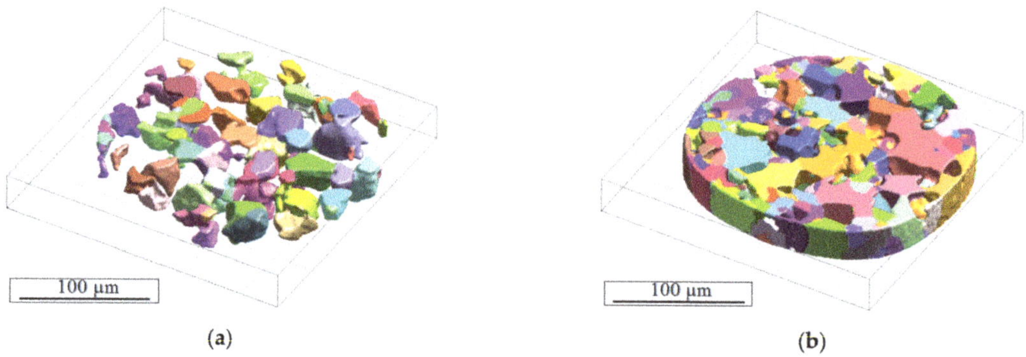


**Figure 7.** Reconstruction of (**a**) the austenite and (**b**) the ferrite grain microstructure.

By means of the phase contrast tomography technique and the commercial 3D data visualization software AVIZO FIRE, the crack shown in Figure 8a was examined nondestructively and represented three-dimensionally after different numbers of load cycles. In order to execute the tomography investigations, the miniature sample had to be removed intermittently from the ultrasonic test rig and precisely placed and mounted into the synchrotron beam. Following each tomography experiment, the miniature sample was installed back again in the ultrasonic fatigue testing device in order to continue the cyclic loading, and so forth. The loading parameter values of the ultrasonic fatigue equipment were thoroughly kept constant during the whole experiment.

The crack front after a predefined numbers of loading cycles is displayed in the view direction parallel to the sample axis in Figure 8b. Clearly, the crack propagates roughly in a half-elliptical shape. However, the crack propagation rate obviously shows significant local differences. The crack is presented in Figure 8c after $3.75 \cdot 10^5$ loading cycles in the view direction perpendicular to the sample surface. Furthermore, Figure 8d shows the crack profile along the dashed white line delineated in Figure 8b. Both images (Figure 8c,d) clearly show the characteristic zigzag path that is typical of propagation of microstructurally short fatigue cracks, which propagate along single slip planes under predominant local single-slip conditions. The changes of the propagation direction at grain or phase boundaries are caused by changes in the crystallographic orientation (not shown here).

A three-dimensional microstructure of the investigated duplex stainless steel is presented in Figure 9. The ferrite phase is shown at the left side and the corresponding austenite phase is depicted at the right side. The geometry of the grains and the phase distribution was determined by means of the combination of DCT and PCT (see Section 2). The geometry information was discretized by tetrahedral finite elements by means of the commercial image processing software AVIZO FIRE. The experimentally determined phase affiliation of each grain and the individual crystallographic grain orientation were considered. Moreover, anisotropic elasticity and lattice-type-specific crystal plasticity were taken into account in the stress distribution calculations (Section 3). The volume fraction of both phases was confirmed to be about 50% each. In order to consider the behavior of the surrounding of the depicted volume of the microstructure in a reasonable approach, this section was embedded in a frame consisting of finite elements with isotropic-elastic material behavior (Young's modulus $E$ = 197 MPa and Poisson's ratio $\nu$ = 0.3). This surrounding frame is not shown in Figure 9 for the sake of clarity.

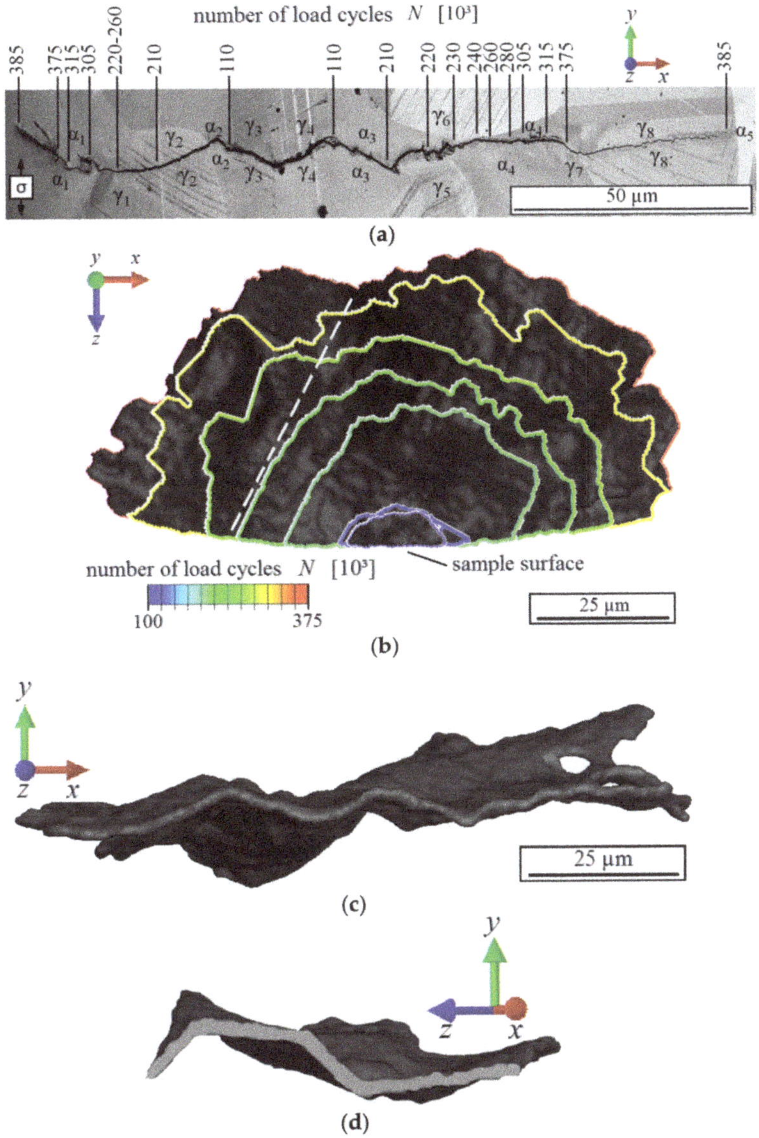


**Figure 8.** (**a**) Short fatigue crack propagation at sample surface. (**b**) Fracture surface of the fatigue crack shown in a) as a function of the number of load cycles (view in loading direction). (**c**) 3D geometry of the fatigue crack (view perpendicular to sample surface). (**d**) Profile of the fatigue crack along the dashed line in (**b**).

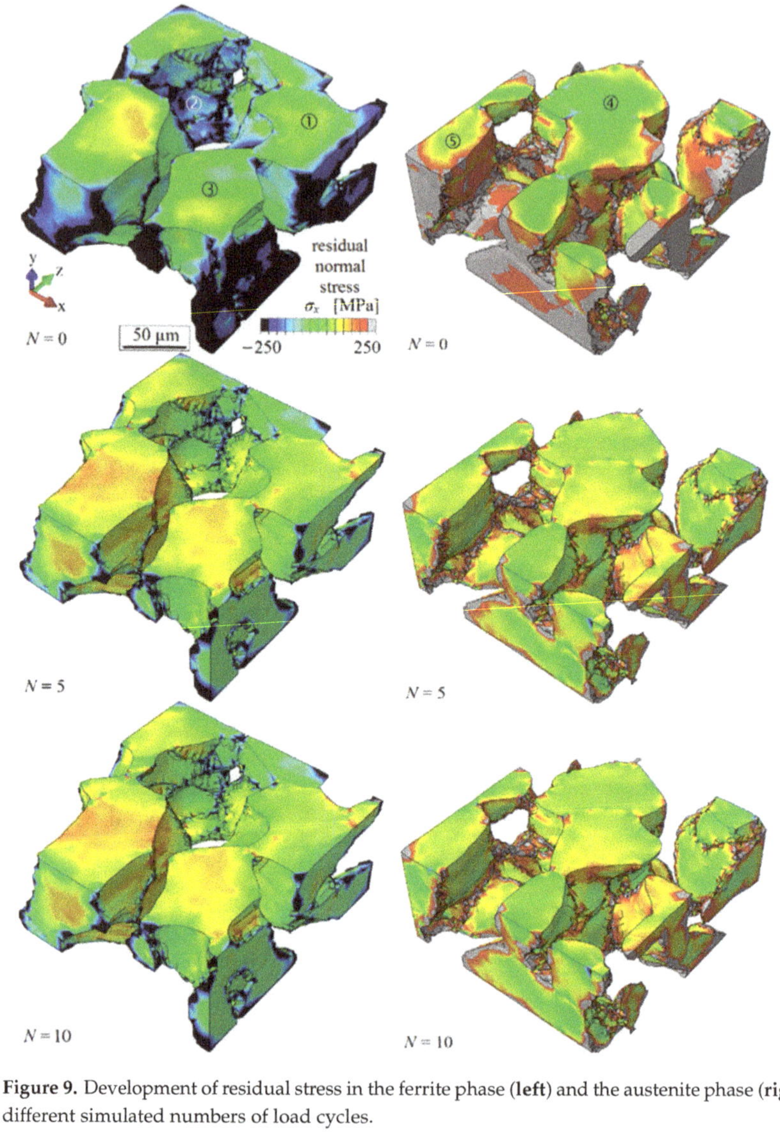


**Figure 9.** Development of residual stress in the ferrite phase (**left**) and the austenite phase (**right**) after different simulated numbers of load cycles.

The nodes at the left face of the model were fixed in x-direction. One node at the upper left corner of the model was fixed in z-direction and one node at the lower left corner was additionally fixed in z- and y-direction to avoid rigid body motions. To define the starting condition, the residual stress distribution resulting from the cooling at the end of the initial annealing process was calculated. For this purpose, the quenching process from 1050 °C to room temperature was numerically simulated applying temperature-dependent elastic constants $E_{11}$, $E_{12}$ and $E_{44}$; a temperature-dependent thermal expansion coefficient $\alpha$; temperature-dependent microstructural frictional shear stresses of both phases; and the temperature-dependent Young's modulus $E$ for the isotropic-elastic frame (details are provided in [2]). Furthermore, the first order residual stresses, which were experimentally determined by conventional X-ray diffraction measurements, were considered in the microstructure section of Figure 9 by applying a load of 50.2 MPa in tension at the nodes at the right face of the model and 21 MPa in compression at the upper and lower face of the model. The simulation of the quenching process was followed by a simulation of the effect of load cycles. For this, a cyclic normal stress amplitude of 350 MPa was applied at the nodes at the right face of the model.

Because of the higher coefficient of thermal expansion of the austenite phase as compared to the ferrite phase, the starting condition after quenching was characterized by dominating compressive internal stresses in the ferrite and tensile stresses in the austenite. The distribution of the simulated residual normal stress in load direction $\sigma_x$ in the ferrite phase after different numbers of load cycles $N$ is presented at the left side of Figure 9. Some grains showed no change of the initial residual stresses with increasing numbers of load cycles (e.g., position 1). However, initial compressive residual stresses in some other grains were significantly reduced during the first loading cycles, but less during further cycles (e.g., position 2). Furthermore, significant tensional residual stresses built up during the first loading cycles in some grains, which reached a saturation state during the following load cycles—for example position 3.

At the right side of Figure 9, the simulated residual normal stress in load direction $\sigma_x$ is presented for the austenite phase after different numbers of load cycles $N$. Some grains showed no significant change of residual stress with increasing numbers of loading cycles (e.g., position 4). However, a decrease of initial tensile residual stress was observed in the other grains (e.g., position 5).

Basically, the expected equalization of the residual stresses by means of cyclic plastic loading could be confirmed. However, in some ferritic grains, a localized stress increase took place that seemed to promote crack initiation. It should be emphasized that this type of calculations were found to explain the location of crack initiation [3]. In this respect, 3D-calculations on the basis of an extruded microstructure, which do not take the third-dimension correctly into account, are much less unerring.

## 5. Conclusions

Advanced diffraction contrast tomography (DCT) and phase contrast tomography (PCT) were combined and successfully applied to an austenitic–ferritic duplex stainless steel. It was shown that a three-dimensional microstructure of a metallic material that contained two phases with different crystal structures could be reconstructed nondestructively. Besides the information on the phase distribution and grain microstructure of the material, this sophisticated technique additionally provided crystallographic orientation data of the grains.

The obtained data were used to create realistic three-dimensional finite-element models of the microstructure. By means of a crystal plasticity finite-element analysis, experimentally determined first and second order residual stresses caused by the manufacturing process of the material were simulated. The results of subsequent cyclic deformation simulations were found to be in good agreement with experimental results on the evolution of initial residual stress on the microstructural level and strongly support the identification of the critical fatigue crack nucleation sites. In the duplex stainless steel investigated, solely in ferrite grains, residual stresses built up locally due to cyclic deformation promoting crack nucleation in this phase in accordance to the experimental observations.

By applying a suitable fatigue damage parameter (e.g., the one proposed in ref. [2]), crystal plasticity finite-element simulations based on three-dimensional microstructures, which were obtained nondestructively by means of DCT and PCT, provide a powerful instrument to analyze and quantitatively judge separately the effect of microstructural parameters regarding their influence on the fatigue behavior of a material. Furthermore, cyclic loading experiments intermitted by DCT and PCT analyses enable a 3D visualization and re-modeling of crack propagation behavior, being, in particular, useful for an improved understanding of the mechanisms which control the propagation of microstructural short fatigue cracks. The results of the present work clearly demonstrate the significance of phase and grain boundaries as microstructural barriers for short fatigue crack growth, both at the surface and in the interior of multi-phase alloys.

**Author Contributions:** Conceptualization, U.K., H.-J.C., M.S., A.K.H. and U.P.; Methodology, W.L.; investigation, B.D., M.S. and A.K.H.; writing—original draft preparation, B.D.; writing—review and editing, H.-J.C.; supervision, U.P., H.-J.C., U.K. and W.L.; funding acquisition, U.K., U.P. and W.L. All authors have read and agreed to the published version of the manuscript.

**Funding:** This research was funded by Deutsche Forschungsgemeinschaft (German Science Foundation) in the framework of the Priority Program SPP 1466: Life—Infinite Life of Cyclically Loaded High Performance Materials.

**Institutional Review Board Statement:** Not applicable.

**Informed Consent Statement:** Not applicable.

**Data Availability Statement:** The data presented in this study are available on request from the corresponding author.

**Acknowledgments:** The authors thank Deutsche Forschungsgemeinschaft (DFG, German Science Foundation) for financial support in the framework of the priority program "Life" (SPP1466).

**Conflicts of Interest:** The authors declare no conflict of interest.

## References

1. Shiozawa, K.; Matsushita, H. Crack initiation and small fatigue crack growth behavior of beta Ti-15V-3Cr-3Al-3Sn alloy. In Fatigue '96: Proceedings of the Sixth International Fatigue Congress, Berlin, Germany, 6–10 May 1996; Lütjering, G., Nowack, H., Eds.; Elsevier: Oxford, UK; p. 301.
2. Dönges, B.; Fritzen, C.-P.; Christ, H.-J. Experimental investigation and simulation of the fatigue mechanisms of a duplex stainless steel under HCF and VHCF conditions. *Key Eng. Mater.* **2015**, *664*, 267–274. [CrossRef]
3. Dönges, B.; Fritzen, C.-P.; Christ, H.-J. Fatigue mechanisms and its modelling of an austenitic-ferritic duplex stainless steel under HCF and VHCF loading conditions. In *Fatigue of Materials at Very High Number of Loading Cycles—Experimental Techniques, Mechanisms, Modeling and Fatigue Life Assessment*; Christ, H.-J., Ed.; Springer Spektrum, Springer Fachmedien Wiesbaden GmbH: Wiesbaden, Germany, 2018; pp. 111–131.
4. Groeber, M.; Ghosh, S.; Uchic, M.D.; Dimiduk, D.S. A framework for automated analysis and simulation of 3D polycrystalline microstructures: Part 1: Statistical characterization. *Acta Mater.* **2008**, *56*, 1257–1273. [CrossRef]
5. McGarrity, K.S.; Sietsma, J.; Jonbloed, G. Characterisation and quantification of microstructural banding in dual-phase steels; Part 2—Case study extending to 3D. *Mater. Sci. Technol.* **2012**, *28*, 903–910. [CrossRef]
6. Calcagnotto, M.; Ponge, D.; Demir, E.; Raabe, D. Orientation gradients and geometrically necessary dislocations in ultrafine grained dual-phase steels studied by 2D and 3D EBSD. *Mater. Sci. Eng. A* **2010**, *527*, 2738–2746. [CrossRef]
7. Cloetens, P.; Ludwig, W.; Baruchel, J.; Van Dyck, D.; Van Landuyt, J.; Guigay, J.P.; Schlenker, M. Holotomography: Quantitative phase tomography with micrometer resolution using hard synchrotron radiation X-rays. *Appl. Phys. Lett. B* **1999**, *75*, 2912–2914. [CrossRef]
8. Dönges, B.; Istomin, K.; Söker, M.; Schell, N.; Krupp, U.; Pietsch, U.; Fritzen, C.-P.; Christ, H.-J. Experimental investigation and numerical description of the damage evolution in a duplex stainless steel subjected to VHCF-loading. *Mater. Sci. Eng. A* **2015**, *646*, 8–18. [CrossRef]
9. Dönges, B.; Giertler, A.; Krupp, U.; Fritzen, C.-P.; Christ, H.-J. Significance of crystallographic misorientation at phase boundaries for fatigue crack initiation in a duplex stainless steel during high and very high cycle fatigue loading. *Mater. Sci. Eng. A* **2014**, *589*, 146–152. [CrossRef]
10. Istomin, K.; Dönges, B.; Schell, N.; Christ, H.-J.; Pietsch, U. Analysis of VHCF damage in a duplex stainless steel using hard X-ray diffraction techniques. *Int. J. Fatigue* **2014**, *66*, 177–182. [CrossRef]
11. Düber, O. Untersuchungen zum Ausbreitungsverhalten Mikrostrukturell Kurzer Ermüdungsrisse in Zweiphasigen Metallischen Werkstoffen am Beispiel eines Austenitisch-Ferritischen Duplexstahls. Ph.D. Thesis, Universität Siegen, Düsseldorf, Germany, 2007.
12. Syha, M.; Trenkle, A.; Lödermann, B.; Graff, A.; Ludwig, W.; Weygand, D.; Gumbsch, P. Validation of three-dimensional diffraction contrast tomography reconstructions by means of electron backscatter diffraction characterization. *J. Appl. Crystallogr.* **2013**, *46*, 1145–1150. [CrossRef] [PubMed]
13. Herbig, M.; King, A.; Reischig, P.; Proudhon, H.; Lauridsen, E.M.; Marrow, J.; Buffière, J.Y.; Ludwig, W. 3-D growth of a short fatigue crack within a polycrystalline microstructure studied using combined diffraction and phase-contrast X-ray tomography. *Acta Mater.* **2011**, *59*, 590–601. [CrossRef]

14. Reischig, P.; King, A.; Nervo, L.; Vigano, N.; Guilhem, Y.; Palenstijn, W.J.; Batenburg, K.J.; Preuss, M.; Ludwig, W. Advances in X-ray diffraction contrast tomography: Flexibility in the setup geometry and application to multiphase materials. *J. Appl. Crystallogr.* **2013**, *46*, 297–311. [CrossRef]
15. Mayer, H. Fatigue crack growth and threshold measurements at very high frequencies. *Inter. Mater. Rev.* **1999**, *44*, 1–34. [CrossRef]
16. Bathias, C.; Paris, P.C. *Gigacycle Fatigue in Mechanical Practice*; Marcel Dekker: New York, NY, USA, 2005.
17. Huang, Y. *A User-Material Subroutine Incorporating Single Crystal Plasticity in the ABAQUS Finite Element Program. Mech Report 178*; Division of Engineering and Applied Sciences, Harvard University: Cambridge, MA, USA, 1991.
18. Asaro, R.J. Micromechanics of crystals and polycrystals. *Adv. Appl. Math. Mech.* **1983**, *23*, 1–115.
19. Peirce, D.; Shih, C.F.; Needleman, A. A tangent modulus method for rate dependent solids. *Comput. Struct.* **1984**, *18*, 875–887. [CrossRef]
20. Kysar, J.W. *Abbendum to 'A User-Material Subroutine Incorporating Single Crystal Plasticity in the ABAQUS Finite Element Program, Mech Report 178'*; Division of Engineering and Applied Sciences, Harvard University: Cambridge, MA, USA, 1997.

materials

Article

# Machine-Learning-Based Atomistic Model Analysis on High-Temperature Compressive Creep Properties of Amorphous Silicon Carbide

Atsushi Kubo * and Yoshitaka Umeno

Institute of Industrial Science, The University of Tokyo, Tokyo 113-8654, Japan; umeno@iis.u-tokyo.ac.jp
* Correspondence: kubo@ulab.iis.u-tokyo.ac.jp

**Abstract:** Ceramic matrix composites (CMCs) based on silicon carbide (SiC) are used for high-temperature applications such as the hot section in turbines. For such applications, the mechanical properties at a high temperature are essential for lifetime prediction and reliability design of SiC-based CMC components. We developed an interatomic potential function based on the artificial neural network (ANN) model for silicon-carbon systems aiming at investigation of high-temperature mechanical properties of SiC materials. We confirmed that the developed ANN potential function reproduces typical material properties of the single crystals of SiC, Si, and C consistent with first-principles calculations. We also validated applicability of the developed ANN potential to a simulation of an amorphous SiC through the analysis of the radial distribution function. The developed ANN potential was applied to a series of creep test for an amorphous SiC model, focusing on the amorphous phase, which is expected to be formed in the SiC-based composites. As a result, we observed two types of creep behavior due to different atomistic mechanisms depending on the strain rate. The evaluated activation energies are lower than the experimental values in literature. This result indicates that an amorphous region can play an important role in the creep process in SiC composites.

**Keywords:** silicon carbide; ceramic matrix composites; creep properties; high-temperature strength; molecular dynamics; artificial neural network

**Citation:** Kubo, A.; Umeno, Y. Machine-Learning-Based Atomistic Model Analysis on High-Temperature Compressive Creep Properties of Amorphous Silicon Carbide. *Materials* **2021**, *14*, 1597. https://doi.org/10.3390/ma14071597

Academic Editor: Matthias Posselt

Received: 27 February 2021
Accepted: 22 March 2021
Published: 25 March 2021



## 1. Introduction

From the viewpoint of structural materials, silicon carbide (SiC) is used as the matrices and reinforcing fibers in the ceramics matrix composites (CMCs). The CMC whose matrix and fiber consist of SiC is called SiC/SiC. The SiC/SiC composite has prominent mechanical properties such as light weight, high elastic modulus, high strength, and high fracture toughness [1,2], and these features makes the SiC/SiC composites applicable to the hot section in turbines [3,4]. For such application of SiC-based structural materials, the mechanical properties at high temperature are of crucial importance and have been investigated by experiments (e.g., the temperature dependence of tensile strength and creep properties for a SiC fiber [5,6], and high-temperature fatigue behavior of a SiC/SiC composite [7]). Since such experiments require a lot of time and special experimental equipment, assistance by numerical simulation approaches is necessary for evaluation and analysis of the mechanical properties. The molecular dynamics (MD) simulation can be a powerful tool to investigate the essential mechanisms of deformation and fracture, if a suitable interatomic potential function is available. Moreover, ideally, the material properties obtained by the MD simulations can be used as the material parameters for larger-scale simulations; e.g., the activation energies of creep and diffusion are applied to the phase field simulations, the finite element method (FEM) analysis, etc.

The reliability of MD simulation is mainly dependent on the quality of the applied potential function, and thus the potential function should be chosen (or developed) properly according to the purpose of the simulation. Thus far many potential functions have been developed for the Si-C system based on various physical models such as the bond-order model [8,9], the modified embedded atom model [10], the ionic model [11,12], and the charge-transfer model [13]. Nevertheless, there are still some non-trivial difficulties in development of an interatomic potential function for the high-temperature analysis of SiC, as follows:

1. In SiC fibers and matrices, the ratio of Si and C atoms is *not* necessarily stoichiometric but may be varied. For example, Hi-Nicalon, a typical SiC fiber, consists of Si and C of 39 at% and 60.4 at%, respectively [6]. Therefore, a constant-charge model, such as Vashishta et al. [11] and Kubo et al. [12], is not applicable any longer.
2. Unlike a low-temperature condition, the local atomic structure can easily change at a high temperature via, e.g., creep and diffusion processes. In such situation, the bond-order type potential functions may cause qualitative and quantitative errors, because in general such potential functions tend to overestimate the bonding energy or critical force of bond breaking.
3. It is also unclear what interactions are dominant on the mechanical properties of SiC at high temperature. In SiC, both ionic and covalent interactions are competitive and thus play an important role in deformation and fracture in SiC. That is true even in the case of a perfect crystal of SiC at 0 K, where cleavage, slip, and phase transition were observed with a slight difference in the loading condition in a first-principles (FP) analysis [14]. This complex feature of the interaction in SiC requires a highly flexible and versatile formulation in the potential model.
4. The practical SiC materials include other types of atoms (e.g., B, O, N, Al, etc.) as impurities and/or dopants [15,16] (Relatedly, ceramic fibers consisting of Si, B, N, and C also have been produced [17]). Therefore, the potential function for the Si-C systems should be extendable to a many-species system for further investigation. The existence of such impurities and/or dopants also requires a flexible formulation of the potential model because the additional atoms can be metallic, covalent, or ionic.

In short, MD simulations of the high-temperature mechanics of SiC require the interatomic potential model to be highly transferable, flexible and extendable. It is difficult to address all these problems by means of conventional interatomic potential models (Even if possible, it may be accompanied by a substantial modification of the function forms). In contrast, the interatomic potential models based on machine-learning approaches can be a suitable solution to overcome these problems because of their systematic formulations. Several types of machine-learning potentials have been thus far proposed [18], e.g., the artificial neural network (ANN) model [19,20], the Gaussian approximation potential [21], the gradient-domain machine learning [22], and other approaches [23]. Especially the artificial neural network (ANN) potential models have been widely applied because of its flexibility. In a mathematical sense, an ANN can mimic any function with a given accuracy [24–26]. In a physical sense, an ANN has no physical background and thus can be applied to any type of materials including metals [27,28], metal oxides [20,29], carbon polymorphs [30,31], and organic molecules and complexes [32]. In addition, application of the ANN models to atomistic analyses is not limited to use as a potential function. Very recently, several ANN applications have been reported, where ANNs are used for prediction of electronic properties in the atomic systems [33–35]. Thanks to a lack of the physical background in ANN models, a framework of the ANN potential function is applicable to other objects, e.g., a prediction of the electronic density of state, without any special modification in the ANN architecture itself, as was demonstrated by Umeno and Kubo [33].

In this study, we develop an ANN potential function for the Si-C systems, for the purpose of investigation of high-temperature mechanical properties in the SiC structural materials. While real SiC-based ceramics contain other atomic species, we omit such impurity or dopant atoms and focus on pure Si-C systems for simplicity. Some mechanical properties are evaluated for SiC single crystals for the purpose of validation of the ANN potential. After that, the ANN potential function is applied to high-temperature deformation analyses. On performing MD simulations, it is required to set a reasonable simulation model of atomic structure because a whole CMC system cannot be examined by molecular simulations owing to the limitation of the length scale. According to experimental observations, the grain boundary region plays an important role in the high-temperature mechanical behavior such as creep in ceramics [6,36]. Thus, we mainly examine the amorphous-phase structures to mimic the non-crystalline grain boundary regions instead of directly investigating a CMC structure or a polycrystal model for simplicity and efficiency of simulation; i.e., we regard the amorphous model as the limit of small-grain. The effect of oxygen atoms, which is a major cause of forming the amorphous phase, is also omitted for simplicity. The creep properties of amorphous SiC is evaluated as a relationship among the strain rate, stress, and temperature. The mechanisms of creep in the amorphous SiC are discussed by analyzing the structural change during deformation.

This study is organized as follows. In Section 2, we introduce the potential function based on the ANN model and explain the procedure of parametrization. In Section 3, the developed ANN potential is validated through several simple MD simulations. Typical material properties such as the lattice constant are compared with the FP calculations and experiments in literature. In Section 4, we apply the ANN potential to practical analyses to evaluate the high-temperature mechanical properties, especially the creep properties under a variety of loading conditions. Lastly, in Section 5, we summarize the results and make the concluding remarks.

## 2. Construction of Artificial Neural Network Potential Model

### *2.1. Formulation*

We adopted the framework of the ANN potential model proposed by Behler and Parrinello [19], where the potential energy for each atom is evaluated as a function of the local atomic structure within the cutoff radius. The ANN architecture is schematically shown in Figure 1. An ANN model is composed by the input layer (layer 0), internal layers (layers 1, ..., $N-1$) and the output layer (layer $N$). Each layer consists of nodes, whose state is expressed by a real number. The state of the nodes in the input layer is determined from local atomic structure by the basis functions. The output layer has a single node and, its state is returned as the potential energy. The state of the $\alpha$-th node in the $n$-th layer, $x_\alpha^n$ ($n > 0$), is determined with the nodes in the previous layer $n-1$ as follows:

$$x_\alpha^n = f_\mathrm{a}^n\left(\sum_{\beta=1}^{M_{n-1}} w_{\alpha\beta}^n x_\beta^{n-1} + w_{\alpha 0}^n\right), \tag{1}$$

where $w_{\alpha\beta}^n$ denotes the weight parameter, and $w_{\alpha 0}^n$ is the bias weight parameter independent of the state of the previous layer. Those weight parameters are to be optimized through machine learning. The function $f_\mathrm{a}^n$ is the activation function for the $n$-th layer. The controllable architectural parameters in this model are the number of the internal layers $N$, the number and types of the basis functions, and the number of nodes in each layer $M_n$ for $n = 1, ..., N-1$ (the number of nodes in the input layer, $M_0$, is equal to the number of the basis functions). In addition, also the type of the activation function $f_\mathrm{a}^n$ ($n = 1, ..., N-1$) can be changed.

While the original Behler-Parrinello model applies an exponential-type basis function set, we adopted the basis function set formulated by Artrith et al. [32] This basis set includes the two-body interaction $\varphi_\mu(r)$ and the three-body interaction $\psi_\mu(\theta)$:

$$\phi_\mu(r) = T_\mu\left(\frac{2r}{r_\mathrm{c}} - 1\right), \tag{2}$$

$$\psi_\mu(\theta) = T_\mu(\cos\theta), \tag{3}$$

where the series of $T_\mu(x)$ ($x\in[-1, 1]$, $\mu = 0, 1, 2, ...$) is the Chebyshev polynomials of the first kind defined recursively as

$$\begin{cases} T_0(x) = 1, \\ T_1(x) = x, \\ T_{\mu+1}(x) = 2xT_\mu(x) - T_{\mu-1}(x). \end{cases} \quad (4)$$

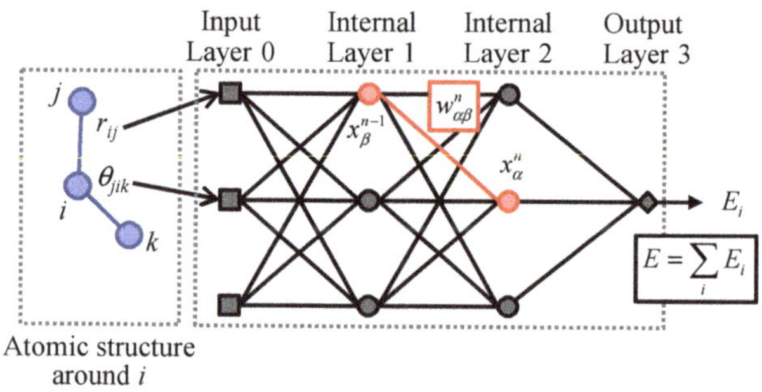


**Figure 1.** Schematic illustration of ANN potential model. This figure shows an exemplified network model with two internal layers (layers 1 and 2) and three nodes each in the input and internal layers.

Using these basis sets, the state of the nodes in the input layer, $x^0{}_\alpha$, is determined as follows:

$$x_\alpha^{0\ (\mathrm{pair})} = \sum_j \phi_{\alpha-1}(r_{ij}) f_c(r_{ij}), \quad (5)$$

$$x_\alpha^{0\ (\mathrm{angle})} = \sum_{j,k} \psi_{\alpha-1}(\theta_{jik}) f_c(r_{ij}) f_c(r_{ik}), \quad (6)$$

where $f_c$ is the cutoff function to truncate the basis functions at the cutoff radius $r_c$. In this formulation, each of two-body and three-body basis functions has only two empirical parameters, i.e., the number of terms for expansion, $N_2$ and $N_3$, and the cutoff radius $r_c$. This simple feature enables a systematic expansion of the basis function.

The activation function $f_a^n$ is given as

$$f_a^n(x) = \begin{cases} 1.7159\tanh\left(\frac{2x}{3}\right) + 0.1x & (n = 1, \ldots, N-1) \\ x & (n = N) \end{cases}. \quad (7)$$

While the cutoff function $f_c$ is given by a cosine function in the original model [32], we modified the cutoff function to the following form:

$$f_c(r) = \begin{cases} \frac{1}{2}\left[1 + \sin\left(\frac{\pi}{2}\cos\left(\frac{\pi r}{r_c}\right)\right)\right] & (r \le r_c) \\ 0 & (r > r_c) \end{cases}. \quad (8)$$

The main purpose of this modification is to make the basis functions smoother at the cutoff radius $r_c$.

An individual ANN model is assigned for each atomic species, i.e., Si and C, with the identical architecture. The architectural parameters of the basis functions are summarized in Table 1. The number of nodes in the input layer is equal to the number of terms of basis functions, i.e., 44 (16 terms for two pair types and 4 terms for three triplet types). Each ANN possesses two internal layers with 20 nodes each. Therefore, each ANN model has 1341 parameters ($w_{\alpha\beta}^n$ and $w_{\alpha 0}^n$) to be optimized.

**Table 1.** Architectural parameters of basis function set. X in the combination type indicates the atomic species of centering atom (X = Si, C).

| | 2-Body | 3-Body |
|---|---|---|
| Combination type | X-Si, X-C | Si-X-Si, Si-X-C, C-X-C |
| Number of terms per combination | 16 | 4 |
| Cutoff radius $r_c$ [Å] | 8.0 | 6.5 |

### 2.2. Optimization Procedure and Reference Data

The internal parameters in the ANN potential were optimized in order that the ANN model can reproduce the relationship between the atomic structure and the potential energy in the target material system. We refer to this relationship or mapping as the reference data, and the atomic structures for the reference data are called the reference structures. We obtained the reference data by the FP calculation based on the density functional theory (DFT) [37], which can reproduce the basic material properties such as the lattice constants and elastic constants for the Si-C systems [12,14] and deal with relatively large number of atoms compared with other FP approaches.

The optimization was carried out in the following way: Firstly, the initial reference data set was obtained for basic atomic structures. We conducted DFT calculations for the stable atomic structures of SiC (2H, 3C), Si (diamond), and C (diamond, graphene). In addition, we collected the reference data for other typical structures of high symmetry (e.g., bcc, fcc, etc.) and the strained structures, where affine deformation is applied up to 20% in typical modes. The structures and deformation modes are shown in Appendix A.

Secondly, the ANN parameters were optimized to develop a provisional ANN potential function. Here, the architecture of the ANN model (the number of layers, the number of nodes, etc.) was not changed during optimization. We chose 90% of the reference data at random and used them as the training data. The remaining 10% was left as the test data for validation.

Thirdly, several MD simulations were performed using the provisional ANN potential to confirm its validity in qualitative and quantitative aspects. If the ANN potential can reproduce all the material properties of interest, the optimization process is over. However, an atomic structure often falls into an unrealistic structure during MD simulation, especially at an early stage of the optimization cycle. If impermissible errors or unphysical behavior (e.g., unrealistic phase transition) are found in the ANN results, the atomic structures with such a disagreement are added to the reference data set, and the ANN potential is optimized again with the additional reference data. This series of development cycle (addition of reference data, optimization, and validation) was repeated until the resultant ANN potential reaches an acceptable level of quality. This method can selectively detect atomic structures for which the ANN returns an erroneous result, and thus enables an efficient optimization of the potential function [12,38]. For this purpose, we also examined untypical atomic structures that are obtained from the MD simulations under ultimate conditions, such as clusters consisting of a small number of atoms, structures fused at an extremely high temperature, etc. These data can enhance the transferability of the resultant ANN potential. Exemplified atomic structures obtained via this process are shown in Figure 2.

Note that the potential energy tends to be underestimated in the atomic structures obtained by the MD simulation with a provisional ANN potential. That is simply because an atomic structure is unlikely to appear during MD simulation if its potential energy is overestimated. This trend inevitably causes a bias in the reference structures and indicates a limitation of the abovementioned approach (this problem is *not* peculiar to the ANN potential model or other machine-learning potentials but common to any type of the potential models). To address this problem, we also collected the reference structures in another way. We picked up several atomic structures resulting from the MD simulations, and those structures were relaxed by the DFT calculation. The fully relaxed structure and transient structures during relaxation were used as the reference structures, and the potential energy of those structures is likely to be overestimated by the provisionally developed ANN potential. Since this approach requires more computational cost for relaxation by DFT calculation, relatively fewer reference data were generated by this method.

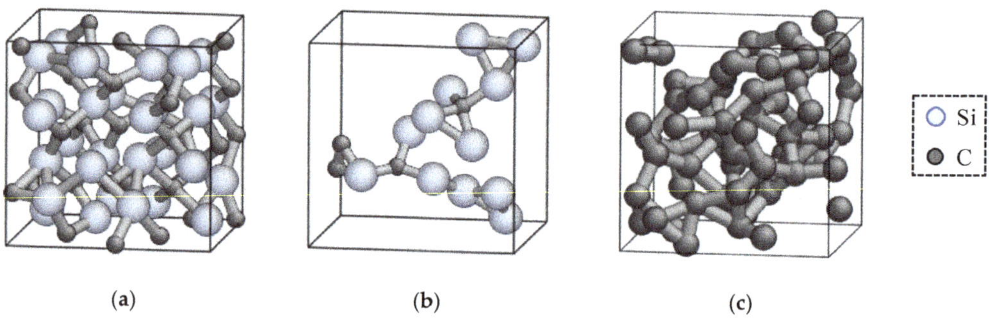


**Figure 2.** Example of reference structures obtained by MD simulation using provisional ANN potential functions. (**a**) 3C-SiC crystal near the melting point; (**b**) non-stoichiometric sparce Si-C system; (**c**) fused C system.

All the DFT calculations were conducted by Vienna Ab-initio Simulation Package (VASP) [39,40] with the generalized gradient approximation (GGA) by Perdew et al. [41] The core electrons were dealt with by the projector augmented wave (PAW) method [42]. The ANN potential was developed by the Atomic Energy Network (ænet) package [20] with the limited-memory Broyden-Fletcher-Goldfarb-Shanno method [43]. The MD simulations were carried out by the LAMMPS code [44,45] on which the ænet library was implemented by Mori [46]. The numbers of the reference structures for training and testing were 18,571 and 2063 in total, respectively.

## 3. Validation of ANN Potential Function

### 3.1. Result of Optimization

Figure 3 shows the comparison of the potential energies of the reference structures calculated by the DFT calculation and the developed ANN potential. Each data point corresponds to one reference structure. If the potential energy of a reference structure is consistent with the ANN and DFT calculations, the corresponding data point locates near the diagonal line, $y = x$. Almost all the data points in Figure 3 are found to locate near the diagonal line, and thus the developed ANN potential is expected to be able to reproduce the DFT results for various atomic structures. The ANN potential overestimates the potential energy of some reference structures, which are amorphous structures composed by carbon. Thus, the ANN potential may cause not a little error for the simulations with such structures. Although this issue should be addressed eventually, such error is still acceptable for the present purpose, because (i) the amorphous phases of pure carbon is out of our scope and not close to our target, and (ii) overestimation of the potential energy is much less likely to cause a fatal problem than underestimation, which can induce an unrealistic phase transition.

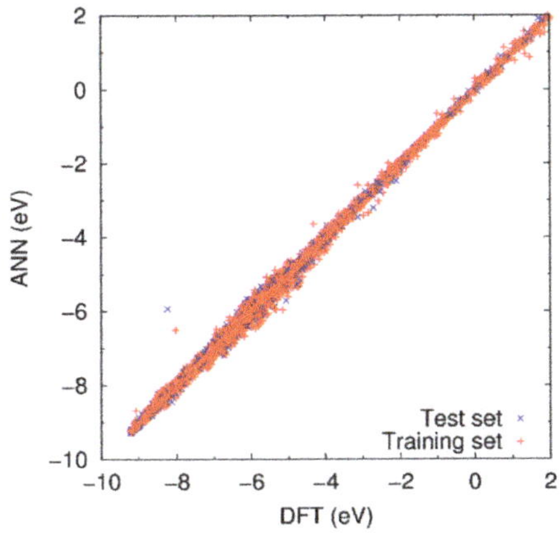


**Figure 3.** Potential energy of reference structures obtained by ANN potential and DFT calculation. Note that there are data points with the potential energy beyond +2.0 eV, which are not shown here because of less importance.

### 3.2. Material Properties at Equilibrium States

Table 2 lists the lattice constants, potential energies, and elastic constants at equilibrium state of the typical crystal structures of SiC, Si, and C obtained by the ANN potential and DFT calculation.

**Table 2.** Lattice constants $(a, c)$, potential energy $(E)$, and elastic constants $(C_{ij})$ at equilibrium state of typical structures of SiC, Si, and C. Parameter $d$ for C-graphene indicates the bond length. The column of "Exp." and "MM" indicates the values obtained by experiments [47–51] and molecular mechanics simulation with conventional potential models (Tersoff (T94) [8], Erhart and Albe (EA) [9]), respectively.

| | | | ANN | DFT | Exp. | MM T94 [a]/EA |
|---|---|---|---|---|---|---|
| SiC | 3C (zincblende) | $a$ [Å] | 4.371 | 4.379 | 4.3596 [b] | 4.280/4.359 |
| | | $E$ [eV/atom] | −7.540 | −7.532 | - | −6.434/−6.340 |
| | | $C_{11}$ [GPa] | 425 | 384 | 390 [b] | 447/382 |
| | | $C_{12}$ [GPa] | 189 | 127 | 142 [b] | 138/145 |
| | | $C_{44}$ [GPa] | 190 | 233 | 256 [b] | 293/240 |
| | 2H (wurtzite) | $a$ [Å] | 3.082 | 3.091 | 3.076 [c] | - |
| | | $c$ [Å] | 5.103 | 5.073 | 5.048 [c] | - |
| | | $E$ [eV/atom] | −7.539 | −7.530 | - | - |
| | | $C_{11}$ [GPa] | 593 | 498 | - | - |
| | | $C_{33}$ [GPa] | 592 | 537 | - | - |
| | | $C_{12}$ [GPa] | 226 | 98 | - | - |
| | | $C_{13}$ [GPa] | 94 | 49 | - | - |
| | | $C_{44}$ [GPa] | 183 | 153 | - | - |
| | 4H | $a$ [Å] | 3.085 | - | 3.080 [d] | - |
| | | $c$ [Å] | 10.196 | - | 10.081 [d] | - |
| | 6H | $a$ [Å] | 3.088 | - | 3.080 [d] | - |
| | | $c$ [Å] | 15.294 | - | 15.098 [d] | - |
| Si | diamond | $a$ [Å] | 5.486 | 5.469 | 5.429 [e] | 5.432/5.429 |
| | | $E$ [eV/atom] | −5.418 | −5.424 | - | −4.63/−4.63 |
| | | $C_{11}$ [GPa] | 137 | 154 | 168 [e] | 143/167 |
| | | $C_{12}$ [GPa] | 66 | 57 | 65 [e] | 75/65 |
| | | $C_{44}$ [GPa] | 135 | 74 | 80 [e] | 119/60 |

**Table 2.** *Cont.*

| | | | ANN | DFT | Exp. | MM T94 [a]/EA |
|---|---|---|---|---|---|---|
| C | diamond | $a$ [Å] | 3.572 | 3.572 | 3.567 [f] | 3.556/3.566 |
| | | $E$ [eV/atom] | −9.127 | −9.096 | - | −7.473/−7.373 |
| | | $C_{11}$ [GPa] | 1336 | 1052 | 1081 [f] | 1010/1082 |
| | | $C_{12}$ [GPa] | 663 | 126 | 125 [f] | 169/127 |
| | | $C_{44}$ [GPa] | 785 | 551 | 579 [f] | 545/635 |
| | graphene | $d$ [Å] | 1.436 | 1.424 | 1.42 [f] | 1.555/1.475 |
| | | $E$ [eV/atom] | −9.261 | −9.230 | - | −5.314/−7.374 |

[a] Calculated by Halicioglu [52]. [b] Lambrecht et al. [47]. [c] Merz and Adamsky [48]. [d] Lundqvist [49]. [e] Ref. [50]: Properties-Of-Silicon. [f] Ref. [51]: Numerical-Data.

It is found that the ANN potential can evaluate the lattice constants and the potential energy of typical crystal structures in a good agreement with the DFT calculation. Remarkably, the ANN potential can successfully reproduce the small energy gaps between the 3C and 2H structures of SiC, and the diamond and graphene structures of C. This result indicates that the developed ANN potential is applicable to a variety of atomic environment. On the other hand, the elastic constants are found to have relatively large error from the DFT results. This is mainly because the elastic constants are related to the second derivative of the potential energy surface, which is not directly taken into account in the target function of machine learning. Note that such a large error in the elastic constants is also found in the analytic potential models (For example, Erhart and Albe compared the elastic constants of SiC, Si, and C obtained by various analytic potential models [9]). The accuracy of the elastic constants may be improved by including physical quantities related to the derivative of the potential energy in the reference data, e.g., as is adopted in the force-matching method [53]. Note that the ANN potential still has an advantage, even if the conventional potential functions can reproduce the structural and mechanical properties with a comparable accuracy. This is because ANN potentials are much more flexible and extendable than the flamework of the conventional potential models. For example, one can easily extend ANN potentials for Si-C systems to multi-species systems (e.g., Si-C-Al-B) without any special care or consideration, while conventional models are expected to require a substantial effort to determine a suitable function form for such multi-species systems.

### 3.3. *Phase Stability of Crystalline SiC*

We examined the phase stability of crystalline SiC by conducting a melting test. A crystalline SiC model with 3C structure consisting of 512 atoms was prepared and heated at various heating rate $dT/dt$ from 20 K/ps to 200 K/ps, where the cell size was adjusted in order that the thermal stress on the simulation cell was relaxed. Figure 4 shows the relationship between temperature and the potential energy per atom at 20 and 200 K/ps. It is found that the potential energy exhibits a discontinuous increase at certain temperature range ($T \approx$ 3500–4000 K), which is interpreted as the melting point $T_\mathrm{m}$. Note that $T_\mathrm{m}$ is dependent on the heating rate, i.e., $T_\mathrm{m} = T_\mathrm{m}(dT/dt)$, because of high heating rates in the MD simulation. Therefore, the *real* $T_\mathrm{m}$ should be evaluated by extrapolating as $dT/dt \to 0$. By approximating the $T_\mathrm{m}$-$dT/dt$ relationship with a quadratic function (See Appendix B), we estimated $T_\mathrm{m} \to \approx$3500 K. This value is in a fairly good agreement with the experimental decomposition temperature ($\approx$3000 K) [54].

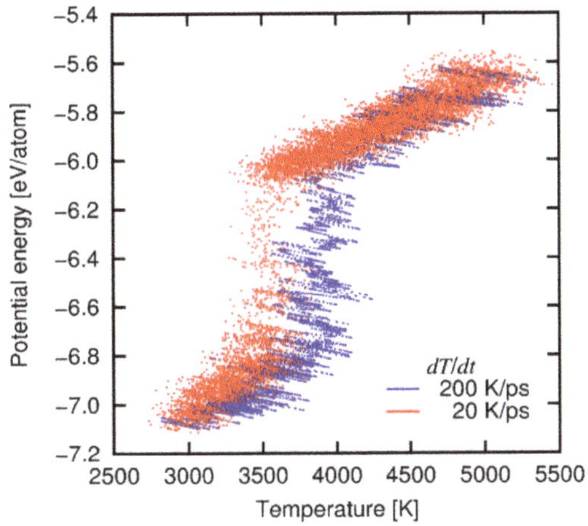


**Figure 4.** Relationship between potential energy and temperature at heating rates of 20 and 200 K/ps.

### 3.4. Structural Property of Amorphous SiC

To confirm the applicability of the ANN potential to the analysis of amorphous SiC, we prepared for an amorphous SiC structure model by the MD simulation of the melt-quench process. A 3C-SiC model with 512 atoms was fused at 4500 K and cooled to 300 K at the cooling rate $dT/dt = 4.2$ K/ps. Figure 5 shows the radial distribution function (RDF) of the amorphous SiC structure model at 300 K obtained by the melt-quench simulation. The profile of RDF is in a qualitatively good agreement with experiments [55,56], and typical peak profiles are well reproduced. Note that we observed partial recrystallization during the cooling process, and thus the RDF is dependent on the condition of the heating and cooling processes (e.g., an annealing process promotes recrystallization [55]).

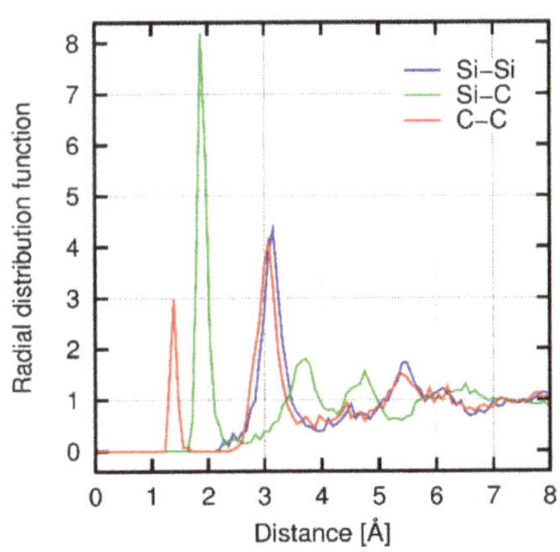


(**a**)

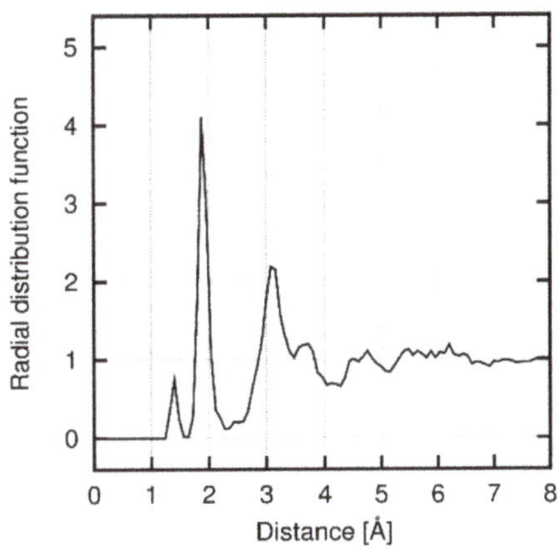


(**b**)

**Figure 5.** Radial distribution functions (RDFs) of amorphous SiC where (**a**) each contribution is shown separately, and (**b**) all combinations are included. All the RDFs are normalized by the global number density of atoms (regarding the corresponding pair type) and thus converge to 1 with the distance $r \to \infty$.

# 4. Temperature-Dependence of Creep Properties of Amorphous SiC

## *4.1. Preparation of Simulation Cells*

The amorphous structures for the creep tests were prepared for each target temperature. We set the target temperatures of $T$ = 1000–3000 K, but the creep tests were actually examined only at $T$ = 1000 K and 1500 K for the reason explained at the end of this subsection. While the real amorphization process is expected to be involved with the oxygen atoms, here we created the amorphous structures by the melt-and-quench method for simplicity. The structures were made in the following procedure (schematically shown in Figure 6): (i) A 3C-SiC supercell was prepared with 4 unit cells in each direction (512 atoms in total); (ii) The simulation cell was heated from 10 K to 5000 K (>$T_m$) at a constant heating rate, $dT/dt$ = 499 K/ps; (iii) The atomic structure was fully shuffled at 5000 K for 10 ps; (iv) The cell was quenched at a constant cooling rate, $dT/dt$ = 499 K/ps, to the target temperature; (v) The cells were relaxed for 500 ps at the target temperature. During this process, isotropic stress was controlled to zero by adjusting the cell size. Note that the obtained structures are not completely amorphous because the structure is recrystallized during the quenching and relaxation processes to some extent. Especially, most part of the simulation cell was recrystallized during the relaxation process at 2000 K and 2500 K. Figure 7 shows the relationship between the potential energy of the relaxed structures and the temperature. From these results, we set the temperature range for the creep tests to $T$ = 1000–1500 K, where the amorphous phase is likely to be kept.

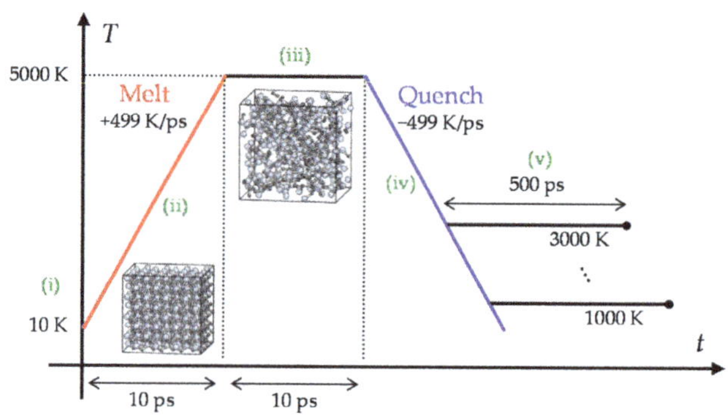


**Figure 6.** Procedure of preparation for amorphous structures.

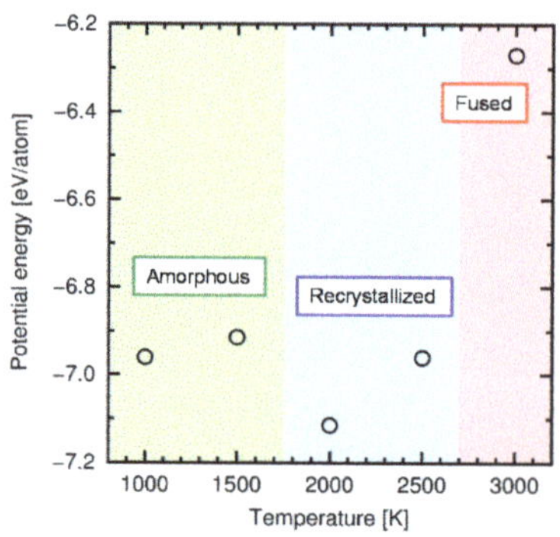


**Figure 7.** Potential energy of the quenched and relaxed structure at various temperatures (relaxation for 500 ps). Three regions indicate the resultant atomic structure at each temperature.

### 4.2. Deformation Condition

We evaluate the creep properties of amorphous SiC, especially, the relationship between the stress and the strain rate. We considered two types of deformation conditions, i.e., constant-strain-rate and constant-stress conditions (hereafter $d\varepsilon/dt$- and $\sigma$-constant conditions, respectively). While most experimental creep tests were conducted under a tensile loading condition, we applied compressive strain or stress to avoid void nucleation and brittle fracture. For the $d\varepsilon/dt$-constant condition, true compressive strain was applied up to $\varepsilon = 0.5$ at constant true strain rates with the range of $2 \times 10^{-3}$–$5 \times 10^{-1}$ ps$^{-1}$. We conducted three sets of creep test along three different loading directions (i.e., $x$, $y$, and $z$) for each condition. For the $\sigma$-constant condition, we applied the compressive stress of 5–8 GPa at 1000 K and 4–7 GPa at 1500 K for 10 ps. As well as the case of the $d\varepsilon/dt$-constant condition, we conducted the simulations three times along different loading directions for each condition. For both conditions, the normal stress components perpendicular to the loading direction were fully relaxed.

### 4.3. Results and Discussion

Figure 8 show the stress-strain relationship obtained under the $d\varepsilon/dt$-constant conditions at $T = 1000$ K and 1500 K. For all the cases, the stress-strain relationship consists of two regions, i.e., the nearly elastic region at small strain and the steady-state region at large strain, and yielding occurs between those regions at $\varepsilon \approx 0.1$. Relatively large fluctuation of stress in the steady-state region is attributed to the limitation of the simulation cell size. Figure 9 shows the atomic structure before and after deformation, obtained at $T = 1500$ K and $d\varepsilon/dt = 2 \times 10^{-3}$ ps$^{-1}$. This result indicates that the atomic structure is kept amorphous during the deformation test, which will be also confirmed by the RDFs later in this section. We evaluated the steady-state stress as the average value of stress over this region $\varepsilon = 0.2$–$0.5$. Figure 10 shows the relationship between stress and strain rate at the steady state. It is found that there are two regions at the low and high strain rates. The trend in both the regions follows the power law, which is qualitatively consistent with the experiments and thus supports the validity of the developed ANN potential. Hereafter we refer to the regions of low strain rates ($d\varepsilon/dt = 2 \times 10^{-3}$–$2 \times 10^{-2}$ ps$^{-1}$) and high strain rates ($d\varepsilon/dt = 5 \times 10^{-2}$–$5 \times 10^{-1}$ ps$^{-1}$) as Regions I and II, respectively.

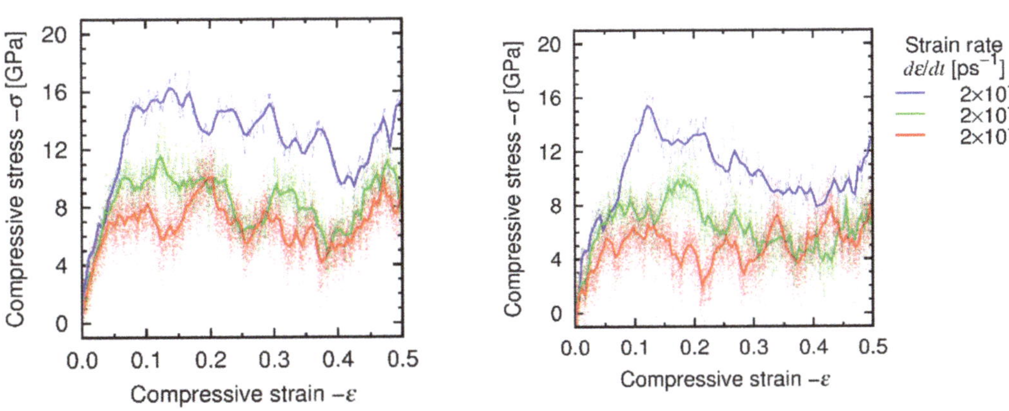


(**a**) (**b**)

**Figure 8.** True stress-strain relationships under the $d\varepsilon/dt$-constant conditions at (**a**) $T = 1000$ K and (**b**) $T = 1500$ K. The solid curves indicate the Bezier approximation.

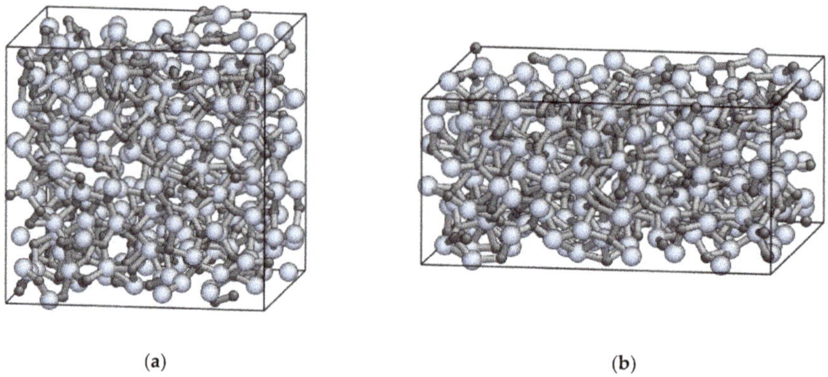


**Figure 9.** Atomic structure obtained at deformation test under the $d\varepsilon/dt$-constant condition at $T$ = 1500 K and $d\varepsilon/dt = 2 \times 10^{-3}$ ps$^{-1}$; (**a**) $\varepsilon$ = 0.0 (before deformation), (**b**) $\varepsilon$ = 0.5 (after deformation).

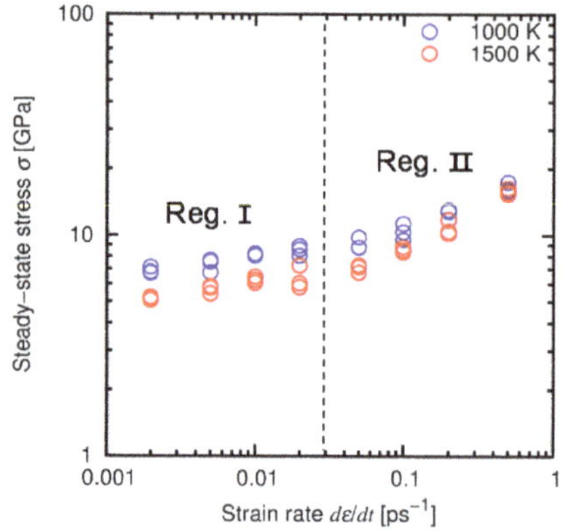


**Figure 10.** Relationship between stress and strain rate at the steady-state creep. The scales are shown in a logarithmic form. The vertical dashed line indicates the border of Regions I and II.

Figure 11 shows typical temporal developments of strain under the $\sigma$-constant conditions. Creep deformation was observed in all the cases, and we calculated the creep strain rates $d\varepsilon/dt$ from the $\varepsilon$-$t$ relationship for 20–100 ps for each stress and temperature. Figure 12 shows the relationship between the stress and the strain rate at the steady state (the data of $d\varepsilon/dt$-constant conditions are also shown for comparison and clarity). The results of the $\sigma$-constant condition are basically consistent with the $d\varepsilon/dt$-constant results and found to belong to Region I. This result also shows that the $\sigma$-$d\varepsilon/dt$ relationship satisfies the power law in a wide range of $d\varepsilon/dt$.

The creep behaviors at Regions I and II are expected to be based on different atomistic mechanisms. To reveal those mechanisms, we characterized the trend of change in the atomistic structures under deformation through the mean-square displacement (MSD) and the radial distribution function (RDF), which are related to the diffusion of atoms and local structural change, respectively. Figure 13 shows the temporal development of MSD and creep strain at Region I under the $\sigma$-constant condition, and Figure 14 shows the RDFs obtained under the $d\varepsilon/dt$-constant conditions at Regions I and II before and after deformation. A clear correspondence between the trends in the MSD and the creep strain is found (Figure 13), while the RDFs at Region I remain nearly unchanged after deformation (Figure 14a). Therefore, the creep mechanism at Region I is attributed to diffusion of atoms in the amorphous structure. Figure 15 shows the temporal development of the individual MSDs for Si and C (decomposed from the MSD in Figure 13), which indicates that both the Si and C atoms are equally involved with the creep process. At Region II, in contrast, the profiles of the RDFs are partly affected by deformation (Figure 14b); especially, the peak at $r$ = 3 Å of the Si-Si pair is significantly broadened after deformation (See Appendix C in detail). This result reveals that at Region II the deformation is faster than (or comparable to) relaxation of the local structure, and thus the creep behavior at Region II is characterized by the relaxation process in the local atomic structure. This type of deformation only occurs at a very high strain rate and thus is presumably negligible under a realistic condition. Note that the atomic structures are kept amorphous at both Regions I and II through the examination, as indicated by RDF profiles before and after deformation (Figure 14).

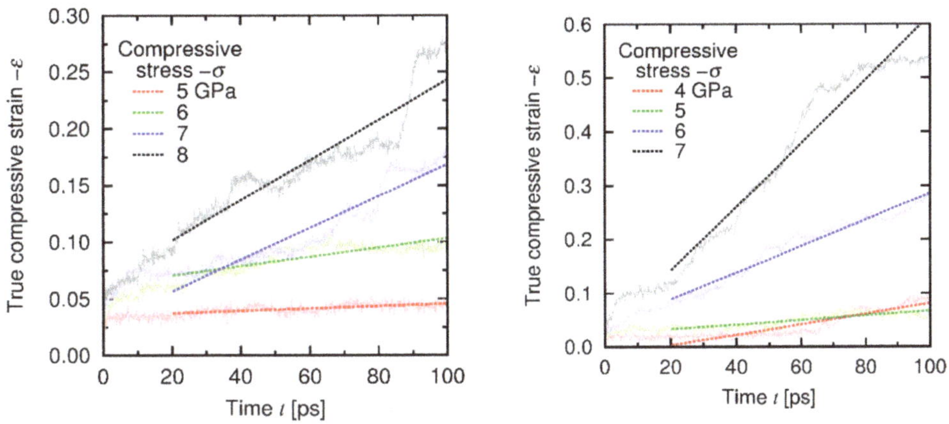


(**a**) (**b**)

**Figure 11.** Temporal development of true compressive strain under the $\sigma$-constant conditions at (**a**) 1000 K and (**b**) 1500 K. The dashed lines indicate the linear approximations, and its slope is equal to the strain rate $d\varepsilon/dt$ at the steady state.

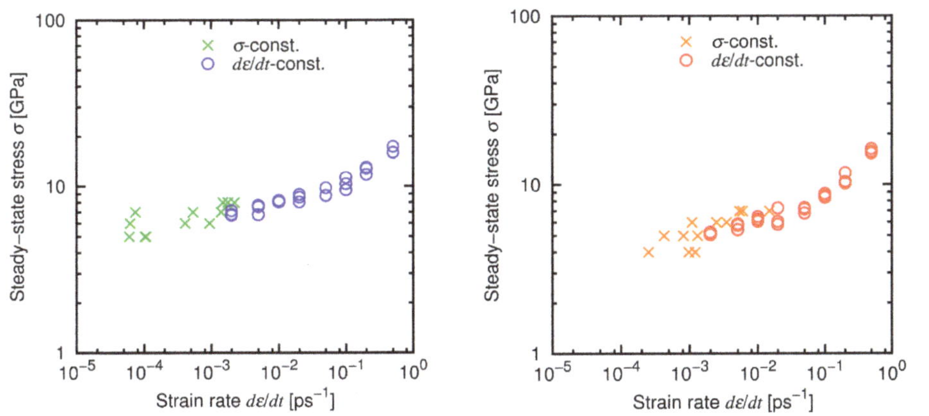


(**a**) (**b**)

**Figure 12.** Relationship between the steady-state stress and the strain rate obtained under the $\sigma$- and $d\varepsilon/dt$-constant conditions at (**a**) 1000 K and (**b**) 1500 K.

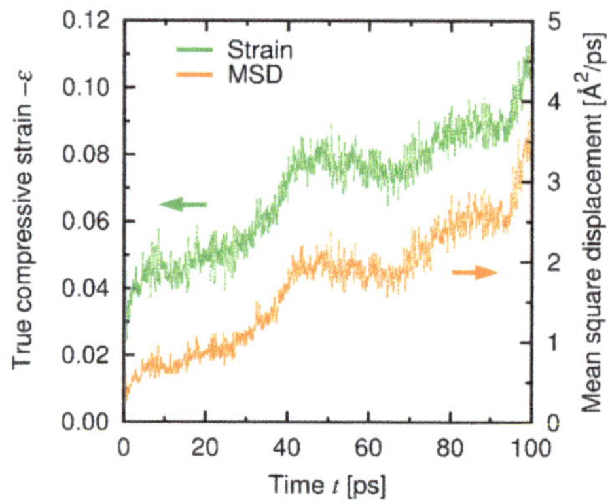


**Figure 13.** Temporal development of MSD and creep strain at Region I under the $\sigma$-constant condition ($\sigma$ = 7 GPa, $T$ = 1000 K). The MSD includes the contributions of both Si and C.

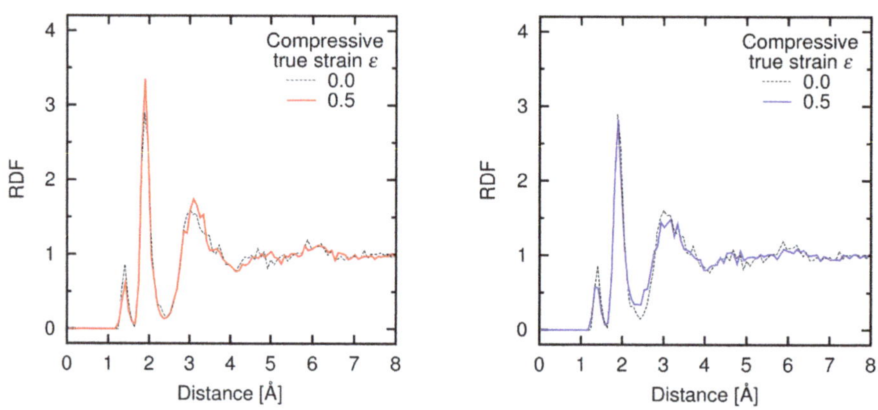


(**a**) (**b**)

**Figure 14.** RDFs under $d\varepsilon/dt$-constant conditions for all contributions (Si-Si, Si-C, and C-C) at (**a**) $d\varepsilon/dt = 2 \times 10^{-3}$ ps$^{-1}$ (Region I) and (**b**) $d\varepsilon/dt = 2 \times 10^{-1}$ ps$^{-1}$ (Region II) before and after deformation.

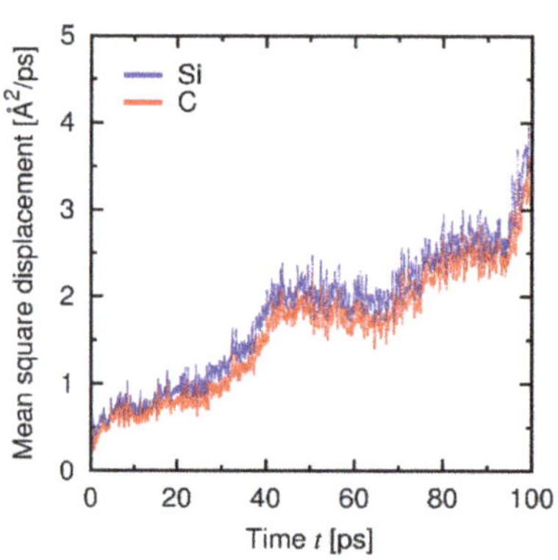


**Figure 15.** Temporal development of the individual MSDs for Si and C ($\sigma$ = 7 GPa, $T$ = 1000 K).

According to the experiments, the relationship between the stress and the strain rate at the steady state is described by a nonlinear viscoelastic model as follows [6,36]:

$$\frac{d\varepsilon}{dt} = A\sigma^n \exp\left(\frac{-Q}{RT}\right), \tag{9}$$

where $A$ is a material parameter. The parameters $n$ and $Q$ are the stress exponent and the activation energy, respectively. The constant $R$ is the gas constant ($R$ = 8.314 (J/mol)/K). The parameters $n$ and $Q$ are to be obtained via the present simulations. The $\sigma$-$d\varepsilon/dt$-$T$ relationship was fitted with this equation for each region from the data of the $d\varepsilon/dt$-condition (Note that the data of the $\sigma$-condition were not used for approximation because of a relatively large deviation). The resultant fitting parameters are listed in Table 3 for each region, compared with the experimental results [5,6]. The stress exponents $n$ are relatively large (especially at Region I) compared with the experimental values (typically $n$ = 1–6 [36]) but within a reasonable range. The evaluated activation energies $Q$ are significantly small compared with an experimental evaluation for a SiC fiber (Hi-Nicalon) [6] and the known elementary processes in SiC such as self-diffusion and grain-boundary diffusion [5]. The activation energy of creep in the SiC fiber is an apparent (nominal) value affected by various type of factors and elementary processes, and the diffusion in the amorphous phase can play an important role in those processes owing to its low activation energy. Note that the creep mechanism in the SiC fiber cannot be explained only via the known elementary processes mentioned above (i.e., self-diffusion and grain-boundary diffusion) because the activation energies of those processes are considerably higher than that of the SiC fiber, while those processes are presumably major factors. Thus, it is implied that the existence of a creep mechanism with a quite low activation energy, and the diffusion process in the amorphous phase is a possible cause of such low activation energy. It is a likely scenario that stress concentration at amorphous regions near the grain boundary results in a considerable creep deformation, owing to a quite large stress exponent $n$.

**Table 3.** The fitted parameters for the relationship of the stress and the strain rate.

| | | $A$ [GPa$^n$·ps$^{-1}$] [a] | $n$ | $Q$ [kJ/mol] |
|---|---|---|---|---|
| This work | Reg. I | $7.7 \times 10^{-7}$ | 7.5 | 50 |
| - | Reg. II | $2.7 \times 10^{-4}$ | 3.2 | 13 |
| Exp. [6] | Hi-Nicalon | - | 2–3 | 193–423 |
| Exp. [5] | self-diffusion (Si) | - | - | 912 |
| - | GB-diffusion (Si/C) | - | - | 564/841 |

[a] The unit depends on the parameter $n$.

## 5. Conclusions

We developed an interatomic potential function based on the ANN model for the purpose of evaluating the high-temperature mechanical properties of Si-C system. The internal parameters in the model were optimized in order to reproduce the first principles results. It was confirmed by a series of verification simulations that the developed potential function can reproduce basic material properties such as the lattice constants and the stability of typical atomic structures. The elastic properties have a relatively large discrepancy, which should be improved in the future work by adopting advanced fitting approaches such as the force-matching method. The potential function was applied to the creep test of amorphous SiC under various loading conditions. As a result, it was found that there exist two regimes with different behaviors according to the strain rate. The mechanisms of creep deformation at the slow and fast deformation regions were attributed to atomic diffusion in the amorphous phase and relaxation of the local atomic structure, respectively. The activation energy of creep obtained from the present simulation is significantly lower than those of self-diffusion and grain-boundary diffusion. This result indicates the importance of diffusion in the amorphous phase.

While we dealt with a pure Si-C binary system, practical SiC/SiC composites include other various atomic species. To take them into account, it is necessary to develop a potential function for the multi-species system, based on the present ANN potential. In addition, a SiC/SiC component requires the environmental barrier coating system, which prevents SiC from chemical reaction with high-temperature vapor. Thus, it is also needed to investigate the mechanical properties of the interface between the SiC/SiC component and the coating layer, because a stress concentration and/or a thermal stress can be exerted near such a dissimilar interface. The MD results (material properties and/or deformation mechanisms) can be used for the larger-scale simulations, such as the phase-field and finite element analyses. Those subjects are left as future works.

**Author Contributions:** Conceptualization, A.K. and Y.U.; methodology, A.K.; validation, A.K. and Y.U.; formal analysis, A.K.; resources, A.K. and Y.U.; writing—original draft preparation, A.K. and Y.U.; writing—review and editing, A.K. and Y.U.; visualization, A.K.; supervision, Y.U.; funding acquisition, A.K. and Y.U. All authors have read and agreed to the published version of the manuscript.

**Funding:** This research was funded by Research Grant of The Japan Refractories Association and The Technical Association of Refractories, Japan. The APC was funded by Research Grant of The Japan Refractories Association and The Technical Association of Refractories, Japan and Institute of Industrial Science, The University of Tokyo.

**Institutional Review Board Statement:** Not applicable.

**Informed Consent Statement:** Not applicable.

**Data Availability Statement:** Data (the developed ANN potential) available on request.

**Acknowledgments:** Authors acknowledge Hideki Mori (College of Industrial Technology, Japan) for providing a computational code (LAMMPS with implementation of ænet code) for the present MD simulations.

**Conflicts of Interest:** The authors declare no conflict of interest.

## Appendix A. Typical Reference Crystal Structures

The typical reference crystal structures are listed in Table A1 (several non-crystal molecules are also included, e.g., dimer, trimer, etc.). The $x$, $y$, and $z$ axes in "Deformation Mode" indicate the [100], [010], and [001] directions for the cubic structures, and the $[2\bar{1}\bar{1}0]$, $[01\bar{1}0]$ and [0001] directions for the hexagonal structures, respectively. The deformation mode "$XY$" indicates that the $X$ component is deformed in the $Y$ direction (thus, the case of $X = Y$ means a normal strain, and otherwise shear strain). The combination of $XX$ and $YY$ indicates the coupling of two deformation modes. The mode "iso" is the isotropic deformation, i.e., $xx + yy + zz$. The deformation modes with the notation of "$d = A$–$B$ Å" indicate that the interatomic distance from the first neighbors was varied instead of changing the strain (mainly applied to non-crystal systems). The atomic structures that are not typical or that we defined are visualized in Figure A1.

**Table A1.** Typical crystal structures and deformation modes considered as reference data. The structures with an asterisk ("*") are visualized in Figure A1.

| System | Structure | Deformation Mode |
|---|---|---|
| SiC | zincblende (3C) | iso, $xx$, $xx+yy$, $xx-yy$, $xy$ |
| | wurtzite (2H) | iso, $xx$, $yy$, $zz$, $xx+yy$, $xx-yy$, $zx$, $zy$ |
| | rock salt | iso, $xx$, $xx+yy$, $xx-yy$, $xy$ |
| | cesium chloride | iso, $xx$, $xx+yy$, $xx-yy$, $xy$ |
| | hexagonal boron nitride * | iso, $xx$, $yy$, $zz$, $xx+yy$, $xx-yy$, $zx$, $zy$ |
| | monolayer boron nitride * | $d$ = 0.80–3.00 Å |
| | fluorite ($SiC_2$, $CSi_2$) | iso, $xx$, $xx+yy$, $xx-yy$, $xy$ |
| | tungsten carbide * | iso, $xx$, $yy$, $zz$, $xx+yy$, $xx-yy$, $zx$, $zy$ |
| | dimer | $d$ = 0.80–4.00 Å |
| Si | diamond | iso, $xx$, $xx+yy$, $xx-yy$, $xy$ |
| | graphene | $d$ = 0.80–3.00 Å |
| | bcc | iso, $xx$, $xx+yy$, $xx-yy$, $xy$ |
| | fcc | iso, $xx$, $xx+yy$, $xx-yy$, $xy$ |
| | sc | iso, $xx$, $xx+yy$, $xx-yy$, $xy$ |
| | simple hexagonal * | iso, $xx$, $yy$, $zz$, $xx+yy$, $xx-yy$, $zx$, $zy$ |
| | octahedra * | iso, $xx$, $xx+yy$, $xx-yy$, $xy$ |
| | 2D-square * | $d$ = 0.90–3.00 Å |
| | 2D-triangle * | $d$ = 0.90–3.00 Å |
| | chain (straight) * | $d$ = 0.90–4.00 Å |
| | dimer | $d$ = 1.00–4.00 Å |
| C | graphene | $d$ = 0.80–3.00 Å |
| | graphite (α, β) | iso, $xx$, $yy$, $zz$, $xx+yy$, $xx-yy$, $zx$, $zy$ |
| | diamond | iso, $xx$, $xx+yy$, $xx-yy$, $xy$ |
| | lonsdaleite * | iso, $xx$, $yy$, $zz$, $xx+yy$, $xx-yy$, $zx$, $zy$ |
| | bcc | iso, $xx$, $xx+yy$, $xx-yy$, $xy$ |
| | fcc | iso, $xx$, $xx+yy$, $xx-yy$, $xy$ |
| | hcp | iso, $xx$, $yy$, $zz$, $xx+yy$, $xx-yy$, $zx$, $zy$ |
| | sc | iso, $xx$, $xx+yy$, $xx-yy$, $xy$ |
| | simple hexagonal * | iso, $xx$, $yy$, $zz$, $xx+yy$, $xx-yy$, $zx$, $zy$ |
| | octahedra * | iso, $xx$, $xx+yy$, $xx-yy$, $xy$ |
| | 2D-square * | $d$ = 0.80–3.00 Å |
| | 2D-triangle * | $d$ = 0.80–3.00 Å |
| | chain (straight) * | $d$ = 0.80–4.00 Å |
| | chain (triangle) * | $d$ = 0.80–3.00 Å |
| | dimer | $d$ = 0.80–4.00 Å |
| | 3-mer (triangle) * | $d$ = 0.80–3.00 Å |
| | 4-mer (tetrahedron) * | $d$ = 0.80–3.00 Å |
| | 6-mer (octahedron) * | $d$ = 0.80–3.00 Å |
| | 8-mer (cube) * | $d$ = 0.80–3.00 Å |
| | 8-mer (fcc-like) * | $d$ = 0.80–3.00 Å |

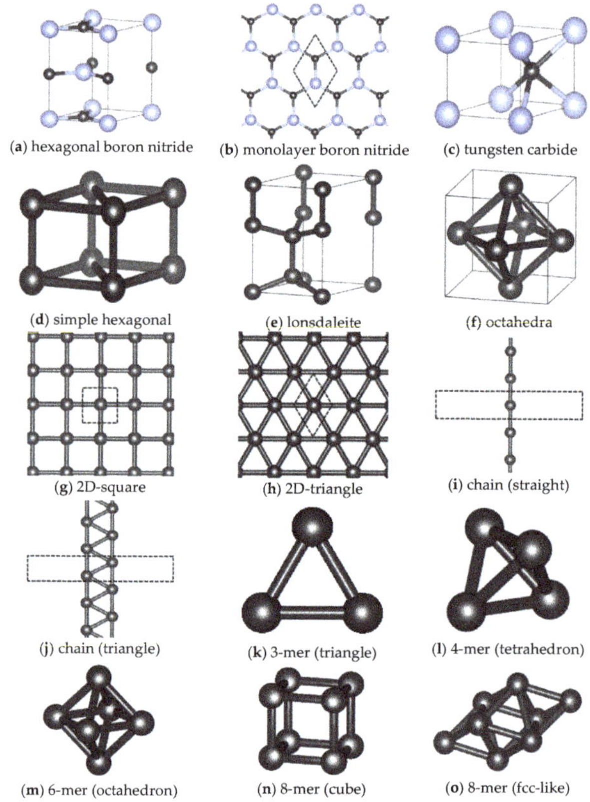


**Figure A1.** Atomic structures (untypical) used as reference structure. Thin solid and dashed lines indicate the unit cells.

## Appendix B. Approximation of Melting Point

Figure A2 shows the melting point $T_m$ as a function of temperature rate $dT/dt$ for 3C SiC. The data points were fitted with a quadratic function, i.e., $T_\mathrm{m} = A{\cdot}(dT/dt)^2 + B{\cdot}(dT/dt) + C$. Then the fitting parameter $C$ is equal to $T_\mathrm{m}$ at $dT/dt \to 0$.

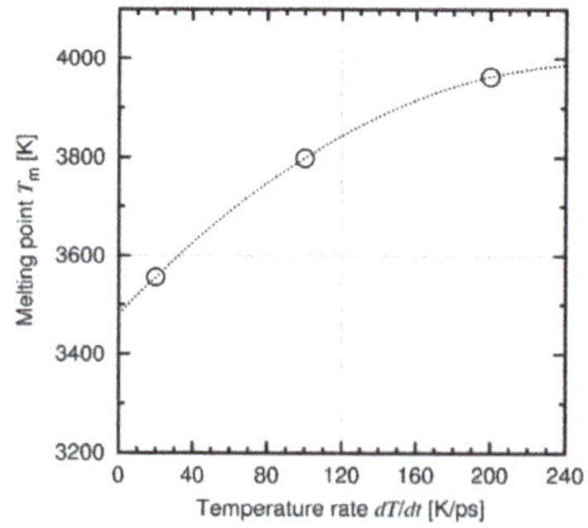


**Figure A2.** Melting temperature as a function of heating rate. The dashed curve indicates a quadratic approximation.

## Appendix C. Radial Distribution Function during Deformation

Figure A3 shows the RDFs obtained under the $d\varepsilon/dt$-constant conditions at Regions I and II before and after deformation for each atomic pair (decomposed from Figure 14). The distribution of the Si-Si pair is affected by deformation at the high strain rate (Reg. II; Figure A3(a-ii)), while all the distributions are nearly unchanged at the low strain rate (Reg. I; Figure A3(a-i)–(c-i)). This result implies relaxation of the local structure as a major contributor of creep deformation at Reg. II (see main text).

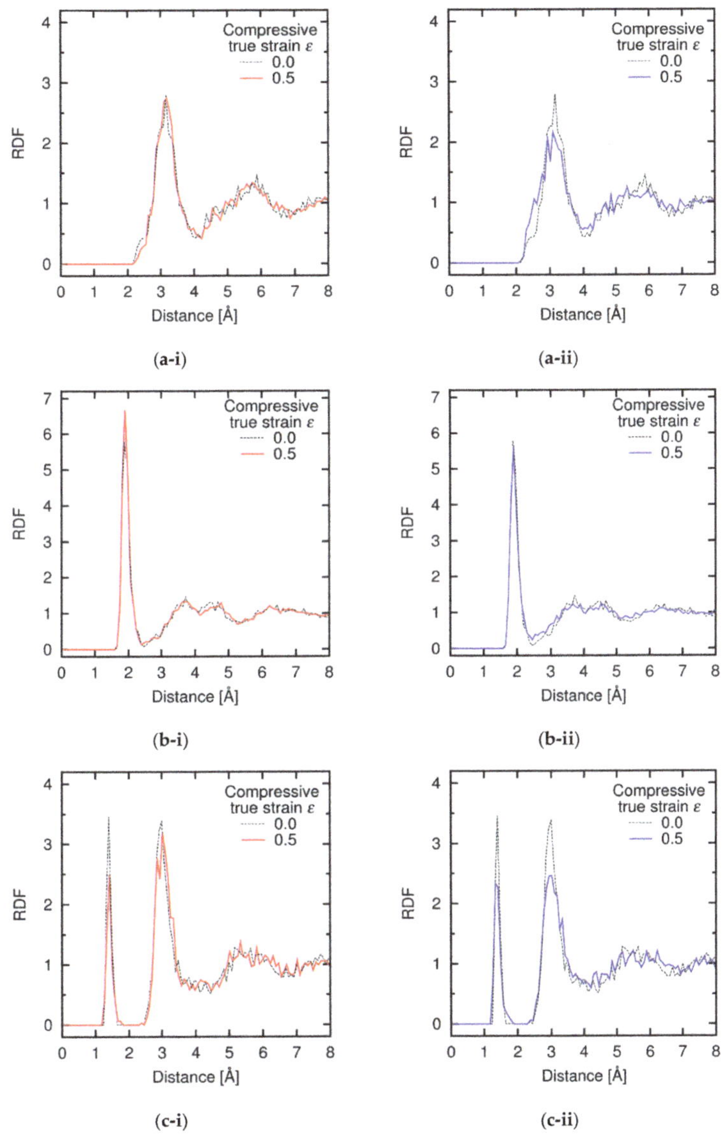


**Figure A3.** RDFs under $d\varepsilon/dt$-constant conditions for (**a**) Si-Si, (**b**) Si-C, and (**c**) C-C at (**i**) $d\varepsilon/dt = 2 \times 10^{-3}$ ps$^{-1}$ (Region I) and (**ii**) $d\varepsilon/dt = 2 \times 10^{-1}$ ps$^{-1}$ (Region II) before and after deformation.

## References


1. Droillard, C.; Lamon, J. Fracture toughness of 2-D woven SiC/SiC CVI-composites with multilayered interphases. *J. Am. Ceram. Soc.* **1996**, *79*, 849–858. [CrossRef]
2. Xu, Y.; Cheng, L.; Zhang, L.; Yin, H.; Yin, X. High toughness, 3D textile, SiC/SiC composites by chemical vapor infiltration. *Mater. Sci. Eng. A* **2001**, *318*, 183–188. [CrossRef]
3. Padture, N.P. Advanced structural ceramics in aerospace propulsion. *Nat. Mater.* **2016**, *15*, 804–809. [CrossRef] [PubMed]
4. Perepezko, J.H. The hotter the engine, the better. *Science* **2009**, *326*, 1068–1069. [CrossRef]
5. DiCarlo, J.A. Creep of chemically vapor deposited SiC fibres. *J. Mater. Sci.* **1986**, *21*, 217–224. [CrossRef]
6. Bodet, R.; Bourrat, X.; Lamon, J.; Naslain, R. Tensile creep behaviour of a silicon carbide-based fibre with a low oxygen content. *J. Mater. Sci.* **1995**, *30*, 661–677. [CrossRef]
7. Ruggles-Wrenn, M.B.; Lee, M.D. Fatigue behavior of an advanced SiC/SiC ceramic composite with a self-healing matrix at 1300 °C in air and in steam. *Mater. Sci. Eng. A* **2016**, *677*, 438–445. [CrossRef]
8. Tersoff, J. Chemical order in amorphous silicon carbide. *Phys. Rev. B* **1994**, *49*, 16349–16352. [CrossRef] [PubMed]
9. Erhart, P.; Albe, K. Analytical potential for atomistic simulations of silicon, carbon, and silicon carbide. *Phys. Rev. B* **2005**, *71*, 035211. [CrossRef]
10. Huang, H.; Ghoniem, N.M.; Wong, J.K.; Baskes, M.I. Molecular dynamics determination of defect energetics in β-SiC using three representative empirical potentials. *Modelling Simul. Mater. Sci. Eng.* **1995**, *3*, 615–627. [CrossRef]
11. Vashishta, P.; Kalia, R.K.; Nakano, A.; Rino, J.P. Interaction potential for silicon carbide: A molecular dynamics study of elastic constants and vibrational density of states for crystalline and amorphous silicon carbide. *J. Appl. Phys.* **2007**, *101*, 103515. [CrossRef]
12. Kubo, A.; Nagao, S.; Umeno, Y. Molecular dynamics study of deformation and fracture in SiC with angular dependent potential model. *Cumput. Mater. Sci.* **2017**, *139*, 89–96. [CrossRef]
13. Takamoto, S.; Yamasaki, T.; Nara, J.; Ohno, T.; Kaneta, C.; Hatano, A.; Izumi, S. Atomistic mechanism of graphene growth on a SiC substrate: Large-scale molecular dynamics simulations based on a new charge-transfer bond-order type potential. *Phys. Rev. B* **2018**, *97*, 125411. [CrossRef]
14. Umeno, Y.; Kubo, A.; Nagao, S. Density functional theory calculation of ideal strength of SiC and GaN: Effect of multi-axial stress. *Comput. Mater. Sci.* **2015**, *109*, 105–110. [CrossRef]
15. Lipowitz, J.; Barnard, T.; Bujalski, D.; Rabe, J.; Zank, G. Fine-diameter polycrystalline SiC fibers. *Compos. Sci. Technol.* **1994**, *51*, 167–171. [CrossRef]
16. Chen, D.; Zhang, X.-F.; Ritchie, R.O. Effects of grain boundary structure on the strength, toughness, and cyclic-fatigue properties of a monolithic silicon carbide. *J. Am. Ceram. Soc.* **2000**, *83*, 2079–2081. [CrossRef]
17. Baldus, P.; Jansen, M.; Sporn, D. Ceramic fibers for matrix composites in high-temperature engine applications. *Science* **1999**, *285*, 699–703. [CrossRef] [PubMed]
18. Behler, J. Perspective: Machine learning potentials for atomistic simulations. *J. Chem. Phys.* **2016**, *145*, 170901. [CrossRef] [PubMed]
19. Behler, J.; Parrinello, M. Generalized neural-network representation of high-dimensional potential-energy surfaces. *Phys. Rev. Lett.* **2007**, *98*, 146401. [CrossRef]
20. Artrith, N.; Urban, A. An implementation of artificial neural-network potentials for atomistic materials simulations Performance for $TiO_2$. *Comput. Mech. Sci.* **2016**, *114*, 135–150. [CrossRef]
21. Bartók, A.P.; Payne, M.C.; Kondor, R.; Csányi, G. Gaussian approximation potentials: The accuracy of quantum mechanics without electrons. *Phys. Rev. Lett.* **2010**, *104*, 136403. [CrossRef]
22. Chmiela, S.; Tkatchenko, A.; Sauceda, H.E.; Poltavsky, I.; Schütt, K.T.; Müller, K.-R. Machine learning of accurate energy-conserving molecular force fields. *Science Adv.* **2017**, *3*, e1603015. [CrossRef] [PubMed]
23. Rupp, M.; Tkatchenko, A.; Müller, K.R.; von Lilienfeld, O.A. Fast and accurate modeling of molecular atomization energies with machine learning. *Phys. Rev. Lett.* **2012**, *108*, 058301. [CrossRef] [PubMed]
24. Cybenko, G. Approximation by superpositions of a sigmoidal function. *Math. Control. Signals Syst.* **1989**, *2*, 303–314. [CrossRef]
25. Hornik, K. Approximation capabilities of multilayer feedforward networks. *Neural Netw.* **1991**, *4*, 251–257. [CrossRef]
26. Leshno, M.; Lin, V.Y.; Pinkus, A.; Schocken, S. Multilayer feedforward networks with a nonpolynomial activation function can approximate any function. *Neural Netw.* **1993**, *6*, 861–867. [CrossRef]
27. Mori, H.; Ozaki, H. Neural network atomic potential to investigate the dislocation dynamics in bcc iron. *Phys. Rev. Mater.* **2020**, *4*, 040601. [CrossRef]
28. Eshet, H.; Khaliullin, R.Z.; Kühne, T.D.; Behler, J.; Parrinello, M. Ab initio quality neural network potential for sodium. *Phys. Rev. B* **2010**, *81*, 184107. [CrossRef]
29. Artrith, N.; Hiller, B.; Behler, J. Neural network potentials for metals and oxides–First applications to copper clusters at zinc oxide. *Phys. Status Solidi B* **2013**, *250*, 1191–1203. [CrossRef]
30. Khaliullin, R.Z.; Eshet, H.; Kühne, T.D.; Behler, J.; Parrinello, M. Graphite-diamond phase coexistence study employing a neural-network mapping of the *ab initio* potential energy surface. *Phys. Rev. B* **2010**, *81*, 100103. [CrossRef]
31. Khaliullin, R.Z.; Eshet, H.; Kühne, T.D.; Behler, J.; Parrinello, M. Nucleation mechanism for the direct graphite-to-diamond phase transition. *Nat. Mater.* **2011**, *10*, 693–697. [CrossRef] [PubMed]

32. Artrith, N.; Urban, A.; Ceder, G. Efficient and accurate machine-learning interpolation of atomic energies in compositions with many species. *Phys. Rev. B* **2017**, *96*, 014112. [CrossRef]
33. Umeno, Y.; Kubo, A. Prediction of electronic structure in atomic model using artificial neural network. *Comput. Mater. Sci.* **2019**, *168*, 164–171. [CrossRef]
34. Chandrasekaran, A.; Kamal, D.; Batra, R.; Kim, C.; Chen, L.; Ramprasad, R. Solving the electronic structure problem with machine learning. *NPJ Comput. Mater.* **2019**, *5*, 1–7. [CrossRef]
35. Yeo, B.C.; Kim, D.; Kim, C.; Han, S.S. Pattern learning electronic density of states. *Sci. Rep.* **2019**, *9*, 5879. [CrossRef]
36. Cannon, W.R.; Langdon, T.G. Creep of ceramics. *J. Mater. Sci.* **1983**, *18*, 1–50. [CrossRef]
37. Kohn, W.; Sham, L.J. Self-consistent equations including exchange and correlation effects. *Phys. Rev.* **1965**, *140*, A1133–A1138. [CrossRef]
38. Kubo, A.; Wang, J.; Umeno, Y. Development of interatomic potential for Nd-Fe-B permanent magnet and evaluation of magnetic anisotropy near the interface and grain boundary. *Modelling Simul. Mater. Sci. Eng.* **2014**, *22*, 065014. [CrossRef]
39. Kresse, G.; Hafner, J. Ab initio molecular dynamics for liquid metals. *Phys. Rev. B* **1993**, *47*, 558–561. [CrossRef]
40. Kresse, G.; Furthmüller, J. Efficiency of ab-initio total energy calculations for metals and semiconductors using a plane-wave basis set. *Comput. Mater. Sci.* **1996**, *6*, 15–50. [CrossRef]
41. Perdew, J.P.; Burke, K.; Ernzerhof, M. Generalized gradient approximation made simple. *Phys. Rev. Lett.* **1996**, *77*, 3865–3868. [CrossRef]
42. Blöchl, P.E. Projector augmented-wave method. *Phys. Rev. B* **1994**, *50*, 17953–17979. [CrossRef]
43. Byrd, R.H.; Lu, P.; Nocedal, J.; Zhu, C. A limited memory algorithm for bound constrained optimization. *SIAM J. Sci. Comput.* **1995**, *16*, 1190–1208. [CrossRef]
44. Plimpton, S. Fast parallel algorithms for short-range molecular dynamics. *J. Comput. Phys.* **1995**, *117*, 1–19. [CrossRef]
45. LAMMPS Web Page. Available online: https://lammps.sandia.gov (accessed on 24 March 2021).
46. HidekiMori-CIT, Aenet-Lammps. Available online: https://github.com/HidekiMori-CIT/aenet-lammps (accessed on 24 March 2021).
47. Lambrecht, W.R.L.; Segall, B.; Methfessel, M.; van Schilfgaarde, M. Calculated elastic constants and deformation potentials of cubic SiC. *Phys. Rev. B* **1991**, *44*, 3685–3694. [CrossRef] [PubMed]
48. Merz, K.M.; Adamsky, R.F. Synthesis of the wurtzite form of silicon carbide. *J. Am. Chem. Soc.* **1959**, *81*, 201–251. [CrossRef]
49. Lundqvist, D. On the crystal structure of silicon carbide and its content of impurities. *Acta. Chem. Scand.* **1948**, *2*, 177–191. [CrossRef]
50. Harris, G.L. *Properties of Silicon, Emis Datareviews Series No. 4*; INSPEC: London, UK, 1988.
51. Every, A.G.; McCurdy, A.K. *Numerical Data and Functional Relationships in Science and Technology*; New Series, Group III; Ullmaier, H., Landolt-Börnstein, Eds.; Springer: Heidelberg, Germany, 1991; Volume 29, Pt. A.
52. Halicioglu, T. Comparative study on energy-and structure-related properties for the (100) surface of β-SiC. *Phys. Rev. B* **1995**, *51*, 7217–7223. [CrossRef]
53. Ercolessi, F.; Adams, J.B. Interatomic potentials from first-principles calculations: The force-matching method. *Europhys. Lett.* **1994**, *26*, 583–588. [CrossRef]
54. Daviau, K.; Lee, K.K.M. Decomposition of silicon carbide at high pressures and temperatures. *Phys. Rev. B* **2017**, *96*, 174102. [CrossRef]
55. Ishimaru, M.; Bae, I.-T.; Hirotsu, Y.; Matsumura, S.; Sickafus, K.E. Structural relaxation of amorphous silicon carbide. *Phys. Rev. Lett.* **2002**, *89*, 055502. [CrossRef] [PubMed]
56. Ishimaru, M.; Hirata, A.; Naito, M.; Bae, I.-T.; Zhang, Y.; Weber, W.J. Direct observation of thermally induced structural changes in amorphous silicon carbide. *J. Appl. Phys.* **2008**, *104*, 033503. [CrossRef]

materials

MDPI

*Article*

# Effect of Wave Process of Plastic Deformation at Forging on the Fatigue Fracture Mechanism of Titanium Compressor Disks of Gas Turbine Engine

**Andrey A. Shanyavskiy [1,2,*], Alexey P. Soldatenkov [1] and Alexandr D. Nikitin [2]**

[1] Aviation Register for Russian Federation, Airport Sheremetievo-1, PO Box 54, 141426 Moscow Region, Chimkinskiy State, Russia; 106otdel@gmail.com
[2] Institute of Computer Aided Design, 2nd Brestskaya street 19/18, 123056 Moscow, Russia; nikitin_alex@bk.ru
* Correspondence: 106otdel@mail.ru

**Abstract:** The low-cycle fatigue behavior of the VT3-1 titanium alloy (Ti–6Al–3Mo–2Cr alloy) under loading with a triangular and trapezoidal shape of cycle waveform was studied on round specimens prepared from forged compressor disks of a gas turbine engine. The filament type structure after forging has alternating filaments with the ductile and quasi-brittle state of the metal as a result of the wave process of plastic deformation during the metal forging process. The crack propagation, regardless of the cyclic waveform shape, occurs by the crack meso-tunneling mechanism: initially, the cracks propagate along the filaments by a quasi-brittle mechanism with the formation of a facetted pattern relief on the fracture surface reflecting the two-phase structure of the titanium alloy, and then, the bridge between the meso-tunnels is fractured with the formation of fatigue striations. The part of the crack growth duration $N_p/N_f$ in the durability $N_f$ is determined on the basis of measuring the fatigue striation spacing, and it depends on the crack path with respect to the material filaments. The growth of a fatigue crack in the case of in-service failure of a compressor disk of a gas turbine engine is considered, taking into account the crack meso-tunneling effect, and the fatigue crack growth duration in the disk is determined on the basis of quantitative fractography.

**Keywords:** forged titanium alloy; low-cycle fatigue; cycle wave form; durability; crack meso-tunneling; fractography; crack growth duration

**Citation:** Shanyavskiy, A.A.; Soldatenkov, A.P.; Nikitin, A.D. Effect of Wave Process of Plastic Deformation at Forging on the Fatigue Fracture Mechanism of Titanium Compressor Disks of Gas Turbine Engine. *Materials* **2021**, *14*, 1851. https://doi.org/10.3390/ma14081851

Academic Editor: Jaroslav Pokluda

Received: 26 February 2021
Accepted: 1 April 2021
Published: 8 April 2021



## 1. Introduction

Compressor disks of gas turbine engines (GTE) are manufactured from various titanium alloys and operate under high-stress conditions. Therefore, the main criterion for determining their service life is low-cycle fatigue (LCF) [1,2]. A high-stress level of the disks is considered in repetitive cycles of variable loads, which represent a load block for a flight or a cycle of starting and stopping the engine. It should be emphasized that the LCF regime is considered in terms of the number of loading cycles within $4 \times 10^4$ engine start and stop cycles in service (FLC—flight loading cycle). In this case, the stress level in zones with a high-stress concentration does not exceed the value of $\sigma_{max} = 0.8 \cdot \sigma_{0.2}$, where $\sigma_{0.2}$ is the yield stress of the material. Therefore, according to fatigue S-N curves, the service life of more than $4 \times 10^4$ FLC can be realized. This requirement reflects the fact that the transition to the LCF regime according to the criteria of the physics of metals means that the yield stress of the metal is reached, which is associated with a change in the shape of the object, and such a case is not permissible for the compressor disk in the structure.

Plastic strains appear to be localized, but not leading to a change in the shape of the structure on the macro-scale level. Macroplastic strains, characterizing the LCF regime for the specimens, are not observed during the LCF regime of disks.

The problem of the fatigue crack development in compressor disks of gas turbine engines arose in the mid-1980s [1,3–6]. It is generally associated with the structural features of two-phase ($\alpha + \beta$) titanium alloys and the pronounced occurrence of their sensitivity to dwell-time under load [6–8]. Cracks can appear both in the rim [4] and in the hub part of the disks [3,5,6]. Therefore, the fracture of disks is determined not by the presence of material defects but by the sensitivity of the material to operating conditions (long dwell-time under load) or the arising anomalous high-stress state, in most cases leading to the initiation and propagation of cracks in the LCF regime.

Analysis of crack growth regularities in fractured disks, mainly made of VT3-1 titanium alloy (Ti–6Al–3Mo–2Cr alloy), showed that the main feature of the fracture surface formation is expressed in a combination of two crack growth mechanisms [5,6]. Firstly, a quasi-brittle fracture occurs with the formation of a facetted pattern that reflects the geometry of the two-phase structure of the metal (globular or lamellar), and secondly, fatigue striations are formed, reflecting the ductile development of fracture typical of a material [5,6,9–11]. It is obvious that the combination of the two fracture mechanisms reflects a radical difference in the reaction of the material structure to the realized loading conditions of the disks during the full cycle of operation from start to stop of the engine. During the specified period of engine operation, the disk is affected by a sequence of loads varying in level and the number of unit cycles associated with the sequence of flight stages: taxiing the aircraft along the runway, takeoff, cruising, landing, and taxiing the aircraft. It is worth emphasizing that during the full flight loading cycle, the disks experience long-time loading with a constant level of stress, for example, in the cruising mode, which leads to an intensification of the fracture process, which was expressed in the formation of a facetted pattern.

Since the engines operate at modes with various numbers of revolutions due to different stages of a flight, not one but several cycles of material damage are considered. Hence, the actual number of cycles damaging the disks is increased up to five unit cycles for one FLC [4,12]. Therefore, when estimating the crack growth duration in terms of the number of engine start and stop cycles, it is necessary to take into account the actual number of damaging cycles based on the schematized block of cyclic loads.

Tests of one of the GTE compressor disks in air and at room temperature, as well as of specimens prepared from the compressor disk, using the triangular and trapezoidal shape of the loading cycle showed [4] that with an increase in the duration of the loading cycle and dwell-time up to 1 min, the durability and the crack growth duration decrease simultaneously. The decrease in durability is accompanied by a change in the mechanism of material fracture from ductile to quasi-brittle. In the first case, fatigue striations are predominantly formed, and in the second case, a facetted pattern of the fracture surface, reflecting the two-phase ($\alpha + \beta$) lamellar or globular structure of the material, is formed [5].

Thus, in the existing technological cycle of disk manufacturing, when the mechanical characteristics of the material specified in the drawing and the chemical composition of the material meet the requirements imposed on them, the material may turn out to be both less and more sensitive to the conditions of its loading in operation.

To eliminate the inhomogeneity in the material structure formed during forging, the technology of the multiple and multidirectional disk forging process was proposed [13]. However, as shown by fractographic studies, in this case, even with a durability of 800 cycles for smooth specimens, cracks originate under the surface, which excludes their control in service [14].

Reducing the stress state of the disk or limiting their resource does not exclude the occurrence of new fractures due to the natural scatter of material properties in the rim or the hub part of the disks, as well as variations in the stress state of the material as a result of natural clearances in the area of the blade positions. Unfortunately, this leads to the appearance of fatigue cracks in the disks of different compressor stages at different operating times.

At the same time, the experience of studying similar compressor disks shows that the process of fatigue crack propagation in them is characterized by general regularities, which are expressed in the following: there is a pronounced process of alternating fracture surface areas with a facetted pattern, which reflects the structure of a two-phase titanium alloy, as a rule, lamellar or mixed lamellar-globular type; these areas alternate with areas of fracture surface with fatigue striations, the spacing of which can sharply change over a short fracture surface length; the areas are oriented in a certain direction, which is set by the orientation of filaments in the texture of the deformed material.

In order to introduce more advanced technological solutions into the disk manufacturing process, it became necessary to study in detail the mechanism of fracture development using the specimens from a titanium compressor disk. The results of a comprehensive study of specimens that were prepared from titanium compressor disks and subsequently tested in the LCF regime as well as a discussion of the fracture mechanism and an estimation of the crack growth duration in the titanium compressor disk of the GTE of the An-124 aircraft are presented below in a generalized form.

## 2. Materials and Methods

At the first stage of the research, the specimens that were prepared from the first-stage disks of the engine low-pressure compressor were tested. The disk operated for about 8000 flight cycles (FC) without violating the operating conditions. The specimen dimensions were calculated in accordance with the ASTM E606 [15]. The specimens have a uniform-gage test section with a diameter of 6 mm and gage length of 20 mm. In the grips area, the specimens had a larger diameter (12 mm) and threads to prevent them from slipping out of the grips.

Before preparing the specimens, all disks were identified as corresponding in terms of mechanical characteristics and chemical composition to the two-phase VT3-1 titanium alloy in accordance with the requirements of the drawing (Table 1). The chemical element content was evaluated by the method of an atomic emission spectroscopy.

**Table 1.** The results of the atomic emission spectroscopic analysis of the disk specimen in comparison with the main alloying elements of the grade composition in wt %.

| Element/ Specimen | Ti | Al | Mo | Cr | Si | Fe |
|---|---|---|---|---|---|---|
| Disk | balance | 6.0 | 2.5 | 2.0 | 0.4 | 0.5 |
| VT3-1 grade | balance | 5.5–7.0 | 2.0–3.0 | 0.8–2.3 | 0.15–0.40 | 0.2–0.7 |

The fatigue tests were performed on specimens according to the loading scheme with constant maximum strain at three levels 0.0065, 0.0073, and 0.008 with two waveforms of the loading cycle: (1) triangular, with a frequency of 0.5 Hz; (2) trapezoidal, with both loading and unloading times of 5 s and a dwell-time of 20 s at a maximum strain level. A total amount of 72 specimens were tested at the stress ratio $R = 0$.

After testing, all fractured specimens were subjected to metallographic, spectroscopic, and fractographic analyses.

The metallographic analysis confirmed the correspondence of the specimen material to a two-phase ($\alpha + \beta$) titanium alloy with a mixed type of globular and lamellar structure (Figure 1).

Energy dispersive X-ray analysis was carried out to identify the chemical composition for individual phases. Fractographic analysis of the fractured specimens was performed using the scanning electron microscope EVO-40 (Carl Zeiss GmbH, Oberkochen, Germany).

At the second stage of the study, a fractographic analysis of the regularities of fatigue crack propagation in the rim of the disk of the intermediate-pressure compressor (IPC) of the GTE was carried out.

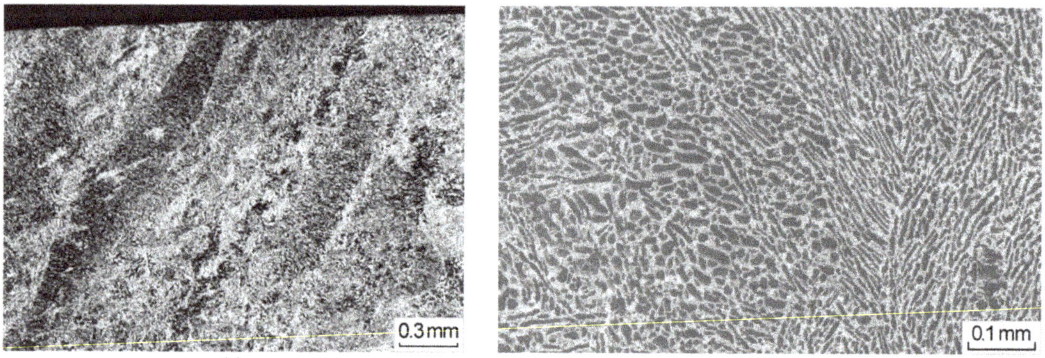


(**a**) (**b**)

**Figure 1.** The microstructure of the two-phase ($\alpha + \beta$) titanium alloy with a mixed type of globular and lamellar structure at (**a**) 0.3 mm magnification and (**b**) 0.1 mm magnification.

During the operation of the An-124 aircraft with the number RA-82043, a surge of the GTE of the power plant No. 4 occurred in the climb regime, with a fire alarm and automatic activation of the first stage of the fire extinguishing system. Subsequently, the engine No. 4 was automatically shut down. The aircraft landed safely at the departure airport. Inspection on the ground of the failed engine established its non-localized fracture due to the release of fragments of the titanium second-stage disk of the IPC (Figure 2).

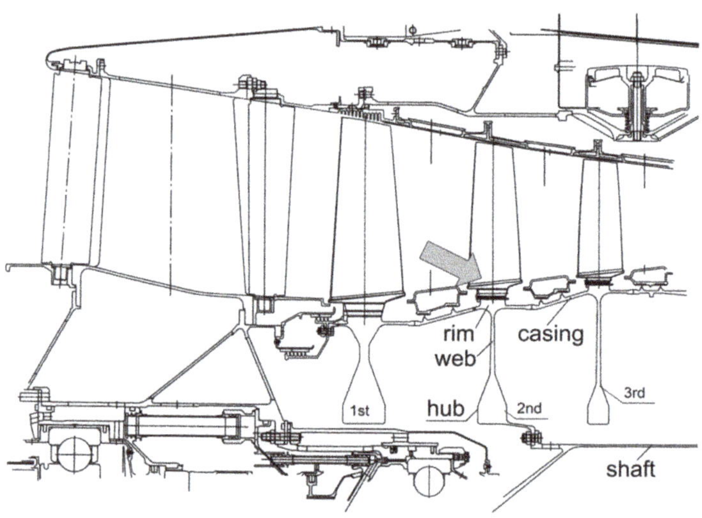


**Figure 2.** Fragment of the scheme of the GTE intermediate-pressure compressor. The arrow indicates the location of the initial fracture of the second-stage disk.

The engine operated for 14,986 h (3299 FC) since the beginning of the operation, including 11,191 h (2486 FC) after the last overhaul.

For all specimens and the disk fracture surface, the duration of fatigue crack growth was estimated based on measuring the spacing of fatigue striations in the direction of crack propagation, regardless of the striation orientation [1,14].

## 3. Results and Discussion of Fatigue Tests

The test results are shown in Figure 3. In the specimens tested with different shapes of the cycle waveform, different fracture surface patterns were formed, in which the behavior of the material corresponded to the first, second, or third type according to the classification considered below.

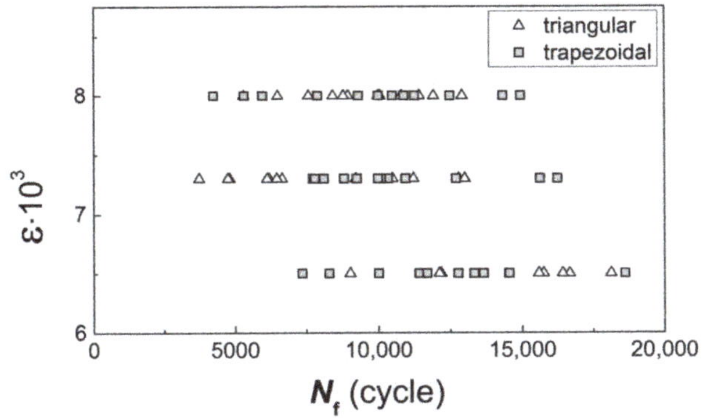


**Figure 3.** Fatigue ε-$N$ curve plotted for tested specimens with different material behavior. The specimens with a life $N_f$ of 7000 cycles and less showed sensitivity to loading conditions.

The cracks in all the specimens initiated from the scores (<5 μm) made on the surface during the manufacturing procedure of the specimens. Subsequent crack propagation occurred along, across, or at an angle of less than 90° to the material filaments.

A fundamental difference in the durability of the specimens due to the presence of scores with different shapes and depths was not obtained. For example, for a specimen with high durability, the depth of the score differs a little from a specimen that has fractured with significantly lower durability with the same type of fracture surface pattern.

The feature in the behavior of the material, which is sensitive to the loading conditions, was revealed in a durability decrease; in some cases, the durability of such specimens was almost three times less than for the majority of specimens tested with the same strain range (see Figure 3). The decrease in durability was related to the dominance of the quasi-brittle fracture surface pattern with the formation of a predominantly facetted pattern of a fracture surface and few local zones with fatigue striations.

This concerns primarily specimens tested with the triangular cycle waveform at a maximum strain level of 0.0073. In the fracture surfaces, the filament type structure of the material was revealed, inherited from the forging of disks. A similar situation was found in some specimens tested with the trapezoidal shape of the cycle waveform at the same maximum strain level of 0.0073 (Figure 4). The crack propagated in the specimen along the filaments with pronounced tunneling and the formation of a predominantly quasi-brittle fracture surface relief. The connection of meso-tunnels along the bridge between them led to the formation of local areas with groups of fatigue striations oriented almost perpendicular to the main direction of crack growth (Figure 4b). It follows that in those zones of crack propagation where the quasi-brittle fracture occurred, the material had a reduced resistance to crack growth. During loading, the crack initially propagated along the filaments on the uploading part of a cycle, while during the formation of fatigue striations, the crack grew on the unloading part of a cycle [16].

Depending on the loading conditions and the stress state of the alloy, three types of fatigue crack propagation are possible under cyclic loading conditions.

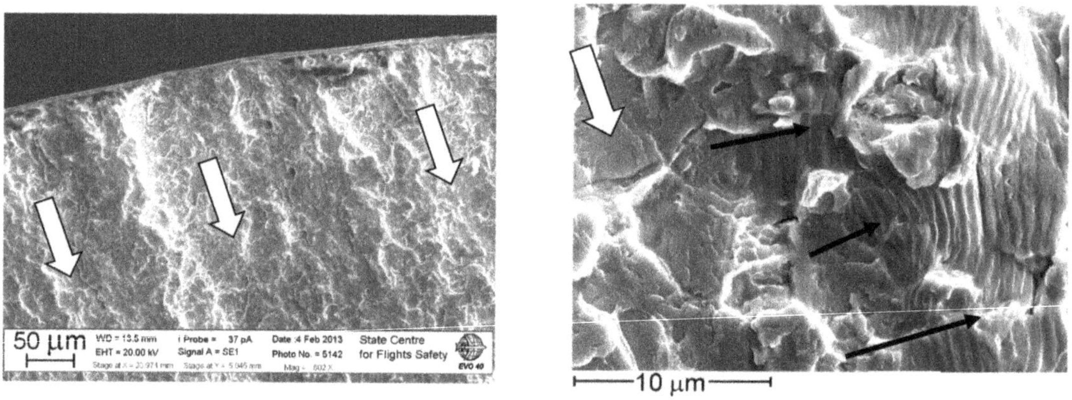


(a) (b)

**Figure 4.** Meso-tunnels (indicated by hollow arrows) near the fracture origin (**a**) and at a several distance (**b**) in the crack growth direction for specimen tested at the trapezoidal shape of cycle waveform with $\varepsilon_{max}$ = 0.0073 and $N_f$ = 10,902 cycles. Dark arrows indicate the direction of crack growth during the formation of fatigue striations.

First type. The crack initially propagates along the brittle filaments with the formation of a facetted pattern of fracture surface and meso-tunnels, as shown in Figure 5. Then, the crack develops across the ductile filaments by the bridge fracture with the formation of fatigue striations.

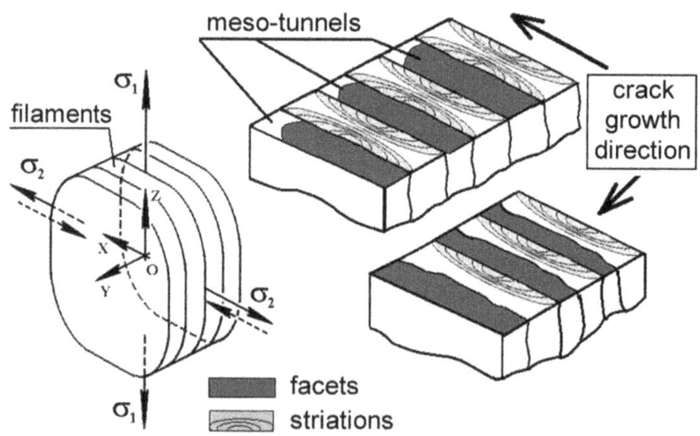


**Figure 5.** Scheme of the biaxial stress state of the material after forging and the difference in fracture surface formation along brittle and across ductile filaments at different directions of crack growth.

Second type. The crack propagates perpendicular to the filaments with an alternating waved pattern of a fracture surface and zones with fatigue striations (see Figure 5).

Third type. The crack grows at an angle to the elongated zones with filaments, which leads to an alternation of the fracture process according to the first and second variants.

Nevertheless, fatigue striations were observed on all fracture surfaces with a different fraction of the whole fracture surface area.

The performed local X-ray analysis showed that in all zones of fracture surface, the similar homogeneity in the distribution of alloying elements is systematically repeated. On average, the distribution of chemical elements in both phases of material meets the requirements of the technical specifications for the VT3-1 titanium alloy.

The distribution of chemical elements in phases within the slice is such that Mo is not detected within an accuracy of 0.5% in the α-phase, and its content in the β-phase is higher than the grade composition, which meets the requirements of the alloy grade composition, since Mo is a β-phase stabilizer. Cr is typically present in the β-phase and may not be observed in the α-phase.

The indicated distribution of chemical elements does not affect the regularity of the formation of a particular fracture relief along the α-phase (Figure 6). No fundamental differences in the chemical composition of the alpha phase are observed during the formation of both facetted patterns and relief with striations. Therefore, there is no phase difference in chemical composition for different types of fracture surface relief. The difference in the crack growth mechanisms is determined only by the local plastic characteristics of the material state.

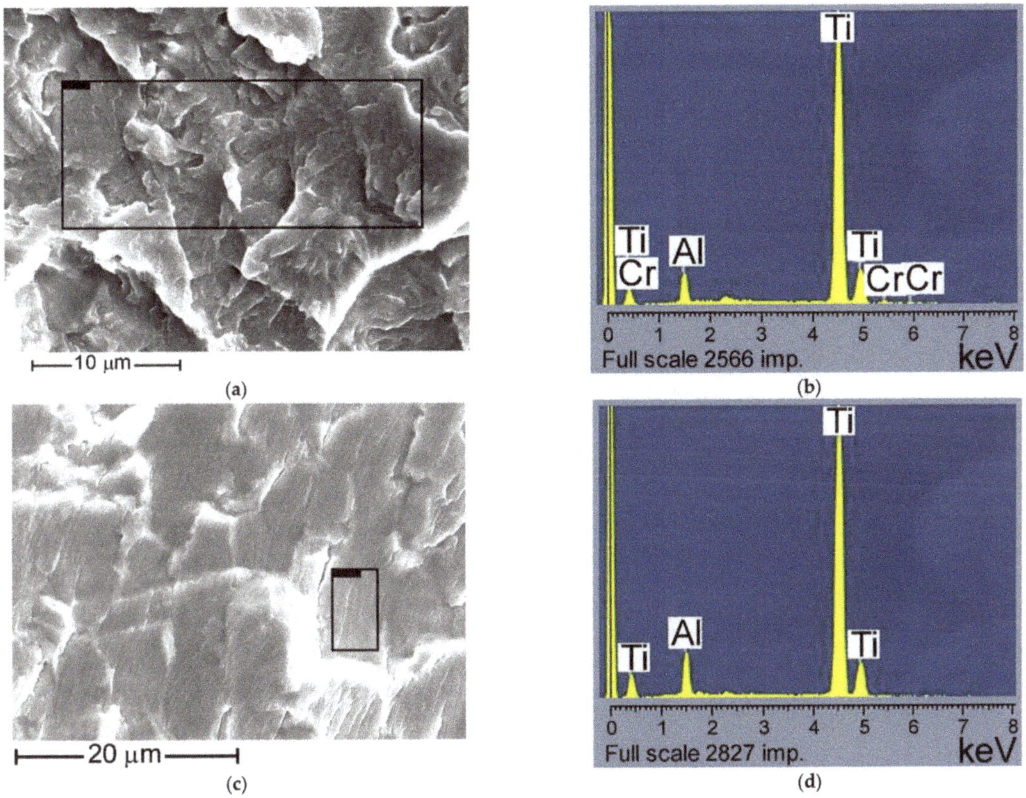


**Figure 6.** (**a**) Region of fracture surface with quasi-brittle pattern; (**b**) the spectrum of the main alloy elements distribution over the rectangular area of fracture surface on (a); (**c**) region of fracture surface with striations; (**d**) the spectrum of the main alloy elements distribution over the rectangular area of fracture surface on (c).

Fractographic studies made it possible to conclude that the material of the tested specimens had a high inhomogeneity in resistance to the action of loads with various cycle shapes. The obtained durability scatter reveals the influence of two factors on material behavior.

First, in all specimens on the surface, there are scores of different depths (<5 μm) created during the manufacturing process. This creates for all specimens statistically the same inhomogeneity in the stress state of the surface in the crack initiation zones and affects the scattering of durability by the criterion of crack initiation, underestimating its value in relation to the behavior of smooth specimens.

Second, in a small group of specimens, a sharp decrease in durability is observed since the material sensitivity to loading conditions has a dominant influence on durability.

The data on fatigue tests of fractographically investigated specimens illustrates the conclusions (see Figure 3).

A group of specimens with the durability of 7000 cycles and less should be characterized by a single fatigue curve, the parameters of which indicate significantly lower durability of the material at the considered strain levels. A group of specimens with a life of more than 10,000 cycles should be characterized by a different fatigue curve. In fact, a bimodal distribution of fatigue life is considered when a small part of specimens in which the material is sensitive to loading conditions determines the minimum allowable resource for the entire set of tested specimens. Therefore, for in-service disks, there is a risk of failure with a low operating time of single instances, the material of which is in an unsatisfactory state from the point of view of the criterion under consideration; i.e., the sensitivity to loading conditions is revealed and consisted in predominant fracture with the formation of a facetted pattern even for a triangular shape of the cyclic load waveform.

The results obtained indicate the statistical inhomogeneity in the material behavior within one disk from one zone to another and also from one disk to another, since the specimens studied were made from different disks as well as from randomly selected zones of one disk.

Thus, the material sensitivity to the in-service loading conditions is not a characteristic of manufacturing a particular disk. In different areas of the disk, the sensitivity is revealed differently for the same forging process. Therefore, in the existing technology of disk manufacturing, disk samples can be obtained with different zones according to the material sensitivity to loading conditions, when the durability of the disks can be more than three times reduced as compared to most other disks that do not have the specified sensitivity.

Based on the performed statistical analysis of the chemical element distribution over all investigated fracture surfaces, one can conclude that the realized mechanisms of the facetted pattern or the fatigue striation formation in the $\alpha$-phase is not associated with the local redistribution of chemical elements, but it is determined by the plastic properties of the metal after its deformation (i.e., during manufacturing). In local regions where the plastic deformation process can occur in the material, fatigue striations are formed. In the zones where the plastic deformation has been exhausted at the stage of manufacturing the disk, the formation of the facetted pattern relief on the fracture surface is realized.

As a result of measuring the distance between the fatigue striations, an ambiguous relationship between the crack growth duration and the specimen durability was revealed. First of all, this is argued by the difference in the critical crack lengths at which the final fracture is realized.

In specimens where the stable crack growth with the formation of fatigue striations was realized almost for the total cross-section, the behavior of the material should be considered as the most favorable from the viewpoint of its structural (stress-strain) state. However, in several specimens, even with the predominant formation of fatigue striations, there was a sharp transition from the zone of stable crack growth to the fast fracture. As a rule, in the specimens, the cracks grew initially across the material filaments following propagation along the filaments. Such a sharp transition is associated with the fact that the crack began to develop not only within the plane of the main fracture but also along the filaments of the material, almost perpendicular to the plane of the fracture. This indicates that in some filaments, the material behaves as quasi-brittle, and the propagation of a crack through them is not accompanied by the formation of fatigue striations (Figure 7). In other filaments, the material is more ductile, and the propagated crack leads to the formation of fatigue striations.

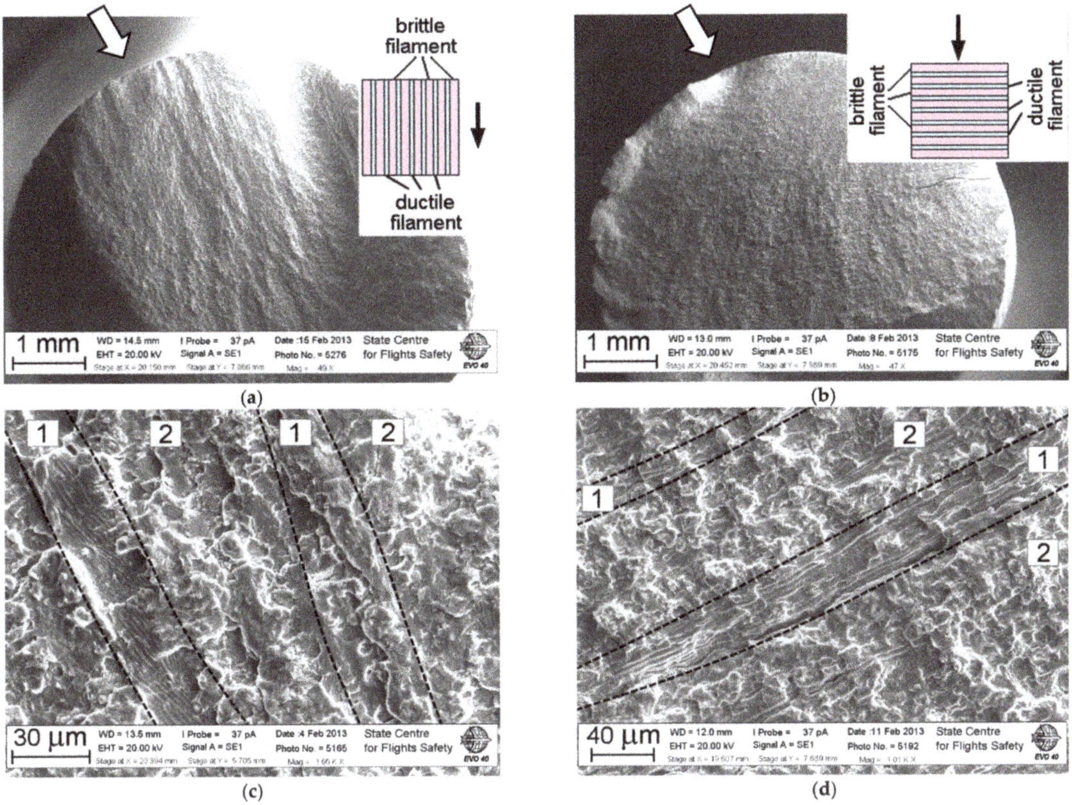


**Figure 7.** Fracture surfaces with the sketch of crack propagation (**a**,**c**) along and (**b**,**d**) across the filaments in a two-phase ($\alpha$ + $\beta$) structure of specimens tested with a triangular shape of the cycle waveform and the durability $N_f$ of 16,566 cycles and 64,860 cycles, respectively. The dark arrow on the sketch indicates the direction of crack growth. The fatigue crack initiation zone is shown by the hollow arrow. Regions 1 and 2 correspond to fracture of ductile and brittle filaments, respectively.

In fact, in the process of forging, the disk material is formed as inhomogeneous with alternating filaments, within which the plastic deformation is significantly exhausted, and filaments can still realize significant plastic deformation in the process of fatigue crack growth. In the process of forging, individual grains of the material can be self-organized in volume with the formation of a filament type texture in such a way that it is in the grains that a facilitated sliding will be realized along the system of planes along which the subsequent crack development will predominantly occur. In this case, the material is already prepared for the implementation of the quasi-brittle fracture mechanism in the $\alpha$-phase along the slip planes.

It is also necessary to take into account the natural variation in the crystallographic orientation of the $\alpha$-phase within the filament group with relation to the realized loading [17]. In the case of significant exhaustion of plastic strain in the material along a particular group of filaments with an unfavorable orientation regarding the applied force, the faster initiation of a fatigue crack occurs. Then, transition to the fast fracture zone along the boundaries of the metal filament texture appears. This determines the scatter in the critical crack lengths.

Thus, during the disk forging process, an inhomogeneous wave process of plastic deformation occurs [18], which leads to the creation of a composite material with different levels of residual plastic strain along the formed layers (Figure 8). The deformation waves create periodically alternating filaments with a high and low degree of strain. The crack development along the filaments or a change in the orientation of the crack growth retains the difference in the fracture mechanisms for brittle and ductile filaments. However, due to the change in the crack orientation in the field of the biaxial stress state, the fracture of the brittle filaments occurs more ductile and without the formation of a facetted pattern [9,13,19], namely: for a positive $\lambda$-ratio, the fracture is more brittle than for a negative $\lambda$-ratio, where $\lambda$ is the ratio of the first and the second principal stresses. Therefore, in the case of deformation of a two-phase titanium alloy during manufacturing in the material, the process of self-organized distribution of the residual stress level occurs due to the wave nature of the distribution of the intensity of the deformation effect on individual structural elements created during forging as a filament-type structure. Subsequent heat treatment does not lead to a uniformly prepared structure and the removal of stress state inhomogeneity in the material, which is due to a distinct difference in the plasticity of the formed individual filaments. In fact, the introduction of a dwell time under cyclic loading conditions for an inhomogeneous material creates conditions for its self-organization, which are expressed in the implementation of various fracture mechanisms.

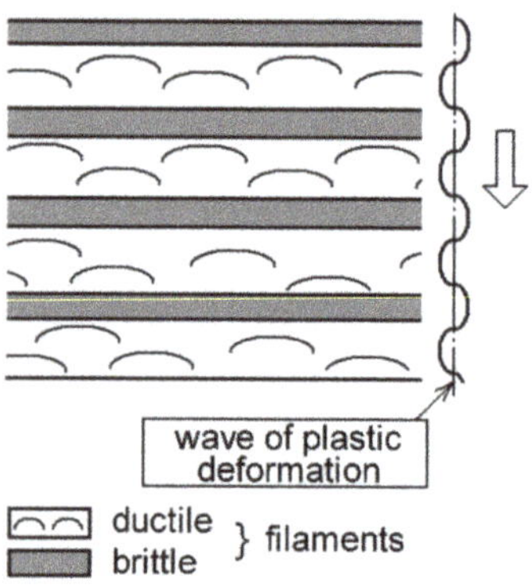


**Figure 8.** Scheme of the wave process of material deformation during forging with the formation of alternating filaments with a brittle and ductile material state.

Due to the existing difference in the material response to cyclic loading at the stage of crack nucleation and growth, with the same strain level and different orientation of initial crack propagation according to one of the above-mentioned types, it becomes necessary to estimate the crack growth duration as a component of the part in durability. For this purpose, the $N_p/N_f$ ratio between the crack growth duration $N_p$ and the durability $N_f$ was used.

In alloys on different bases, a stable and unambiguous relationship between the values of $N_p/N_f$ and $N_f$ in the case of a statistically homogeneous state of the material exists [20–22]. In the case of specimens with a pronounced anisotropy in the material structure in the form of filaments, this relationship may be ambiguous due to a scatter of durability.

Therefore, initially, for each strain level and the triangular cycle waveform, the research results for specimens with minimum and maximum durability from the group located at the boundary of the scatter region are considered (Figure 9). It was established that the specimens tested at the same shape of cycle waveform belong to two different groups, for which the regularities of the change in the crack growth duration in relation to the realized specimen durability are different. One group of specimens in its behavior fits into the regularity of the crack growth duration, which was obtained earlier for the investigated specimens of the VT3-1 titanium alloy [1]. This analyzed dependence belongs to the largest values of the crack growth duration (Curve 2 in Figure 9). In this case, the crack propagated across the filaments, and ductile fracture with fatigue striations was mainly realized.

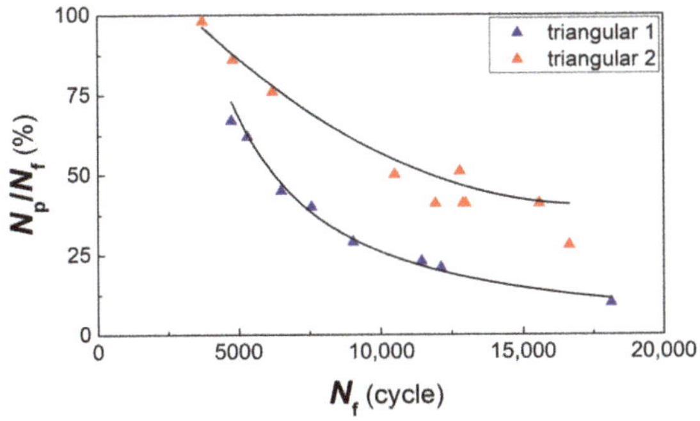


(**a**)

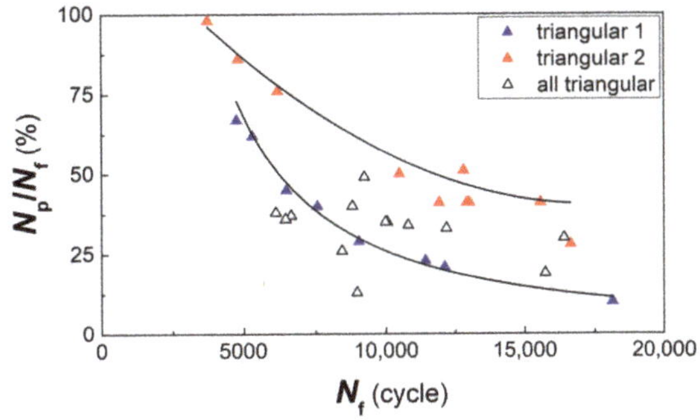


(**b**)

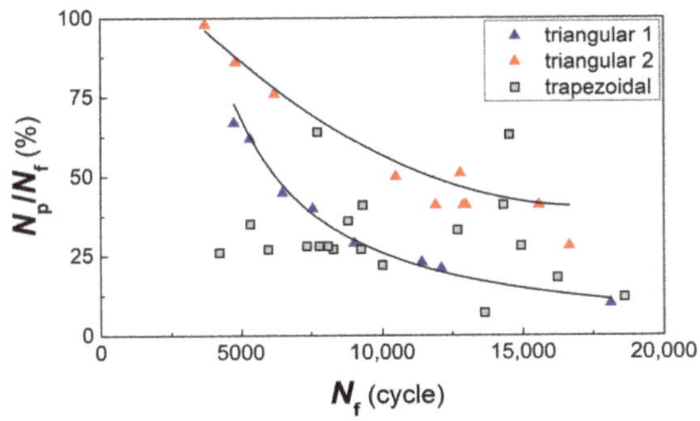


(**c**)

**Figure 9.** (**a**) Two groups of dependences 1 and 2 of the relative part of the crack growth duration $N_p/N_f$ on the durability $N_f$ for specimens tested with the triangular shape of the cycle waveform, and the general dependence of changes in the relative fatigue crack growth duration on the durability for all specimens tested with (**b**) triangular and (**c**) trapezoidal shape of cycle waveform.

Another group of specimens indicates a smaller part of the crack growth duration in specimen durability (see Curve 1 in Figure 9). The surface of these specimens was possibly more hardened during their manufacturing than the surface of other specimens. Nevertheless, at the stage of crack propagation, faster fracture as compared to other specimens with approximately the same durability occurred due to the low resistance of the material to fatigue crack growth. It means that these specimens belonged to the first type of crack growth mechanism with the formation of meso-tunnels.

With a decrease in durability, starting from 7000 cycles, the data on relative crack growth duration approach each other. This indicates that at a high-stress level and fatigue life of 4000 cycles and less, the differences in the material structure of the specimens are not revealed significantly in the relative part of the crack growth duration.

Accounting for all the specimens tested with the triangular shape of the cycle waveform showed a significant scatter of the experimental data relative to the two dependences presented (see Figure 9b).

For a group of specimens tested with a trapezoidal waveform of the cycle, a similar analysis was performed. Some part of the specimens appears to have a similar difference in their behavior, and the relative part of the crack growth duration is grouped around the two dependencies on durability identified for the case of the triangular cycle waveform (see Figure 9c). This is quite natural, since the specimens were prepared under the same conditions, and their surface condition is statistically uniform.

It is necessary to point out a group of specimens with a life range of 4000–8000 cycles (on Figure 9c), which does not fit into the presented regularities. For this group of specimens, the relationship between the relative part of crack growth and specimen durability is not observed. It is the specimens that showed their sensitivity to the loading conditions, and at the stage of crack growth, a predominantly facetted pattern of quasi-brittle fracture with pronounced meso-tunneling of fatigue cracks was formed.

Two obtained dependencies (see Figure 9) confirm the fact that the tested specimens demonstrate a set of material states, one of which is bad, when already with a triangular waveform of the cycle, rapid nucleation and a short crack growth stage occur. Specimens showing reduced fatigue resistance during the crack growth stage should be classified as medium or also bad specimens. The arising wave process of plastic deformation [18], the scheme of which is shown in Figure 8, is responsible for such a material behavior. A good state will be observed when, in any direction of crack propagation, even in the material with weakly revealed filaments, the fracture mechanism associated with the formation of fatigue striations will dominate.

The established regularities of the metal texture formation during forging the disks allowed us to explain the observed regularities of the in-service compressor disks' fracture process, which are demonstrated below on the instance of the GTE compressor disk fracture.

## 4. Fracture of the In-Service IPC Disk

### *4.1. Crack Growth Mechanisms*

Based on the disk fracture surface analyses, it was established that the main fracture occurred in the radial direction from the rim part toward the disk hub (Figure 10). The initial zone of crack initiation and growth within about 5 mm from the lateral surface of the rim (zone I in Figure 10) is located along the slot for the blade. The fracture surface boundaries are oriented according to the scores from machining the surface along the radius transition of the bottom of the slot for the dovetail blade to the lateral surface of the inter-slot volume. Therefore, in the crack initiation zone along the slot for the blade along the radius transition, a combination of radial and tangential loads took place. Analysis of the crack initiation zone shape, its location, and changes in the path of fracture development at the subsequent stage shows that the crack initiation in the disk is generally due to the local stress concentration in the shape of the radius transition of the slot for the blade, and not to the total stress intensity of the disk.

It is established by the analysis of the fracture surface relief in the region I that the fracture has a multi-origin fatigue nature with the formation of a meso-tunnels' cascade (Figure 11). In the fracture surface origins, a predominantly facetted pattern was observed. Between them, the blocks of fatigue striations were formed during the fracture of the bridges between the meso-tunnels. The uniformity of the fatigue striation formation has such a manner that within their block, the spacing remains almost the same. However, in some regions of the fracture surface (Figure 11b) within one meso-tunnel, there was a sharp change in the block of fatigue striation spacing of the same size to a block with a wider spacing. This reflects the specifics of the changing stress state of the material in the bridges between the meso-tunnels.

The observed nature of crack propagation with the formation of two types of fracture surface relief is typical for a material with a pronounced filament texture when the crack growth process is initially realized as a result of crack extension by the mechanism of a facet formation. Then, the bridges between the previously formed fracture surface regions are connected with the formation of fatigue striations (see, for example, Figure 4). In case of fracture from multiple origins, the duration of crack propagation in the direction 1 (see scheme in Figure 10) should be evaluated only starting from a distance of about 5 mm, when the crack emerged from the radius transition of the slot bottom under the blade (region II in Figure 10). This approach is based on the observation that on fracture surface region I in the direction 1, fatigue striations are formed toward each other (Figure 11b).

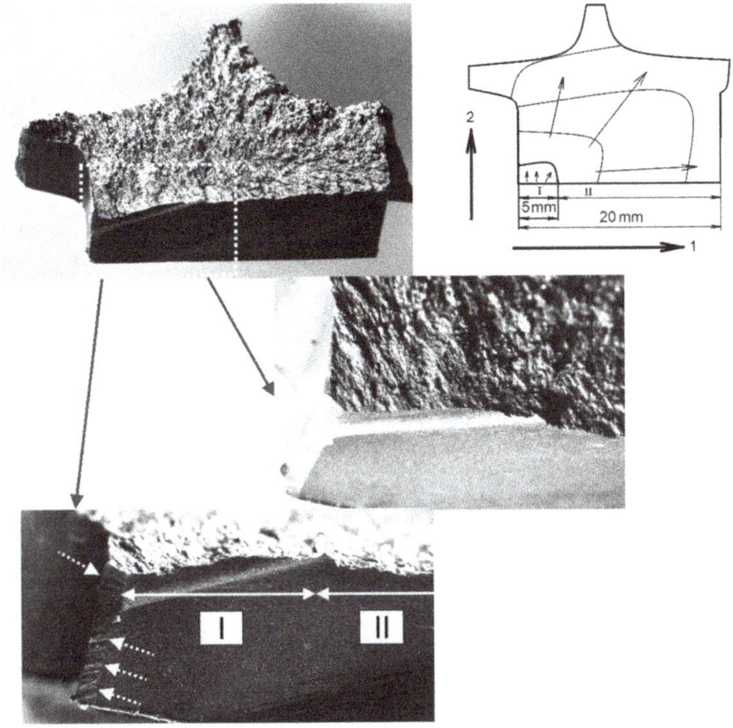


**Figure 10.** Initial fracture zone of the second-stage disk and its schematic representation indicating the orientation of the fracture surface formation from the region of the slot for the blade along the radius transition with the origin of the fatigue crack (indicated by a rectangle) in the disk. Dashed arrows indicate a region with high roughness along the boundary of the transition of the bottom of the slot for the blade to the rim surface. Roman numerals designate successively formed different zones of the fracture surface.

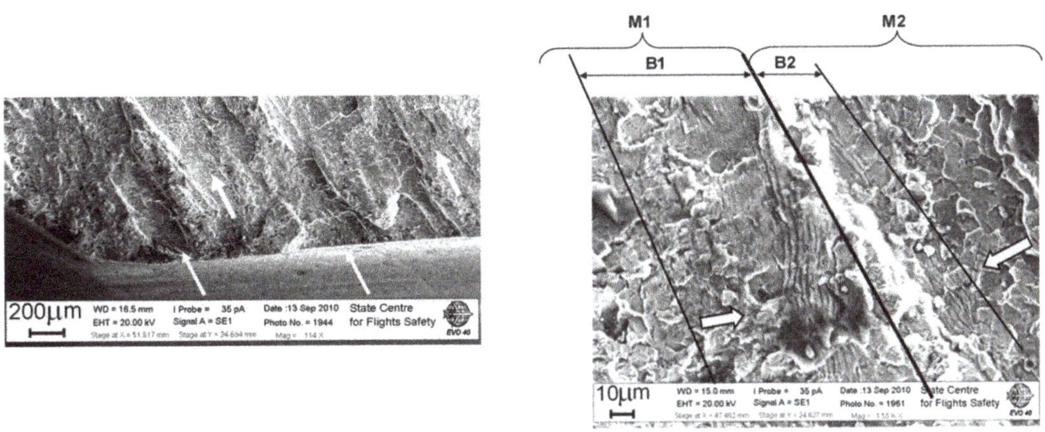


(**a**) (**b**)

**Figure 11.** (**a**) Features on the zone I (see Figure 10) of fracture surface in the bridge between the initial meso-tunnels and (**b**) two meso-tunnels M1 and M2 with a facetted pattern. Blocks of fatigue striations formed simultaneously to meet each other are designated by B1 and B2, and the directions of fracture development are shown by hollow arrows.

In the directions 1 and 2, the crack development took place with different regularity in the formation of fatigue striation spacing. The spacing of striations already in the zone I of crack initiation (see Figure 10) is 0.6 μm, which characterizes the process of initiation and development of disk fracture in the LCF regime. Along the initial fracture region with a width of about 5 mm in the direction 1, the fatigue striation spacing is almost constant. Further along the length of the crack in the direction 1, an increase in the striation spacing is observed. However, at a crack length of about 12 mm in the direction 1, the process of dimple formation increases, which indicates a transition to the unstable fracture zone of the disk material.

The crack has passed through a longer distance in direction 1 than in direction 2 during the same duration of disk fracture development. The size of the fracture zone in the direction 2 was about 10 mm, which is two times less than in the direction 1. Therefore, in the direction 1, the crack in the range of 12–20 mm grew at a higher rate than in the direction 2 for the same number of disk loading cycles.

The analysis of the regularities of the formation of cracks revealed during non-destructive testing in other slots for the blades showed the following. Fracture surfaces of the opened cracks in the rim part are formed similarly as in the above-described main fracture zone with the location of origins in the zones of the radius transition of the slot bottom for the blade to the surface of the inter-slot volume. In all fracture surfaces, the quasi-brittle fracture relief elements were observed, which are similar to the main fracture. However, the fracture origins are located only along the radius of transition from the surface of the slot for the blade to the inter-slot volume.

The observed relief elements indicate the development of secondary cracks in radius transitions along the inter-slot volume of the disk according to the LCF mechanism. These cracks appeared, most probably, after the LCF crack began to develop along the main fracture zone and had spread to a considerable depth. This is indicated by the fracture surface pattern along secondary cracks with a pronounced fast fracture process, the small length of the fatigue fracture zones, as well as the abrupt boundary of the transition to the fast fracture zone, formed in-service. In addition, fatigue striations in the zones were hardly pronounced.

According to the scheme of the main fracture surface formation (see Figure 10), the crack growth duration was estimated along the above-indicated directions 1 and 2. In direction 1, the estimation of the crack duration was important, because it was in this area that the crack extended to the greatest length over the surface (about 20 mm). However, as shown by the results of fractographic studies, it was in this area that the crack developed in such a manner that several meso-tunnels were formed simultaneously over a length of 5 mm, which were interconnected with the formation of fatigue striation blocks. This is indicated by the slightly different fatigue striations spacing, being in the range of 0.6–1.0 μm. The noted feature of relief formation is additionally illustrated by a fragment of a fracture surface with a pronounced process of crack development with almost the same fatigue striation spacing in the bridge between the meso-tunnels at the initial stage of the fracture zone formation (see Figure 11). From the facts considered above, it follows that the initial fracture zone with a length of 5 mm in the direction 1 should be excluded from the estimation of the crack growth duration, which underestimates slightly the total number of cycles during which the fatigue crack propagated. Such an underestimation goes into the safety margin and is acceptable for the practical implementation of non-destructive testing.

Therefore, in the direction 1, the regularity of fatigue striation formation was analyzed with the recalculation of the established number of unit cycles into the number of flight cycles on the length range of 5–20 mm. For example, the method of determining the crack growth duration by the fatigue striation spacing is shown in [14]. The fatigue striation spacing was measured, the number of loading cycles for the rim part of the disk during the fatigue crack propagation was calculated, and the results are shown in Figure 12. The increase in the fatigue striation spacing along the crack length was established to be insignificant. Stable formation of areas with fatigue striations occurred in the crack length range of 5–11 mm. Then, along the considered direction, an accelerated, repeated-static fracture of the disk was realized with the formation of areas with the facetted patterns and dimples. The crack growth duration resulted from measurements of the fatigue striation spacing at a length of 5–11.5 mm is about 5000 cycles and an additional 1900 cycles at a length of 11.5–20 mm according to the approximation of the spacing dependence on the length obtained at the interval of 5–11.5 mm.

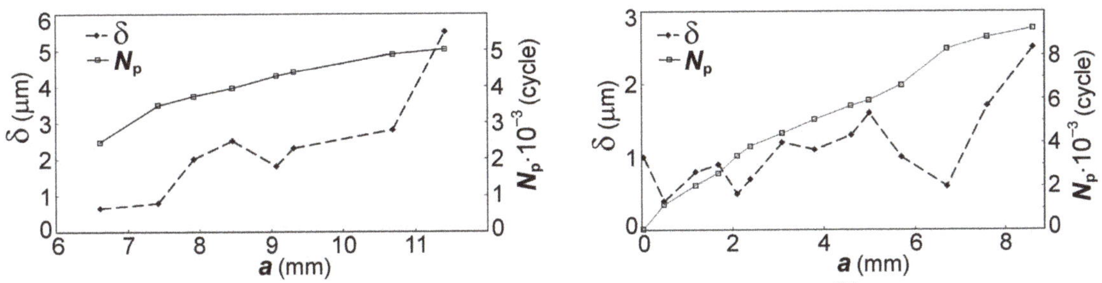


(a) (b)

**Figure 12.** Dependences of fatigue striation spacing δ and the number of cycles for the crack propagation $N_P$ on the crack length $a$ in the rim of the disk for direction (**a**) 1 and (**b**) 2.

As indicated above, the presented estimation is not complete, since it does not include the crack growth within the first 5 mm at the stage of origination zone formation. Therefore, the regularities of crack growth in the direction 2 (see Figure 10) were analyzed.

The results of the performed measurements showed that in this direction, from the corner of the fracture surface and before the transition to the fast crack growth zone with the formation of facetted patterns and dimples on the fracture surface, the crack development was obtained for a longer duration (Figure 12b). This indicates the fact that for the crack length range of 0–5 mm in the zone I, the formation of multiple fracture origins occurred and the initial stage of crack growth was quite long. Comparison of the crack growth duration values obtained for the two considered directions 1 (6900 cycles) and 2 (9600 cycles) showed the following. The duration of the fracture process for the region of 0–5 mm was at least 30% of the total crack growth duration from the initiation up to the critical length.

### *4.2. The Number of Aircraft Flights with a Developing Crack*

The results obtained for evaluating the crack growth duration in terms of unit loading cycles allow one to proceed to the analysis of the crack growth duration in flight cycles. To make the conversion, let us consider the disk loading when the engine is running according to the stages of a typical flight of the An-124 aircraft (Figure 13). The disk loading should be considered at least in five stages of changing the engine operating modes, resulting in the propagation of a fatigue crack for a flight: starting and entering the operating mode, changing the mode of operation when maneuvering, as well as the stages of changing the mode of operation before landing. This is consistent with the previously performed analysis of schematized loading cycles for other types of engines with the introduction of recalculation of the crack growth duration in terms of unit cycles into flight cycles [1,4]. The above estimation of the crack growth duration in terms of unit cycles should be reduced by a factor of 5.

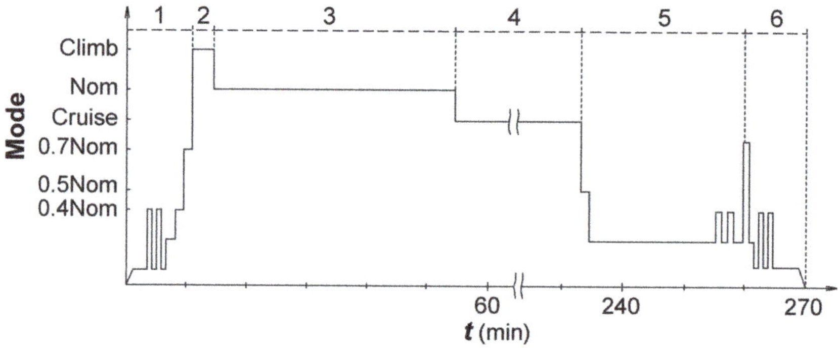


**Figure 13.** The schematized flight cycle of the GTE concerning with engine speed of revolutions relative to flight stages: 1—start-up, warm-up, taxiing (10 min); 2—take-off (3.5 min); 3—climb (40 min); 4—cruise (180 min); 5—descending (27 min); 6—reverse, engine cooling, landing, taxiing, engine shut-down (10 min). "Nom" is the nominal speed of revolutions.

In addition, it is necessary to take into account the delay in the development of the crack estimated by both the fatigue striations and the crack growth rate along the disk surface. This discrepancy leads to the usage of a factor of 1.6 and reduces the calculation results based on measuring the striation spacing when recalculating into the number of flight cycles [1,4,12]. Therefore, the recalculation of the number of unit cycles $N_p$ into flight cycles $n_p$ should be carried out by using the following expression: $n_p = 5 \times 1.6\, N_p = 8.0 \cdot N_p$.

The validity of considering several loading cycles of the disk, causing the crack propagation during flight, is confirmed by the results of fractographic studies, presented, for example, in Figure 14. Within the fracture surface region length of about 40 μm, there are two blocks of fatigue striations "a" and "b" with an average spacing of 0.36 and 1.2 μm, respectively. In some fracture zones, pronounced blocks of fatigue striations formed without changing their spacing within sequences of striations belonging to one block (8–14 striations) were observed. These sequences can be a result of material fracture during the implementation of transient loading modes, as follows from the scheme of engine operating conditions for a typical flight (see Figure 13).

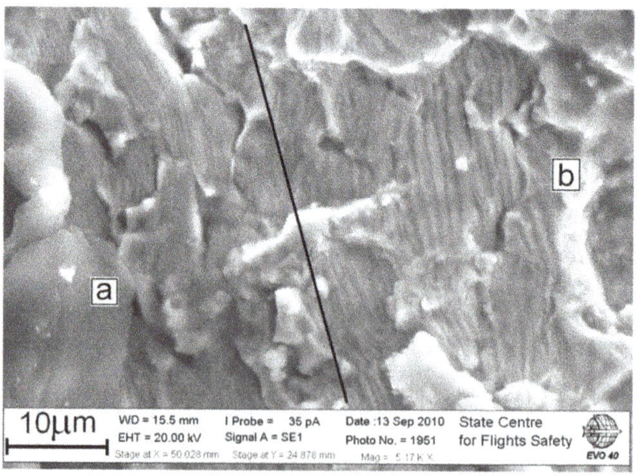


**Figure 14.** Fracture surface with two blocks "a" and "b" of fatigue striations with similar average spacing in each block, but differing by more than three times for the blocks "a" and "b".

The performed recalculation and the described regularities of disk fracture allowed obtaining the following results (Figure 15): (1) the crack growth duration along the direction 2 was about 1200 flight cycles; (2) the fracture durations in the directions 2 and 1 correspond to each other; and (3) within the range of first 5 mm along the surface of the slot for the blade, $n_p$ of about 340 flight cycles should be considered (i.e., $n_p$ = (9600–6900)/8). The dependencies shown in Figure 15 allow determining the frequency of disk inspections, starting with the crack length determined by the resolution of the inspection tool and the accessibility of the zone where the crack can be detected.

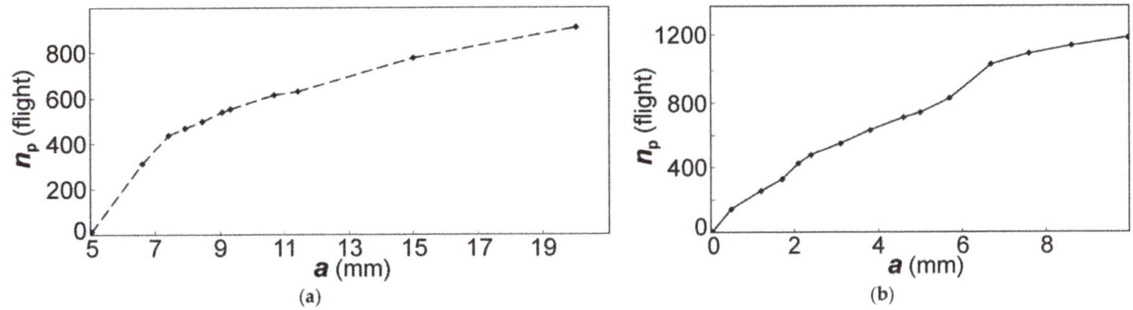


**Figure 15.** Dependences of the fatigue crack propagation duration in terms of flights $n_p$ on the crack length $a$ in the second-stage disk of the intermediate-pressure compressor (IPC) for directions (**a**) 1 and (**b**) 2 (shown in Figure 10).

Thus, the obtained results of estimating the crack growth duration in terms of flight cycles using a ratio of eight unit cycles per one flight are quite reliable.

### 4.3. Summarizing the Results of the Disk Study

The results obtained in the study of the disk material quality and the established regularities of a fatigue crack growth indicate that the initiation of fatigue cracks occurred in a material that did not possess a deviation in the manufacturing quality of the second-stage disk of the IPC. Nevertheless, the concentration of the crack initiation zone localized within the radius transition, the orientation of the initial fracture zone boundary along the surface of the radius transition, and the LCF corresponding fracture surface peculiarities are evidence of the high-stress intensity of the disk in the fracture initiation zone. To identify the nature of such stress intensity, the history of engine operation and overhaul was analyzed.

Analysis of data on the operation and overhaul of the engine, including the second-stage disk of IPC under investigation, showed the following. At the initial stage of engine operation, the first-stage disk was replaced, which could affect the stress state of the second-stage disk, since they are welded into one drum of the compressor. The replacement could not fully reproduce the stress-state conditions that were realized during the initial manufacturing of the drum with the second-stage disk of the IPC. Later, during the second overhaul, a cascade of fluorescence was observed in the second-stage disk after an operating time of 2858 h and 5 min (or 539 FC). Fluorescence was detected in the radius transition to the bottom of the dovetail slot along a length of 5 mm, as well as from the side of the entrance and an acute angle in many locations of the disk. This implies the certain relaxation processes of stress state in the second-stage disk, which was created in the disk by welding to the cascade of the first-stage disk of the IPC in the first overhaul.

During the overhaul, a thin layer of disk material was removed to eliminate the fluorescence (assuming material inhomogeneities). On the one hand, this led to the elimination of possible material inhomogeneities, but, on the other hand, it did not remove the total changed stress state of the disk caused by welding of the first-stage disk of the IPC.

The analysis of regularities of fatigue crack initiation and growth in the second-stage disk of IPC showed that in the region of crack initiation, the fatigue striation spacing was reached 0.6 μm (see Figure 12). This indicates a high-stress intensity of the disk precisely in the initial fracture zone, where the multi-focal origin of the fracture occurred. The subsequent fracture development occurred under varying disk stress state conditions that follow from the analysis of the regularity of the fatigue crack propagation. Despite the significant intensity of the initial crack growth stage, further fracture development took place with insignificant acceleration, resulting in a long crack growth duration. This fact implies that the disk stress intensity decreased in the direction of crack growth. The geometry of the initial fracture zone confirms the existence of the extremely high-stress state of the disk material in the zone that is oriented along the radius transition to the slot bottom within 5 mm. Then, the crack kinked, and its orientation became almost perpendicular to the radial force (see Figure 10), i.e., to the total stress state realized in the disk during in-service loading. Moreover, the analysis of the boundary between the fracture zone with the region of radius transition to the slot bottom showed that it is oriented along the machining scores of the slot surface. At a high-stress intensity of the material, the radius transition plays a key role in creating a stress concentration, rather than the scores along this surface that are insignificant in depth (<5 μm). Their main influence was revealed in the formation of the crack initiation zone along the scores on the radius transition surface; i.e., there was a slight increase in the stress concentration along the specified surface.

Thus, it can be concluded that the failure of the second-stage disk of the IPC of the gas turbine engine has resulted from its high-stress intensity in the region of the radius transition to the slot bottom. Most probably, the indicated stress state resulted from welding of the first-stage disk of IPC after an operating time of 1985 h and 20 min (342 FC). At the operating time of 2858 h and 5 min (539 FC), the initial origins of the multiple non-uniform states in the second-stage disk detected during non-destructive testing were removed in the overhaul. This led to a decrease in the stress intensity in the region of the radius transition zone to the bottom of the slot but did not eliminate it.

The recommended inspection intervals for a disk with a developing fatigue crack were evaluated in accordance with the existing standard [1]. It follows from the standard that when introducing periodic monitoring, it is necessary to calculate the duration of the operation period between two adjacent inspections $n_{\mathrm{insp}}$ based on the fractographic assessment of the crack growth duration $n_{\mathrm{p}}$ in terms of flight cycles by using the expression:

$$n_{\mathrm{insp}} = \frac{2}{3} \times \frac{n_{\mathrm{p}}}{K_{\mathrm{N}}}, \quad (1)$$

where the value of factor $K_{\mathrm{N}}$ is varied from 2–4. Only in special cases, with a statistically reliable substantiation of the crack propagation duration $n_{\mathrm{p}}$ based on several cases of disk failure, the value of the factor can be reduced up to 2. In the considered situation with a single case of disk failure, a statistically reliable estimation of the crack growth duration is impossible. Therefore, the inspection intervals for the second-stage disk of the IPC should be considered with the application of $K_{\mathrm{N}} = 4$ as a reduction factor, and the value of the inspection interval $n_{\mathrm{insp}} = (2/3) \times (900/4) = 150$ FC with an inspection method resolution of about 2 mm (see Figure 15). For an average flight time of 5 h and a flight frequency of once every three days, the duration of engine operation between inspections is of 750 h or 450 days, which guarantees a stable operation for a long period. With an increase in test-sensitivity, the inspection interval can be extended.

Thus, based on fractographic analysis of the regularities of fatigue crack growth, taking into account the effect of fatigue crack meso-tunneling, the inspection intervals for disks in service on the entire fleet of An-124 aircraft are established and substantiated, with ensuring the required level of flight safety, which are at least 150 flights. As information concerning the in-service inspections of disks, the crack detections and the estimation of the crack growth duration will be accumulated, the in-service duration between two adjacent inspections of the disk can be increased.

## 5. Conclusions

1. A VT3-1 titanium alloy with an inhomogeneous filament type texture obtained as a result of forging was studied in the LCF regime on specimens manufactured from compressor disks under loading conditions with triangular and trapezoidal cycle waveform. The considered structure of the material is formed at forging the compressor disks as a result of the wave process of plastic deformation, leading to the creation of alternating layers of metal with low and sufficiently high plasticity.
2. It is established that depending on the orientation of the fatigue crack propagation, a different fracture surface relief can be realized under loading conditions: in the case of crack growth perpendicular to the filaments, alternating regions of fatigue striations and quasi-brittle waved pattern are formed; in the case of crack growth along the material filaments, the process of crack meso-tunneling occurs.
3. Meso-tunneling of a crack is expressed in the initial fracture of the material along the brittle filaments with the formation of a facetted pattern relief on the fracture surface evidencing the two-phase structure of the titanium alloy, and then the bridges between the adjacent meso-tunnels are fractured with the formation of a block of fatigue striations.
4. The part of the fatigue crack growth duration $N_p/N_f$ is characterized by different dependences on the durability $N_f$ in the case of crack propagation along or across the filaments of a forged titanium alloy.
5. Based on the analysis of the fracture process of the titanium second-stage disk of the intermediate-pressure compressor of the engine, it is shown how the effect of a crack meso-tunneling determines the regularities of fatigue crack propagation under implemented in-service conditions.
6. The duration of the fatigue crack propagation was estimated by measuring the fatigue striation spacing, and the consideration of the requirement to introduce a reduction factor when recalculating the number of unit loading cycles of the disk into the number of flight loading cycles was discussed, and a recommendation on a duration of the inspection intervals for the disk in service was given as well.
7. The reason for the initiation of a fatigue crack in the compressor disk is considered on the basis of data on its operation and taking into account the welded structure of the rotor of the IPC.

**Author Contributions:** Conceptualization, A.A.S., A.P.S. and A.D.N.; Methodology, A.A.S. and A.D.N.; Formal Analysis, A.A.S., A.P.S. and A.D.N.; Investigation, A.A.S., A.P.S. and A.D.N.; Writing–Original Draft Preparation, A.A.S. and A.P.S.; Writing–Review and Editing, A.A.S. and A.D.N. All authors have read and agreed to the published version of the manuscript.

**Funding:** The work was supported by RSF, grant N19-19-00705.

**Institutional Review Board Statement:** Not applicable.

**Informed Consent Statement:** Not applicable.

**Data Availability Statement:** The data presented in this study are available on request from the corresponding author.

**Conflicts of Interest:** The authors declare no conflict of interest.

## References

1. Shanyavskiy, A. *Tolerance Fatigue Failures of Aircraft Components. Synergetics in Engineering Applications*; Monografy: Ufa, Russia, 2003. (In Russian)
2. Williams, J.C.; Starke, E.A., Jr. Progress in Structural Materials for Aerospace Systems. *Acta Mater* **2003**, *51*, 5775–5799. [CrossRef]
3. McEvily, A.J. Failures in inspection procedures: case studies. *Eng. Fail. Anal.* **2004**, *11*, 167–176. [CrossRef]
4. Shanyavskiy, A.A.; Stepanov, N.V. Fractographic Analysis of Fatigue Crack Growth in Engine Compressor Disks of Ti-6Al-3Mo-2Cr Titanium Alloy. *Fatigue Fract. Engng Mater. Struct.* **1995**, *18*, 539–550. [CrossRef]
5. Shanyavskiy, A.A.; Losev, A.I. Fatigue Crack Growth in Aroengine Compressor Disks Made from Titanium Alloy. *Fatigue Fract. Engng. Mater. Struct.* **1999**, *22*, 949–966.

6. Shanyavskiy, A.A.; Losev, A.I.; Banov, M.D. Development of Fatigue Cracking in Aircraft Engine Compressor Disks of Titanium Alloy Ti-6Al-3Mo-2Cr. *Fatigue Fract. Eng. Mater. Struct.* **1998**, *21*, 297–313. [CrossRef]
7. Wanhill, R.J.H.; Oldersma, A. Fatigue and Fracture in an Aircraft Engine Pylon. In *Engineering Against Fatigue, Sheffield, UK, 17–21 March 1997*; Beynon, J.H., Brown, M.W., Lindley, T.C., Smith, R.A., Tomkins, B., Eds.; A.A. Balkema: Rotterdam, the Netherlands, 1999; pp. 721–727.
8. Howard, I.C. Fracture of an Aircraft Horizontal Stabilizer. In *Case Histories Involving Fatigue and Fracture Mechanics, Charleston, SC, USA, 21–22 March 1985*; Hudson, C., Rich, T., Eds.; ASTM International: West Conshohocken, PA, USA, 1986; pp. 259–276.
9. Pilchak, A.L.; Williams, J.C. Observations of Facet Formation in Near-α Titanium and Comments on the Role of Hydrogen. *Metall. Mater. Trans. A* **2010**, *42*, 1000–1027. [CrossRef]
10. Kocańda, D.; Kocańda, S.; Tomaszek, H. Probabilistic Description of Fatigue Crack Growth in a Titanium Alloy Notched Member. In *Notch Effects in Fatigue and Fracture. NATO Science Series II: Mathematics, Physics and Chemistry*; Pluvinage, G., Gjonaj, M., Eds.; Springer: Dordrecht, The Netherlands, 2001; pp. 239–255.
11. Pilchak, A.L. Fatigue Crack Growth Rates in Alpha Titanium: Faceted vs. Striation Growth. *Scr. Mater.* **2013**, *68*, 277–280. [CrossRef]
12. Shanyavskiy, A.A.; Losev, A.I. Synergistic Problem of Introduction of Tolerance Damage Service of Titanium Disks of Aircraft Engines. In *Progress in Mechanical Behaviour of Materials. Vol. III: Advanced Materials and Modelling of Mechanical Behaviour, Proceedings of the Eighth International Conference on the Mechanical Behaviour of Materials (ICM8), Victoria, BC, Canada, 16–21 May 1999*; Ellyin, F., Provan, J.W., Eds.; Fleming Printing Ltd.: Victoria, BC, Canada, 1999; pp. 1227–1232.
13. Woodfield, A.P.; Gorman, M.D.; Corderman, R.R.; Sutliff, J.A. Effect of Microstructure on Dwell Fatigue Behavior of Ti-6242. In Proceedings of the Eighth World Conference on Titanium (Titanium '95: Science and Technology), Birmingham, UK, 22–26 October 1995; Blenkinsop, P.A., Evans, W.J., Flower, H.M., Eds.; Institute of Materials: London, UK, 1996; pp. 1116–1123.
14. Shanyavskiy, A.A. The Effects of Loading Waveform and Microstructure on the Fatigue Response of Ti–6Al–2Sn–4Zn–2Mo Alloy. *Fatigue Fract. Eng. Mater. Struct.* **2005**, *28*, 195–204. [CrossRef]
15. *ASTM E606 / E606M-19e1, Standard Test Method for Strain-Controlled Fatigue Testing*; ASTM International: West Conshohocken, PA, USA, 2019.
16. Shanyavskiy, A.A.; Burchenkova, L.M. Mechanism for Fatigue Striations as Formed under Variable Negative R-ratio in Al-based Structural Alloys. *Int. J. Fatigue* **2013**, *50*, 47–56. [CrossRef]
17. Jha, S.K.; Larsen, J.M. Random Heterogeneity Scales and Probabilistic Description of the Long-Lifetime Regime of Fatigue. In Proceedings of the Fourth International Conference on Very High Cycle Fatigue (VHCF-4), University of Michigan, Ann Arbor, MI, USA, 19–22 August 2017; Allison, J.E., Jones, J.W., Larsen, J.M., Ritchie, R.O., Eds.; TMS: Warrendale, PA, USA, 2007; pp. 385–396.
18. Panin, V.E.; Egorushkin, V.E.; Panin, A.V. The Plastic Shear Channeling Effect and the Nonlinear Waves of Localized Plastic Deformation and Fracture. *Phys. Mesomech.* **2010**, *13*, 215–232. [CrossRef]
19. De Los Rios, E.; Kandi, A.; Miller, K.J.; Brown, M.W. A Metallographic Study of Multiaxial Creep-fatigue Behavior in 316 Stainless Steel. In *Multiaxial Fatigue. ASTM STP 853*; Miller, K., Brown, M., Eds.; ASTM International: West Conshohocken, PA, USA, 1985; pp. 669–687.
20. Miller, K. *Creep and Fracture*; Metallurgy: Moscow, Russia, 1986.
21. Couper, M.J.; Neeson, A.E.; Griffiths, J.R. Casting Defects and the Fatigue Behaviour of an Aluminium Casting Alloy. *Fatigue Fract. Eng. Mater. Struct.* **1990**, *13*, 213–227. [CrossRef]
22. Takahashi, I.; Yoshii, T.; Iidaka, H.; Fujii, E.; Matsuoka, K. Fatigue Strength of Non-load-carrying Fillet Welded Joints: Effects of Weld Residual Stresses and Stress Concentration. *Fatigue Fract. Eng. Mater. Struct.* **1993**, *16*, 37–51. [CrossRef]

materials

MDPI

Article

# Usability of Ultrasonic Frequency Testing for Rapid Generation of High and Very High Cycle Fatigue Data

**Michael Fitzka [1], Bernd M. Schönbauer [1], Robert K. Rhein [2], Niloofar Sanaei [2], Shahab Zekriardehani [2], Srinivasan Arjun Tekalur [2], Jason W. Carroll [2] and Herwig Mayer [1,*]**

1 Department of Material Sciences and Process Engineering, Institute of Physics and Materials Science, University of Natural Resources and Life Sciences (BOKU), 1190 Vienna, Austria; michael.fitzka@boku.ac.at (M.F.); bernd.schoenbauer@boku.ac.at (B.M.S.)

2 Center for Materials and Manufacturing, Eaton Corporation, Southfield, MI 48076, USA; RobKRhein@eaton.com (R.K.R.); NiloofarSanaei@eaton.com (N.S.); ShahabZekriardehani@eaton.com (S.Z.); SrinivasanArjunTekalur@Eaton.com (S.A.T.); JasonWCarroll@eaton.com (J.W.C.)

* Correspondence: herwig.mayer@boku.ac.at

**Abstract:** Ultrasonic fatigue testing is an increasingly used method to study the high cycle fatigue (HCF) and very high cycle fatigue (VHCF) properties of materials. Specimens are cycled at an ultrasonic frequency, which leads to a drastic reduction of testing times. This work focused on summarising the current understanding, based on literature data and original work, whether and how fatigue properties measured with ultrasonic and conventional equipment are comparable. Aluminium alloys are not strain-rate sensitive. A weaker influence of air humidity at ultrasonic frequencies may lead to prolonged lifetimes in some alloys, and tests in high humidity or distilled water can better approximate environmental conditions at low frequencies. High-strength steels are insensitive to the cycling frequency. Strain rate sensitivity of ferrite causes prolonged lifetimes in those steels that show crack initiation in the ferritic phase. Austenitic stainless steels are less prone to frequency effects. Fatigue properties of titanium alloys and nickel alloys are insensitive to testing frequency. Limited data for magnesium alloys and graphite suggest no frequency influence. Ultrasonic fatigue tests of a glass fibre-reinforced polymer delivered comparable lifetimes to servo-hydraulic tests, suggesting that high-frequency testing is, in principle, applicable to fibre-reinforced polymer composites. The use of equipment with closed-loop control of vibration amplitude and resonance frequency is strongly advised since this guarantees high accuracy and reproducibility of ultrasonic tests. Pulsed loading and appropriate cooling serve to avoid specimen heating.

**Keywords:** ultrasonic fatigue; frequency effect; strain rate effect; environmental effect; size effect; high cycle fatigue; very high cycle fatigue

**Citation:** Fitzka, M.; Schönbauer, B.M.; Rhein, R.K.; Sanaei, N.; Zekriardehani, S.; Tekalur, S.A.; Carroll, J.W.; Mayer, H. Usability of Ultrasonic Frequency Testing for Rapid Generation of High and Very High Cycle Fatigue Data. *Materials* **2021**, *14*, 2245. https://doi.org/10.3390/ma14092245

Academic Editor: Jaroslav Pokluda

Received: 31 March 2021
Accepted: 23 April 2021
Published: 27 April 2021



## 1. Introduction

Ultrasonic fatigue testing is a powerful method to investigate the fatigue properties of materials in high cycle fatigue (HCF) and very high cycle fatigue (VHCF) regimes. Specimens are stimulated by resonance vibrations at a frequency close to 20,000 Hz rather than being stressed by external forces as in conventional (e.g., servo-hydraulic, rotating bending, resonance tester) fatigue tests. The high cycling frequency shortens testing times and makes fatigue investigations at extremely high numbers of cycles possible, in a regime that is hardly accessible with conventional testing due to excessive testing times. Thus, ultrasonic fatigue testing is the most appropriate technique to study fatigue properties of materials in the VHCF regime [1].

One of the main questions involved with ultrasonic testing is whether the measured fatigue data would be similarly obtained using conventional testing methods working at much lower frequencies [2–5]. Fatigue tests with servo-hydraulic test frames, resonance testing apparatus or rotating bending equipment are typically performed at cycling frequencies below 100 Hz. Moreover, technical components are typically stressed at cycling frequencies far below the ultrasonic range, so that cycling frequencies in conventional tests are much more comparable to those prevailing in actual applications. It is important for high-frequency testing to consider possible frequency effects on the progress of fatigue damage and the measured fatigue lifetimes. Additionally, there are generally accepted standards for how to perform servo-hydraulic tests (i.e., measurement and control of the applied forces) or rotating bending tests (i.e., application of a defined bending moment). Although there have been individual attempts (e.g., WES 1112 [6] requiring control of displacement amplitude), a similar degree of standardisation is not available for ultrasonic tests. Different setups for ultrasonic fatigue testing are being used, which influence the accuracy and replicability of tests and the measured data.

This paper gives an overview of the existing knowledge on the applicability of ultrasonic fatigue testing for material qualification and collection of fatigue data based on the experimental data generated by the authors, as well as data gathered from the literature. An ultrasonic testing setup is described that enables reproducible and high-precision fatigue tests, avoids specimen heating, and controls, monitors and logs specimen loading. It was shown that extrinsic influences, such as size effects can influence the measured cyclic properties. The effect of a high strain rate on plastic deformation and crack initiation [7,8] and the time-dependent influence of the testing environment [9,10] are considered as the two, in principle, possible influences of the testing frequency on fatigue damage.

Fatigue performance of aluminium alloys, steels and titanium alloys as well as nickel alloys, magnesium alloys, fibre-reinforced polymer composites and graphite measured with conventional and ultrasonic equipment were compared based on the data available in the literature. Moreover, original fatigue data generated by the authors on a Ti6Al4V and 2024-T351 aluminium alloy by both conventional and ultrasonic fatigue were used in this study for discussion and comparison purposes. The present work aims to review and compare ultrasonic and conventional fatigue data, discuss the potential variations between these results, and offer insights into how these data correlate for each material.

## 2. Materials and Methods

### 2.1. Ultrasonic Fatigue Testing Procedure

In an ultrasonic test, an appropriately designed specimen is stimulated to resonance vibrations at close to 20 kHz. The displacement movement of the specimen is sinusoidal with time with maximum vibration amplitudes at both ends of the specimen. Cyclic loading is determined by the amplitude and frequency of the ultrasonic vibration. Strain amplitudes are highest in the centre of the specimen, where a vibration node is formed. Strain gauges are commonly used to measure the cyclic strain amplitude and to calibrate the ultrasonic fatigue test. Alternatively, several non-contact methods to measure strains at high frequencies have been employed in the past (e.g., optical interferometry based strain/displacement gauge [11], or laser vibrometry [12]). Particularly high temperatures may limit the applicability of strain gauges. They are also not suitable to be used for the actual control of the strain amplitude during a test. Cyclic stresses (or cyclic forces) cannot be measured directly but they can be calculated from the measured strains. In the high and very high cycle fatigue range, where ultrasonic fatigue testing is used, many materials show approximately linear elastic behaviour. Then, cyclic stresses, $\sigma$ are calculated from the measured strains, $\varepsilon$ with Young's modulus, $E$ and using Hooke's law (Equation (1)).

$$\sigma = E \times \varepsilon \quad (1)$$

No widely accepted standard exists for how to perform ultrasonic fatigue tests. A successfully used method to control ultrasonic loading is to measure the vibration amplitude and to control it in a closed-loop [5]. The vibration amplitude of the specimen's ends and strain amplitudes in the specimen's centre are directly proportional. Therefore, the measured vibration amplitude is a suitable signal to control cyclic loading. Vibration amplitude can, for example, be measured with an induction coil, with a laser vibrometer or with a gap sensor. Strain gauges attached to the specimen's centre also deliver a signal proportional to the cyclic load. However, strain gauges will fatigue and fracture during the high strains and high numbers of cycles that occur during an ultrasonic test and, therefore, are not suitable for control purposes.

By using the measured displacement signal in a closed-loop circuit, the vibration amplitude can be controlled with high accuracy. The maximum deviation between selected and actually realised vibration amplitudes can be limited to $\pm 1\%$. In a second closed-loop circuit, the stimulation frequency is controlled and kept very close to the resonance frequency of the load train within $\pm 1$ Hz. Control of stimulation frequency is necessary, since the actual resonance frequency of the system can change during the test due to slight variations of temperature or due to the initiation of a crack, for example. The vibration amplitude can be measured and the magnitude and number of cycles applied can be stored using a computer unit for backtracking the specimen's loading after fracture [5].

The measured fatigue data is strongly influenced by the chosen method of ultrasonic fatigue test. One of the most basic problems related to ultrasonic fatigue cycling is the heat generated in the specimen due to internal friction. Cyclic plastic deformation as well as the movement of interstitial atoms (i.e., Snoek effect in carbon steel) leads to heat dissipation and can increase the specimen's temperature [13,14]. If specimens are loaded continuously at ultrasonic frequencies, their temperature may increase beyond acceptable values. For about 40 years, intermittent loading with periodic sequences of pulses and cooling pauses has been known as the appropriate method to avoid specimen heating [15]. Additionally, the specimen may be cooled with forced air. Both intermittent loading and forced air-cooling are appropriate methods to limit temperature increase by a maximum of 5 °C to 10 °C above room temperature during tests in ambient air, for example. Specimen temperature can be monitored with thermocouple or contact-free infrared thermography.

Intermittent loading requires precise closed-loop control of the vibration amplitude. Otherwise, overshoot of the amplitude at the beginning of the pulse can lead to undefined specimen loading and inaccurate tests. The rapid increase of the nominal vibration amplitude without overshoot cannot be guaranteed by all ultrasonic fatigue testing systems. Some simpler testing systems do not employ feedback control of cyclic displacement, but rather rely on controlling power output to the ultrasonic converter, which yields inferior loading amplitude accuracy. Measured ultrasonic fatigue data are influenced by the way the test is performed, the accuracy of specimen loading and the method of data evaluation. Therefore, differences between ultrasonic and conventional fatigue tests may not only be the result of intrinsic frequency influences but may also be a consequence of an improper experimental setup and an unsuitable testing procedure.

### 2.2. Servo-Hydraulic Fatigue Testing Procedure

In contrast to ultrasonic fatigue tests, testing with servo-hydraulic equipment is highly standardised regarding testing procedure, closed-loop control of static and dynamic force, data-acquisition systems and accuracy requirements. The choice of specimen geometry is not limited by the resonance criterion, also enabling investigations with specimens with large testing volumes and component testing.

Original fatigue data generated by the authors on Ti6Al4V and 2024-T351 aluminium alloys were used in this study. These tests were performed using MTS 810.10 servo-hydraulic fatigue testing equipment (MTS Systems Co., Eden Prairie, MN, USA) with a load capacity of 50 kN. The maximum testing frequency was $\leq 70$ Hz. Tests were performed in force-control mode with active peak-value compensation to guarantee at least $\pm 1\%$ peak value accuracy.

### 2.3. Specimen Geometry

Specimen geometries are constrained mainly by the resonance criterion at approximately 20 kHz for ultrasonic fatigue tests, the self-heating properties of the investigated material, the maximum power, and achievable displacement amplitude of the used setup. Specimen geometries can be chosen more freely in conventional fatigue tests. However, if the comparability between conventional and ultrasonic fatigue tests is investigated, it is important to have a comparable material loaded under comparable conditions.

It is necessary that the loaded volume of specimens used in both testing series is identical. If a significantly larger testing volume would be used for conventional tests, for example, lifetimes would be shifted to lower values due to the size effect. It is, therefore, sensible to use the same gauge section geometry in low-frequency and ultrasonic tests to make the results comparable.

Besides the influence of testing volume, the stress distribution in the specimens must be considered. The specimen geometry can affect the stress distribution along the specimen's length and over the load-bearing cross-section for conventional and ultrasonic resonance loading differently. Geometries featuring rectangular cross-sections, for example, exhibit a distinct stress profile across their cross-sections under resonance loading, where strain amplitudes can vary by as much as 10%. It is, therefore, necessary not only to exclude a size effect, but also to guarantee comparable stress distributions over the loaded cross-sections for low and ultrasonic tests.

For the present study, authors tested fatigue performance of Ti6Al4V and 2024-T351 aluminium alloys using the same specimen geometry, as shown in Figure 1, by both servo-hydraulic and ultrasonic tests. The strain distribution along the specimen's length was calculated with AUTODESK Inventor Nastran 2021 with the *Normal Modes* and the *Linear Static* simulation types, respectively. A sufficiently high mesh resolution of 0.2 mm was chosen over the gauge section. Figure 1 shows that the cylindrical gauge section of 10 mm length experiences almost uniform strain (variation of strain amplitude is less than 1% in the modal, and 0% in the static loading case, respectively). Additionally, the stress distribution over the load-bearing cross-section (not shown) identifies less than 1% deviation from nominal stress for both loading conditions.

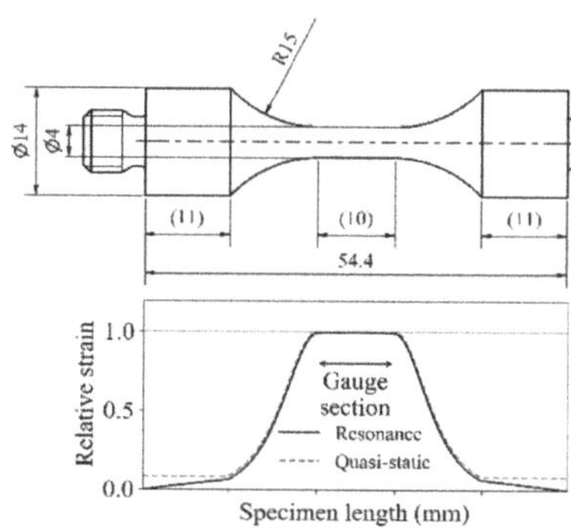


**Figure 1.** Specimen shape used for ultrasonic and servo-hydraulic fatigue tests (all dimensions in mm); strain distribution along the specimen's length under ultrasonic resonance loading (solid line) and quasi-static (servo-hydraulic) loading (dashed line) is shown.

### 2.4. Materials

Ti6Al4V and 2024-T351 aluminium alloys were tested with displacement-controlled ultrasonic fatigue testing equipment in the range 19–20 kHz and with force-controlled servo-hydraulic or resonance fatigue tests at frequencies below 100 Hz.

The age-hardened aluminium alloy 2024 was tested under T351 conditions (i.e., solution annealed, water quenched, cold worked and age hardened). The chemical composition of the material was (in% mass): Cu 4.46, Mg 1.42, Mn 0.61, Zn 0.18, Si 0.06, Fe 0.14, Cr 0.005, Ti 0.025, Pb 0.0020, Ni 0.0058, Sn 0.055 and Al balance. The material was obtained in rolled plates. Mechanical properties are as follows: tensile strength 473 ± 3 MPa, 0.2% proof stress 364 ± 4 MPa, elongation 18 ± 1% and Vickers hardness 141 ± 2 HV.

Dumbbell-shaped specimens with cylindrical gauge sections were machined with their longitudinal dimension aligned to the rolling direction of the plates. The gauge sections were ground parallel to the specimen's longitudinal dimension with abrasive paper of grade 1000 to obtain well-defined surface conditions. The same specimen shape was used for fatigue testing with ultrasonic and servo-hydraulic testing equipment. Specimen shape and strain distribution along the specimen's length under servo-hydraulic and ultrasonic fatigue loading are shown in Figure 1.

The titanium alloy Ti6Al4V was tested in mill annealed condition. The chemical composition was (in% mass): Al 6.2, V 4.1, Fe 0.06, O 0.17, N 0.004, H 0.0013, C 0.014, Y < 0.001, other impurities total <0.04, Ti balance. Mechanical properties of the material are as follows: tensile strength 993 MPa, 0.2% proof stress 965 MPa, elongation 21% and Vickers hardness 331 HV.

Specimens were manufactured from rods with a diameter of 15 mm. The same specimen shape was used for testing Ti6Al4V as well as for 2024-T351 (Figure 1). The surface in the gauge section is ground parallel to the specimen's longitudinal dimension with elastic grinding wheels of grade 150 (approximately equivalent to abrasive paper grade 1000). The same specimen geometry with the same surface preparation was used for servo-hydraulic and ultrasonic tests.

Fatigue tests were performed with thin sheets made of 18Ni maraging steel with nitrided surfaces [16]. The chemical composition was (in% mass): Ni 18, Co 5, Mo 5, Al 1, Cr 1 and Fe (balance). Sheet specimens were manufactured from coil material in solution-annealed condition, followed by precipitation hardening (480 °C, 2.5 h) and gas nitriding. The material showed a tensile strength of 2000 MPa and a yield strength of 1800 MPa. Vickers hardness of the nitrided surface layer was 1000 HV and 560 HV in the core of the sheet.

Two different specimen shapes were used in the tests, which are named large specimens and small specimens in the following [16]: The thickness of both specimens was 0.435 mm. Large specimens had constant cross-sections in the centre with a width of 14 mm and a length of 20 mm with a 170 mm shoulder fillet radius. The small specimens had a width of 2.4 mm in the centre. The width increased towards both ends with a radius of 20 mm. The maximum stress in small specimens was solely in the centre, whereas it was constant over a length of 20 mm in large specimens. Consequently, the stressed volumes were different. The material volume subjected to more than 95% of the nominal stress was 3.4 mm$^3$ for the small specimens and 260 mm$^3$ for the large specimens i.e., the testing volume was larger by a factor of 76 for the large specimens.

## 3. Conventional and Ultrasonic Fatigue Testing

### 3.1. Aluminium Alloys

Most aluminium alloys do not show a fatigue limit and they fail even at very high numbers of cycles [17]. However, for some cast aluminium alloys, a fatigue limit is observed. At stress amplitudes below the fatigue limit, fatigue cracks can initiate at casting porosity in these materials; however, the cracks are non-propagating and do not lead to fracture even if cycled to the VHCF regime [18,19]. In automotive applications, engine components, wheels and chassis components may be stressed with several hundred million cycles during service. The fatigue behaviour in the VHCF regime cannot be reliably extrapolated from HCF data, and accelerated testing techniques are therefore of great interest. Different types of aluminium-based alloys have already been investigated with ultrasonic fatigue testing, including wrought alloys, cast alloys, metal matrix composites, foams, laminates and additively manufactured (AM) alloys. Due to the low damping properties of most aluminium alloys, heat generation is low and fatigue tests can be performed with high effective frequency (i.e., test frequency considering that the specimen is not loaded continuously but in pulsed mode). Therefore, ultrasonic fatigue investigations up to $10^9$ cycles or beyond are possible within one or a few days.

Fatigue lifetimes for 2024-T351 in the HCF and VHCF regimes have been the subject of several investigations in the author's laboratory, including constant and variable amplitude loading at different load ratios at low and ultrasonic frequencies [20–22] as well as ultrasonic cyclic torsion tests [23]. Fatigue performance of 2024-T351 specimens tested at $R = 0.1$ and 0.5 load ratios with both displacement-controlled ultrasonic testing equipment and with force-controlled servo-hydraulic testing equipment, is shown in Figure 2. Specimens with the shape shown in Figure 1 from the same batch and with the same surface preparation were used in both testing series. Tests with servo-hydraulic equipment were performed at cycling frequencies between 8 Hz and 70 Hz and ultrasonic tests at 20 kHz. Mean lifetimes measured at 20 kHz at $R = 0.1$ were $1.7 \times 10^5$ cycles at 145 MPa, $6.7 \times 10^5$ cycles at 120 MPa, and $1.1 \times 10^7$ cycles at 100 MPa. In servo-hydraulic tests, mean lifetimes measured at these stress amplitudes were $1.6 \times 10^5$ cycles, $4.8 \times 10^5$ cycles and $1.0 \times 10^7$ cycles. At load ratio 0.5 and stress amplitude 80 MPa, the mean lifetime is $1.2 \times 10^7$ cycles at low and $2.2 \times 10^7$ cycles at ultrasonic frequencies.

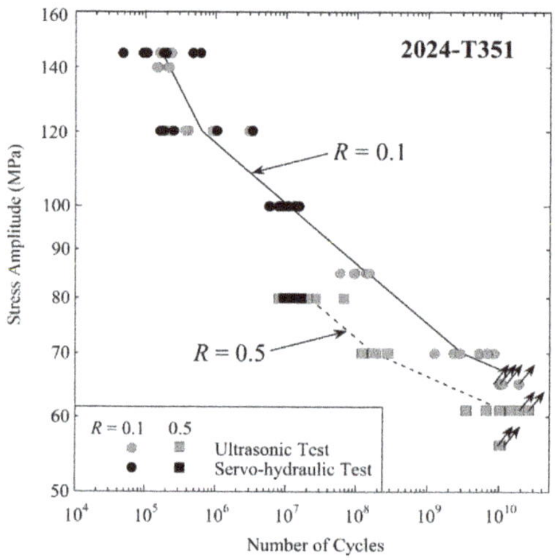


**Figure 2.** Fatigue data of aluminium alloy 2024-T351 measured at load ratio $R = 0.1$ (circles) and $R = 0.5$ (squares) with servo-hydraulic equipment at cycling frequencies between 8 Hz and 70 Hz (black symbols) and with ultrasonic equipment at 20 kHz (grey symbols).

Considering all four testing series, the mean fatigue lifetimes were a factor 1.3 higher in ultrasonic than in servo-hydraulic tests. However, this must be compared to the scatter of fatigue lifetimes. Assuming a log-normal distribution of cycles to failure, lifetimes for 90% failure probability were factor 4 higher than those for 10% fracture probability for ultrasonic and servo-hydraulic tests. Thus, the deviation of mean lifetimes measured in ultrasonic and servo-hydraulic tests was much smaller than the scatter of lifetimes, and no statistically significant deviation of lifetimes was found.

Fracture surfaces at the places of fatigue crack initiation after fracturing at 145 MPa in the ultrasonic test and in the servo-hydraulic test are shown in Figure 3a,b, respectively. Cycles to failure were $9.1 \times 10^4$ cycles in the test at low frequency and $1.8 \times 10^5$ cycles in the ultrasonic frequency test. Images were obtained in backscatter electron mode to improve material contrast. Both figures show crack initiation at fractured secondary phase particles. Energy-dispersive X-ray analysis suggests that these are $Al_7Cu_2$(Fe, Mn) particles. Crack growth close to the particle was trans-granular, mainly quasi-cleavage fracture with a brittle appearance of the fracture surfaces. No obvious difference was found for crack initiation at low and ultrasonic frequencies indicating that the mechanism of fatigue damage is not influenced by the drastic increase of cycling frequency.

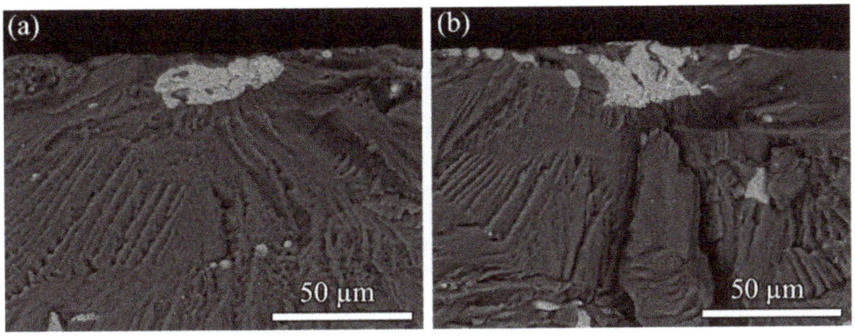


**Figure 3.** Fatigue crack initiation in 2024-T351 at a fractured secondary phase particle, probably $Al_7Cu_2$(Fe, Mn), cycling at stress amplitude 145 MPa (**a**) in ultrasonic test ($1.8 \times 10^5$ cycles to failure) and (**b**) in servo-hydraulic test ($9.1 \times 10^4$ cycles to failure).

A wrought aluminium alloy similar to 7075 (i.e., alloy AlZnMgCu1.5) was tested with displacement controlled ultrasonic testing equipment and with a resonant testing machine working at 100 Hz [24]. Fatigue data in the regime from $10^5$ to $10^7$ cycles were measured at both frequencies, and comparable lifetimes were found.

Investigation of the fatigue performance of cast aluminium alloy E319-T7 with ultrasonic and servo-hydraulic equipment by Zhu et al. [9] delivered a different result. The measurements were also performed with displacement controlled ultrasonic fatigue testing equipment, and specimens with comparable geometry and stress distribution were used in tests at 20 kHz and 75 Hz.

The tests were performed at 20 °C, 150 °C and 250 °C, and results obtained at room temperature are shown in Figure 4. The mean number of cycles to failure was increased by a factor of 5 to 10 in ultrasonic tests at all three temperatures. The extended lifetimes in ultrasonic tests are explained by the environmental influence of air humidity, which is weaker at ultrasonic than at low cycling frequency. Fatigue loading leads to early crack initiation at porosity in the cast alloy. Water vapour of ambient air diffuses to the crack tip and causes material embrittlement. At ultrasonic frequencies, the crack is not open long enough for full coverage. Thus, the chemical processes caused by water vapour act more severely in the servo-hydraulic than in the ultrasonic test, and lifetimes are prolonged. Performing ultrasonic fatigue tests in distilled water significantly decreased the lifetimes and led to an *S-N* curve close to the *S-N* curve at 75 Hz in ambient air [10].

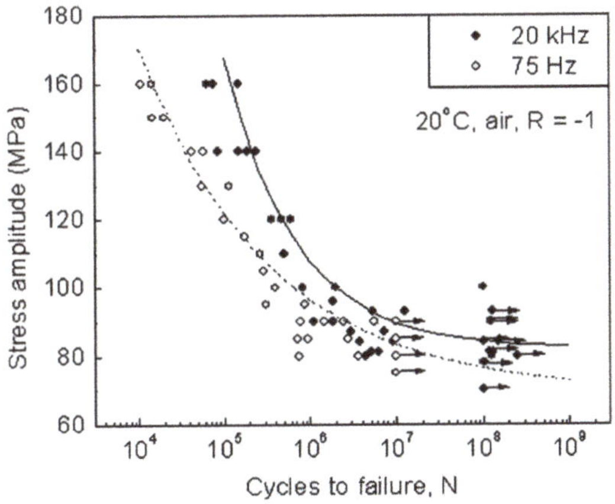


**Figure 4.** Fatigue data of cast aluminium E319-T7 with servo-hydraulic equipment at 75 Hz and with ultrasonic equipment at 20 kHz; reprinted with permission from Springer Nature [9].

Caton et al. [19] investigated cyclic properties of W319-T7 cast aluminium alloy using displacement controlled ultrasonic equipment and compared these results with servo-hydraulic tests. Microstructures with different secondary dendrite arm spacings were studied. In the range of $10^5$ to $10^7$ cycles, fatigue data were produced at 40 Hz and 20 kHz cycling frequency and no frequency influence on fatigue lifetime was found. An interesting result, which could have solely been found with ultrasonic fatigue testing, was the demonstration of a fatigue limit in this cast aluminium.

The fatigue properties of Al1Mg0.6Si foam cylinders were tested at 1–10 Hz with a servo-hydraulic testing machine in the low and high cycle fatigue regime, and at 20 kHz using ultrasonic equipment in the HCF and VHCF regimes [25]. Fatigue date in the overlapping regime showed no significant influence of frequency and testing procedure. The foam shows an endurance limit, where cracks initiated internally in the closed-cell structure, at holes or pre-existing cracks, but these cracks did not propagate to fracture.

The powder metallurgical aluminium–silicon alloy DISPAL® S232-T6x was tested with displacement controlled ultrasonic equipment at the authors' laboratory at 20 °C and 150 °C [26]. In parallel, servo-hydraulic fatigue tests were performed at IWK Aachen, Germany. In the HCF regime, where fatigue lifetimes were measured with both methods, no influence of the cycling frequency was found at room or elevated temperature. Additional ultrasonic fatigue tests in distilled water delivered similar lifetimes to the ultrasonic tests in ambient air.

Additive manufactured (AM) AlSi10Mg produced by selective laser melting were studied by Awd et al. [27] with servo-hydraulic equipment at 20 Hz and with ultrasonic equipment at 20 kHz. They concluded that the influence of testing at ultrasonic frequencies on the measured fatigue data was not significant. Pores were found as preferential crack initiation locations at low and high cycling frequencies. Due to the possibility for rapid generation of HCF and VHCF data, the ultrasonic method is a powerful method to study AM aluminium alloys and was used to rate different alloys [28] and to investigate the influence of process parameters on cyclic strength [29].

Investigations of aluminium alloys show comparable lifetimes in ultrasonic and servo-hydraulic tests for some alloys, whereas others show prolonged lifetimes at high frequencies. Face-centred cubic (fcc) lattice structures are relatively insensitive to strain rate influences [7] which makes strain rate influences in aluminium alloys improbable. This is supported by fatigue crack growth investigations in a vacuum, where similar growth rates and threshold stress intensity factors were found at 20 Hz and 20 kHz [30,31]. In ambient air, humidity causes chemical processes at the crack tip that accelerate crack growth and deteriorate the cyclic properties. The ratio of water vapour partial pressure and cycling frequency, the $p/f$ determines the diffusion rate of water vapour to the crack tip and consequently its deleterious effect. The diffusion rate is too low to allow full coverage of newly created surfaces with aluminium-hydroxide during an ultrasonic cycle [9,10], whereas this is possible for ambient air conditions and conventional cycling frequencies [32,33]. Increasing the water vapour pressure in ultrasonic experiments decreases the difference measured in low and ultrasonic frequency tests. Water partial pressure can be increased by a factor of about 1000, cycling the specimen in (distilled) water instead of humid air, i.e., the $p/f$ for cycling a specimen at ultrasonic frequencies in water is in the same regime as the $p/f$ for cycling a specimen at low frequency in humid air [9,34]. Experiments in water can, therefore, be used to assess whether an alloy is sensitive to air humidity, and frequency effects are possible or if the alloy is insensitive and ultrasonic tests and servo-hydraulic tests deliver comparable lifetimes.

### 3.2. Steels

The failure of high-strength steels beyond the conventional fatigue limit at $10^7$ cycles was one of the main findings that triggered interest in the very high cycle fatigue (VHCF) regime [35–37]. Fatigue cracks preferentially initiate at the surface in the HCF regime whereas internal inclusions are a starting point for cracks leading to VHCF failure. In the latter case, the fracture surface in the close vicinity of the crack initiating inclusion shows a typical rough appearance, referred to as the optically dark area (ODA) [38,39] or fine granular area (FGA) [40] in tests performed at negative and slightly positive load ratios (i.e., $R = 0.1$). An example of VHCF failure from an interior non-metallic inclusion with a distinct FGA is shown in Figure 5. Crack propagation rates inside the FGA are two or more decades below one Burgers vector per cycle and nearly exclusively determine VHCF lifetime, and failures can be found up to extremely high numbers of cycles such as $>2 \times 10^{10}$ cycles [41]. Since such high numbers of cycles can only be reached with a high-frequency testing technique, ultrasonic fatigue has become a widely used testing technique for VHCF investigations of high-strength steels.

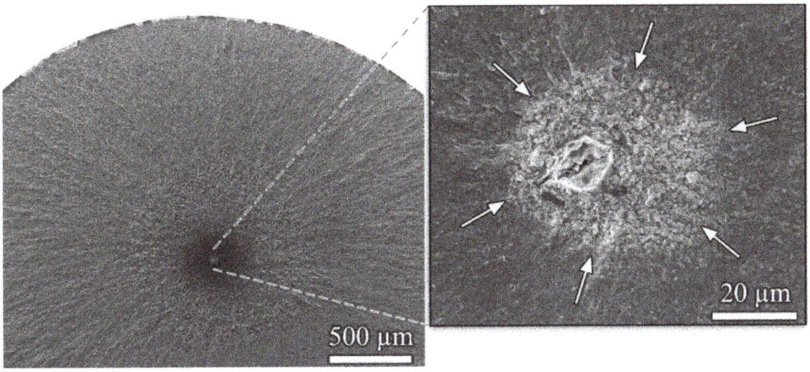


**Figure 5.** Fatigue crack initiation in 100Cr6 bearing steel at an interior non-metallic inclusion after ultrasonic fatigue testing at $\sigma_a = 875$ MPa, $R = -1$, $N_f = 2.4 \times 10^9$ cycles; the border of FGA is marked with arrows.

An investigation into frequency influences on fatigue properties of high-strength steels was performed with 18Ni maraging steel [16]. Figure 6 shows fatigue data of maraging steel sheet specimens obtained in three testing series: large specimens (testing volume 260 mm$^3$) tested at 95 Hz, small specimens (testing volume 3.4 mm$^3$) tested at 95 Hz, and small specimens tested at 20 kHz.

The result of the investigation shows that there is a strong size effect in this maraging steel. Mean lifetimes measured with large specimens are about one order of magnitude lower than mean lifetimes of small specimens. This is found when cycling the small specimen at ultrasonic frequencies (Figure 6a) or cycling both specimens at the same low frequencies (Figure 6b). In contrast, when cycling small specimens at low and ultrasonic frequencies, the difference in mean lifetimes is within the range of scatter (Figure 6c). The mean size of inclusions in large specimens is 9.9 µm, whereas it is 4.4 µm in small specimens. For the fatigue crack initiating at an internal inclusion, this means that the starting crack length was longer in large specimens, which shortened the crack propagation period and decreased the fatigue lifetime.

Cracks initiated exclusively at aluminates in large specimens. Crack initiation was at aluminate and Zr(N,C)-inclusions in small specimens for cycling with both 95 Hz and 20 kHz. Fracture of the interface between aluminates and matrix and particle fracture of the (smaller) Zr(N,C)-inclusions are the two crack initiating mechanisms. Due to the small highly stressed volume in small specimens, about half of the samples do not contain aluminate inclusions large enough to initiate a crack, and failure is caused by the smaller Zr(N,C) inclusions.

Japanese Researchers from the National Institute of Materials Science performed ultrasonic fatigue and multiple-axis, cantilever-type, rotating bending fatigue testing [42], up to $10^{10}$ cycles which took more than 3 years at 100 Hz. No frequency effects were found for the two investigated high-strength steels (spring steel SUP7 and low-alloy steel

SCM440) as shown in Figure 7. The same research group further demonstrated that the size of the specimens significantly affects the fatigue lifetimes even in the VHCF regime [43,44] due to a lower probability of large inclusions in small testing volumes—as discussed for 18Ni maraging steel sheets.

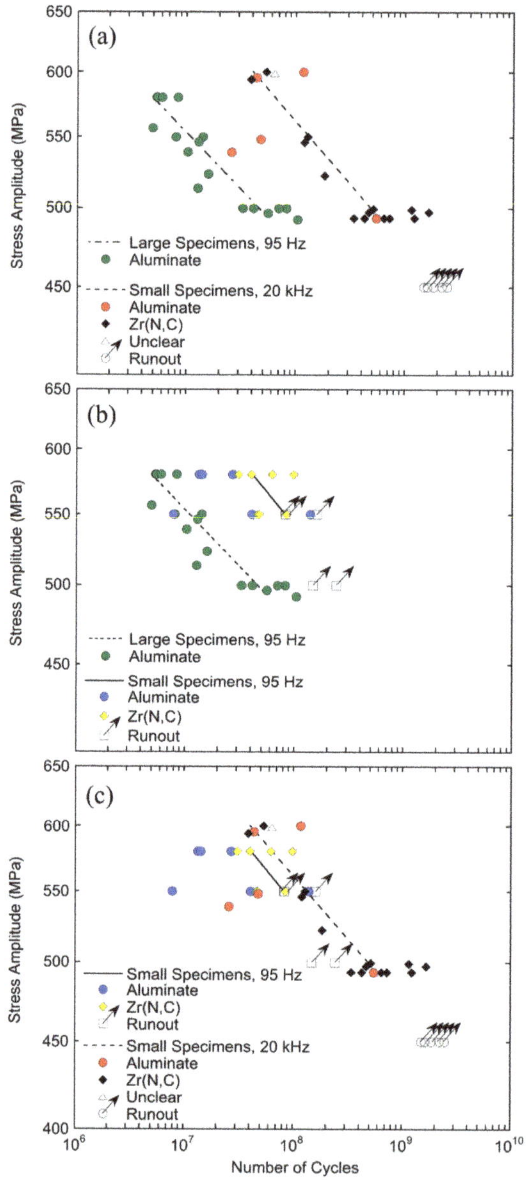


**Figure 6.** Comparison of *S-N* data at load ratio $R = 0.1$ (**a**) of large specimens tested at 95 Hz and of small specimens tested at 20 kHz; (**b**) of large and small specimens tested at 95 Hz; (**c**) of small specimens tested at 95 Hz and 20 kHz; crack initiation was at internal aluminate inclusions (circles) or at internal Zr(N,C) inclusions (diamonds); runouts are marked with arrows; reprinted with permission from Elsevier [16].

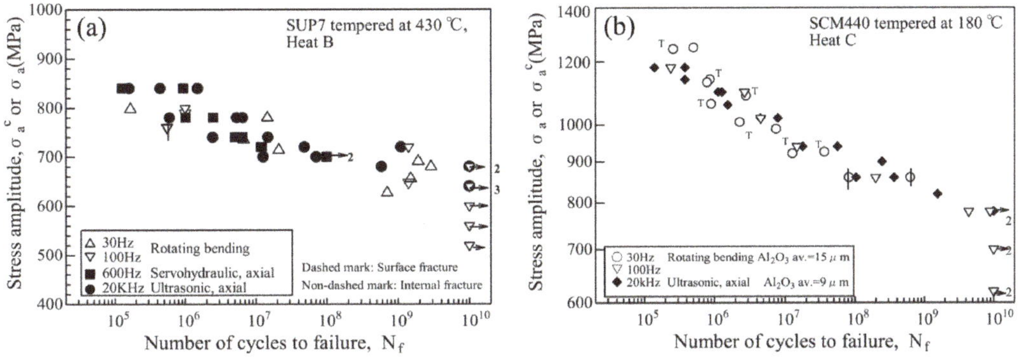

**Figure 7.** Fatigue data of high-strength spring steel SUP7 (**a**) and low-alloy steel SCM440 (**b**) with rotating-bending equipment (30 Hz and 100 Hz), servo-hydraulic equipment (600 Hz), and ultrasonic equipment at 20 kHz [42].

In addition to the examples demonstrated above, several investigations of the VHCF properties of martensitic stainless steels [41,45–47] have shown that ultrasonic frequencies do not affect the fatigue lifetimes if the failure occurs due to inherent defects such as non-metallic inclusions.

In contrast to martensitic steels, where the inclusion size is large compared to the characteristic microstructural size (block size) [48], austenitic steels may rather fail due to crack initiation at surface slip bands. In this case, it can be assumed that the austenite grain size determines the fatigue strength. Bending fatigue tests at 160 Hz to 200 Hz and ultrasonic fatigue tests at 20 kHz were performed with tubular specimens made of austenitic stainless steel AISI904L by Carstensen et al. [49] (Figure 8), where no frequency effect was observed. Ultrasonic and resonance pulsating fatigue tests at 150 Hz were also performed with austenitic stainless steel AISI316L steel [50]. A significantly increased fatigue limit and prolonged fatigue lifetimes were observed at 20 kHz. The strong damping property of austenitic steel required to use specimens with much smaller testing volume in ultrasonic tests, and the difference in measured strength was attributed to frequency as well as scale effects [50].

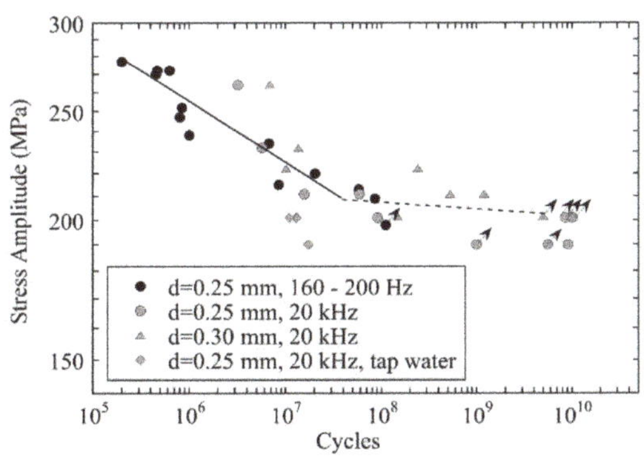

**Figure 8.** Fatigue data of tubes made of austenitic stainless steel AISI 904L measured with rotating bending at 160 Hz–200 Hz and ultrasonic fatigue equipment at 20 kHz; adapted with permission from John Wiley and Sons [49].

In contrast to the fcc austenite, it is well known that a body-centred cubic (bcc) ferritic microstructure exhibits strain-rate dependent mechanical properties [7,8]. Not only yield and tensile strength increase with a higher strain rate [51] but also the fatigue properties are exaggerated at higher testing frequencies. Frequency effects in low carbon steels, where fatigue crack initiation occurs in the ferrite phase, were systematically investigated by Tsutsumi et al. [52] employing conventional tension-compression tests at 10 Hz and ultrasonic-fatigue testing. As shown in Figure 9, fatigue lifetimes and fatigue limit were increased at 20 kHz for both annealed (Vickers hardness of $HV$ = 123) and 10% pre-strained ($HV$ = 160) specimens. This was explained by larger cyclic yield stress and a reduction in plastic zone size at the crack tip in ultrasonic fatigue tests [52]. The latter was corroborated by the observation of abundant slip bands around fatigue cracks under conventional fatigue tests that were rarely visible after ultrasonic fatigue testing (Figure 9).

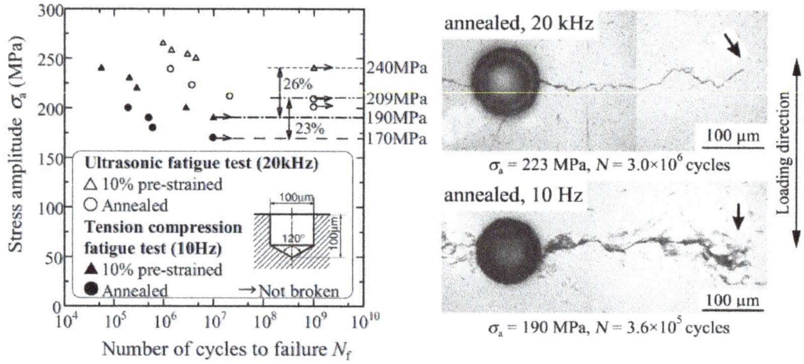


**Figure 9.** Servo-hydraulic and ultrasonic tension-compressions fatigue test results with annealed and 10% pre-strained low carbon steel specimens containing 1-hole defects; reprinted with permission from John Wiley and Sons [52].

Extended fatigue lifetimes in ultrasonic fatigue tests were reported in several investigations of mild steels [53–55], and the strain rate sensitivity of ferrite, decreased plastic deformation at high strain rates and retardation in crack formation in the bcc phase are held responsible. Attempts have been made to consider the strain rate sensitivity of ferrite in the presentation of ultrasonic fatigue data. It has been shown [55] that if lifetimes are presented versus normalised stress amplitudes, i.e., the ratio of stress amplitude and tensile strength at room temperature (conventional testing) and the ratio of stress amplitude and tensile strength at 350 °C (ultrasonic testing), low and high-frequency data show good agreement. Other investigations [56,57] have considered ways to control the fatigue tests, i.e., that ultrasonic experiments are displacement controlled whereas servo-hydraulic tests are force controlled. Considering the strain rate dependence of yield strength, ultrasonic stresses are re-calculated using the cyclic stress–strain curve, which then leads to a good agreement between conventional and ultrasonic fatigue data.

Surprising observations were recently made with the martensitic stainless steels 17-4PH: while no frequency effects were apparent from the results obtained by servo-hydraulic tension-compression, rotating-bending or ultrasonic fatigue testing [41,58], fatigue lifetimes were significantly prolonged under cyclic torsional loading at ultrasonic frequencies [59] (Figure 10). This can be explained by the presence of elongated δ ferrite grains where shear cracks can more easily initiate under low-frequency cycling due to the strain-rate sensitivity of the bcc microstructure. Under uniaxial and rotating bending loading, δ ferrite grains were oriented parallel to the maximum principal stress direction. In this case, the relevant grain size is similar to the diameter of non-metallic inclusions, and failure was more likely to occur from the latter [41], regardless of the testing frequency. Under torsional loading, however, δ ferrite grains were inclined by approximately 45° to the direction of the major principal stress and, thus, were oriented parallel to the maximum shear direction. This facilitated the initiation of shear cracks with a length exceeding the size of inherent defects, such as inclusions of uncritical size. As shown in Figure 10, the influence of testing frequency under torsional loading disappeared when artificial defects larger than a critical size were introduced in the test specimens.

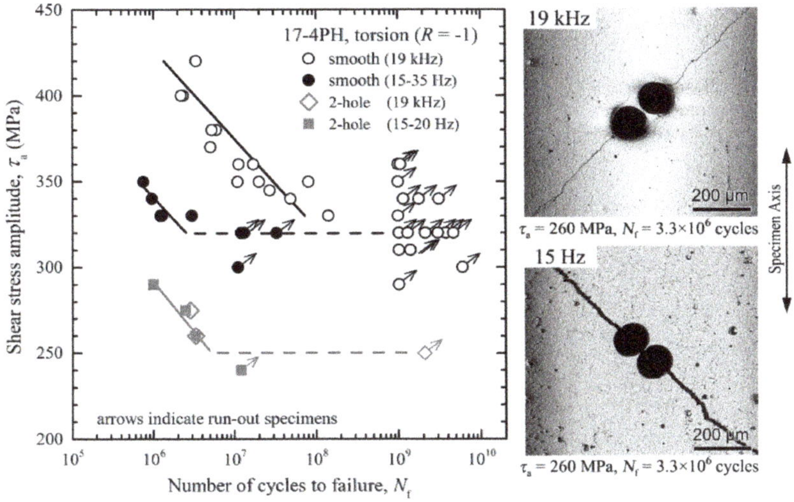


**Figure 10.** Servo-hydraulic and ultrasonic torsional fatigue tests with martensitic stainless steel 17-4PH (smooth and 2-hole defect containing specimens) [59].

Based on these research results it can be concluded that frequency effects can be neglected if non-metallic inclusions or other defects are the origins of the fatigue fracture. Small defects are the source of the fatigue fracture in high-strength steels and determine their cyclic strength [60]. Several ultrasonic and conventional investigations of the HCF and the VHCF regimes have shown the absence of a frequency effect. A strong size effect due to the increasing probability of encountering large defects with increasing testing volumes must be considered when comparing fatigue data measured with different methods. If the fatigue cracks in steels initiate in a ferritic phase, however, higher fatigue limits and lifetimes are observed in ultrasonic fatigue experiments due to the strain-rate sensitivity of the bcc microstructure. This leads to reduced plastic deformation, a prolonged crack initiation period and an extended crack propagation period. In special cases, this effect may even be relevant in high-strength steels, for example, if ferritic grains are embedded in a martensitic matrix and if they are of sufficient size in the maximum shear direction. In austenitic stainless steels, no frequency effects are expected due to the fcc microstructure. However, comparative tests with conventional and ultrasonic fatigue tests with austenitic steels are limited, and further investigations are necessary.

### 3.3. Titanium Alloys

Titanium alloys are commonly used in aerospace industries, where their high strength and relatively low weight allow for significant weight savings. The additional high corrosion resistance makes them favourable for applications where good performance under high frequencies and at high numbers of load cycles regime is required. HCF-related gas turbine engine failures observed by the US Air Force led to increased interest in fatigue of Ti6Al4V in the early 1990s, resulting in the "High Cycle Fatigue Science And Technology Program" by the US Air Force, starting in 1994 (e.g., [61]). Today, the interest in titanium alloys extends to additively manufactured (AM) applications, where a methodology for the rapid characterisation of fatigue properties for the many different processes, build conditions and post-processing procedures could greatly speed up material development.

Several studies investigated the fatigue performance of Ti6Al4V at $R = -1$ with conventional and ultrasonic fatigue (Figure 11). Generally, wrought Ti6Al4V tested in a high vacuum showed the best performance followed by wrought Ti6Al4V tested in ambient air. The HIPed AM specimens had shown fatigue performance as good as the wrought material and even better than cast material. Finally, non-HIPed AM specimens showed the poorest fatigue performance. As-built electron beam melting (EBM) processed specimens and selective laser melting (SLM) processed and heat-treated specimens showed almost equal fatigue performance, since they have been shown to be similar in terms of defect sizes and distribution.

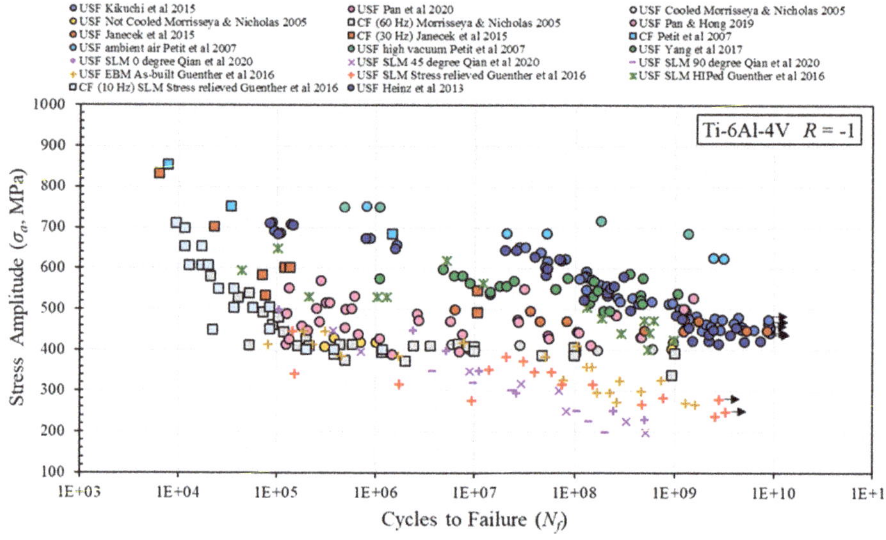


**Figure 11.** Fatigue data for Ti6Al4V across numerous investigations, generated by conventional and ultrasonic fatigue testing at $R = -1$ [62–71].

Most studies, however, only present either data produced with the conventional testing technique or data from ultrasonic tests, respectively (Figure 11). Direct comparison between conventional and ultrasonic fatigue data across different investigations is not possible, due to the large number of stock material conditions (cast, wrought, AM), heat treatments (affecting, e.g., primary $\alpha$ grain sizes), testing methodologies and specimen geometries. This makes it difficult to draw meaningful conclusions regarding only the effect of testing frequency. It is necessary to test material of the same lot with the same gauge section geometry in both conventional and ultrasonic equipment to only look at the effect of test frequency.

The authors measured the fatigue performance of mill annealed Ti6Al4V using both servo-hydraulic and displacement-controlled ultrasonic fatigue testing equipment. Specimens from the same stock with identical geometry (Figure 1) and identical surface preparation were fatigued in ambient air (24 °C, RH 50%) at $R = 0.1$ at 55 Hz cycling frequency (servo-hydraulic, force-controlled) and ~20 kHz (ultrasonic, displacement controlled), respectively. The resulting lifetimes are shown in Figure 12. Nine samples were cycled with servo-hydraulic equipment with stress amplitude, $\sigma_a$ between 350 MPa and 400 MPa, and eight samples with ultrasonic equipment between 290 MPa and 375 MPa. For 350 MPa and 375 MPa, lifetimes measured with both methods are available. Measured lifetimes were very similar, i.e., the mean lifetimes only deviate by factor 1.1. Assuming a log-normal distribution of cycles to failure, lifetimes for 90% failure probability were a factor 2.2 higher than for 10% fracture probability for ultrasonic and servo-hydraulic tests. Approximating data with a power fit, a mean lifetime of $10^7$ cycles was found at $\sigma_a = 359.7$ MPa for servo-hydraulic and $\sigma_a = 356.3$ MPa for ultrasonic tests, respectively, i.e., the difference is less than 1%. Thus, the deviation of mean lifetimes measured in ultrasonic and servo-hydraulic tests is much smaller than the scatter of lifetimes, and no statistically significant deviation of lifetimes is found.

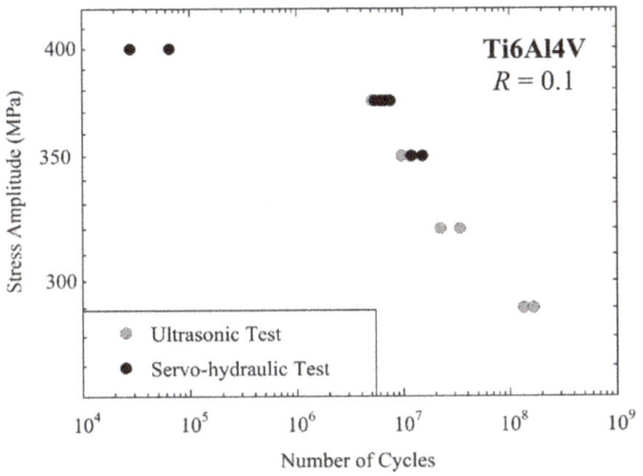


**Figure 12.** Fatigue data of Ti6Al4V ELI (mill annealed) measured at $R = 0.1$ with servo-hydraulic equipment at 55 Hz cycling frequency (black) and with ultrasonic equipment at 20 kHz (grey).

The fractographic investigation confirmed similar surface morphologies for both testing series: Figure 13 shows fracture surfaces of specimens that were cycled at $\sigma_a = 350$ MPa and that failed at approximately $10^7$ cycles. Both consistently show agglomerations of primary $\alpha$ grains at the crack initiation locations in the back-scatter images (right column). There the material is weakened due to the softer $\alpha$ phase, and cracks are preferentially initiated.

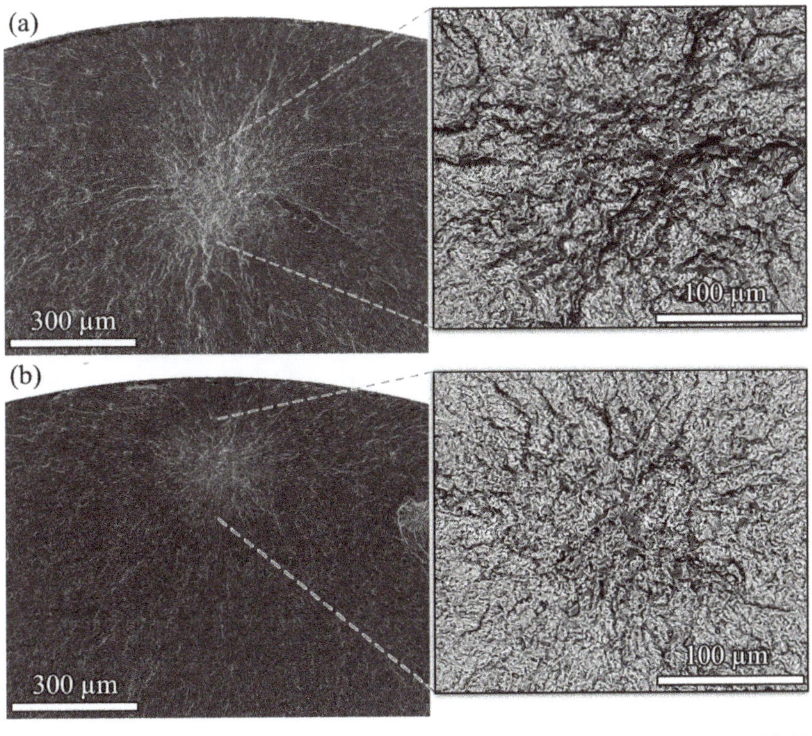


**Figure 13.** Fracture surfaces from specimens that were cycled at $\sigma_a = 350$ MPa; at (**a**) 20 kHz, $N_f = 1.18 \times 10^7$ and at (**b**) 55 Hz, $N_f = 1.51 \times 10^7$.

The Ti6Al4V fatigue data from the literature verify these findings. Morrissey and Nicholas [63,72] performed fully reversed fatigue tests with Ti6Al4V (tensile strength 968 MPa) with servo-hydraulic equipment at 60 Hz and with ultrasonic equipment at 20 kHz. *S-N* data are available from both testing methods in a wide range between $3 \times 10^5$ and $10^8$ cycles. Very consistent lifetimes were found at both frequencies, indicating that no frequency influence is present in the investigated conditions.

Similarly, Takeuchi et al. [73] tested Ti6Al4V from three different manufacturers (tensile strengths between 906 MPa and 967 MPa; all three heats satisfied ASTM specifications) at 120 Hz, 600 Hz and 20 kHz at $R = -1$. Two of the three heats showed internal crack initiation above approximately $5 \times 10^6$ cycles. Fatigue lifetimes measured for these two heats showing internal crack initiation in the long lifetime regime were similar for all three testing frequencies. However, the third heat showed solely surface crack initiation and no failures above $5 \times 10^6$ cycles as well as the significant effect of the testing method and significantly higher cyclic strength when tested at the higher frequencies.

Furuya and Takeuchi [74] studied Ti6Al4V from the same three suppliers at $R = 0$ and $R = 0.3$ as well as with fixed maximum stress and cycling frequencies of 120 Hz and 20 kHz. With superimposed tensile mean loads, all three heats showed failures above $5 \times 10^6$ cycles with crack initiation in the interior. No effect of the cycling frequency was found in the HCF and VHCF regimes for nine testing series overall.

Günther et al. [67] investigated the impact of EBM and SLM additive manufacturing on the fatigue life of Ti6Al4V in the HCF and VHCF regimes. Cylindrical fatigue specimens in differently post-treated conditions were tested at 10 Hz with servo-hydraulic equipment and at 20 kHz using ultrasonic equipment. Data from the SLM batch shows that the lifetimes and failure modes (i.e., crack initiation locations) are closely similar for both testing methods.

The fatigue properties of laser-additive manufactured Ti6Al4V were investigated by Wycisk et al. [75]. Specimens in stress relieved as well as HIPed conditions were analysed for crack initiation site, mean stress sensitivity and overall fatigue performance. Fatigue tests in tension-compression loading at 59 Hz and 20 kHz in the HCF and VHCF regimes until $10^9$ cycles revealed a clear shift of crack initiation from surface to internal initiation with increasing cycles to failure, regardless of the testing principle. They also stated that no effects of test frequency on life span could be determined.

Other than Ti6Al4V, Papakyriacou et al. [76] tested Ti6Al7Nb that is used for medical applications, with rotating bending equipment at 100 Hz and with ultrasonic equipment at 20 kHz. Fatigue lifetimes between $10^5$ and $2 \times 10^8$ cycles were measured with both methods, and no frequency effect on fatigue lifetimes was found for this alloy.

Szczepanski et al. [77] characterised the fatigue behaviour of the alpha-beta titanium alloy Ti6Al2Sn4Zr6Mo in the HCF regime using conventional testing at 20 Hz and in the VHCF regime using ultrasonic equipment at 20 kHz, respectively, at a load ratio of 0.05. The lifetimes they found in both testing series showed a clear separation between earlier failures with crack initiation at the surface and longer lifetimes with crack initiation occurring below the surface. They concluded that, even though the stress levels of the employed testing methods do not overlap, the data follow the same trend as expected for a typical *S-N* curve. Together with a similar fractographic appearance, the authors interpret this as an indication that there is no appreciable frequency effect on fatigue lifetime between testing frequencies of 20 kHz and 20 Hz in the investigated Ti6Al2Sn4Zr6Mo alloy.

These results show that titanium alloys may be considered insensitive to frequency effects. Similar lifetimes are found with conventional and ultrasonic equipment, and crack initiation occurs preferentially in the interior at $\alpha$ grains in the VHCF regime for both low and high-frequency testing. The absence of a frequency effect may be expected, irrespective of the way titanium alloys are produced. However, the actual production process of a given titanium alloy can strongly affect the microstructure and consequently the measured fatigue properties.

### 3.4. Nickel Alloys

Components made of nickel-based superalloys such as turbine blades are typically exposed to a very high number of load cycles, and investigations of the VHCF regime are, therefore, of major importance. In the absence of detrimental defects, e.g., process-related flaws, such as casting pores or lack of fusion, polycrystalline nickel-based alloys typically fail due to favourably oriented large grains located at the surface or in the interior [78,79].

Stöcker et al. [80] performed fatigue tests with nickel alloys over a wide range of lifetimes, from low cycle fatigue to very high cycle fatigue, and obtained a single curve (Figure 14). Other investigations using ultrasonic fatigue testing similarly showed no frequency effect [81,82]. Due to their practical importance for gas turbine engines, the high-temperature fatigue properties of nickel alloys are important. This led to interesting further developments of the ultrasonic fatigue testing technique to be used up to 1000 °C [82,83].

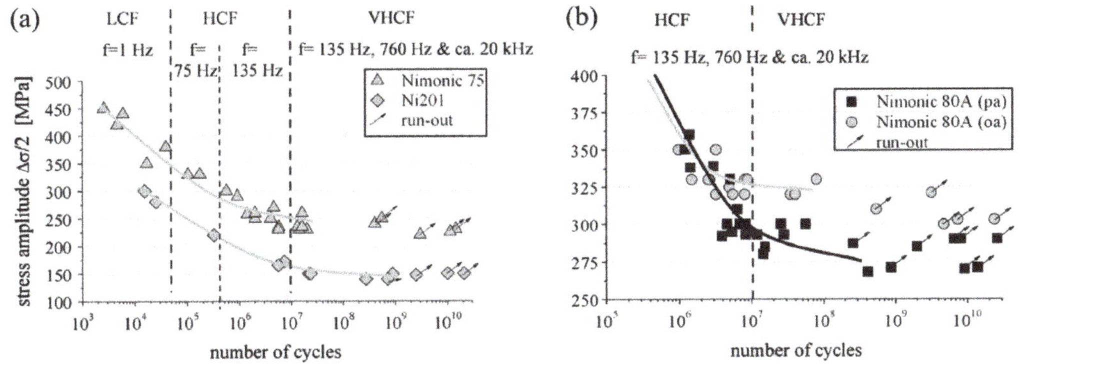


**Figure 14.** Fatigue data of different Ni-based alloys obtained with servo-hydraulic, resonance electromechanical and ultrasonic fatigue testing; (**a**) pure nickel (Ni201), Nimonic 75 and (**b**) peak-aged (pa) and overaged (oa) Nimonic 80A; reprinted with permission from Elsevier [80].

### 3.5. Magnesium Alloys

The three high-pressure die-cast magnesium alloys AZ91 hp, AM60 hp and AE42 hp, were investigated using ultrasonic equipment and servo-hydraulic equipment [18]. As an example, Figure 15 shows the fatigue data of AE42 hp measured with both methods. Fatigue cracks initiate at shrinkage and gas porosity in these high-pressure die-cast materials. Fatigue lifetimes show pronounced scatter due to the different sizes of crack-initiating defects in the different specimens. Within the ranges of scatter, similar lifetimes were measured for the three alloys at both 50 Hz and 20 kHz, respectively. Non-propagating cracks were found at porosities in runout specimens. Porosity can be considered an initial crack, and the fatigue limit can be correlated with the minimum stress intensity factor required to propagate the crack to fracture [18].

### 3.6. Fibre-Reinforced Polymers Composites

The number of applications of polymers and their fibre-reinforced composites has been growing rapidly during the past few decades, due to their low density, excellent chemical resistance, design flexibility, ease of manufacturing and robust mechanical properties. Such materials are used in the aerospace and automotive industries where components may be subjected to high numbers of load cycles. However, cyclic properties of fibre-reinforced polymer composites (FRPC) are mostly investigated in the regime below $10^6$ cycles. Except for ultrasonic studies, only one gigacycle fatigue investigation can be found in the literature where a carbon fibre-reinforced composite was tested with a resonance tester [84]. VHCF investigations with conventional methods are time-consuming, and therefore attempts have been made to use ultrasonic equipment for fatigue testing of FRPC.

This is challenging due to the viscoelasticity of polymers and the associated heat generation at ultrasonic frequencies. Due to the low thermal conductivity coefficient of polymers (~0.2–0.5 W/m/K) compared to metals (~20–300 W/m/K), great care must be taken to avoid excessive heating during high-frequency cycling.

Testing glass fibre-reinforced polymer (GFRP) in pulsed mode, with forced air-cooling and with control of specimen surface temperature, it was shown that ultrasonic fatigue testing could be an appropriate technique for gathering valid fatigue data of FRPC [85]. Ultrasonic tests were performed with a static load superimposed to the resonance vibration to realise load ratio $R = 0.1$. Figure 16 shows the results of ultrasonic and servo-hydraulic fatigue tests of quasi-unidirectional GFRP (90% of the fibres aligned in the main direction while 10% of them are woven transversally for stabilisation) with a fibre volume fraction of 60%. Fatigue strengths measured with both techniques agree well, which indicates the suitability and reliability of the developed ultrasonic testing method.

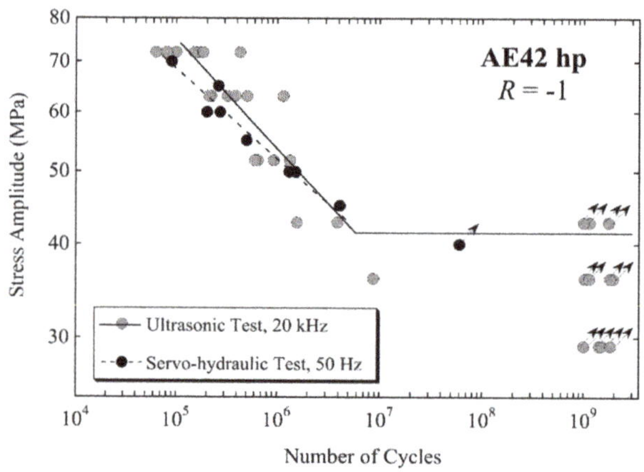


**Figure 15.** Fatigue data of magnesium alloy AE42 hp produced by high-pressure die-casting; data measured with servo-hydraulic (black circles) and ultrasonic equipment (grey circles) are shown; adapted with permission from Elsevier [18].

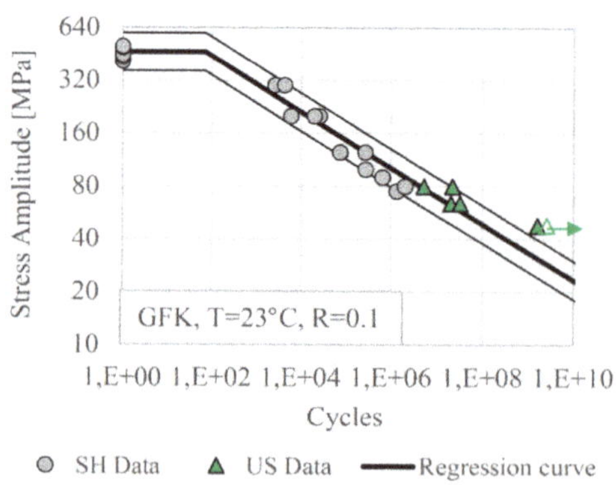


**Figure 16.** Fatigue data of a quasi-unidirectional glass fibre-reinforced polymer measured at load ratio $R = 0.1$ with servo-hydraulic equipment at 10 Hz and ultrasonic equipment at 20 kHz; reprinted with permission from Elsevier [85].

Cyclic properties of GFRP were also studied by Lee et al. [86] using ultrasonic and hydraulic equipment working at 3 Hz, respectively. Tests were performed under fully reversed loading conditions, and shorter lifetimes were found in the ultrasonic tests. Since the results of ultrasonic tests are influenced by numerous parameters (e.g., calibration procedure, calculation of cyclic stress amplitude, the accuracy of equipment, the temperature of the specimen) and different specimen shapes were used in low and ultrasonic tests, a number of reasons for shorter lifetimes other than the frequency are possible.

Several ultrasonic studies showed the benefits of high-frequency testing for understanding the process of fatigue damage in FRPCs in the HCF and VHCF regimes. A three-point bending setup, where a specimen was stimulated to bending resonance vibrations at ultrasonic frequencies, was used by Backe et al. [12,87] to investigate the VHCF properties of carbon fibre-reinforced polymer (CFRP). These studies demonstrated the progress of fatigue damage as a multi-stage process starting with the debonding of fibre and matrix, then initiation of transverse cracks in the 90° layers, micro delamination between 0° and 90° fibre roving, followed by macro delamination and finally fracture. Cui et. al. [88] performed similar three-point bending ultrasonic tests of CFRP and showed differences in failure form in the low, high and very high cycle fatigue regime. Ding and Cheng studied the benefit of incorporating nano-silica [89] and multiwall carbon nanotubes [90] into carbon fibre-reinforced composites and reported great improvement in mechanical properties and fatigue resistance of composite materials in presence of nanoparticles.

### 3.7. Graphite

Fatigue properties of isotropic polycrystalline porous graphite (mean grain size 10 μm, porosity 13 vol%) were investigated in the regime between $10^3$ and $10^9$ cycles. Cyclic tension-compression tests with ultrasonic equipment at 20 kHz and fully-reversed cyclic bending tests at 25 Hz were performed [91]. The tensile strength of graphite is lower than its bending strength. Comparing the results of both testing series, this is considered by presenting lifetimes versus normalised stress amplitudes. The normalised stress amplitude is the ratio of tension-compression stress amplitude and tensile strength in ultrasonic tests, and the ratio of bending stress amplitude and bending strength in cyclic bending tests.

Figure 17 shows fatigue lifetimes of graphite measured at 25 Hz and 20 kHz. Ultrasonic fatigue data are slightly shifted towards higher lifetimes compared with cyclic bending data; however, this difference is well within the range of scatter. It can be concluded, therefore, that the ultrasonic testing technique is appropriate for the rapid collection of fatigue data of graphite. Moreover, ultrasonic fatigue testing was also used to demonstrate the beneficial effect of infiltration of graphite with AlSi7Mg alloy cyclic strength [91].

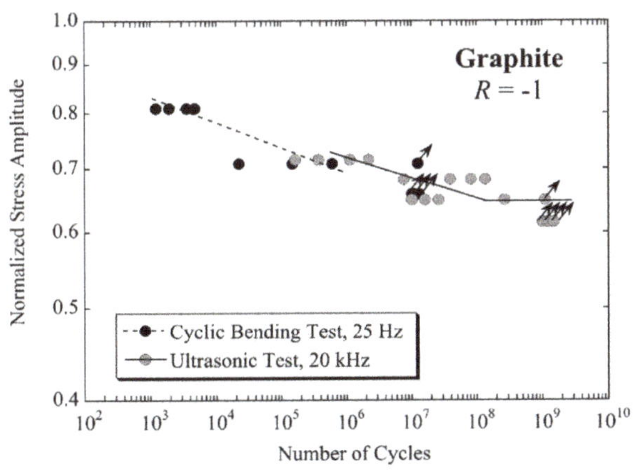


**Figure 17.** Fatigue data of polycrystalline graphite measured under fully reversed loading condition for cyclic bending (black circles) and ultrasonic tension–compression loading (grey circles), respectively; fatigue lifetimes are presented versus normalised stress amplitudes; adapted with permission from Elsevier [91].

### 3.8. Other Materials

In addition to the aforementioned material systems, several others have been successfully tested with ultrasonic equipment.

Fatigue crack initiation, slow crack propagation and conditions for crack arrest were studied in the aluminium laminates ARALL [92] and GLARE [93]. These materials are used in aerospace applications, where crack initiation at notches and slow fatigue crack growth are of great practical interest. Particularly, GLARE with glass fibres preloaded in the tension demonstrated strongly improved cyclic properties compared to aluminium alloy laminates [93].

In a specifically developed setup, thin wires (Ø0.1 mm) of the high-entropy alloy CoNiCr alloy MP35N were tested in the VHCF regime [94]. In this special application, the wire specimen does not vibrate in resonance, but is cyclically stressed in a quasi-static fashion at ultrasonic frequencies. The usability of the newly developed method was demonstrated by measuring lifetimes that are comparable to conventional tests [94].

Cyclic compression loading was purpose-developed for VHCF testing of concrete. With this technique, the damage mechanism in the very long lifetime regime could be demonstrated for concrete used in wind power plant foundations [95].

Fundamental studies on the fatigue mechanisms have been extensively performed with pure copper. It was shown, for example, that new persistent slip bands are formed even after $5 \times 10^8$ cycles [96].

Phase transformation and associated fatigue damage were studied in the shape memory alloy Nitinol [97]. In situ synchrotron experiments successfully demonstrated the occurrence of the forward and reverse transformation between the material's austenitic and martensitic phases in a super elastic state at ultrasonic frequencies.

## 4. Conclusions

Based on the literature and original data, this paper summarises the current understanding of whether and how fatigue properties measured with ultrasonic and conventional equipment are comparable.

- While the orders of magnitude are faster than conventional fatigue testing, no generally accepted standard for ultrasonic fatigue testing exists. The use of equipment with closed-loop control of vibration amplitude and the resonance frequency is strongly advised since this guarantees high accuracy and reproducibility of ultrasonic tests. Pulsed loading and appropriate cooling are necessary to avoid specimen heating.
- Depending on the material, frequency influences can be caused by strain rate influences on plastic deformation, and by time-dependent influences of the testing environment. The size effect must be considered for all materials, if low and ultrasonic frequency data are compared.
- Several aluminium alloys show comparable lifetimes in ultrasonic and conventional tests. Some alloys tested in ambient air show prolonged lifetimes at high frequency, due to a reduced influence of air humidity. Ultrasonic tests in high humidity or in distilled water can better approximate the environmental conditions acting at low cycling frequencies.
- Frequency effects can be neglected in high-strength steels, where non-metallic inclusions or other defects are preferential crack initiation locations in the regime of long lifetimes. Ultrasonic tests of steels lead to prolonged lifetimes if fatigue cracks initiate in a ferritic phase. Austenitic stainless steels are less prone to frequency effects.
- Titanium alloys may be considered insensitive to frequency effects, as suggested by data from both the original work and the literature. Similar lifetimes and similar crack initiation locations are found with conventional and ultrasonic equipment.
- Ultrasonic tests with nickel alloys showed no frequency effect. On the basis of limited data, the same conclusion can be drawn for cast magnesium alloys and graphite.
- Ultrasonic tests of a glass fibre-reinforced polymer delivered comparable fatigue lifetimes to servo-hydraulic tests, i.e., high-frequency testing is in principle applicable

to testing fibre-reinforced polymer composites. However, further research is needed to better understand the influence of the experimental procedure on the measured data.

**Author Contributions:** M.F.: methodology, formal analysis, validation, investigation, data curation, writing—original draft, visualization; B.M.S.: formal analysis, data curation, writing—original draft, visualization; R.K.R.: data curation, writing—review and editing; N.S.: data curation, writing—review and editing, visualisation; S.Z.: data curation, writing—review and editing; S.A.T.: supervision, funding acquisition, writing—review and editing; J.W.C.: supervision, funding acquisition; H.M.: conceptualisation, methodology, formal analysis, writing—original draft, supervision, project administration. All authors have read and agreed to the published version of the manuscript.

**Funding:** This research received no external funding.

**Institutional Review Board Statement:** Not applicable.

**Informed Consent Statement:** Not applicable.

**Data Availability Statement:** The data are not publicly available as further investigations are currently ongoing.

**Acknowledgments:** We would like to pay our gratitude and our respects to our late colleague G. Pfersmann, who has worked with us for many years and who is, in large, responsible for the design, development and realisation of the ultrasonic fatigue testing equipment at BOKU. Without him, this and many preceding works would not have been possible.

**Conflicts of Interest:** The authors declare no conflict of interest.

## References

1. Stanzl-Tschegg, S. Very high cycle fatigue measuring techniques. *Int. J. Fatigue* **2014**, *60*, 2–17. [CrossRef]
2. Willertz, L.E. Ultrasonic Fatigue. *Int. Met. Rev.* **1980**, *2*, 65–78.
3. Roth, L.D. Ultrasonic Fatigue Testing. In *ASM Handbook*; Newby, J.R., Davis, J.R., Refsnes, S.K., Dietrich, D.A., Eds.; ASTM: Philadelphia, PA, USA, 1992; Volume 8, pp. 240–258.
4. Mayer, H. Fatigue crack growth and threshold measurements at very high frequencies. *Int. Mater. Rev.* **1999**, *44*, 1–36. [CrossRef]
5. Mayer, H. Recent developments in ultrasonic fatigue. *Fatigue Fract. Eng. Mater. Struct.* **2016**, *39*, 3–29. [CrossRef]
6. Standards, T.J.W.E. *Standard Method for Ultrasonic Fatigue Test in Metallic Materials*; The Japan Welding Engineering Society Standards: Tokyo, Japan, 2017; Volume WES 1112.
7. Laird, C.; Charsley, P. Strain Rate Sensitivity Effects in Cyclic Deformation and Fatigue Fracture. In *Ultrasonic Fatigue (Proc. 1st Int. Conf. on Fatigue and Corrosion Fatigue Up to Ultrasonic Frequencies)*; Wells, J.M., Buck, O., Roth, L.D., Tien, J.K., Eds.; Transactions of the Metallurgical Society of AIME: Philadelphia, PA, USA, 1982; pp. 187–205.
8. Mughrabi, H.; Herz, K.; Stark, X. Cyclic Deformation and Fatigue Behaviour of a-Iron Mono- and Polycrystals. *Int. J. Fract.* **1981**, *17*, 193–220. [CrossRef]
9. Zhu, X.; Jones, J.W.; Allison, J.E. Effect of frequency, environment, and temperature on fatigue behavior of E319 cast aluminum alloy: Stress-controlled fatigue life response. *Met. Mater. Trans. A* **2008**, *39*, 2681–2688. [CrossRef]
10. Zhu, X.; Jones, J.W.; Allison, J.E. Effect of frequency, environment, and temperature on fatigue behavior of E319 cast aluminum alloy: Small crack propagation. *Met. Mater. Trans. A* **2008**, *39*, 2666–2680. [CrossRef]
11. Sharpe, W.N. *A 20 kHz Optical Strain Gauge*, 4th ed.; In International Conference on Very High Cycle Fatigue (VHCF-3); Ann, A.M.I., Allison, J.E., Jones, J.W., Larsen, J.M., Ritchie, R.O., Eds.; TMS: Warrendale, PA, USA, 2007; pp. 341–346.
12. Backe, D.; Balle, F.; Eifler, D. Fatigue testing of CFRP in the Very High Cycle Fatigue (VHCF) regime at ultrasonic frequencies. *Compos. Sci. Technol.* **2015**, *106*, 93–99. [CrossRef]
13. Papakyriacou, M.; Mayer, H.; Plenk, H.; Tschegg, S. Cyclic plastic deformation of tantalum and niobium at very high numbers of cycles. *Mater. Sci. Eng. A* **2002**, *325*, 520–524. [CrossRef]
14. Zettl, B.; Mayer, H.; Ede, C.; Stanzl-Tschegg, S. Very high cycle fatigue of normalized carbon steels. *Int. J. Fatigue* **2006**, *28*, 1583–1589. [CrossRef]
15. Stanzl, S.E.; Tschegg, E.K.; Mayer, H.R. Lifetime Measurements for Random Loading in the Very High Cycle Fatigue Range. *Int. J. Fatigue* **1986**, *8*, 195–200. [CrossRef]
16. Fitzka, M.; Pennings, B.; Karr, U.; Schönbauer, B.; Schuller, R.; Tran, M.-D.; Mayer, H. Influence of cycling frequency and testing volume on the VHCF properties of 18Ni maraging steel. *Eng. Fract. Mech.* **2019**, *216*, 106525. [CrossRef]
17. Stanzl-Tschegg, S.E.; Mayer, H.R. Fatigue and fatigue crack growth of aluminium alloys at very high numbers of cycles. *Int. J. Fatigue* **2001**, *23*, 231–237. [CrossRef]
18. Mayer, H.; Papakyriacou, M.; Zettl, B.; Stanzl-Tschegg, S.E. Influence of porosity on the fatigue limit of die cast magnesium and aluminium alloys. *Int. J. Fatigue* **2003**, *25*, 245–256. [CrossRef]

19. Caton, M.J.; Jones, J.W.; Mayer, H.; Stanzl-Tschegg, S.; Allison, J.E. Demonstration of an Endurance Limit in Cast 319 Aluminum. *Met. Mater. Trans. A* **2003**, *34*, 33–41. [CrossRef]
20. Mayer, H.; Schuller, R.; Fitzka, M. Fatigue of 2024-T351 aluminium alloy at different load ratios up to $10^{10}$ cycles. *Int. J. Fatigue* **2013**, *57*, 113–119. [CrossRef]
21. Mayer, H.; Fitzka, M.; Schuller, R. Variable amplitude loading of Al 2024-T351 at different load ratios using ultrasonic equipment. *Int. J. Fatigue* **2014**, *60*, 34–42. [CrossRef]
22. Fitzka, M.; Mayer, H. Constant and variable amplitude fatigue testing of aluminum alloy 2024-T351 with ultrasonic and servo-hydraulic equipment. *Int. J. Fatigue* **2016**, *91*, 363–372. [CrossRef]
23. Mayer, H. Ultrasonic torsion and tension-compression fatigue testing: Measuring principles and investigations on 2024-T351 aluminium alloy. *Int. J. Fatigue* **2006**, *28*, 1446–1455. [CrossRef]
24. Mayer, H.; Papakyriacou, M.; Pippan, R.; Stanzl-Tschegg, S. Influence of Loading Frequency on the High Cycle Fatigue Properties of AlZnMgCu1.5 Aluminium Alloy. *Mater. Sci. Eng. A* **2001**, *314*, 51–57. [CrossRef]
25. Zettl, B.; Mayer, H.; Stanzl-Tschegg, S.E. Fatigue properties of Al1Mg0.6Si foam at low and ultrasonic frequencies. *Int. J. Fatigue* **2001**, *23*, 565–573. [CrossRef]
26. Stanzl-Tschegg, S.E.; Mayer, H.; Schuller, R.; Przeorski, T.; Krug, P. Fatigue properties of spray formed hypereutectic aluminium silicon alloy DISPAL® S232 at high and very high numbers of cycles. *Mater. Sci. Eng. A* **2012**, *538*, 327–334. [CrossRef]
27. Awd, M.; Siddique, S.; Johannsen, J.; Emmelmann, C.; Walther, F. Very high-cycle fatigue properties and microstructural damage mechanisms of selective laser melted AlSi10Mg alloy. *Int. J. Fatigue* **2019**, *124*, 55–69. [CrossRef]
28. Awd, M.; Siddique, S.; Walther, F. Microstructural damage and fracture mechanisms of selective laser melted Al-Si alloys under fatigue loading. *Appl. Fract. Mech.* **2020**, *106*, 102483. [CrossRef]
29. Rhein, R.K.; Shi, Q.; Arjun Tekalur, S.; Wayne Jones, J.; Carroll, J.W. Effect of direct metal laser sintering build parameters on defects and ultrasonic fatigue performance of additively manufactured AlSi10Mg. *Fatigue Fract. Eng. Mater. Struct.* **2021**, *44*, 295–305. [CrossRef]
30. Holper, B.; Mayer, H.; Vasudevan, A.K.; Stanzl-Tschegg, S.E. Near Threshold Fatigue Crack Growth in Aluminium Alloys at Low and Ultrasonic Frequency: Influences of Specimen Thickness, Strain Rate, Slip Behaviour and Air Humidity. *Int. J. Fatigue* **2003**, *25*, 397–411. [CrossRef]
31. Holper, B.; Mayer, H.; Vasudevan, A.K.; Stanzl-Tschegg, S.E. Near Threshold Fatigue Crack Growth at Positive Load Ratio in Aluminium Alloys at Low and Ultrasonic Frequency: Influences of Strain Rate, Slip Behaviour and Air Humidity. *Int. J. Fatigue* **2004**, *26*, 27–38. [CrossRef]
32. Wei, R.P. Rate Controlling Processes and Crack Growth Response. In *Hydrogen Effects in Metals*; Bernstein, I.M., Thompson, A.W., Eds.; TMS: Warrendale, PA, USA, 1980; pp. 677–689.
33. Wei, R.P.; Pao, P.S.; Hart, G.; Weir, T.W.; Simmons, G.W. Fracture Mechanics and Surface Chemistry Studies of Fatigue Crack Growth in an Aluminium Alloy. *Met. Trans. A* **1980**, *11*, 151–185. [CrossRef]
34. Engler-Pinto, C.C.; Frisch, R.J.; Lasecki, J.V.; Mayer, H.; Allison, J.R. *Effect of Frequency and Environment on High Cycle Fatigue of Cast Aluminum Alloys, 4th International Conference on Very High Cycle Fatigue*; Ann, A.M.I., Allison, J.E., Jones, J.W., Larsen, J.M., Ritchie, R.O., Eds.; TMS: Warrendale, PA, USA, 2007; pp. 421–427.
35. Atrens, A.; Hoffelner, W.; Duerig, T.W.; Allison, J.E. Subsurface crack initiation in high cycle fatigue in Ti6A14V and in a typical martensitic stainless steel. *Scr. Met. Mater.* **1983**, *17*, 601–606. [CrossRef]
36. Naito, T.; Ueda, H.; Kikuchui, M. Fatigue behavior of carburized steel with internal oxides and nonmartensitic micro-structure near the surface. *Met. Trans. A* **1984**, *15*, 1431–1436. [CrossRef]
37. Asami, K.; Sugiyama, Y. Fatigue strength of various surface hardened steels. *J. Heat Treat. Technol. Assoc.* **1985**, *25*, 147–150.
38. Murakami, Y.; Nomotomo, T.; Ueda, T.; Murakami, Y. On the mechanism of fatigue failure in the superlong life regime (>$10^7$ cycles). Part I: Influence of hydrogen trapped by inclusions. *Fatigue Fract. Engng. Mater. Struct.* **2000**, *23*, 893–902. [CrossRef]
39. Murakami, Y.; Nomotomo, T.; Ueda, T.; Murakami, Y. On the mechanism of fatigue failure in the superlong life regime (>$10^7$ cycles). Part II: A fractographic investigation. *Fatigue Fract. Eng. Mater. Struct.* **2000**, *23*, 903–910. [CrossRef]
40. Sakai, T.; Sato, Y.; Oguma, N. Characteristic S-N properties of high-carbon-chromium-bearing steel under axial loading in long-life fatigue. *Fatigue Fract. Engng. Mater. Struct.* **2002**, *25*, 765–773. [CrossRef]
41. Schönbauer, B.M.; Fitzka, M.; Karr, U.; Mayer, H. Variable amplitude very high cycle fatigue of 17–4PH steel with a stepwise S-N curve. *Int. J. Fatigue* **2021**, *142*, 105963. [CrossRef]
42. Furuya, Y.; Hirukawa, H.; Takeuchi, E. Gigacycle fatigue in high strength steels. *Sci. Technol. Adv. Mat.* **2019**, *20*, 643–656. [CrossRef] [PubMed]
43. Furuya, Y. Specimen size effects on gigacycle fatigue properties of high-strength steel under ultrasonic fatigue testing. *Scr. Mater.* **2008**, *58*, 1014–1017. [CrossRef]
44. Furuya, Y. Notable size effects on very high cycle fatigue properties of high-strength steel. *Mater. Sci. Eng. A* **2011**, *528*, 5234–5240. [CrossRef]
45. Li, W.; Sakai, T.; Li, Q.; Luc, L.T.; Wang, P. Effect of loading type on fatigue properties of high strength bearing steel in very high cycle regime. *Mater. Sci. Eng. A* **2011**, *528*, 5044–5052. [CrossRef]
46. Kovacs, S.; Beck, T.; Singheiser, L. Influence of mean stresses on fatigue life and damage of a turbine blade steel in the VHCF-regime. *Int. J. Fatigue* **2013**, *49*, 90–99. [CrossRef]

47. Schmid, S.; Hahn, M.; Issler, S.; Bacher-Hoechst, M.; Furuya, Y.; Mehner, A.; Bomas, H.; Zoch, H.W. Effect of frequency and biofuel E85 on very high cycle fatigue behaviour of the high strength steel X90CrMoV18. *Int. J. Fatigue* **2014**, *60*, 90–100. [CrossRef]
48. Morito, S.; Yoshida, H.; Maki, T.; Huang, X. Effect of block size on the strength of lath martensite in low carbon steels. *Mater. Sci. Eng. A* **2006**, *438–440*, 237–240. [CrossRef]
49. Carstensen, J.; Mayer, H.; Bronsted, P. Very high cycle regime fatigue of thin walled tubes made from austenitic stainless steel. *Fatigue Fract. Eng. Mater. Struct.* **2002**, *25*, 837–844. [CrossRef]
50. Grigorescu, A.C.; Hilgendorff, P.M.; Zimmermann, M.; Fritzen, C.P.; Christ, H.J. Cyclic deformation behavior of austenitic Cr–Ni-steels in the VHCF regime: Part I—Experimental study. *Int. J. Fatigue* **2016**, *93*, 250–260. [CrossRef]
51. Manjoine, M.J. Influence of Rate of Strain and Temperature on Yield Stresses of Mild Steel. *J. Appl. Mech.* **2021**, *11*, A211–A218. [CrossRef]
52. Tsutsumi, N.; Murakami, Y.; Doquet, V. Effect of test frequency on fatigue strength of low carbon steel. *Fatigue Fract. Eng. Mater. Struct.* **2009**, *32*, 473–483. [CrossRef]
53. Guennec, B.; Ueno, A.; Sakai, T.; Takanashi, M.; Itabashi, Y. Effect of the loading frequency on fatigue properties of JIS S15C low carbon steel and some discussions based on micro-plasticity behavior. *Int. J. Fatigue* **2014**, *66*, 29–38. [CrossRef]
54. Nonaka, I.; Setowaki, S.; Ichikawa, Y. Effect of load frequency on high cycle fatigue strength of bullet train axle steel. *Int J. Fatigue* **2014**, *60*, 43–47. [CrossRef]
55. Torabian, N.; Favier, V.; Dirrenberger, J.; Adamski, F.; Ziaei-Rad, S.; Ranc, N. Correlation of the high and very high cycle fatigue response of ferrite based steels with strain rate-temperature conditions. *Acta Mater.* **2017**, *134*, 40–52. [CrossRef]
56. Geilen, M.B.; Klein, M.; Oechsner, M.; Kaffenberger, M.; Störzel, K.; Melz, T. A method for the strain rate dependent correction for control type of fatigue tests. *Int. J. Fatigue* **2020**, *138*, 105726. [CrossRef]
57. Geilen, M.B.; Schönherr, J.A.; Klein, M.; Leininger, D.S.; Giertler, A.; Krupp, U.; Oechsner, M. On the Influence of Control Type and Strain Rate on the Lifetime of 50CrMo4. *Metals* **2020**, *10*, 1458. [CrossRef]
58. Schönbauer, B.M.; Yanase, K.; Endo, M. VHCF properties and fatigue limit prediction of precipitation hardened 17–4PH stainless steel. *Int. J. Fatigue* **2016**, *88*, 205–216. [CrossRef]
59. Schönbauer, B.M.; Yanase, K.; Chehrehrazi, M.; Endo, M.; Mayer, H. Effect of microstructure and cycling frequency on the torsional fatigue properties of 17–4PH stainless steel. *Mater. Sci. Eng. A Struct.* **2021**, *801*, 140481. [CrossRef]
60. Murakami, Y. Material defects as the basis of fatigue design. *Int. J. Fatigue* **2012**, *41*, 2–10. [CrossRef]
61. Morrissey, R.J.; McDowell, D.L.; Nicholas, T. Frequency and stress ratio effects in high cycle fatigue of Ti-6Al-4V. *Int. J. Fatigue* **1999**, *21*, 679–685. [CrossRef]
62. Kikuchi, S.; Heinz, S.; Eifler, D.; Nakamura, Y.; Ueno, A. Evaluation of Very High Cycle Fatigue Properties of Low Temperature Nitrided Ti-6Al-4V Alloy Using Ultrasonic Testing Technology. *Key Eng. Mater.* **2016**, *664*, 118–127. [CrossRef]
63. Morrissey, R.J.; Nicholas, T. Fatigue strength of Ti–6Al–4V at very long lives. *Int. J. Fatigue* **2005**, *27*, 1608–1612. [CrossRef]
64. Janecek, M.; Novy, F.; Harcuba, P.; Strasky, J.; Trsko, L.; Mhaede, M.; Wagner, L. The Very High Cycle Fatigue Behaviour of Ti6Al4V Alloy. *Acta Phys. Polonica A* **2015**, *128*, 497–502. [CrossRef]
65. Petit, J.; Sarrazin-Baudoux, C.; Martinez, J.; Stanzl-Tschegg, S.; Mayer, H. *Very High Cycle Fatigue Behavior of a Ti6246 Alloy in air and In High Vacuum, 4th Internation Conference on Very High Cycle Fatigue*; Allison, J.E., Jones, J.W., Larsen, J.M., Ritchie, R.O., Eds.; TMS: Warrendale, PA, USA, 2007; pp. 399–408.
66. Qian, G.; Li, Y.; Paolino, D.S.; Tridello, A.; Berto, F.; Hong, Y. Very-high-cycle fatigue behavior of Ti-6Al-4V manufactured by selective laser melting: Effect of build orientation. *Int. J. Fatigue* **2020**, *136*, 105628. [CrossRef]
67. Günther, J.; Krewerth, D.; Lippmann, T.; Leuders, S.; Tröster, T.; Weidner, A.; Biermann, H.; Niendorf, T. Fatigue life of additively manufactured Ti–6Al–4V in the very high cycle fatigue regime. *Int. J. Fatigue* **2017**, *94*, 236–245. [CrossRef]
68. Pan, X.; Hong, Y. High-cycle and very-high-cycle fatigue behaviour of a titanium alloy with equiaxed microstructure under different mean stresses. *Fatigue Fract. Eng. Mater. Struct.* **2019**, *42*, 1950–1964. [CrossRef]
69. Pan, X.; Qian, G.; Wu, S.; Fu, Y.; Hong, Y. Internal crack characteristics in very-high-cycle fatigue of a gradient structured titanium alloy. *Sci. Rep.* **2020**, *10*, 4742. [CrossRef]
70. Heinz, S.; Balle, F.; Wagner, G.; Eifler, D. Analysis of fatigue properties and failure mechanisms of Ti6Al4V in the very high cycle fatigue regime using ultrasonic technology and 3D laser scanning vibrometry. *Ultrasonics* **2014**, *53*, 1433–1440. [CrossRef]
71. Yang, K.; He, C.; Huang, Q.; Huang, Z.Y.; Wang, C.; Wang, Q.; Liu, Y.J.; Zhong, B. Very high cycle fatigue behaviors of a turbine engine blade alloy at various stress ratios. *Int. J. Fatigue* **2017**, *99*, 35–43. [CrossRef]
72. Morrissey, R.J.; Nicholas, T. Staircase testing of a titanium alloy in the gigacycle regime. *Int. J. Fatigue* **2006**, *28*, 1577–1582. [CrossRef]
73. Takeuchi, E.; Furuya, Y.; Nagashima, N.; Matsuoka, S. The effect of frequency on the giga-cycle fatigue properties of a Ti–6Al–4V alloy. *Fatigue Fract. Engng. Mater. Struct.* **2008**, *31*, 599–605. [CrossRef]
74. Furuya, Y.; Takeuchi, E. Gigacycle fatigue properties of Ti–6Al–4V alloy under tensile mean stress. *Mater. Sci. Eng. A Struct.* **2014**, *598*, 135–140. [CrossRef]
75. Wycisk, E.; Siddique, S.; Herzog, D.; Walther, F.; Emmelmann, C. Fatigue Performance of Laser Additive Manufactured Ti–6Al–4V in Very High Cycle Fatigue Regime up to $10^9$ Cycles. *Frontiers* **2015**, *2*, 72.
76. Papakyriacou, M.; Mayer, H.; Pypen, C.H.P., Jr.; Stanzl-Tschegg, S. Influence of Loading Frequency on High-Cycle Fatigue Properties of b.c.c. and h.c.p. Metals. *Mater. Sci. Eng. A* **2001**, *308*, 143–152. [CrossRef]

77. Szczepanski, C.J.; Jha, S.K.; Larsen, J.M.; Jones, J.W. Microstructural Influences on Very-High-Cycle Fatigue-Crack Initiation in Ti-6246. *Met. Mater. Trans. A* **2008**, *39*, 2841–2851. [CrossRef]
78. Miao, J.; Pollock, T.M.; Wayne Jones, J. Crystallographic fatigue crack initiation in nickel-based superalloy René 88DT at elevated temperature. *Acta Mater.* **2009**, *57*, 5964–5974. [CrossRef]
79. Okazaki, S.; Takakuwa, O.; Ogawa, Y.; Okita, K.; Funakoshi, Y.; Yamabe, J.; Matsuoka, S.; Matsunaga, H. Effect of defects on the fatigue limit of Ni-based superalloy 718 with different grain sizes. *Fatigue Fract. Eng. Mater. Struct.* **2019**, *42*, 1203–1213.
80. Stöcker, C.; Zimmermann, M.; Christ, H.-J. Localized cyclic deformation and corresponding dislocation arrangements of polycrystalline Ni-base superalloys and pure Nickel in the VHCF regime. *Int. J. Fatigue* **2011**, *33*, 2–9. [CrossRef]
81. Shyam, A.; Torbet, C.; Jha, S.; Larsen, J.; Caton, M.; Szczepanski, C.; Pollock, T.; Jones, J. Development of Ultrasonic Fatigue for Rapid, High Temperature Fatigue Studies in Turbine Engine Materials. *Proc. Int. Symp. Superalloys* **2004**, *259*, 268.
82. Cervellon, A.; Cormier, J.; Mauget, F.; Hervier, Z.; Nadot, Y. Very High Cycle Fatigue of Ni-Based Single-Crystal Superalloys at High Temperature. *Met. Mater. Trans. A* **2018**, *49*, 3938–3950. [CrossRef]
83. Yi, J.Z.; Torbet, C.J.; Feng, Q.; Pollock, T.M.; Jones, J.W. Ultrasonic fatigue of a single crystal Ni-base superalloy at 1000 °C. *Mater. Sci. Eng. A Struct.* **2007**, *443*, 142–149. [CrossRef]
84. Michel, S.A.; Kieselbach, R.; Martens, H.J. Fatigue strength of carbon fibre composites up to the gigacycle regime (gigacycle-composites). *Int. J. Fatigue* **2006**, *28*, 261–270. [CrossRef]
85. Flore, D.; Wegener, K.; Mayer, H.; Karr, U.; Oetting, C.C. Investigation of the high and very high cycle fatigue behaviour of continuous fibre reinforced plastics by conventional and ultrasonic fatigue testing. *Compos. Sci. Technol.* **2017**, *141*, 130–136. [CrossRef]
86. Lee, C.S.; Kim, H.J.; Amanov, A.; Choo, J.H.; Kim, Y.K.; Cho, I.S. Investigation on very high cycle fatigue of PA66-GF30 GFRP based on fiber orientation. *Compos. Sci. Technol.* **2019**, *180*, 94–100. [CrossRef]
87. Backe, D.; Balle, F. Ultrasonic fatigue and microstructural characterization of carbon fiber fabric reinforced polyphenylene sulfide in the very high cycle fatigue regime. *Compos. Sci. Technol.* **2016**, *126*, 115–121. [CrossRef]
88. Cui, W.; Chen, X.; Chen, C.; Cheng, L.; Ding, J.; Zhang, H. Very High Cycle Fatigue (VHCF) Characteristics of Carbon Fiber Reinforced Plastics (CFRP) under Ultrasonic Loading. *Materials* **2020**, *13*, 908. [CrossRef] [PubMed]
89. Ding, J.; Cheng, L. Ultra-high three-point bending fatigue performance of nano-silica-reinforced CFRP. *Int. J. Fatigue* **2021**, *145*, 106085. [CrossRef]
90. Ding, J.; Cheng, L. Ultra-high three-point bending fatigue fracture characteristics of CFRP modified by MWCNTs and fatigue life data analysis. *Compos. Struct.* **2021**, *259*, 113468. [CrossRef]
91. Mayer, H.; Papakyriacou, M. Fatigue behaviour of graphite and interpenetrating graphite–aluminium composite up to 109 load cycles. *Carbon* **2006**, *44*, 1801–1807. [CrossRef]
92. Stanzl-Tschegg, S.E.; Papakyriacou, M.; Mayer, H.R.; Schijve, J.; Tschegg, E.K. High Cycle Fatigue Crack Growth Properties of Aramid Reinforced Aluminum Laminates. In *Composite Materials: Fatigue and Fracture, Vol. IV*; Stinchcomb, W.W., Ashbaugh, N.E., Eds.; ASTM: Warminster, PS, USA, 1993; Volume ASTM STP 1156, pp. 637–652.
93. Papakyriacou, M.; Schijve, J.; Stanzl-Tschegg, S.E. Fatigue Crack Growth Behaviour of Fiber-Metal Laminate Glare 1 and Metal Laminate 7475 with Different Blunt Notches. *Fatigue Fract. Eng. Mater. Struct.* **1997**, *20*, 1573–1584. [CrossRef]
94. Fitzka, M.; Catoor, D.; Irrasch, D.; Reiterer, M.; Mayer, H. Fatigue testing of thin CoNiCr wire up to $10^{10}$ cycles. *Int. J. Fatigue* **2017**, *98*, 92–100. [CrossRef]
95. Karr, U.; Schuller, R.; Fitzka, M.; Denk, A.; Strauss, A.; Mayer, H. Very high cycle fatigue testing of concrete using ultrasonic cycling. *Mater. Test.* **2017**, *59*, 438–444. [CrossRef]
96. Stanzl-Tschegg, S.E. Influence of material properties and testing frequency on VHCF and HCF lives of polycrystalline copper. *Int. J. Fatigue* **2017**, *105*, 86–96. [CrossRef]
97. Fitzka, M.; Rennhofer, H.; Catoor, D.; Reiterer, M.; Lichtenegger, H.; Checchia, S.; di Michiel, M.; Irrasch, D.; Gruenewald, T.A.; Mayer, H. High Speed In Situ Synchrotron Observation of Cyclic Deformation and Phase Transformation of Superelastic Nitinol at Ultrasonic Frequency. *Exp. Mech.* **2020**, *60*, 317–328. [CrossRef]

materials

MDPI

Article

# Effect of Underload Cycles on Oxide-Induced Crack Closure Development in Cr-Mo Low-Alloy Steel

Pavel Pokorný [1], Tomáš Vojtek [1,2,*], Michal Jambor [1], Luboš Náhlík [1] and Pavel Hutař [1]

1 Institute of Physics of Materials, Czech Academy of Sciences, Žižkova 22, 616 00 Brno, Czech Republic; pokorny@ipm.cz (P.P.); jambor@ipm.cz (M.J.); nahlik@ipm.cz (L.N.); hutar@ipm.cz (P.H.)
2 Central European Institute of Technology (CEITEC), Brno University of Technology, Purkyňova 123, 612 00 Brno, Czech Republic
* Correspondence: vojtek@ipm.cz

**Abstract:** Underload cycles with small load amplitudes below the fatigue crack growth threshold are dominantly considered as insignificant cycles without any influence on fatigue lifespan of engineering structural components. However, this paper shows that in some cases these underload cycles can retard the consequent crack propagation quite significantly. This phenomenon is a consequence of oxide-induced crack closure development during cyclic loading below the threshold. The experimentally described effect of fatigue crack growth retardation was supported by measurement of the width and the thickness of the oxide debris layer using the EDS technique and localized FIB cuts, respectively. Both the retardation effect and the amount of oxide debris were larger for higher number and larger amplitudes of the applied underload cycles. Crack closure measurement revealed a gradual increase of the closure level during underload cycling. Specimens tested in low air humidity, as well as specimens left with the crack open for the same time as that needed for application of the underload cycles, revealed no retardation effect. The results can improve our understanding of environmental effects on fatigue crack propagation and understanding the differences between the results of laboratory testing and the fatigue lives of components in service.

**Keywords:** underload cycles; crack closure; threshold; fatigue crack growth; oxidation; EA4T steel

**Citation:** Pokorný, P.; Vojtek, T.; Jambor, M.; Náhlík, L.; Hutař, P. Effect of Underload Cycles on Oxide-Induced Crack Closure Development in Cr-Mo Low-Alloy Steel. *Materials* **2021**, *14*, 2530. https://doi.org/10.3390/ma14102530

Academic Editor: Carlos Garcia-Mateo

Received: 28 February 2021
Accepted: 26 April 2021
Published: 13 May 2021



## 1. Introduction

Fatigue crack propagation is an important stage of the fatigue failure of engineering components, especially in applications where the damage-tolerance design is employed. In these applications, components are regularly inspected for the presence of fatigue cracks using non-destructive techniques, which brings about the need for a high precision of predictions of residual fatigue life (RFL). The near-threshold regime is the most significant area to study, since the majority of RFLs are associated with the beginning of crack propagation, where the stress intensity factors are small. In addition, the majority of loading cycles in the spectra of variable amplitude loading usually has small amplitudes. Moreover, this area is characterized by many unresolved theoretical questions. In current research, it is desirable to improve our understanding of mechanisms of resistance to fatigue crack growth (FCG) in the near-threshold regime, where the crack closure effects represent a significant part of the resistance, and they are very difficult to predict due to a lack of quantitative models. Crack closure effects are also responsible for the generally large scatter of reported threshold values for particular metallic materials.

The crack closure phenomenon was firstly described in [1] and is a crack tip shielding mechanism [2] diminishing the crack driving force. Due to premature contact of fracture surfaces, the cyclic plastic zone is reduced and, therefore, crack propagation is slower than in the case of the absence of crack closure effects (such as at high load ratios). The most commonly considered mechanisms of crack closure are due to plastically deformed materials in the crack wake, roughness of fracture surfaces, and the presence of enhanced oxide debris at fracture surfaces, as illustrated schematically in Figure 1. More details about crack closure mechanism can be found in the literature, e.g., [3–5]. While plasticity-induced crack closure (PICC) is present in both the near-threshold and the Paris regime, roughness-induced crack closure (RICC) the oxide-induced crack closure (OICC) are significant only in the near-threshold regime owing to small cyclic crack opening displacements.

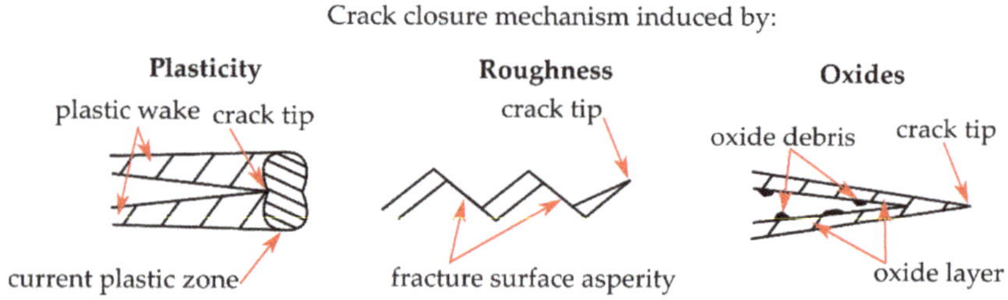


**Figure 1.** Mechanisms of crack closure due to plastic deformation in the crack wake, fracture surfaces roughness and oxide debris produced on fracture surfaces.

The effective (closure-free) threshold in steels is usually about 3 $MPam^{1/2}$, while the measured thresholds in terms of $K_{max}$ are often two or three times larger. The level of plasticity-induced crack closure (PICC) should result in no more than 1/3 of $K_{max}$ in most situations, which means that there is a gap of about 2 to 5 $MPam^{1/2}$ needed to account for RICC and OICC. These two components are difficult to quantify, both experimentally and theoretically. The dependence of the threshold on grain size has been traditionally attributed to the role of RICC. However, it was not clear until recently, what are the proportions between RICC, OICC and the whole threshold value. Experiments in dry air [6] provided one of the first hints on this matter, where the differences between thresholds in dry air and humid air were found to be significant. For railway axle steel EA4T, they were 4 $MPam^{1/2}$ and 7 $MPam^{1/2}$, respectively, in terms of $K_{max}$ for the load ratios $R = 0.1$ and $R = -1$. Thus, the role of OICC was much more significant than the role of RICC in this material. There are experimental indications that such a high significance of OICC for threshold values is also valid for other corroding steels [7] and that there is no physical reason to think the otherwise. Thus, the effect of OICC is a good candidate for an explanation of the differences in threshold values measured under various influencing factors. One of these influences is the load shedding technique, often applied to measure thresholds.

Different fatigue crack propagation rates in the near-threshold area were measured for EA4T steel during load shedding and during subsequent loading with an increasing $\Delta K$, as shown in Figure 2. The low crack growth rates during increasing $\Delta K$ can be explained by the gradual development of an oxide debris layer at fracture surfaces during the load shedding procedure and, consequently, a larger oxide-induced crack closure [6,8,9]. A similar behavior of oxide layer development can be expected for cyclic loading below the threshold.

A majority of laboratory experiments have generally been conducted under constant amplitude or quasi-constant amplitude loading, where only slow gradual changes of $\Delta K$ are applied. However, engineering components in operation are typically subjected to variable amplitude loading (VAL). The load-interaction effects influence RFL in terms of both retardation and acceleration of the crack. The VAL spectra usually have such character that the lowest amplitudes have the highest number of cycles; therefore, not only are the majority of damaging cycles in the near-threshold area, but there is also a large number of non-damaging cycles just below the threshold. The available models of crack propagation under VAL (e.g., [10,11]) include load-interaction effects based on PICC or on the development of residual near-tip stresses. The number of studies on the retardation effect due to OICC is very limited. Only very few publications can be found in the literature on the topic of the effects of underload cycles [12–14].

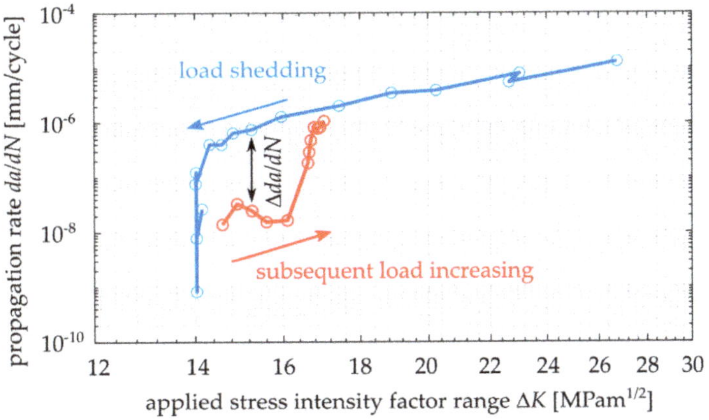


**Figure 2.** Fatigue crack retardation after load shedding procedure in $da/dN$-$K_{max}$ curve, data experimentally obtained for Cr-Mo low-alloy steel, $R = -1$.

Referring to the above-mentioned facts, studying the influence of underload cycles is important for structural components in application, as well as for a deeper understanding of fatigue crack propagation mechanisms. In the literature many papers can be found on the topic of the effects of underloads. However, a majority of them, e.g., [15–18], are related to intense compressive overloads or a reduced mean load (see the underloads of the type A and B in Figure 3a,b). In other words, "underload" means a significant reduction (shift to negative values) of the minimal load. The underload cycles in this work are studied in terms of a reduction of amplitude, while the load ratio is kept constant ($R = -1$). It means that both maximal and minimal values during the loading cycle are shifted closer to the mean load (see the type C underload in Figure 3c). Moreover, the underload cycles considered in this paper are below the threshold of crack propagation, which means that these cycles are non-damaging and are usually considered to have no impact on RFL.

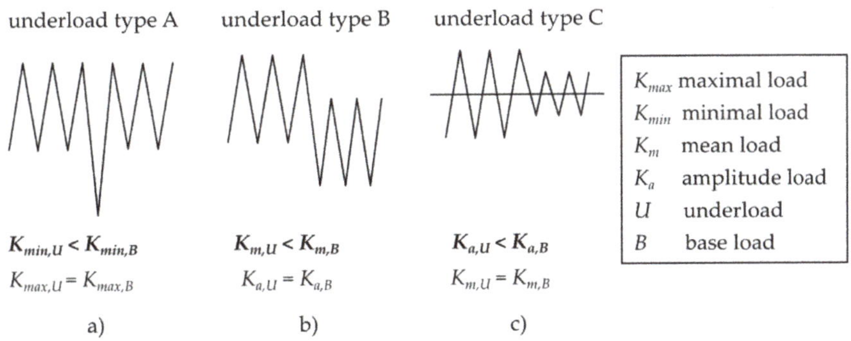


**Figure 3.** The three types of underloads considered in literature: (**a**) a single underload with the shift of $K_{min}$, (**b**) underloads with reduced mean load, and (**c**) underloads with reduced load amplitude.

In [12], it was reported that many of the cycles below threshold contributed to crack growth retardation, when the base load was applied after the underloads. More details about this topic for railway axle steels EA4T and EA1N, tested at load ratios $R = 0.1$ and $-0.5$, are published in [13]. A comprehensive paper by Suresh and Ritchie [14] presented the effects of underloads (type C in Figure 3c) in 2.25Cr-1Mo pressure vessel steel at an ambient temperature and 30% relative humidity. A block of 2,700,000 underload cycles was inserted between base load cycles and the role of the underload cycles on subsequent crack propagation rate was examined (see Figure 4). At loading below the threshold, the crack does not propagate; however, due to cyclic crack opening and closing an oxide debris layer gradually grows.

The presented paper is focused on the role of loading cycles with amplitudes below the threshold on subsequent crack retardation due to the mechanism of OICC in the Cr-Mo low-alloy steel designated EA4T, according to railway standards. The aim is to quantify this retardation effect and the corresponding development of oxide debris layer.

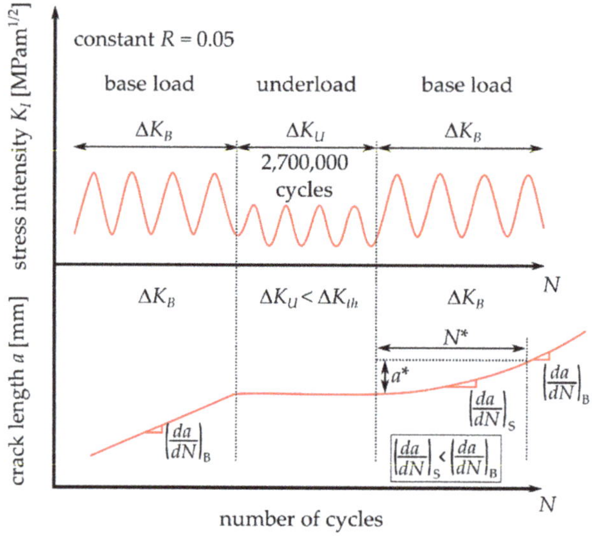


**Figure 4.** A schematic illustration of the loading sequence used by Suresh and Ritchie in [14]. Adapted with permission from ref. [14]. 1981 Elsevier.

## 2. Experimental Procedure and Methodology

All tests were performed on specimens made of the EA4T steel, which is probably the most commonly used material worldwide for the manufacturing of driving railway axles. Therefore, knowledge of its behavior is important for the safety of train operations. Table 1 shows the chemical composition of this steel. The manufacturing process and heat treatment of the axles lead to a microstructure with a dominant portion of bainite. An optical microscopy image of the etched metallography cut of the material is shown in Figure 5.

**Table 1.** Chemical composition of EA4T steel, according to [19].

| Component | C | Si | Mn | P | S | Cr | Cu | Mo | Ni | V | Fe |
|---|---|---|---|---|---|---|---|---|---|---|---|
| min [wt.%] | 0.22 | 0.15 | 0.50 | 0.00 | 0.000 | 0.90 | 0.00 | 0.00 | 0.00 | 0.00 | rest |
| max [wt.%] | 0.29 | 0.40 | 0.80 | 0.02 | 0.015 | 1.20 | 0.30 | 0.30 | 0.30 | 0.06 | |

Table 2 presents selected mechanical properties of the investigated EA4T steel. The parameters related to fatigue crack propagation were experimentally determined using the middle-crack tension (MT) specimen (shown in Figure 6). This type of specimen was also used in experimental investigations of underloads effects in this paper. Note that the specimens were manufactured from the surface part of a real axle in order to more realistically capture the state of the material, see details in [20].

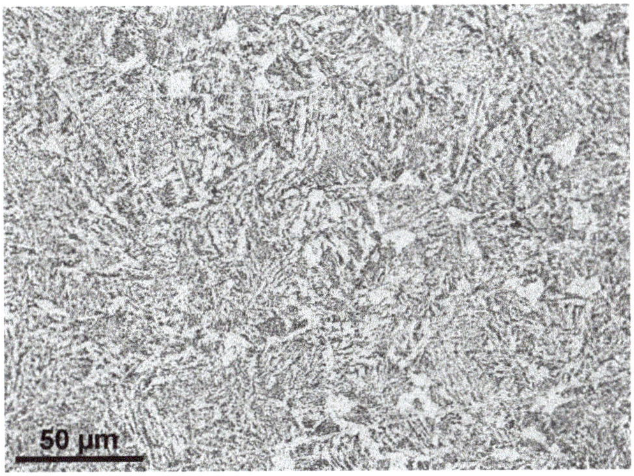


**Figure 5.** Microstructure of the investigated Cr-Mo low-alloy steel (EA4T steel).

**Table 2.** Mechanical properties of the investigated EA4T steel. Values related to fatigue crack behavior were determined on the middle crack tension specimens (MT) with the width of 60 mm, thickness of 5 mm at 23 °C and 50% relative humidity.

| Mechanical Property | Value |
|---|---|
| Yield stress | $S_y$ = 600 MPa |
| Cyclic yield stress | $S_{y,c}$ = 470 MPa |
| Ultimate tensile strength | $S_U$ = 720 MPa |
| Young's modulus | $E$ = 204 GPa |
| Poisson's ratio | $\nu$ = 0.3 |
| Threshold value for $R = -1$ (MT specimen) | $\Delta K_{th}$ = 13.8 MPam$^{1/2}$ |
| Threshold value for $R = 0.8$ (MT specimen) | $\Delta K_{th}$ = 2.9 MPam$^{1/2}$ |
| Paris–Erdogan constant $C$ for $R = -1$ | $C = 1.53 \times 10^{-10}$ for $da/dN$ in [mm/cycle] |
| Paris–Erdogan constant $m$ for $R = -1$ | $m$ = 3.46 |
| Paris–Erdogan constant $C$ for $R = 0.8$ | $C = 1.36 \times 10^{-8}$ for $da/dN$ in [mm/cycle] |
| Paris–Erdogan constant $m$ for $R = 0.8$ | $m$ = 2.72 |

All of the used MT specimens included a sharp central notch produced by electro discharge machining (EDM). A chevron notch was used in order to initiate a symmetrical fatigue crack, (detail B in Figure 6). Experimental tests were carried out on a resonant Schenck PVQ machine with a testing frequency of about 50 Hz (depending on specimen stiffness loading amplitude). The majority of experimental tests were carried out in laboratory air with controlled temperature and humidity. The temperature was set to 23 °C and the absolute humidity was set to 10 g/m$^3$ (relative humidity of ca. 50%). Some of the tests were carried out using the originally designed chamber for tests in very dry air, see Figure 7. This chamber surrounds the central part of the MT specimen and it is filled with silica gel particles to eliminate air humidity (relative humidity is below 15%). Note that such a low humidity significantly retards the development of oxide debris during crack propagation [6,21].

The crack lengths and crack increments were measured optically using two digital Basler acA2040-55um cameras (with the CMOS sensor, resolution of 3.2 Mpx, frame rate 55 fps) and Azure-2514MM objectives (with the format 2/3", manual iris F1.4–F32, focal length 25 mm). The cameras were fit to movable tables and were focused on the fatigue crack tip. The positions of the cameras were measured using Mitutoyo ID-C125XB universal digital indicators (with the resolution of 0.001 mm and the accuracy of 0.003 mm).

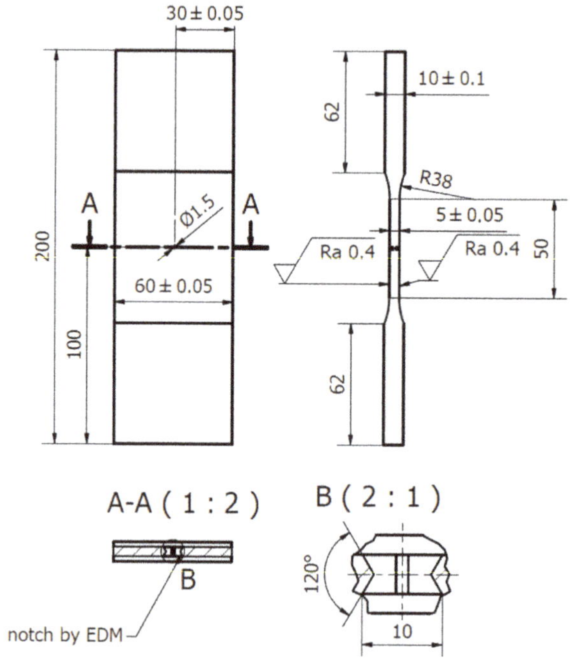


**Figure 6.** Scheme of the MT specimen used for experimental determination of the underload effects in the EA4T steel. (Unit: mm).

Each test was performed at a load ratio $R = -1$. Fatigue cracks were initiated from the machined notch with the length of 10 mm (see Figure 6) via cyclic loading at an amplitude of 23 kN. When a fatigue crack measuring 1 mm in length was initiated on both sides of the specimen, the procedure of load reduction by approx. 5% per crack increment of 0.1 mm was stared. The load shedding procedure continued until the crack growth rate reached $2 \times 10^{-6}$ mm/cycle, which was selected as the growth rate where the production of oxide debris is not very intense yet, but the load is small enough to have easily detectable retardation effect after the enhanced oxide layer is developed during the underload cycles.

Next, a certain number of underload cycles was applied, which ranged from 0 up to 30,000,000 cycles. After that, cyclic loading at the same amplitude as before the underload cycles was started (further referred to as the "base load cycling"), see Figure 8. During that loading, the number of influenced load cycles was investigated. It was assumed that the amount of influenced load cycles $N^*$ corresponds to the number of cycles until the crack growth rate reaches again the base-load rate $(da/dN)_B = 2 \times 10^{-6}$ mm/cycle, see Figure 8. The crack extension during this period was denoted as the influenced zone, $a^*$.

Figure 9 shows a typical difference between the data obtained during load shedding (before application of the underload cycles) and the data obtained after the underload cycles. Note that this behavior is very similar to the behavior depicted in Figure 2, where, after the load shedding, no underloads were applied. However, the underload cycles result in a more pronounced effect of crack retardation.

In order to mathematically express the number of influenced base load cycles after the underload cycles $N^*$ (see Figure 8), the *a-N* data after the underload cycles were plotted and fitted by a polynomial function of the 3rd order:

$$a = k_3N^3 + k_2N^2 + k_1N + k_0 \quad (1)$$

where $k_0$, $k_1$, $k_2$ and $k_3$ are the polynomial constants, $a$ is the crack length and $N$ is the number of cycles after the last underload cycle. The typical evolution of the crack length $a$ with the number of applied base load cycles $N$ is shown in Figure 10.

Estimation of the number of influenced base load cycles $N^*$ after application of the underload cycles was done using the derivative of Equation (1):

$$\frac{da}{dN} = 3k_3N^2 + 2k_2N + k_1 \quad (2)$$

The crack propagation rate before application of the underloads was $2 \times 10^{-6}$ mm/cycle. When this *da/dN* rate is substituted into Equation (2), $N^*$ can be expressed as follows.

$$\begin{gathered} \frac{da}{dN} = 2 \times 10^{-6} = 3k_3N^{*2} + 2k_2N^* + k_1 \\ \rightarrow N^* = \frac{-2k_2 + \sqrt{(2k_2)^2 - 4 \times 3k_3.(k_1 - 2 \times 10^{-6})}}{2 \times 3k_3} \end{gathered} \quad (3)$$

When the number of influenced load cycles after the underload cycles is known, the influenced zone $a^*$ (see Figure 8) can be estimated by substitution of $N^*$ into Equation (1) with a subtraction of the crack length after the underload cycles $a_0$:

$$a^* = k_3N^{*3} + k_2N^{*2} + k_1N^* + k_0 - a_0 \quad (4)$$

where $a^*$ is the influenced zone size, $N^*$ is the number of influenced base load cycles, see Figure 8, and $a_0$ is the crack length before, during and after the underload cycles (the crack does not propagate). Equations (3) and (4) were used for calculation of the data summarized in Table 3.

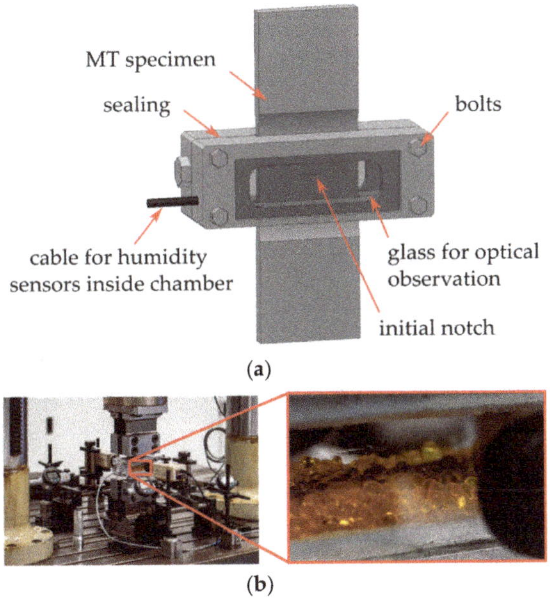


**Figure 7.** Special test setup for tests in dry air: (**a**) scheme of the sealed chamber and (**b**) image of the experiment showing the crack surrounded by silica gel particles.

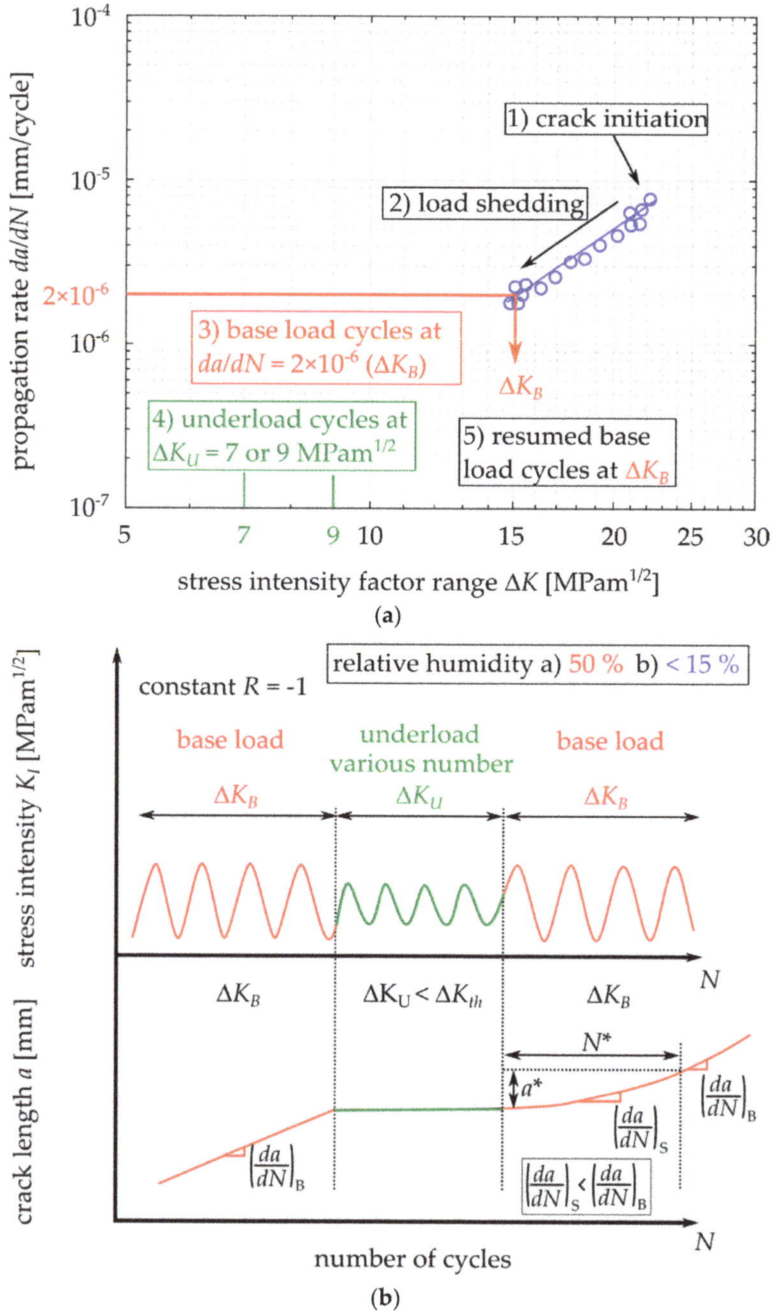


**Figure 8.** Scheme of the conducted experimental procedure. (**a**) Procedure steps in terms of the crack growth rate diagram; (**b**) Loading sequences. Unlike in Figure 4, various numbers of underload cycles were applied and the humidity conditions were varied.

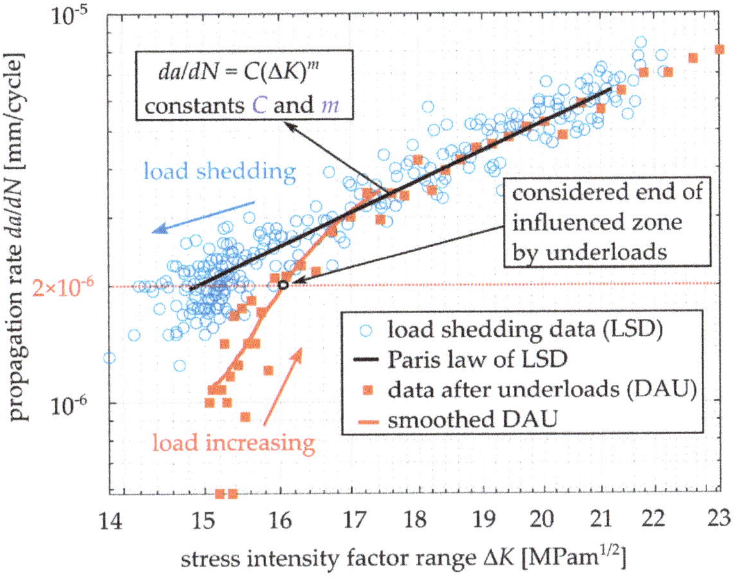


**Figure 9.** Demonstration of different crack propagation rates before and after application of underload cycles.

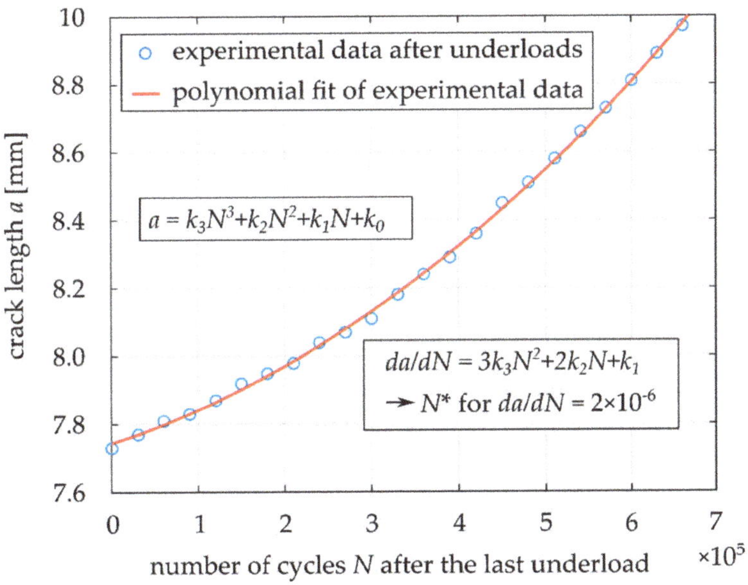


**Figure 10.** Determination of the number of influenced base load cycles after application of the underload cycles based on the fitting equation.

**Table 3.** Experimentally obtained results of the underload effect on crack propagation behavior: $N_U$—number of applied underload cycles, $t_U$—time related to underload cycling or time between the base loading segments, $\Delta K_U$ —level of underloads, *RH*—relative humidity, *N**—number of influenced load cycles according to Equation (3), *a**—influenced zone due to underloads according to Equation (4).

| Specimen | $N_U$ [-] | $t_U$ [h] | $\Delta K_U$ [MPam$^{1/2}$] | *RH* [%] | $N^*$ [-] | $a^*$ [mm] |
|---|---|---|---|---|---|---|
| 1 | 3,000,000 | 17 | 7 | 50 | 375,410 | 0.54 |
| 2 | 3,000,000 | 17 | 9 | 50 | 711,250 | 1.02 |
| 3 [1] | 0 | 17 | static load at *Kmax* | 50 | 0 | 0.00 |
| 4 [2] | 0 | 17 | 0 (no load) | 50 | 0 | 0.00 |
| 5 | 3,000,000 | 17 | 9 | 50 | 706,976 | 1.13 |
| 6 | 3,000,000 | 17 | 7 | 50 | 527,365 | 0.79 |
| 7 [3] | 0 | 0 | no underloads | 50 | 0 | 0.00 |
| 8 | 300,000 | 2 | 9 | 50 | 318,761 | 0.58 |
| 9 | 1,000,000 | 6 | 9 | 50 | 187,009 | 0.33 |
| 10 | 1,000,000 | 6 | 9 | 50 | 75,342 | 0.14 |
| 11 | 3,000,000 | 17 | 9 | 50 | 397,936 | 0.61 |
| 12 | 21,600,000 | 122 | 9 | 50 | 1,084,990 | 1.05 |
| 13 | 16,440,000 | 93 | 9 | 50 | 784,028 | 1.10 |
| 14 | 30,000,000 | 170 | 9 | 50 | 949,323 | 1.31 |
| 15 | 7,220,000 | 41 | 9 | 50 | 827,191 | 1.13 |
| 16 | 3,000,000 | 17 | 9 | <15 | 78,627 | 0.14 |
| 17 [4] | failed | | 9 | <15 | failed | |
| 18 | 15,600,000 | 88 | 7 | <15 | 25,158 | 0.03 |

[1] after reaching $da/dN = 2 \times 10^{-6}$ mm/cycle, specimen 3 was loaded by a static load of $K_{max}$ = 4.5 MPam$^{1/2}$ for 17 h (the same time as specimen 1 and 2), [2] no loading was applied at all (either underloads or static load) for 17 h, then the base load cycling was resumed, [3] in the case of Specimen 7, only the load shedding procedure down to the rate of $2 \times 10^{-6}$ mm/cycle was applied with immediate base load cycling at this level (no loading or pause was applied), [4] Specimen 17 failed during the underload cycles due to low humidity that resulted in a lower threshold than $\Delta K_U$ = 9 MPam$^{1/2}$ (see [21]) and in subsequent crack propagation.

## 3. Results

### 3.1. Fatigue Crack Propagation Rates after Application of Underloads

Table 3 shows the results obtained for 18 MT specimens. In the case of Specimen 1, 3,000,000 underload cycles were applied at a chosen underload level $\Delta K_U$ = 7 MPam$^{1/2}$ ($R = -1$). Application of this number of cycles took ca. 17 h. Subsequently, the base load amplitude was reapplied and the crack growth rate increased to $da/dN = 2 \times 10^{-6}$ mm/cycle after $N^*$ = 375,410 load cycles, which corresponded to the influenced zone $a^*$ = 0.54 mm. In the case of Specimen 2, the procedure was the same, only the level of underload cycles was $\Delta K_U$ = 9 MPam$^{1/2}$. In this case, there were 711,250 influenced load cycles with an influenced zone of 1.02 mm. The difference can be explained by a higher compressive stress at the minimum load during the cycle at $R = -1$. Additionally, a larger crack opening displacement (COD) in the case of loading $\Delta K_U$ = 9 MPam$^{1/2}$ is available to accommodate more oxide debris, compared to the COD at $\Delta K_U$ = 7 MPam$^{1/2}$, allowing a higher level of OICC.

Instead of the underload cycles, Specimen 3 was loaded only by a static load at $K_{max}$ = 4.5 MPam$^{1/2}$ (corresponding to $\Delta K_U$ = 9 MPam$^{1/2}$ at $R = -1$) for 17 h. This aimed to reveal whether the retardation effect is related to the underload cycles or the time that passed without any damaging load cycles. The results showed that even though the crack is opened and accessible for diffusion of oxygen and water vapor, no crack retardation under the subsequent base load occurred (see Table 3). In the case of Specimen 4, the procedure was the same as for Specimen 3, but the load for 17 h was kept at zero. As was expected, Specimen 4 did not reveal any retardation effect either. In the case of Specimen 7, no underloads were applied after load shedding and the base load was immediately restarted. This experiment, again, did not lead reveal any significant retardation effect.

Specimens 16–18 were used for experiments in reduced air humidity, see Figure 7. The low humidity air (relative humidity below 15%) led to an insignificant effect of the underload cycles in comparison with the previous experiments in air with a relative humidity of 50%. This supported the idea that the OICC mechanism was responsible for the observed retardation after underloads. Note that the low humidity air leads to a significantly lower threshold of about $\Delta K_{th}$ = 8.8 $MPam^{1/2}$, see [21]. This fact resulted in failure of Specimen 17, where the "underload" cycles at $\Delta K_U$ = 9 $MPam^{1/2}$ were not really non-damaging. Interestingly, Specimen 16 was also loaded by $\Delta K_U$ = 9 $MPam^{1/2}$ but the crack did not grow during 3,000,000 underload cycles. This showed that the applied $\Delta K_U$ was really very close to the low-humidity threshold.

For a better presentation of the obtained results, Figures 11 and 12 show the dependences of $N^*$ (number of influenced base load cycles) and $a^*$ (influenced zone size) on the number of applied underload cycles $N_u$. The shapes of both diagrams are very similar. There is a relatively large sensitivity up to ca. 500,000 underload cycles and the dependence is quite flat for higher numbers of underloads. However, there was no clear saturation, even after $10^7$ underload cycles, especially in the case of $N^*$ (Figure 12).

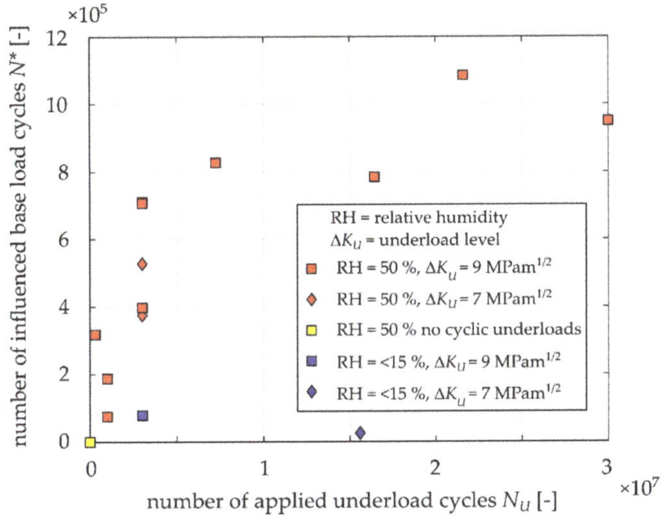


**Figure 11.** Experimentally obtained numbers of influenced load cycles $N^*$ after underloads.

In order to have even better expression of the effect of underloads, the numbers of base load cycles needed for crack elongation by 0.5 mm and 1.0 mm were determined for the tested specimens.

The value $N_{cal}$ represents the number of cycles necessary for the crack extension by 0.5 mm or 1.0 mm without the effect of underloads. The Paris–Erdogan law, with constants given by the load-shedding measurement, was used (see Figure 9). The value $N_{exp}$ represents the experimentally obtained number of base load cycles needed for crack extension by 0.5 mm or 1.0 mm after application of underload cycles. The relative retardation was calculated according to Equation (5).

$$retardation = \frac{N_{exp} - N_{cal}}{N_{cal}} \cdot 100\ [\%] \tag{5}$$

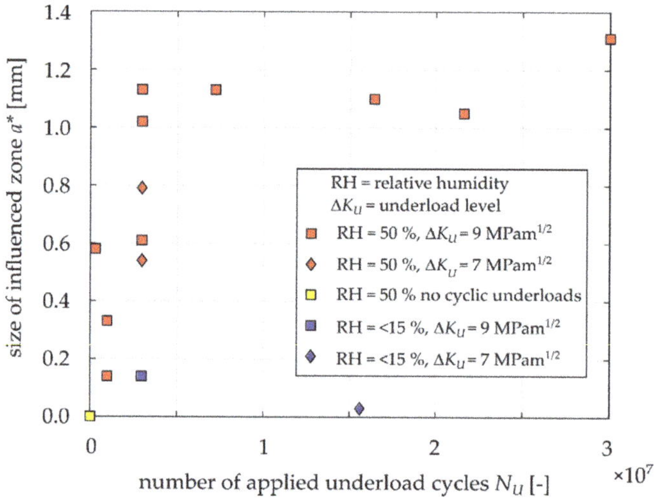


**Figure 12.** Experimentally obtained influenced zones $a^*$ after underloads.

Table 4 summarizes the obtained values for a crack extension of 0.5 mm, while Table 5 summarizes the values for a crack extension of 1.0 mm. One can see that, for a relative humidity of 50%, application of 21.6 or 30 million underload cycles leads to more than 100% retardation in comparison with the case of no underload cycles. For dry air (relative humidity below 15%), the effect of underload cycles almost completely vanished. The retardation effect also vanished when no underload cycles were applied (see Specimens 3, 4 and 7 in Table 4). Note that the retardation values of −2% to 2% should be considered more or less as zero due uncertainty in the data measurement.

**Table 4.** Numbers of base load cycles needed for crack increment of 0.5 mm with ($N_{exp}$) and without ($N_{cal}$) the effect of underload cycles.

| Specimen | $N_U$ [–] | *RH* [%] | $N_{cal}$ [–] | $N_{exp}$ [–] | Retardation |
|---|---|---|---|---|---|
| 1 | 3,000,000 | 50 | 234,187 | 354,545 | 51% |
| 2 | 3,000,000 | 50 | 256,174 | 412,941 | 61% |
| 3 | 0 | 50 | 218,250 | 237,522 | 9% |
| 4 | 0 | 50 | 230,871 | 228,529 | −1% |
| 5 | 3,000,000 | 50 | 255,019 | 350,167 | 37% |
| 6 | 3,000,000 | 50 | 230,516 | 376,923 | 64% |
| 7 | 0 | 50 | 236,181 | 235,000 | −1% |
| 8 | 300,000 | 50 | 238,211 | 275,069 | 15% |
| 9 | 1,000,000 | 50 | 220,487 | 255,732 | 16% |
| 10 | 1,000,000 | 50 | 213,365 | 234,512 | 10% |
| 11 | 3,000,000 | 50 | 234,073 | 307,286 | 31% |
| 12 | 21,600,000 | 50 | 243,481 | 577,167 | 137% |
| 13 | 16,440,000 | 50 | 226,622 | 422,500 | 86% |
| 14 | 30,000,000 | 50 | 212,756 | 443,360 | 108% |
| 15 | 7,220,000 | 50 | 233,103 | 416,842 | 79% |
| 16 | 3,000,000 | <15 | 238,956 | 247,619 | 4% |
| 18 | 15,600,000 | <15 | 246,423 | 245,294 | 0% |

**Table 5.** Numbers of base load cycles needed for crack increment of 1.0 mm with ($N_{exp}$) and without ($N_{cal}$) the effect of underload cycles.

| Specimen | $N_U$ [–] | *RH* [%] | $N_{cal}$ [–] | $N_{exp}$ [–] | Retardation |
|---|---|---|---|---|---|
| 1 | 3,000,000 | 50 | 449,134 | 568,065 | 26% |
| 2 | 3,000,000 | 50 | 495,105 | 703,637 | 42% |
| 3 | 0 | 50 | 410,749 | 447,720 | 9% |
| 4 | 0 | 50 | 433,477 | 441,892 | 2% |
| 5 | 3,000,000 | 50 | 487,622 | 643,033 | 32% |
| 6 | 3,000,000 | 50 | 442,682 | 628,571 | 42% |
| 7 | 0 | 50 | 446,130 | 457,531 | 3% |
| 8 | 300,000 | 50 | 454,380 | 517,891 | 14% |
| 9 | 1,000,000 | 50 | 416,774 | 460,091 | 10% |
| 10 | 1,000,000 | 50 | 401,090 | 438,100 | 9% |
| 11 | 3,000,000 | 50 | 448,896 | 548,273 | 22% |
| 12 | 21,600,000 | 50 | 473,971 | 1,048,283 | 121% |
| 13 | 16,440,000 | 50 | 437,490 | 741,000 | 69% |
| 14 | 30,000,000 | 50 | 412,623 | 795,167 | 93% |
| 15 | 7,220,000 | 50 | 449,260 | 720,000 | 60% |
| 16 | 3,000,000 | <15 | 455,752 | 466,667 | 2% |
| 18 | 15,600,000 | <15 | 468,201 | 459,167 | −2% |

### 3.2. Width of the Enhanced Oxide Debris Area

After the final fracture of the specimens, it was visible even by the naked eye, that some areas were significantly covered by oxide debris. To characterize the oxide layers, a Tescan Lyra 3 scanning electron microscope (SEM) equipped with a focused ion beam (FIB) column and Oxford Ultimax 100 detector for energy-dispersive X-ray spectroscopy (EDS) was used. The fracture surfaces were analyzed via EDS mapping to reveal the distribution of oxygen within the examined region. Parameters, such as accelerating voltage, beam current, working distance, pixel dwell time, etc., were kept unchanged for all tested specimens to ensure the comparability of the obtained results.

Figure 13 shows the fracture surfaces of chosen MT specimens, which were subjected to a block of underload cycles in 50% humidity air. The bright areas in the EDS maps correspond to the formed oxide layers/particles. The largest amount of oxide debris was in the central part of the specimen, while the oxide layers tended to disappear close to the free surfaces, see Figure 13. The maximal width of the oxide layer in the direction perpendicular to the crack front (not to be confused with thickness) was then measured in the maps for all examined specimens. In the case of specimens subjected to low humidity air (relative humidity below 15%), no continuous oxide layer was found, as can be seen for the representative specimen (Specimen 16) presented in Figure 14.

Figure 15 shows the measured widths of the area of oxide debris and the dependence on the applied number of underload cycles. The diagram reveals a clear increasing trend of the oxide layer width with an increasing number of applied underload cycles. After 30,000,000 underload cycles in 50% relative humidity air, the maximal width of the oxide layer was 337 μm. It could be expected from the obtained trend that this value is not a saturated one and that it is likely that this width would further increase with even higher numbers of underload cycles.

For the case of a specimen loaded by a static load of $K_{max}$ = 4.5 MPam$^{1/2}$ instead of the underloads (a possible situation corresponding to the downtime of a train), no continuous oxide layers were detected. Observation of Specimen 16, tested in dry air (relative humidity 6%), did not reveal any oxide layer even after 3,000,000 underload cycles (see the blue point in Figure 15), while the specimen tested in 50% relative humidity air under an equal number of underload cycles exhibited the maximal oxide layer width of 146 μm.

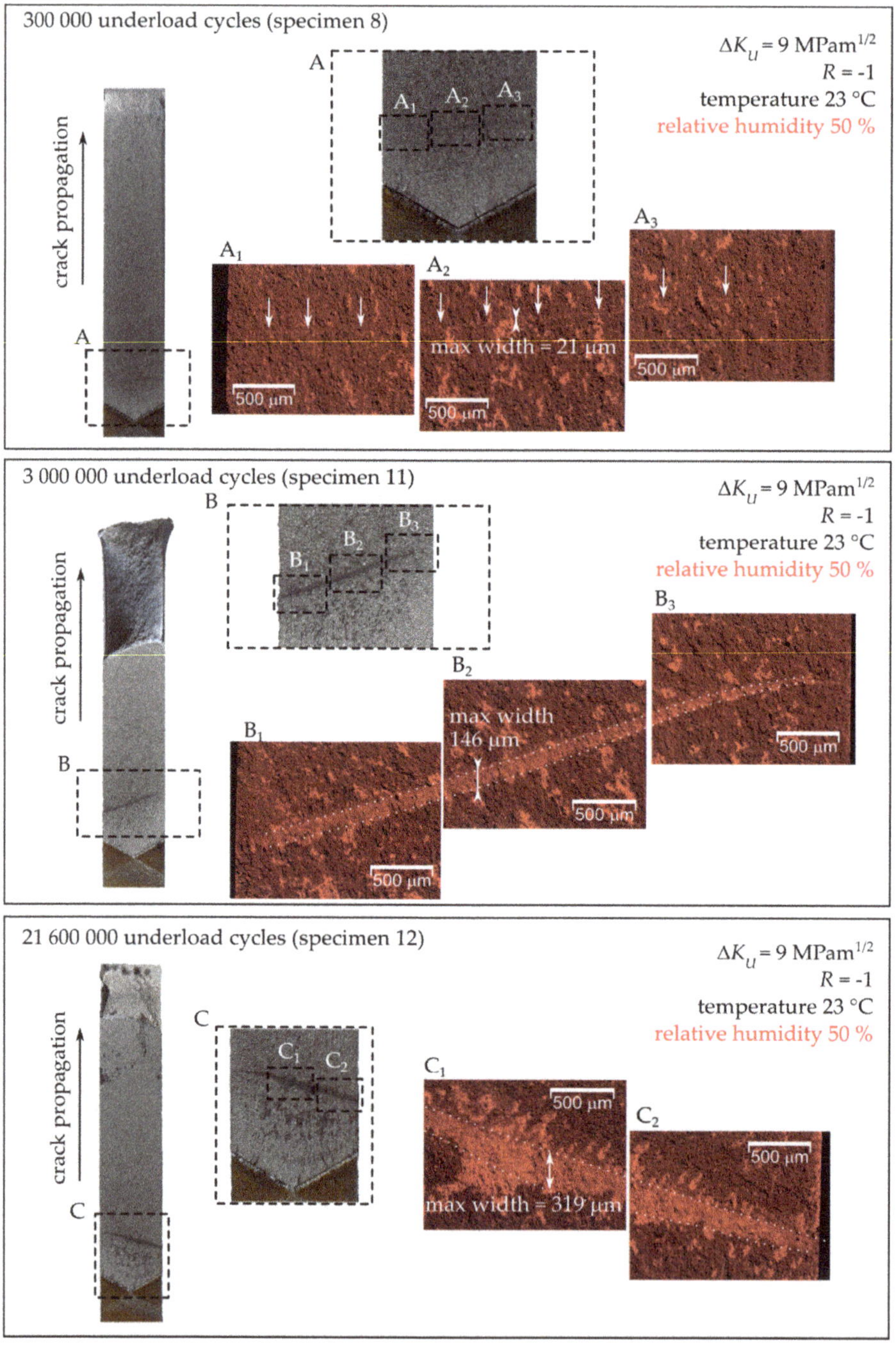


**Figure 13.** Fracture surfaces of Specimens 8, 11 and 12 tested in 50% relative humidity air (see also Table 3). On the left-hand side, the fracture surface of one quarter of the MT specimen with a detail is shown. The right-hand side shows the EDS maps of distribution of oxide particles.

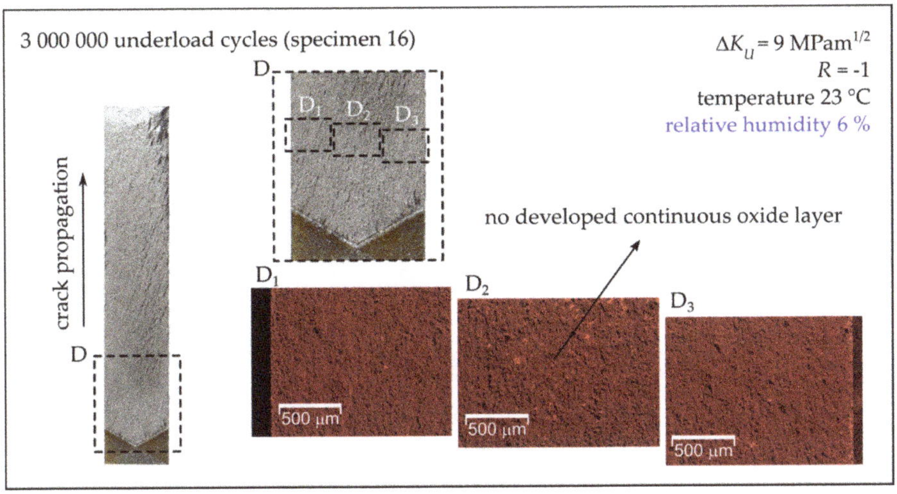


**Figure 14.** Fracture surfaces of Specimens 16 tested in 6% relative humidity air (see also Table 3). On the left-hand side, the fracture surface of one quarter of the MT specimen with a detail is shown. The right-hand side shows the EDS maps of distribution of oxide particles.

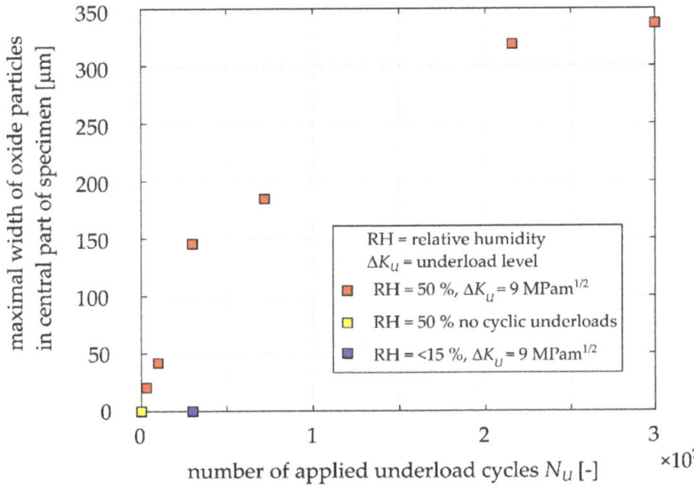


**Figure 15.** Determined maximal widths of oxide layers for the Specimens 3, 4, 7, 8, 10, 11, 12, 14, 15 and 16 based on EDS maps measurements.

### 3.3. Thickness of the Oxide Debris Layer

Apart from the width of oxidized area of the fracture surfaces, the maximal thickness of this layer was evaluated via milling of the cross section by a focused ion beam (FIB). Note that the oxide layers exhibited a large scatter in the thickness within one specimen; hence, the maximal value was considered as representative for mutual comparison. The FIB cut was made in approximately the same location as where the maximal width of the oxide layer was measured. Proper positioning of the FIB-cuts within the oxide layer was ensured by simultaneous observation of the EDS maps. To protect the oxide layer during milling process, a platinum layer was deposited on the top of the area of interest prior to ion milling procedure. Figure 16 shows the FIB cuts for Specimens 8, 11 and 12. The images reveal that the higher number of applied underload cycles, the thicker the oxide layer generated on fracture surfaces. In the case of Specimen 8, loaded by 300,000 underload cycles, the oxide layer thickness was 0.08 μm, whereas Specimen 12, loaded by more than 21,600,000 underload cycles, exhibited a maximal oxide layer thickness of 0.47 μm. The oxide layer thickness was evaluated for all specimens for which the oxidized zone width was measured (see Figure 15). The results are presented in Figure 17. Through comparison of Figures 15 and 17, it can be seen that the oxide layer thickness is about four orders of magnitude smaller than the oxide width. Nevertheless, the trends of both dependencies are very similar. Figure 17 also confirms that the absence of underload cycles, as well as reduced air humidity, lead to suppression of oxide layer development.

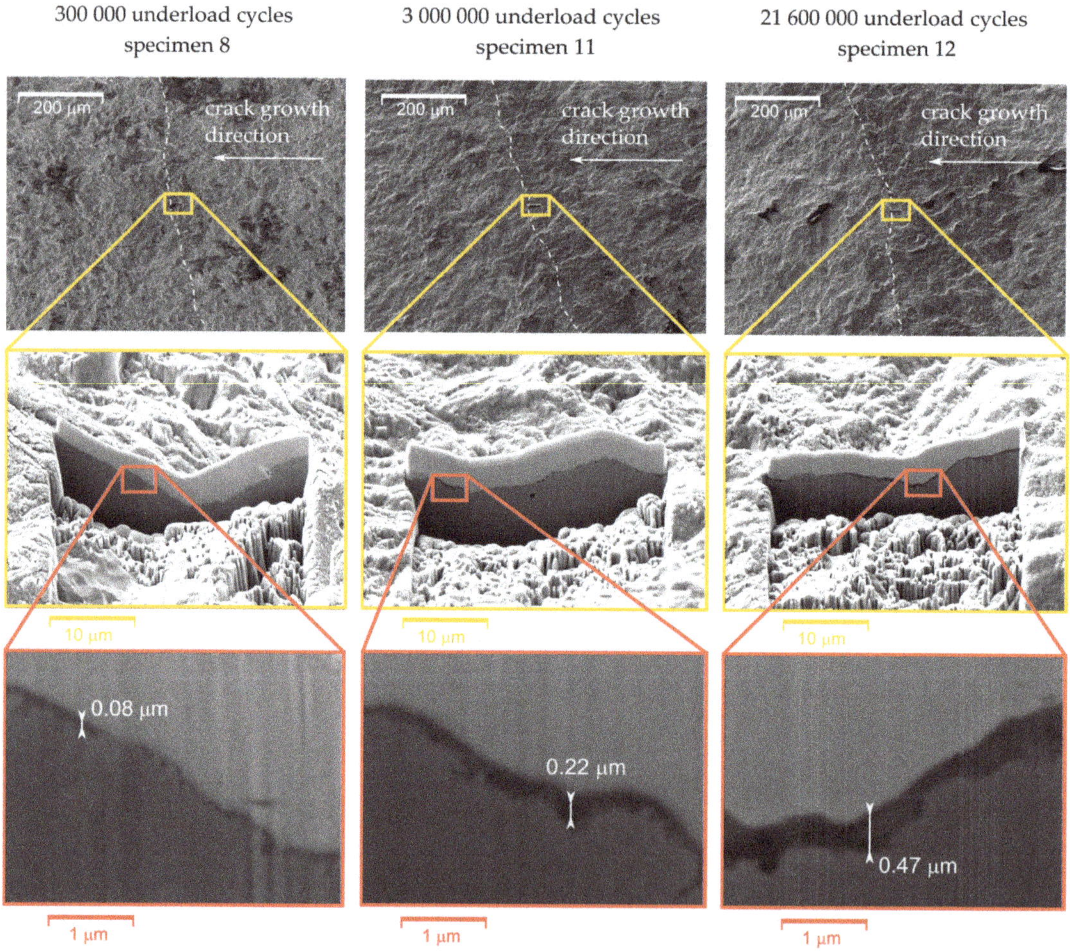


**Figure 16.** Focused ion beam (FIB) cuts of the selected specimens with highlighted thicknesses of the oxide layers. The position of the actual crack front at the time of application of the underload cycles is marked by the white dash-dotted curve in the top images.

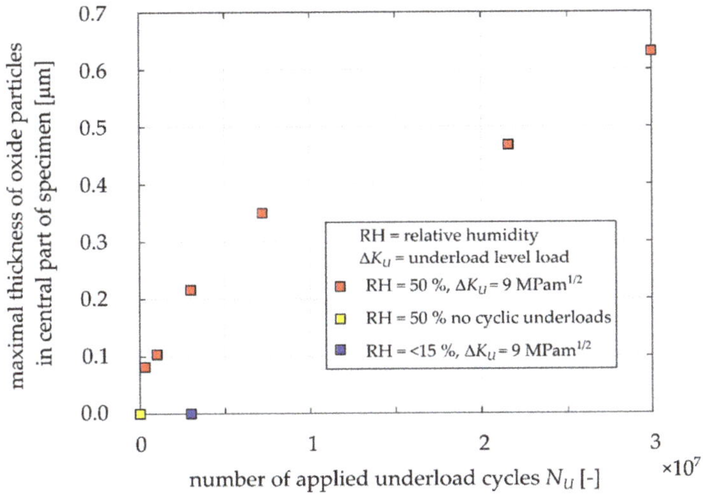


**Figure 17.** Measured thicknesses of the oxide layers for Specimens 3, 4, 7, 8, 10, 11, 12, 14, 15 and 16 based on the FIB cuts imaging.

*3.4. Crack Closure Development during Underload Cycling*

Based on the presented results, it is clear that during the underload cycling the width and thickness of the oxide debris layer gradually increases. Hence, it is expected that oxide-induced crack closure will also increase (see, e.g., [3,6]). Therefore, the level of crack closure was measured in a special experiment. Unfortunately, the MT specimens are not suitable for crack closure measurement during application of many cycles, since the COD gauge cannot be simply used and the strain gauges either provide too weak signals or the period of cycling during which measurement is possible is too short. Therefore, the compact tension (CT) specimen was used to record the evolution of crack closure in dependence on the number of applied underload cycles. The CT specimen with parameters $W = 30$ mm and a thickness of 6 mm, depicted in Figure 18, was tested with an Instron ElectroPuls e3000 machine at a frequency of 60 Hz. Crack closure was determined using the COD gauge (Sandner EXR10-1o) connected to crack mouth of the specimen (also called a CMOD gauge). The compliance change method according to the ASTM standard E647 [22] with an offset of 4% was employed to determine the crack closure level. It should be noted that such determined values correspond to the so called "remote" crack closure data [23,24]. However, such data are sufficient for the purpose of observation of proportional changes of the crack closure level.

The load procedure was the same as in the case of the MT specimen (see Figure 8). When the crack growth rate reached $da/dN = 2 \times 10^{-6}$ mm/cycle, the underload cycles were launched at $R = 0.1$ and $K_{max} = 4.5$ MPam$^{1/2}$, which was the same $K_{max}$ as in the cases of the MT specimen tests. The tests were running in loop sequence, where one loop consisted of the application of 50,000 underload cycles at 60 Hz and 3 underload cycles at 0.2 Hz. The low frequency ensured error reduction of the measured data by the COD gauge.

Each point in Figure 19 represents the determined crack closure level every 50,000 cycles in the range of 0–7,500,000 underload cycles. It is clear that crack closure is permanently increasing and there is no detectable saturation even after such large amount of underload cycles. This result is in agreement with the data shown in Figures 11 and 12, where no saturation can be deduced either.

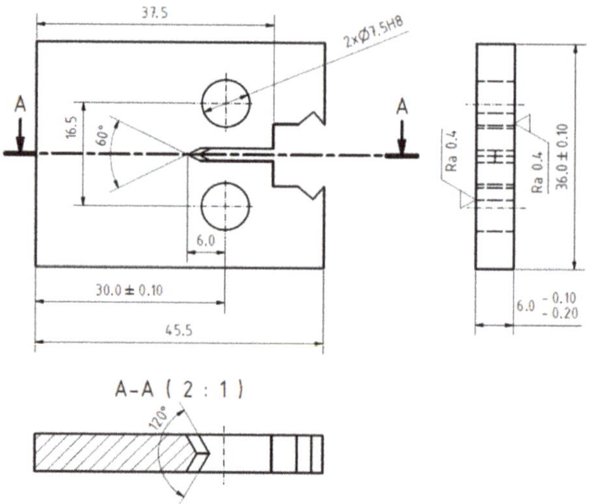


**Figure 18.** CT specimen used for determination of crack closure behavior. (Unit: mm).

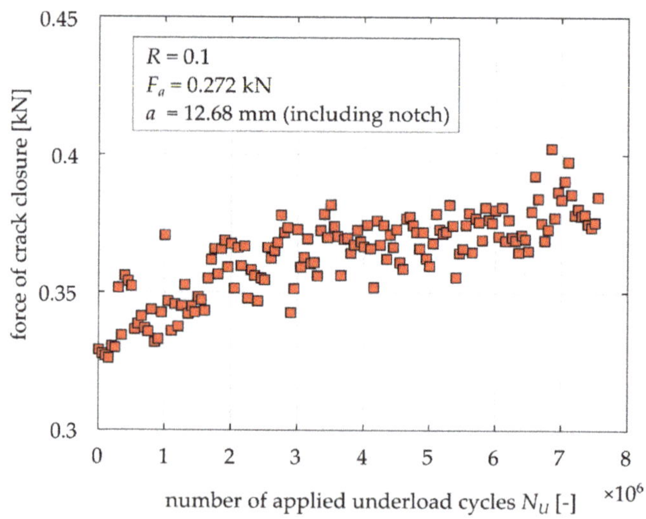


**Figure 19.** Experimental results of crack closure as function of underload cycles.

## 4. Discussion

### *4.1. Factors Influencing the Experimental Results Obtained by Various Methodologies*

In general, crack closure is dominantly caused by the three mechanisms described in the Introduction. The results of the experiments and crack closure measurement showed that OICC is probably a very significant mechanism of crack closure under the investigated loading regimes. On the other hand, the role of PICC can perhaps be neglected during application of the underload cycles, since the crack is not growing and no significant plasticity effect should be expected. The role of RICC during application of the underload cycles could almost be neglected as well, since the crack front stays at the same position. Nevertheless, considering the permanent cycling with fracture surface contact could lead to flattening of the crack flanks, which would decrease the crack closure level. However, Figure 19 shows that there is increase in crack closure during underload cycling, at least for the amount of underload cycles used in the presented experiments. It does not disprove the possibility of change of RICC after application of $10^8$ or $10^9$ cycles, which correspond to the problems in the field of very high cycle fatigue. Therefore, the observed retardation effects can be assigned to the OICC mechanism.

As was pointed out in [14], the oxide layer thickness is proportional to crack opening displacements near the crack tip during the underload cycles. This is also valid for the herein investigated material, EA4T steel (see the shift in data between specimens underloaded at $\Delta K_U$ = 9 MPam$^{1/2}$ and $\Delta K_U$ = 7 MPam$^{1/2}$ in Figures 11 and 12). Nevertheless, this paper also reveals that, in addition to the underload level, the oxide debris thickness also depends on the number of applied underload cycles and the ambient humidity. It was shown that a higher number of underload cycles leads to greater fatigue crack growth retardation due to the development of oxide layer at fracture surfaces during underload cycles. The development is more pronounced in humid air than in dry one. According to Figure 12, the retardation influenced zone can be longer than 1 mm for 5 million or more underload cycles at load level $\Delta K_U$ = 9 MPam$^{1/2}$, $R$ = −1, T = 23 °C and 50% relative humidity. Note that the experimentally obtained size of the influenced zone by oxides is not in agreement with the size estimated by the formula published in [13], even though the investigated material was the same (EA4T steel).

The results obtained in this paper can explain some of the confusing data about oxide thickness at the crack propagation threshold of EA4T, published in [21], where a larger oxide layer thickness of 130 nm was obtained for a relative humidity 30% and a lower oxide layer thickness of 90 nm for a relative humidity 50%. The reason is that there is, not only a dependency on humidity but also on the number of "underload cycles", which are present in the load shedding procedure for the determination of the fatigue propagation threshold. Generally, the specimens are usually subjected to various amounts of "underload cycles" below the threshold at the end of the load shedding procedure before a particular load level is declared as the threshold value.

The results in Tables 4 and 5 lead to similar conclusions, demonstrating that the selection of the parameter of crack extension of 0.5 mm or 1 mm is not so important for the methodology. It should be noted, however, that for railway axles in operation, a crack increment from 1 mm to 2 mm can cover a significant portion of the RFL.

### 4.2. *Significance of the Results for Residual Fatigue Life of Structural Components*

Generally, it can be said that an initial crack length of 1 mm or 2 mm is typically considered for estimation of RFL in the frame of the damage tolerance approach. The size of the considered defect (crack) substantially depends on the acceptable risk and the probability of detection of the crack via non-destructive testing methods [25].

Variability of the loading amplitude in operation is often described by the so-called load spectra, where the multiple of the static load is on the vertical axis and the cumulative number of cycles is on the horizontal axis. Figure 20 shows a typical example of a load spectrum for railway axles made of EA4T material [26], which corresponds to a block of loads repeated every 10,000 km of train operation. This figure also shows a horizontal line corresponding to the threshold for the material with a 1 mm semi-elliptical surface crack. This line splits the load spectrum into two parts. The first part with the loads above the threshold corresponds to the cycles contributing to crack propagation. The second part with the loads below the threshold, which can be considered as an analogy to the underload cycles considered in this work, corresponds to 99.9991% of the total number of loading cycles in the spectrum.

Therefore, at the start of propagation of the 1 mm fatigue crack, the vast majority of the loading cycles from the load spectrum are non-damaging. After some crack extension (or considering a longer initiation crack), the number of damaging cycles increases, however, the non-damaging cycles are still in the majority, unless the crack is so long that the rest of the fatigue life is already negligible in comparison with the total RFL. Hence, the effect of underload cycles studied in this paper is very relevant to consider in applications.

It is common that the newly designed components are tested in laboratories, considering the so-called omission level method, where small loading amplitudes from the load spectrum are excluded from the experiment in order to save on testing time. It is commonly assumed that these small load amplitudes do not contribute to crack growth behavior. However, in the case of the load spectrum shown in Figure 20, the number of damaging cycles is very small and, moreover, crack propagation under these cycles is expected to be retarded due to the underload cycles. Therefore, the difference between the results of the omission level method and the crack behavior under application of the full load spectrum can be very significant. To conclude, omission of the effect of underload cycles due to OICC in both prediction of RFL and the full-scale tests can lead to excessively conservative results.

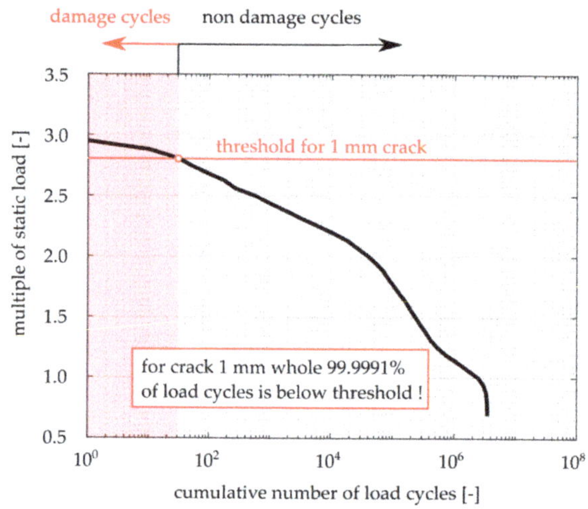


**Figure 20.** Typical load spectrum corresponding to 10,000 km of train operation.

## 5. Conclusions

The influence of underload cycles on subsequent crack growth in the near-threshold regime was experimentally investigated for the railway axle steel, EA4T. It was found that the underload cycles, which are below the crack propagation threshold, lead to formation of an oxide debris layer at fracture surfaces, causing an increase in crack closure and retardation of the following crack propagation. The oxide debris layers were measured in terms of thickness and width (direction perpendicular to the crack front). Both the retardation effect and the amount of oxide debris were larger for higher numbers of applied underload cycles, higher amplitudes of the underload cycles and higher levels of air humidity. The results can be summarized as follows:

- Application of three million underload cycles at $\Delta K_U = 9$ MPam$^{1/2}$ ($R = -1$), corresponding to approx. 65% of the threshold of the investigated steel, resulted in retardation of the subsequent crack growth for about 0.6 to 1.1 mm during 4 to 7 $\times$ $10^5$ base load cycles. The corresponding enhanced oxide layer had a width of 146 μm and thickness of 0.22 μm.
- Application of 30 million underload cycles at the same load level resulted in retardation of the subsequent crack growth for about 1.3 mm during 1 $\times$ $10^6$ base load cycles. The corresponding enhanced oxide layer had a maximal width of 337 μm and maximal thickness of 0.63 μm.
- Application of underload cycles at $\Delta K_U = 7$ MPam$^{1/2}$, corresponding to approx. 50% of the threshold, resulted in thinner oxide layers and shorter influenced crack propagation compared to those of the underloads at $\Delta K_U = 9$ MPam$^{1/2}$ and the same

number of cycles. This can be attributed to lower compressive stress in the negative part of the cycle at $R = -1$ or to a smaller crack opening displacement, which leaves less available space in the crack wake for oxide debris development.
- Measurement of the level of crack closure revealed a gradual increase in crack closure during application of the underload cycles.
- The experiments in dry air showed no or negligible effects of the underload cycles on subsequent crack growth and no enhanced oxide debris layer. This confirmed that the oxide-induced crack closure mechanism was responsible for the retardation effect in normal air humidity.
- A verification experiment was done with a static constant load corresponding to $K_{max}$ of the underload cycles, where the specimen stayed in the testing machine for the same time as what it takes to apply the underload cycles. There was no detectable retardation of the subsequent crack growth and no enhanced oxide layer developed, which confirmed that the effect is caused by the underload cycles and not by any kinds of gas diffusion dynamics or material creep.
- The significance of the effect of underload cycles in applications was demonstrated on a 1 mm surface crack in a railway axle, where 99.9991% of the loading cycles lie below threshold according to the operational load spectrum. Considering the effect of underload cycles in estimation of residual fatigue life of railway axles can lead to better accuracy.

The results contribute to understanding of mechanisms of resistance to fatigue crack growth in the near-threshold area in terms of oxide-induced crack closure in materials prone to corrosion.

**Author Contributions:** Conceptualization, P.P., T.V., M.J., L.N. and P.H.; methodology, P.P., T.V., M.J., L.N. and P.H.; validation, P.P., T.V., M.J., L.N. and P.H.; formal analysis, P.P., T.V., M.J., L.N. and P.H.; investigation, P.P., T.V., M.J., L.N. and P.H.; resources, P.P., T.V., M.J., L.N. and P.H.; data curation, P.P., T.V., M.J., L.N. and P.H.; writing—original draft preparation, P.P., T.V., M.J., L.N. and P.H.; writing—review and editing, P.P., T.V., M.J., L.N. and P.H.; visualization, P.P., T.V., M.J., L.N. and P.H.; supervision, P.P., T.V., M.J., L.N. and P.H.; project administration, P.P., T.V., M.J., L.N. and P.H.; funding acquisition, P.P., T.V., M.J., L.N. and P.H.; All authors have read and agreed to the published version of the manuscript.

**Funding:** This research was funded by the grant No. FV40034 "Development of new design of railway axles with high operational reliability" of the Ministry of Industry and Trade of the Czech Republic.

**Institutional Review Board Statement:** Not applicable.

**Informed Consent Statement:** Not applicable.

**Data Availability Statement:** The data presented in this study are available on request from the corresponding author.

**Acknowledgments:** The equipment and the base of research infrastructure CEITEC were used during the research activities.

**Conflicts of Interest:** The authors declare no conflict of interest.

## References

1. Elber, W. The Significance of Fatigue Crack Closure. *Damage Toler. Aircr. Struct.* **2009**, *486*, 230. [CrossRef]
2. Ritchie, R. Mechanisms of fatigue crack propagation in metals, ceramics and composites: Role of crack tip shielding. *Mater. Sci. Eng. A* **1988**, *103*, 15–28. [CrossRef]
3. Suresh, S. *Fatigue of Materials*, 2nd ed.; Cambridge University Press: Cambridge, UK, 2010; ISBN 0-521-57847-7.
4. Pokluda, J.; Šandera, P. Micromechanisms of Fracture and Fatigue. In *Engineering Materials and Processes*; Metzler, J.B., Ed.; Springer: London, UK, 2010.
5. Newman, J. The merging of fatigue and fracture mechanics concepts: A historical perspective. *Prog. Aerosp. Sci.* **1998**, *34*, 347–390. [CrossRef]
6. Pokorný, P.; Vojtek, T.; Náhlík, L.; Hutař, P. Crack closure in near-threshold fatigue crack propagation in railway axle steel EA4T. *Eng. Fract. Mech.* **2017**, *185*, 2–19. [CrossRef]

7. Vojtek, T.; Pokorný, P.; Oplt, T.; Jambor, M.; Náhlík, L.; Herrero, D.; Hutař, P. Classically determined effective ΔK fails to quantify crack growth rates. *Theor. Appl. Fract. Mech.* **2020**, *108*, 102608. [CrossRef]
8. Suresh, S.; Zamiski, G.F.; Ritchie, D.R.O. Oxide-Induced Crack Closure: An Explanation for Near-Threshold Corrosion Fatigue Crack Growth Behavior. *Met. Mater. Trans. A* **1981**, *12*, 1435–1443. [CrossRef]
9. Suresh, S.; Ritchie, R. On the influence of environment on the load ratio dependence of fatigue thresholds in pressure vessel steel. *Eng. Fract. Mech.* **1983**, *18*, 785–800. [CrossRef]
10. Newman, J.C., Jr.; Brot, A.; Matias, C. Crack-Growth calculations in 7075-T7351 aluminum alloy under various load spectra using an improved crack-Closure model. *Eng. Fract. Mech.* **2004**, *71*, 2347–2363. [CrossRef]
11. Mikheevskiy, S.; Bogdanov, S.; Glinka, G. Analysis of fatigue crack growth under spectrum loading—The UniGrow fatigue crack growth model. *Theor. Appl. Fract. Mech.* **2015**, *79*, 25–33. [CrossRef]
12. Maierhofer, J.; Gänser, H.-P.; Pippan, R. Crack closure and retardation effects—Experiments and modelling. *Procedia Struct. Integr.* **2017**, *4*, 19–26. [CrossRef]
13. Maierhofer, J.; Gänser, H.-P.; Simunek, D.; Leitner, M.; Pippan, R.; Luke, M. Fatigue crack growth model including load sequence effects—Model development and calibration for railway axle steels. *Int. J. Fatigue* **2020**, *132*, 105377. [CrossRef]
14. Suresh, S.; Ritchie, R. On the influence of fatigue underloads on cyclic crack growth at low stress intensities. *Mater. Sci. Eng.* **1981**, *51*, 61–69. [CrossRef]
15. Antunes, F.V.; Paiva, L.; Branco, R.; Borrego, L.P. Effect of Underloads on Plasticity-Induced Crack Closure: A Numerical Analysis. *J. Eng. Mater. Technol.* **2019**, *141*, 1–27. [CrossRef]
16. Zaiken, E.; Ritchie, R. On the role of compression overloads in influencing crack closure and the threshold condition for fatigue crack growth in 7150 aluminum alloy. *Eng. Fract. Mech.* **1985**, *22*, 35–48. [CrossRef]
17. Hassanipour, M.; Verreman, Y.; Lanteigne, J.; Chen, J. Effect of periodic underloads on fatigue crack growth in three steels used in hydraulic turbine runners. *Int. J. Fatigue* **2016**, *85*, 40–48. [CrossRef]
18. Doré, M.J.; Maddox, S.J. Accelerated Fatigue Crack Growth in 6082 T651 Aluminium Alloy Subjected to Periodic Underloads. *Procedia Eng.* **2013**, *66*, 313–322. [CrossRef]
19. Novosad, M.; Fajkoš, R.; Řeha, B.; Řezníček, R. Fatigue tests of railway axles. *Procedia Eng.* **2010**, *2*, 2259–2268. [CrossRef]
20. Pokorný, P.; Dlhý, P.; Poduška, J.; Fajkoš, R.; Vojtek, T.; Náhlík, L.; Grasso, M.; Hutař, P. Influence of heat treatment-induced residual stress on residual fatigue life of railway axles. *Theor. Appl. Fract. Mech.* **2020**, *109*, 102732. [CrossRef]
21. Vojtek, T.; Pokorný, P.; Kuběna, I.; Náhlík, L.; Fajkoš, R.; Hutař, P. Quantitative dependence of oxide-Induced crack closure on air humidity for railway axle steel. *Int. J. Fatigue* **2019**, *123*, 213–224. [CrossRef]
22. *ASTM E 647-08. Standard Test Method for Measurement of Fatigue Crack Growth Rates*; ASTM International: West Conshohocken, PA, USA, 2009.
23. Yamada, Y.; Newman, J.N., Jr. Crack-closure behavior of 2324-T39 aluminum alloy near-threshold conditions for high load ratio and constant Kmax tests. *Int. J. Fatigue* **2009**, *31*, 1780–1787. [CrossRef]
24. Yamada, Y.; Newman, J. Crack closure under high load-ratio conditions for Inconel-718 near threshold behavior. *Eng. Fract. Mech.* **2009**, *76*, 209–220. [CrossRef]
25. Benyon, J.A.; Watson, A.S. The use of Monte Carlo analysis to increase axle inspection interval. In Proceedings of the 13th International Wheelset Congress, Rome, Italy, 17–21 September 2001.
26. Pokorný, P.; Hutař, P.; Náhlík, L. Residual fatigue lifetime estimation of railway axles for various loading spectra. *Theor. Appl. Fract. Mech.* **2016**, *82*, 25–32. [CrossRef]

*materials*

*Article*

# Critical Dynamics of Defects and Mechanisms of Damage-Failure Transitions in Fatigue

**Oleg Naimark *, Vladimir Oborin, Mikhail Bannikov and Dmitry Ledon**

Institute of Continuous Media Mechanics of the Ural Branch, RAS, 614013 Perm, Russia; oborin@icmm.ru (V.O.); mbannikov@icmm.ru (M.B.); ledon@icmm.ru (D.L.)
* Correspondence: naimark@icmm.ru

**Abstract:** An experimental methodology was developed for estimating a very high cycle fatigue (VHCF) life of the aluminum alloy AMG-6 subjected to preliminary deformation. The analysis of fatigue damage staging is based on the measurement of elastic modulus decrement according to "in situ" data of nonlinear dynamics of free-end specimen vibrations at the VHCF test. The correlation of fatigue damage staging and fracture surface morphology was studied to establish the scaling properties and kinetic equations for damage localization, "fish-eye" nucleation, and transition to the Paris crack kinetics. These equations, based on empirical parameters related to the structure of the material, allows us to estimate the number of cycles for the nucleation and advance of fatigue crack.

**Keywords:** fracture; gigacycle fatigue; surface morphology; kinetic equation

**Citation:** Naimark, O.; Oborin, V.; Bannikov, M.; Ledon, D. Critical Dynamics of Defects and Mechanisms of Damage-Failure Transitions in Fatigue. *Materials* **2021**, *14*, 2554. https://doi.org/10.3390/ma14102554

Academic Editor: Jaroslav Pokluda

Received: 28 February 2021
Accepted: 11 May 2021
Published: 14 May 2021



## 1. Introduction

The evaluation of the longevity of critical engineering structures, for example, working parts of aircraft engines, raises new fundamental problems associated with the calculation of service reliability and operating time of materials under cyclic loading during more than $10^9$–$10^{10}$ cycles, generally known as a gigacycle fatigue range [1].

Two problems of a fundamental nature determine the characteristics of fatigue failure at VHCF [1,2]: the problem of localization of damage in conditions of multiscale nucleation and growth of defects, and the problem of the propagation of cracks in the damaged material.

The difficulties in separating the stages of the fatigue crack origin and propagation initiated a fundamentally new problem: a general formulation of damage failure staging including the multiscale kinetics of damage localization, nucleation, and growth of the fatigue crack [3–5]. The duration of these stages is essentially determined by the state of the material structure and the kinetics of the free energy release, revealing the features of critical phenomena [3]. In contrast to HCF, when the fatigue life is associated with the crack advance, the scenario of VHCF damage-failure transition arises in the course of multiscale defects kinetics (PSB, microcracks, pores). Moreover, VHCF reveals a qualitative difference in the nucleation of fatigue cracks in the material, leading to reformulation of the fatigue life problem in which the critical damage-failure transition is estimated with reference to the nonlinearity of the free energy release. The role of the initiation stage is especially important for VHCF loading, which is characterized by the emergence of a "fish-eye" area, the development of small cracks with structure sensitive kinetics that is qualitatively different from the kinetics of crack growth for HCF associated with the Paris law. The correspondence of damage kinetics, the free energy release nonlinearity and specific morphological pattern of the fracture surface [6,7] are the key questions to identifying the VHCF damage-failure transition staging [8,9] discussed in the paper.

Within this fatigue load range, of considerable practical interest is the range corresponding to the number of cycles $N \approx 10^9$ [3]. The behavior of materials in this range is determined by qualitative changes, mechanisms responsible for the initiation and growth of the cracks.

The process of material fracture in the range of gigacycle loading involves several stages, which are classified on account of the damage-induced structural changes occurring on different spatial scales including persistent slip bands (PSBs), fatigue striations, microcracks (formed as a result of PSB crossing), and grain-boundary defects [10,11]. The primary failure is associated with defect scales in the range of 0.1 μm–1 mm, which are considerably smaller than those detected by the standard nondestructive testing methods widely used to assess the fatigue strength of a structure during high cycle fatigue loading.

Today, the quantitative fractography technique is widely used to investigate the role of initial structural heterogeneity, monitoring of defect accumulation at different scales (dislocation ensembles, micropores, microcracks), and determining critical conditions for the damage-failure transition. This method has allowed researchers to distinguish the characteristic stages of fracture (crack nucleation and propagation) and to evaluate the fatigue life of materials and structures under gigacycle loading.

The description of the fracture surface morphology in terms of spatial temporal invariants was first proposed by Mandelbrot [12,13]. This approach is based on the analysis of the fracture surface relief, showing the property of self-affinity, which manifests itself as the invariance of the surface relief characteristics over a broad spectrum of spatial scales. On the other hand, these characteristics reflect the interrelation of defects at different scales evolving into the stages of damage-failure transition and fatigue crack propagation.

The unified laws, governing the relationship between the crack growth rate and a change in the stress intensity factor, are the focus of recent theoretical and experimental studies. The power law, which was first deduced by Paris [1], establishes the subordination of the fatigue crack advance to the self-similar nature of damage accumulation in the vicinity of the crack tip (the process zone). The Paris crack kinetics *da/dN* (a is the crack length, *N* is the number of cycles) obeys the power law with regard to the magnitude of the stress intensity factor, expressed as $\Delta K = K_{max} - K_{min}$, where $K_{max}$ and $K_{min}$ are the maximum and minimum values of stress intensity factors under conditions of cyclic loading:

$$\frac{da}{dN} = C(\Delta K)^m \quad (1)$$

where *C* and m are dependent material constants. For many materials and different crack growth rates under conditions of high cycle fatigue (HCF), the power exponent m is close to 2–4. The fracture of the material under fatigue loading in the Paris regime is defined by the applied stresses and the length of the original crack (and its orientation). In contrast, in the case of small cracks or small stresses, when the structure and damage of the surrounding material are the main contributing factors to the kinetics of crack growth, the traditional formulation of the Paris law needs considerable revision [1].

## 2. Materials and Methods

The successive dynamic and fatigue tests are of particular importance for predicting fatigue strength of gas turbine materials in the conditions of the so-called foreign object damage (FOD) [14]. In [5], the specimens made of aluminum AlMg6 were subjected to dynamic loads using the split Hopkinson pressure bar (SHPB) machine (Perm, Russia) at the strain rate of ~$10^3$ $s^{-1}$ and then subjected to cyclic loading on a Shimadzu USF-2000 ultrasonic loading machine (Tokyo, Japan) at room temperature. Table 1 shows the chemical constitution (percentage by weight) of the AlMg6 alloy. The SHPB machine includes a gas gun and three cylindrical bars (Figure 1), known as the striking bar (SB), incident bar (IB), and transmission bar (TB). The gas gun is used to accelerate the striking bar, which transfers the shock wave into the IB.

**Table 1.** Chemical composition of AlMg6 (percentage by weight).

| Al | Si | Fe | Cu | Mn | Mg | Zn | Ti | Be |
|---|---|---|---|---|---|---|---|---|
| 92.1 | 0.24 | 0.32 | 0.10 | 0.7 | 6.44 | 0.024 | 0.035 | 0.0006 |

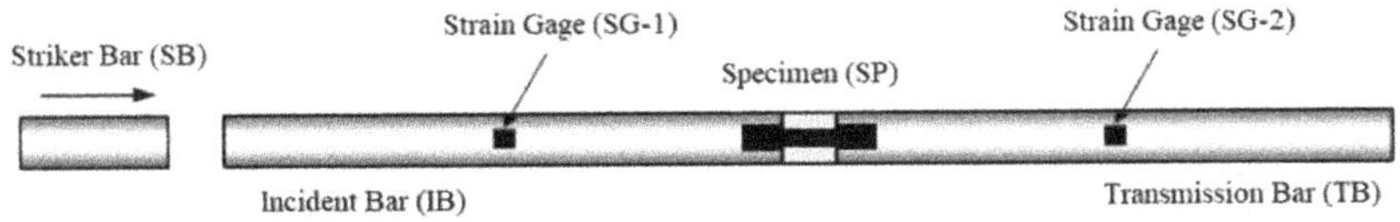


**Figure 1.** Split Hopkinson pressure bar apparatus.

**Figure 1.** Split Hopkinson pressure bar apparatus.

The compressive pulse, which passes through the holder and the specimen, does not produce plastic deformation in the specimen (the main impact of the propagating wave is on the holder, which because of high plastic yield strength shows considerable resistance to the damaging action of the incident bar). As soon as the compressive pulse reaches the free end of this bar, it is reflected as a tensile shock wave, which is just the initial incident wave that causes stretching of the specimen. A part of the tensile pulse after reaching the specimen is transferred through it into the first bar, while another part of this pulse moves back into the second bar. Eventually, this results in plastic deformation of the part of the specimen, which is adjacent to the smallest cross section; the holder, which is disconnected with the bars, does not experience tensile stress.

Then, we performed a series of fatigue tests for 14 initial and nine dynamic preloaded specimens using the Shimadzu USF-2000 ultrasonic fatigue testing machine, which provides cyclic loading ($R = -1$) [2]. The specimens were stressed by a generator, which transformed the frequency of 50 Hz into an ultrasonic electrical sinusoidal signal of frequency 20 kHz with the aid of a piezoelectric transducer. The latter generates longitudinal ultrasonic waves in the frequency range of 20 kHz and the mechanical stress with maximum amplitude at the center of the specimen. The geometry of the specimens is shown in Figure 2. Compressed air cooling of the specimens was used.

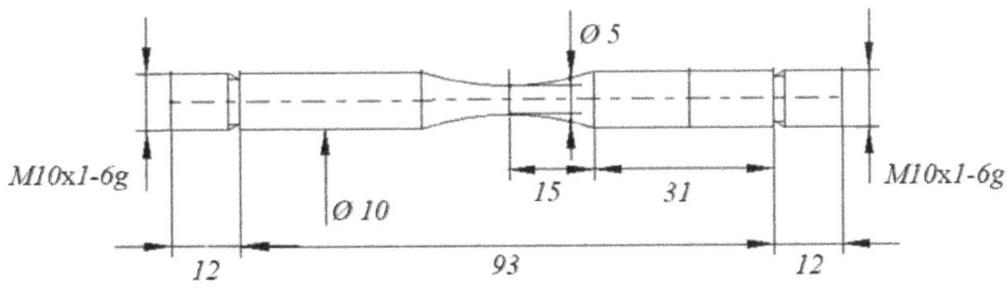


**Figure 2.** The specimen geometry used for dynamic preloading and cycle tests (sizes in mm).

The 0.5 kHz difference in the frequency, which was due to approaching the stage of critical damage, could be viewed as a failure precursor associated with the formation of a crack with a characteristic size of ~2 mm. The applied stresses ranging from 110 to 162 MPa made it possible to investigate the fatigue life under gigacycle loading up to $10^{10}$ cycles. The results of the fatigue testing are shown in Figure 3.

We observed that the cyclic life on the base of $10^9$ cycles for the preloaded AlMg6 alloy decreased from the stress level of 162 MPa to 121–138 MPa (15–25%). There were two specimens (No. 1 and No. 2) that formed fatigue cracks inside the bulk with a characteristic type of fracture called "fish eye". The other ones fractured from the surface.

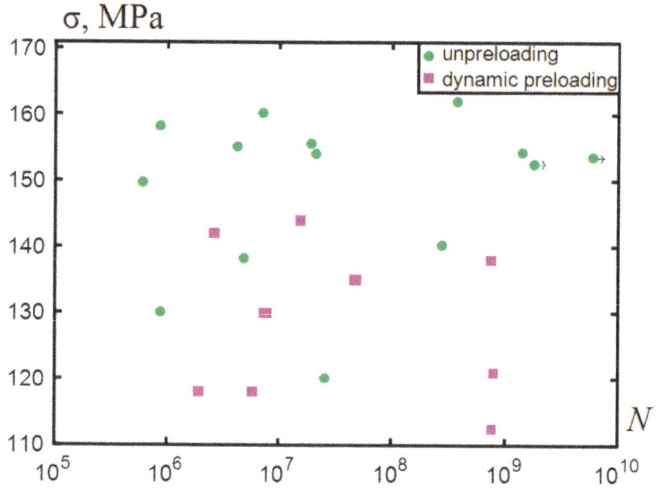


**Figure 3.** The *S–N* curve (Stress in MPa) in the case of dynamic preloading and in the absence of preloading for the AlMg6 alloy.

To determine the mechanism of internal crack formation, the method of amplitude-frequency analysis of changes in the effective elastic properties of materials was used as proposed in [15,16] on the basis of acoustic properties. The resonance testing machine (Shimadzu USF-2000) was combined with a highly sensitive inductive sensor and an analog-to-digital converter system (Figure 4) that measured the amplitudes and frequencies of oscillations of the free end face of the specimen. The software reads the 65,536-point signal every 0.001 s and performs fast Fourier transform to determine the values of fundamental frequency, second and third harmonics and their amplitudes, and wrote them to a file directly during the experiment. By changing the value of these amplitudes, the behavior of the defective structure of the material can be estimated using the nonlinear parameter of the signal $\beta$ [15,16].

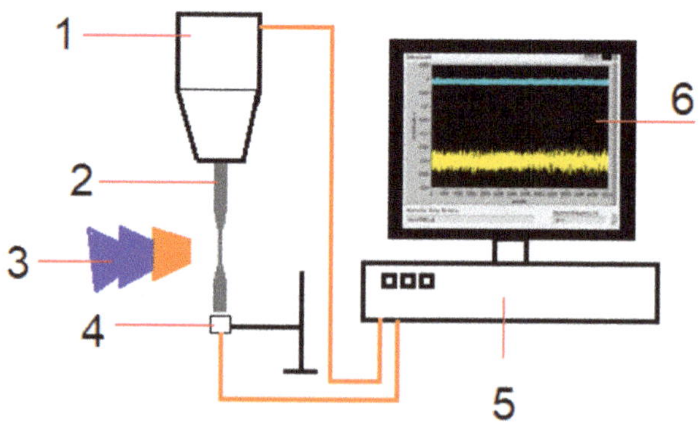


**Figure 4.** The experimental setup: 1—horn, 2—specimen, 3—cooling system, 4—displacement sensor, 5—controlling and analog-digital converter system, 6—analyzing software.

The basis of the damage staging analysis is the registration of current specimen impedance in the presence of harmonic components: the frequency $\omega_0$, associated with forced vibration amplitude $A_1$ and the amplitude $A_2$ of the second harmonics with a frequency $2\omega_0$, and higher harmonics associated with the influence of defects.

The nonlinearity coefficient $\beta$ [15,16] is directly related to the amplitude of the second harmonic. Crack formation and growth lead to a significant increase in the nonlinearity parameter $\beta$ [16]. In this study, we did not need to know the specific value of the nonlinearity parameter, only its qualitative change in time, so it was enough to measure only the amplitude of the second harmonic. The value of amplitude was measured by an inductive displacement sensor with a signal recording frequency of 10 MHz (Figure 4).

To describe the kinetics of crack growth for sizes smaller than the size of "Paris cracks", a phenomenological relation was proposed in [17,18], which along with the macroscopic characteristic of the stress state at the crack tip included the structural parameters of the Burgers vector b and the effective stress intensity factor (Figure 5). The authors of [19] proposed structural parameters $l_{sc}$ (the scale of interaction of defects) and $L_{pz}$, (size of the process zone), which are determined from the fracture surface profile using the correlation function [20–22].

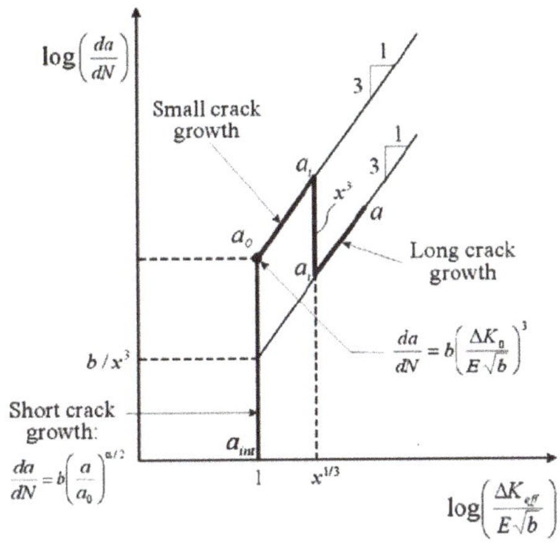


**Figure 5.** Crack advance diagram in HCF [17,18]: b is the Burgers vector, $\Delta K_0$ and $\Delta K_{eff}$ are the stress intensity factors corresponding to the crack lengths $a_0$ and $a_i$.

The self-similar patterns of the nature of fatigue crack growth on specimens loaded in the high and very high-cycle fatigue regimes were studied using the methods of similarity and dimension theory, and as a result, the authors in [21] proposed an equation for crack growth rate, taking into account the structural parameters:

$$\frac{da}{dN} = l_{sc}\left(\frac{\Delta K}{E\sqrt{l_{sc}}}\right)^{\alpha}\left(\frac{L_{pz}}{l_{sc}}\right)^{\beta} \tag{2}$$

where $\alpha$ and $\beta$ are the power exponents reflecting the intermediate asymptotic nature of the crack growth kinetics as a function of dimensionless variables $\Delta K_{eff}/E\sqrt{l_{sc}}, l_{sc}, L_{pz}$. The parameter $\Delta K_{eff} = \Delta K(L_{PZ}/l_{sc})^{a/b}$ is introduced, which allows us to write Equation (2) in a form similar to the Paris law:

$$\frac{da}{dN} = l_{sc}\left(\frac{\Delta K_{eff}}{E\sqrt{l_{sc}}}\right)^{\alpha} \tag{3}$$

Equation (3) can be used to describe both small and large cracks, the kinetics of which are determined by the structural parameters $l_{sc}$, $L_{pz}$, and scaling indices $\alpha$, $\beta$. Parameters $l_{sc}$ and $L_{pz}$, associated with defects, can be determined by fractography of the fracture surface (Figure 6).

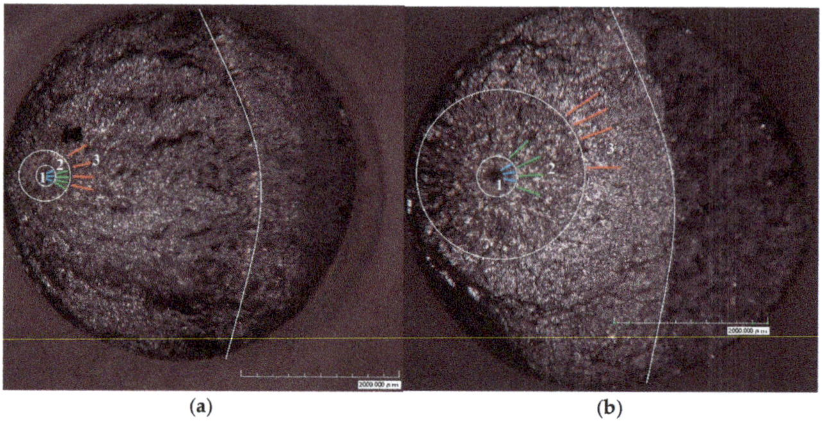

Figure 6. (a) The fracture surface of specimen No. 1 ($\sigma$ = 138 MPa, $N = 7.51 \times 10^8$). (b) The fracture surface of specimen No. 2 ($\sigma$ = 120 MPa, $N = 7.82 \times 10^8$). Colored lines illustrate the directions of analyzed profiles in zones 1 (blue), 2 (green), and 3 (red).

Integrating (3) and setting $\Delta K = \Delta\sigma\sqrt{\pi a}$, we obtain the expression for the number of cycles required for the growth of a fatigue crack from length $a_1$ to $a_2$:

$$N = \frac{2\left(a_1^{1-\frac{\alpha}{2}} - a_2^{1-\frac{\alpha}{2}}\right)}{\alpha - 2} \frac{\left(\frac{L_{pz}}{l_{sc}}\right)^{-\beta} \pi^{\frac{-\alpha}{2}} \left(\frac{\Delta\sigma}{E\sqrt{l_{sc}}}\right)^{-\alpha}}{l_{sc}} \quad (4)$$

## 3. Results

To determine the exact number of cycles required to achieve a certain crack length, we used the methodology to analyze the nonlinear dynamics of oscillations of the free end of the sample described in [15–17]. Three characteristic areas can be distinguished on the fracture surface: 1—fine granular area (FGA); 2—the "fisheye" zone; and 3—the crack growth zone (Figure 6). Sizes of the respective regions were measured using the Hirox optical microscope.

For specimen No. 1 (Figure 6a), the following values of crack length (radius) were taken: zone 1—$a_0$ = 152 μm, zone 2—$a_i$ = 270 μm, and zone 3—$a_k$ = 2679 μm. In all three zones (Figure 6), using the New-View 5010 interferometer-profiler and the procedure described in [13,20], the $l_{sc}$ and $L_{pz}$ values were determined on one-dimensional profiles whose directions coincided with the crack propagation. The morphology of fractured surfaces was investigated using an optical interferometer New-View 5010 (Middlefield, CT, USA) (magnification ×2000), which allows digital three-dimensional surface profiles to be obtained. The interaction of defects with each other during crack formation and growth should leave traces on the fracture surface, and if they exhibit correlated behavior in a wide range of scales, this will be reflected in the form of fractal patterns [19–21].

The surface was scanned near the crack nucleation site (Figure 6a). One-dimensional surface profiles were investigated in the directions of crack growth, starting from zone 1 to zone 3. The profile resolution during scanning was 0.1 nm in the vertical direction and ~0.5 μm in the horizontal direction.

From these profiles, the scale-invariant Hurst parameter [22] was determined, which is calculated from the slope of the correlation function:

$$K(r) = \langle (z(x+r) - z(x))^2 \rangle_x^{1/2} \propto r^H \quad (5)$$

where $K(r)$ is the average difference between the surface elevation values $z(x + r)$ and $z(x)$ in a window of size $r$, and $H$ is the Hurst index (surface roughness index).

The representation of the function *K(r)* in logarithmic coordinates (Figure 7), which should be linear, allowed us to estimate the lower bound of the scaling-scale $l_{sc}$, and to consider the value of the upper bound as the characteristic scale of the process zone $L_{pz}$—the area of the correlated behavior of multiscale defect structures.

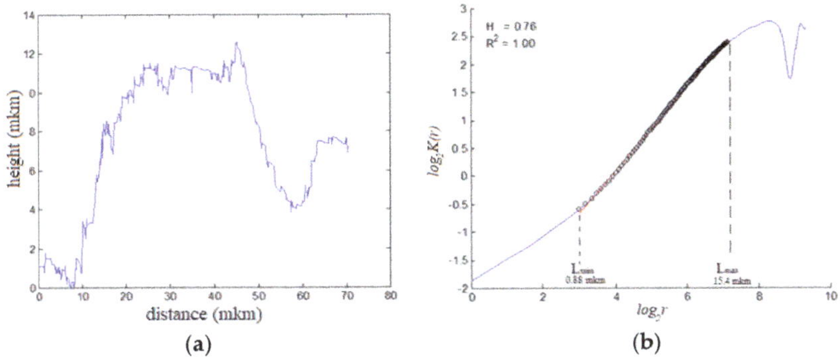

**Figure 7.** Characteristic for "fisheye" zone 1 of sample No. 2: (**a**) one-dimensional profile, (**b**) plot $log_2K(r)$ vs. $log_2(r)$.

The values of $l_{sc}$ and $L_{pz}$ were determined. For sample No. 1: for zone 1: $l_{sc}$ = 0.8 µm, $L_{pz}$ = 11.6 µm; for zone 2: $l_{sc}$ = 2.2 µm, $L_{pz}$ = 28.4 µm; and for zone 3: $l_{sc}$ = 0.4 µm, $L_{pz}$ = 16.6 µm. For sample No. 2: for zone 1: $l_{sc}$ = 0.6 µm, $L_{pz}$ = 17.2 µm; for zone 2: $l_{sc}$ = 0.9 µm, $L_{pz}$ = 26.3 µm; and for zone 3: $l_{sc}$ = 0.5 µm, $L_{pz}$ = 26.2 µm.

The appearance of a fatigue crack caused a major change in the amplitude of the second harmonic and the crack growth, respectively, its monotonic increase (Figure 8). On the basis of data measuring the amplitude of the second harmonic in real time during the fatigue tests, the number of cycles was determined, which was spent on the nucleation and growth of the fatigue crack. The number of cycles required for the nucleation of the fracture site (the first peak in Figure 8) was $N_1 = 7.43 \cdot \times 10^8$. The number of cycles during which the crack grew in zone 2 (Figure 6) was the time between the first and second peaks in Figure 8 (i.e., approximately $N_2 = 1.4 \cdot \times 10^6$ cycles). The remaining $N_3 = 6.6 \cdot \times 10^6$ cycles were the crack growth time in zone 3.

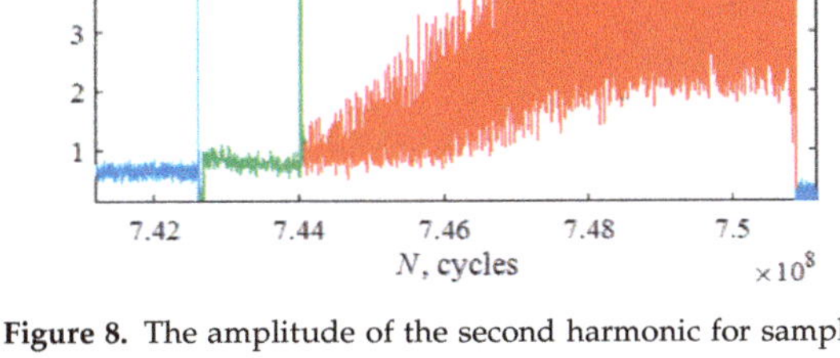


**Figure 8.** The amplitude of the second harmonic for sample No. 1. Blue color ($N = 7.426 \cdot \times 10^8$ cycles)—damage accumulation and nucleation of fatigue crack site. Green color ($N = 1.4 \cdot \times 10^6$ cycles)—crack grew inside zone 2, formation of "fisheye". Red color ($N = 6.7 \cdot \times 10^6$)—crack grown by Paris law.

Then, in Equation (4), two unknown constants remain: $\alpha$ and $\beta$. Their values were determined in the process of solving the minimization problem between the experimental and theoretical number of cycles required for crack growth from 0 to $a_0$ (zone 1), from $a_0$ to $a_i$ (zone 2), and from $a_i$ to $a_k$ (zone 3), respectively. The values of $\alpha$ and $\beta$ were as follows: zone 1: $\alpha = 2.82$, $\beta = 0.45$; zone 2: $\alpha = 3.56$, $\beta = 1.19$; zone 3: $\alpha = 6.22$, $\beta = 1.35$.

The kinetic diagram built according to Equation (3) for a fatigue crack from size $a_0$ to $a_k$ is shown in Figure 9a for sample No. 1 and in Figure 9b for sample No. 2.

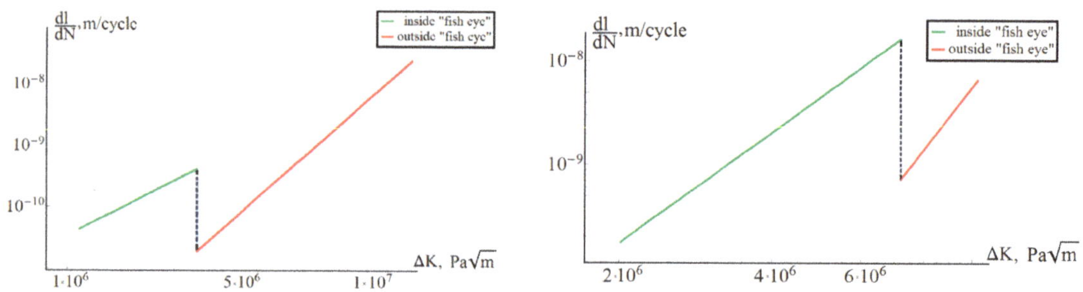


(**a**) (**b**)

**Figure 9.** (**a**) Kinetic diagram for sample No. 1. (**b**) Kinetic diagram for sample No. 2.

Figure 9 shows that the proposed approach allows us to describe the effect of a change in the growth rate of a fatigue crack upon transition from one characteristic region to another. The constants determined for one of the samples (sample No. 1) gave a satisfactory prediction of the number of cycles to failure for other samples. For example, for sample No. 2, the experimental number of cycles to failure was $7.82 \times 10^8$, and the predicted was $6.92 \times 10^8$ (relative error of 11%). The calculation results for sample No. 2 with constants $\alpha$ and $\beta$ obtained on sample No. 1 gave qualitatively consistent results on the number of cycles that went into nucleation and crack growth: $6.91 \cdot \times 10^8$ to nucleation; $0.56 \cdot \times 10^6$ for growth in zone 2; and $0.58 \cdot \times 10^6$ for growth in zone 3. The simulation of the behavior of the second harmonic during the experiment was carried out in [23].

## 4. Discussion

The prediction of the VHCF life time, which includes the definition of fatigue failure as the damage-failure transition problem can be studied by the staging of the multiscale damage localization kinetics, nucleation, and growth of the fatigue crack. The duration of these stages is essentially determined by the state of the material structure that is crucially important for the reliability prediction of the fan blades in the situation of the "foreign object damage". In contrast to HCF, when the fatigue life is associated with the crack advance, the scenario of VHCF damage-failure transition arises in the course of multiscale defects kinetics (PSB, microcracks, pores). At the same time, VHCF reveals a qualitative difference of fatigue crack initiation in the bulk of the material that leads to reformulation of the fatigue life assessment problem as critical damage-failure transition related to the nonlinearity of the free energy release. The role of the initiation stage is especially important for VHCF loading, revealing the specific surface pattern as the "fisheye" area, and the development of small cracks with structure sensitive kinetics that are qualitatively different from the kinetics of crack advance for HCF associated with the Paris law. The correspondence of damage kinetics, the free energy release nonlinearity, and specific morphological pattern of the fracture surface are the key questions to identify the VHCF damage-failure transition staging.

Statistical theory of defects allowed for the development of the phase field approach of damage-failure transition based on specific nonlinear presentation of the free energy release [24,25]. The damage-failure transition staging is linked to the defect induced relaxation properties associated with the generation of collective modes of defects: solitary slip modes and blow-up modes. These modes have the nature of self-similar solutions reflecting singularity features of characteristic morphology areas on the fatigue fracture surface. A special technique was proposed to identify the damage staging during "in situ" measurement of the acoustic impedance of the sample at VHCH loads. Additionally, in classical singularity of the stress field at the crack tip (the stress intensity factor), the presence of singularities of damage accumulation leads to the spatial-temporal kinetics of characteristic areas on the fatigue fracture surface (fisheye, FGA) with pronounced scaling properties of defect induced roughness. Duality of singularities at the fatigue crack process zone (solitary wave PSB image and the stress singularity) explains the power law universality of the Paris law related to the anomaly of the energy absorption in the PSB areas. The self-similar features of damage kinetics were reflected in kinetic equations with material parameters estimated by scaling analysis of the fracture surface. These equations were used for the prediction of the fatigue life-time at the consecutive loading.

The experimental methodology was developed for estimating the ultra-high cycle lifetime with reference to the situation of accidental high-speed collision of solid particles with fan blades and subsequent fatigue failure in the flight cycle conditions, which is common in the practice of operating aircraft engines. We estimated that the fatigue limit of the dynamically preloaded AlMg6 alloy decreased from the stress level of 162 MPa to 121–138 MPa (15–25%), which corresponds to the critical number of cycles ~$10^9$.

## 5. Conclusions

VHCF conditions have both fundamental and applied issues related to qualitative new aspects of fatigue life prediction due to pronounced damage accumulation staging. The definition of stages of initiation and propagation of fatigue cracks is one of the key problems of fatigue failure that can be analyzed by methodology that combined the experimental technique linking the structural and mechanical aspects with the nonlinearity of damage-failure transition. Critical defect kinetics allows one to establish the link of damage staging, fracture surface morphology with self-similar laws of damage kinetics, and fatigue crack advance. Original experiments and scaling analysis of fracture surface were proposed for the formulation of damage kinetic equations and fatigue crack advance.

Summarizing the results, the following conclusions concerning the staging of damage-failure transition in VHCF can be proposed:

(I) Characteristic stages of VHCF damage-failure transition follow the qualitative different mechanisms of crack initiation, crack growth, and crack advance related to the nonlinearity of the free energy release;

(II) There are pronounced quantitative differences of the fracture surface pattern for the initiation, crack origin, and crack advance related to different scaling properties and nonlinearity of the damage induced free energy release.

**Author Contributions:** Conceptualization, O.N.; Methodology, M.B. and V.O.; Software D.L.; Validation, M.B., V.O., and D.L.; Formal analysis, M.B. and O.N.; Investigation, resources, and data curation, V.O. and M.B.; Writing—original draft preparation, M.B. and V.O.; Writing—review and editing, O.N.; Supervision, project administration, and funding acquisition, O.N. All authors have read and agreed to the published version of the manuscript.

**Funding:** This research was supported by the Russian Science Foundation (grant no. 21-79-30041).

**Institutional Review Board Statement:** Not applicable.

**Informed Consent Statement:** Not applicable.

**Data Availability Statement:** All data generated or analyzed during this study are included in the article.

**Conflicts of Interest:** The authors declare no conflict of interest.

## References


1. Bathias, C.; Paris, P.C. *Gigacycle Fatigue in Mechanical Practice*; Marcel Dekker: New York, NY, USA, 2005; p. 305.
2. Bathias, C. Piezoelectric fatigue testing machines and devices. *Int. J. Fatigue* **2006**, *28*, 1438–1445. [CrossRef]
3. Naimark, O.; Palin-Luc, T. Gigacycle fatigue: Non-local and scaling aspects of damage localization, crack initiation and propagation. *Procedia Struct. Integr.* **2016**, *2*, 1143–1148. [CrossRef]
4. Sakai, T. Review and prospects for current studies on very high cycle fatigue of metallic materials for machine structural use. *J. Solid Mech. Mater. Eng.* **2009**, *3*, 425–439. [CrossRef]
5. Hong, Y.; Zhao, A.; Qian, G.; Zhou, C. Fatigue strength and crack initiation mechanism of very-high-cycle fatigue for low alloy steels. *Met. Mater. Trans. A* **2012**, *43*, 2753–2762. [CrossRef]
6. Mughrabi, H. Specific features and mechanisms of fatigue in the ultrahigh-cycle regime. *Int. J. Fatigue* **2006**, *28*, 1501–1508. [CrossRef]
7. Nicholas, T. Critical issues in high cycle fatigue. *Int. J. Fatigue* **1999**, *21*, 221–231. [CrossRef]
8. Mughrabi, H.; Höppel, H.W. Cyclic deformation and fatigue properties of very fine-grained metals and alloys. *Int. J. Fatigue* **2010**, *32*, 1413–1427. [CrossRef]
9. Mughrabi, H. Microstructural fatigue mechanisms: Cyclic slip irreversibility, crack initiation, non-linear elastic damage analysis. *Int. J. Fatigue* **2013**, *57*, 2–8. [CrossRef]
10. Lataillade, J.L.; Naimark, O.B. Mesoscopic and nonlinear aspects of dynamic and fatigue failure (experimental and theoretical results). *Phys. Mesomech.* **2004**, *7*, 55–66.
11. Suresh, S. *Fatigue of Materials*; Cambridge University Press (CUP): Cambridge, UK, 1998; p. 586.
12. Mandelbrot, B.B.; Blumen, A. Fractal geometry: What is it, and what does it do? In *Mathematical and Physical Sciences*; The Royal Society: London, UK, 1989; Volume 423, pp. 3–16.
13. Froustey, C.; Naimark, O.; Bannikov, M.; Oborin, V. Microstructure scaling properties and fatigue resistance of pre-strained aluminium alloys (part 1: Al–Cu alloy). *Eur. J. Mech. A/Solids* **2010**, *29*, 1008–1014. [CrossRef]
14. Cantrell, J.H.; Yost, W.T. Nonlinear ultrasonic characterization of fatigue microstructures. *Int. J. Fatigue* **2001**, *23*, 487–490. [CrossRef]
15. Kumar, A.; Torbet, C.J.; Pollock, T.M.; Jones, J.W. In situ characterization of fatigue damage evolution in a cast Al alloy via nonlinear ultrasonic measurements. *Acta Mater.* **2010**, *58*, 2143–2154. [CrossRef]
16. Kumar, A.; Adharapurapu, R.R.; Jones, J.W.; Pollock, T.M. In situ damage assessment in a cast magnesium alloy during very high cycle fatigue. *Scr. Mater.* **2011**, *64*, 65–68. [CrossRef]
17. Marinesgarcia, I.; Paris, P.; Tada, H.; Bathias, C. Fatigue crack growth from small to long cracks in VHCF with surface initiations. *Int. J. Fatigue* **2007**, *29*, 2072–2078. [CrossRef]
18. Bathias, C.; Paris, P.C. Gigacycle fatigue of metallic aircraft components. *Int. J. Fatigue* **2010**, *32*, 894–897. [CrossRef]
19. Oborin, V.; Bannikov, M.; Naimark, O.; Froustey, C. Long-range-correlation large-scale interactions in ensembles of defects: Estimating reliability of aluminum alloys under dynamic cycling and fatigue loading conditions. *Tech. Phys. Lett.* **2011**, *37*, 241–243. [CrossRef]
20. Oborin, V.; Bannikov, M.; Naimark, O.; Palin-Luc, T. Scaling invariance of fatigue crack growth in gigacycle loading regime. *Tech. Phys. Lett.* **2010**, *36*, 1061–1063. [CrossRef]
21. Oborin, V.A.; Bannikov, M.V.; Bayandin, Y.V.; Naimark, O.B. Lifetime of Amg6 alloy under consecutive shock-wave and gigacycle loading. *PNRPU Mech. Bull.* **2019**, *1*, 121–128. [CrossRef]
22. Bouchaud, E. Scaling properties of cracks. *J. Phys. Condens. Matter* **1997**, *9*, 4319–4344. [CrossRef]
23. Bannikov, M.; Bilalov, D.; Oborin, V.; Naimark, O. Damage evolution in the AMg6 alloy during high and very high cycle fatigue. *Frat. Integrità Strutt.* **2019**, *13*, 383–395. [CrossRef]
24. Naimark, O. Duality of singularities of multiscale damage localization and crack advance: Length variety in Theory of Critical Distances. *Frat. Integrità Strutt.* **2019**, *13*, 272–281. [CrossRef]
25. Naimark, O.; Bayandin, Y.; Uvarov, S.; Bannikova, I.; Saveleva, N. Critical dynamics of damage-failure transition in wide range of load intensity. In *Acta Mechanica*; Springer: Berlin/Heidelberg, Germany, 2021; pp. 1–17. [CrossRef]

materials

MDPI

Article

# About the Role of Interfaces on the Fatigue Crack Propagation in Laminated Metallic Composites

Philip Manuel Pohl [1,2,*], Frank Kümmel [1,3], Christopher Schunk [1,4], Itziar Serrano-Munoz [5], Henning Markötter [5], Mathias Göken [1,2] and Heinz Werner Höppel [1,2]

1 Materials Science & Engineering, Institute I, Friedrich-Alexander-Universität Erlangen-Nürnberg (FAU), Martensstr. 5, 91058 Erlangen, Germany; frank.kuemmel@frm2.tum.de (F.K.); christopher.schunk@fau.de (C.S.); mathias.goeken@fau.de (M.G.); hwe.hoeppel@fau.de (H.W.H.)
2 Joint Institute for New Materials and Processes (ZMP), Friedrich-Alexander-Universität Erlangen-Nürnberg (FAU), Dr.-Mack-Straße 81, 90762 Fürth, Germany
3 Heinz Maier-Leibnitz Zentrum (MLZ), Technische Universität München (TUM), Lichtenbergstr. 1, 85748 Garching, Germany
4 Concept Laser GmbH, An der Zeil 8, 96215 Lichtenfels, Germany
5 Bundesanstalt für Materialforschung und-prüfung (BAM), Unter den Eichen 87, 12205 Berlin, Germany; itziar.serrano-munoz@bam.de (I.S.-M.); henning.markoetter@bam.de (H.M.)
* Correspondence: philip.pohl@fau.de

**Abstract:** The influence of gradients in hardness and elastic properties at interfaces of dissimilar materials in laminated metallic composites (LMCs) on fatigue crack propagation is investigated experimentally for three different LMC systems: Al/Al-LMCs with dissimilar yield stress and Al/Steel-LMCs as well as Al/Ti/Steel-LMCs with dissimilar yield stress and Young's modulus, respectively. The damage tolerant fatigue behavior in Al/Al-LMCs with an alternating layer structure is enhanced significantly compared to constituent monolithic materials. The prevalent toughening mechanisms at the interfaces are identified by microscopical methods and synchrotron X-ray computed tomography. For the soft/hard transition, crack deflection mechanisms at the vicinity of the interface are observed, whereas crack bifurcation mechanisms can be seen for the hard/soft transition. The crack propagation in Al/Steel-LMCs was studied conducting in-situ scanning electron microscope (SEM) experiments in the respective low cycle fatigue (LCF) and high cycle fatigue (HCF) regimes of the laminate. The enhanced resistance against crack propagation in the LCF regime is attributed to the prevalent stress redistribution, crack deflection, and crack bridging mechanisms. The fatigue properties of different Al/Ti/Steel-LMC systems show the potential of LMCs in terms of an appropriate selection of constituents in combination with an optimized architecture. The results are also discussed under the aspect of tailored lightweight applications subjected to cyclic loading.

**Keywords:** laminated metallic composites; toughening mechanisms; interfaces; fatigue crack propagation; fatigue crack growth; large chamber SEM

**Citation:** Pohl, P.M.; Kümmel, F.; Schunk, C.; Serrano-Munoz, I.; Markötter, H.; Göken, M.; Höppel, H.W. About the Role of Interfaces on the Fatigue Crack Propagation in Laminated Metallic Composites. *Materials* **2021**, *14*, 2564. https://doi.org/10.3390/ma14102564

Academic Editor: Jaroslav Pokluda

Received: 31 March 2021
Accepted: 27 April 2021
Published: 14 May 2021



## 1. Introduction

Recent studies [1–4] provide comprehensive overviews on the superior performance of heterostructured metallic materials compared to conventional metallic materials. The superior performance is based on significant synergistic effects associated with strong inter-zone interaction/coupling in the heterostructured materials [1]. The synergistic effects can be attributed to local heterogeneities and local variations of mechanical behavior in, among others, multimodal [5–8], gradient [9–12], or laminated [13–17] structures.

In laminated metal composites (LMCs), these local heterogeneities emerge at the interfaces between layers of different materials. Interface related mechanisms play a significant role regarding the macroscopic properties of laminated structures. For instance, the interface density, which correlates inversely with the layer thickness, was reported to have a major influence on the prevalent deformation mechanisms of laminated metal structures over several length scales of layer thickness [17,18].

Due to their layered architectures, most properties in laminated composites are inherently anisotropic. In terms of mechanical properties associated with crack growth, the most favorable properties are obtained when layers and interfaces are orientated perpendicular to the crack growth direction (crack arrester orientation) for quasistatic [19–22] as well as cyclic [23–26] loading. This can be attributed to the presence of extrinsic toughening mechanisms, resulting in a reduction of local stress intensity at the crack tip and thereby reducing the local crack driving force [27–29]. A comprehensive overview of different toughening mechanisms observed in laminated metal composites is provided by Lesuer et al. [27]. In crack arrester orientation, crack deflection [30–32], crack blunting [33], crack bridging [33], and stress redistribution mechanisms were observed in different LMC systems.

These toughening mechanisms in laminated metal composites are caused by material inhomogeneity effects [34] at the interfaces. Material inhomogeneity effects can be associated for instance with gradients in yield stress (yield stress gradient effect [35]) or elastic properties (elastic inhomogeneity effect [36,37]) at interfaces between dissimilar materials. Several numerical studies have investigated the influence of yield stress gradients on bimetal [31,35,38] as well as interlayer configurations [34,35,39]. A reduction of the local crack driving force at the vicinity of an interface has been reported when the crack approaches the interface from the softer towards the stronger layer. This was attributed to a change of the process zone size ahead of the crack upon interaction with the stronger layer and consequently leads to a crack tip shielding effect [40,41]. An anti-shielding effect was reported for the opposite case when the crack approaches the interface from the stronger to the softer layer. These findings have been confirmed experimentally [30,31,41] in bimetals subjected to constant far-field stress intensity ranges. For the soft/strong transition, the shielding effect was associated with a reduction of the fatigue crack growth rate as well as crack deflection at the vicinity of the interface. The anti-shielding effect was apparent by an increase of the fatigue crack growth rate near the strong/soft transition and a bifurcation of the crack upon entering the soft layer was observed at higher stress intensity ranges in an interlayer system [30]. Regarding the elastic inhomogeneity effect, similar shielding and anti-shielding effects have been reported at the vicinity of interfaces for compliant/stiff and stiff/compliant transitions, respectively, by numerical studies [34,42,43]. Furthermore, different numerical approaches based on cohesive zone models [44] and a strain energy density criterion [45] describe the onset of crack initiation associated with debonding at the interfaces of materials with dissimilar mechanical properties.

A recent study by Kümmel et al. [46] provides a comprehensive overview on the potential of fatigue life enhancement in specifically tailored laminated metal composites by utilizing the inherent material inhomogeneity effects at the interfaces between dissimilar materials. An increase in fatigue life in LMC systems can be accomplished by (a) enhancing resistance against crack initiation based on load transfer from surface layers into adjacent stiffer layers (gradient in elastic properties at interfaces) [15,16,46], and (b) enhancing resistance against crack propagation based on toughening mechanisms at the interfaces (gradient in hardness and elastic properties at interfaces) [15,32,47].

A variation of the stacking sequence of layers with different properties and thus the laminate architecture was reported to significantly influence fatigue life properties in laminated metal composites [15]. Fabrication of laminated metal composites can be achieved by the accumulative roll bonding (ARB) process [48], as this process allows for continuous production [49] of specifically tailored laminates [50] beyond the laboratory scale [51].

The influence of gradients in yield stress and elastic properties on (fatigue) crack growth at interfaces has been studied so far, on the one hand primarily on bimetals and interlayer systems by experimental and numerical approaches, and on the other hand on laminates, where the focus was set on the experimental determination of the fatigue lives. In this study, the influence of gradient effects at the interfaces in laminated metal composites on fatigue crack propagation and the resulting toughening mechanisms at the vicinity of interfaces is emphasized. Additionally, the lightweight application potential of LMCs with optimized architectures is addressed using different laminate architectures, which utilize the effects of gradients in yield stress and elastic properties to different extents in terms of fatigue life and resistance against crack propagation. An improved principal understanding of how these fatigue life-enhancing effects interact in LMCs is essential for an utilization in intelligently designed laminated architectures aiming for a high resistance against fatigue crack propagation.

## 2. Materials and Methods

### 2.1. Processing of the Laminated Metal Composites and Monolithic Materials

Three different systems of laminated metal composites and constituent monolithic materials for reference were produced by means of the accumulative roll bonding (ARB) process in order to study the role of (a) hardness gradients at interfaces and (b) combined gradients in hardness and elastic properties at interfaces on fatigue life and fatigue crack propagation. Regarding the first point, one laminated composite and the constituent monolithic materials were produced to additionally investigate the fatigue crack growth (FCG) properties.

All the investigated laminate and monolithic systems were roll bonded using a four high rolling mill (BW 300, Carl Wezel, Mühlacker, Germany). Prior to roll bonding, the sheet metal surfaces were cleaned with acetone and were wire brushed to remove oxide layers and to achieve sufficient bonding. Subsequently, the treated sheet metal surfaces were stacked and roll bonded at a nominal thickness reduction of 50% per ARB cycle at ambient conditions, air cooled afterwards and halved in length before performing the next cycle. The chemical composition of all sheet metal alloys used in this study is listed in Table 1.

**Table 1.** Chemical composition (wt.%) of the aluminium, steel, and titanium sheet metal, as measured by spark emission spectrometry.

| Alloy | Chemical Composition (wt.%) [1] | | | | | | | | | | |
|---|---|---|---|---|---|---|---|---|---|---|---|
| | Al | Fe | Mg | Zn | Cu | Ti | Si | Cr | Mn | C | O | Others |
| AA1050 | 99.4 | 0.35 | - | - | - | - | 0.15 | - | - | - | - | 0.10 |
| AA2024 | 93.4 | 0.10 | 1.42 | - | 4.37 | - | 0.07 | - | 0.46 | - | - | 0.18 |
| AA5754 | 95.9 | 0.40 | 2.91 | - | - | - | 0.35 | - | 0.31 | - | - | 0.13 |
| AA7075 | 89.4 | 0.12 | 2.72 | 5.77 | 1.53 | - | 0.07 | 0.20 | - | - | - | 0.19 |
| DC05 | - | 99.7 | - | - | - | 0.07 | - | - | 0.10 | 0.01 | - | 0.12 |
| Ti-Grade1 | - | - | - | - | - | 99.8 | - | - | - | - | 0.06 | 0.14 |

[1] Elements with concentration below 0.05 wt.% listed as (-).

In order to study the influence of thin layers of a stronger material embedded in a softer matrix on fatigue life and crack propagation, two different architectures of laminated composites consisting of technically pure aluminium (AA1050) and a solution hardening aluminium alloy (AA5754) as well as the constituent monolithic AA1050 and AA5754 materials were produced using two ARB cycles (N2). All sheet metals had an initial size of 300 mm in length and 100 mm in width. In the first ARB cycle, one sheet of the stronger material (AA5754) was stacked in between two sheets of the softer material (AA1050). The resulting laminate architectures after two ARB cycles with a total thickness of 3.0 mm, respectively, are shown schematically in Figure 1a. For the AA1050/AA5754 composite with a laminated architecture denoted as L1/8, the layer of the stronger material is positioned further towards the surface than for the laminate architecture denoted as L1/4. For both architectures, the volume fraction of the stronger AA5754 material layers accounted for 16.6 vol.% of the total volume of the laminates. Further details on initial stacking sequences for the first ARB cycle (N1) as well as the subsequent second cycle (N2) can be found in previous studies by the authors [15] for LMCs consisting of different constituent materials.

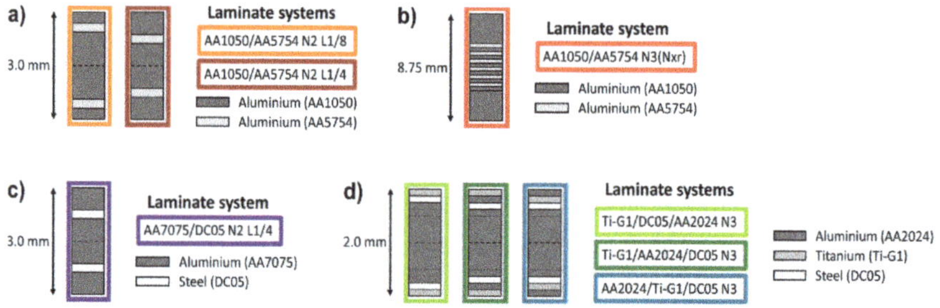


**Figure 1.** Schematic illustration of sheet materials, stacking sequence, and notation of different laminate systems. (**a**) Bi-material systems with hardness gradients at interfaces: AA1050/AA5754 N2 L1/4 and AA1050/AA5754 N2 L1/8; (**b**) bi-material system with hardness gradients at interfaces: AA1050/AA5754 N3(Nxr); (**c**) bi-material system with gradients of hardness and elastic properties at interfaces: AA7075/DC05 N2 L1/4; (**d**) tri-material systems with gradients of hardness and elastic properties at interfaces: Ti-G1/DC05/AA2024 N3, Ti-G1/AA2024/DC05 N3 and AA2024/Ti-G1/DC05.

In order to study the influence of a periodic variation of the soft (AA1050) and strong (AA5754) layers across a laminated structure on fatigue crack growth properties, a AA1050/AA5754 laminate with a total thickness of 8.75 mm was produced. Initially, sheets of AA5754 and AA1050 were 50% cold rolled down to a thickness of 2.5 mm prior to roll bonding. In the first ARB cycle, one sheet of the stronger material (AA5754) and one sheet of the softer material (AA1050) were stacked and rolled to form bimetal sheets. For the second ARB cycle, six of these bimetal sheets were stacked and roll bonded, resulting in a laminate architecture with an alternating sequence of AA1050 and AA5754 layers and a total sheet thickness of 7.5 mm. Subsequently, in the third ARB cycle, this laminate sheet was stacked in between two 2.5 mm thick AA1050 supporting layer sheets and roll bonded. A final laminated architecture for FCG measurements with a nominal layer thickness of 312 μm was achieved. The architecture of the laminated composite is schematically shown in Figure 1b. Monolithic sheet of the constituent AA1050 and AA5754 materials were produced using two ARB cycles. In order to maintain a uniform layer structure across all roll bonding cycles for the laminated composite and better comparability of results with the monolithic materials, all sheets were fully recrystallized prior to the first, second, and if applicable third ARB cycle for 2 h at 365 °C. This ARB processing route is denoted in the following as N3(Nxr) for the laminated composite and as N2(N1r) for monolithic materials in order to account for the intermediate recrystallization steps.

The influence of a combined difference in strength and elastic properties at interfaces between layers on fatigue life and crack propagation was studied (a) on a bi-material laminate system (AA7075/DC05 laminate) and (b) on tri-material laminate systems consisting of AA2024, Ti-Grade 1, and DC05 deep drawing steel.

The ARB processing of the AA7075/DC05 N2 L1/4 architecture followed the same route as described above for the AA1050/AA5754 N2 L1/4 laminate system. The laminate architecture is depicted schematically in Figure 1c. Monolithic AA7075 N2 aluminium and DC05 N2 steel sheets were roll bonded accordingly for reference measurements. To ensure sufficient bonding, all sheets were preheated for 5 min at 280 °C prior to each roll bonding step.

For the AA2024, Ti-Grade 1 and DC05 laminate systems, three roll bonding steps were performed. For the first roll bonding step, a shell structure and a core structure were prepared for each laminate architecture. The core structure consists of four 1 mm AA2024 sheets, which were roll bonded in the first ARB cycle (N1) to form 2.0 mm thick sheets. The shell structures were stacked in three different arrangements of layers of 1.0 mm AA2024, Ti-Grade 1 and DC05 sheets. The nomenclature states the layer materials starting from the surface layer towards the inner most layer. Three different shell structures were stacked in the first ARB cycle: AA2024/Ti-G1/DC05, Ti-G1/DC05/AA2024 and Ti-G1/AA2024/DC05. All shell structures were stacked adding one additional 1.0 mm AA2024 sheet at the inner most position and subsequently roll bonded in the first ARB cycle (N1) to produce 2.0 mm thick laminate sheets. In the second ARB cycle (N2), one core structure and one shell structure were each roll bonded to form asymmetrical laminates. In the ensuing third ARB cycle (N3), two asymmetrical N2 laminate structures were roll bonded at the respective core structures to produce symmetric laminate architectures. The respective stacking sequences (architectures) of the Ti-G1/DC05/AA2024 N3, Ti-G1/AA2024/DC05 N3 and AA2024/Ti-G1/DC05 N3 laminate systems are shown in Figure 1d. Monolithic AA2024 N3, Ti-Grade 1 N3 and DC05 N3 sheets were produced accordingly.

### 2.2. Characterisation of Local Mechanical Properties

The local mechanical properties were measured by means of nanoindentation experiments (Nanoindenter XP, MTS Nano Instruments, Oak Ridge, TN, USA) with the continuous stiffness method [52] using a three-sided Berkovich pyramid. The surfaces of the samples were prepared by grinding down to a grit size of 5 μm and mechanical polishing down to 1 μm grit size using diamond suspension. This was followed up by a chemo-mechanical polishing step using $SiO_2$ polishing suspension (OPS and OPU, Struers, Willich, Germany). In each sample, indentation fields of 50 to 200 indents, depending on the laminate size, were measured across the interfaces. The indentation depth was set to 1000 nm. The distance between indents was chosen to be 25-fold of the indentation depth to avoid influences of the damage zone around previous indents on the current measurement [53].

Uniaxial tensile tests were performed on a universal testing machine (Instron 4505, Instron, Darmstadt, Germany) at room temperature and a constant engineering strain rate of $10^{-3}$ $s^{-1}$. The tensile specimen had a gauge section of 10 mm length (in rolling direction), 4.5 mm width (in traverse direction), and 2.0 mm or 3.0 mm height (in normal direction). For each material, three tensile tests were conducted.

### 2.3. Determination of Fatigue Life

In order to determine fatigue live properties of laminate systems and constituent monolithic materials, fatigue tests were carried out on a vibrophore testing machine (HFP 5100, Roell Amsler, Ulm, Germany) in three-point bending mode. Specimen with dimensions of 20 mm length (in rolling direction), 9.5 mm width (in traverse direction) and 2.0 mm or 3.0 mm height (in normal direction) were machined from the sheets, respectively. The surfaces were prepared by grinding down to a grit size of 5 μm before testing. Laminate samples were tested in crack arrester orientation, where interfaces are oriented perpendicular to the direction of crack propagation. The fatigue tests were conducted in force control with an R-value of 0.1 and a resulting resonance frequency of about 50 to 70 Hz, depending on the stiffness of the sample. The frequency was monitored constantly throughout testing. Experiments were terminated upon a resonance frequency drop of 5 Hz, resulting from loss of stiffness due to (macro-)crack propagation and leading to the value for the numbers of cycles to failure ($N_f$). Post-experiment, the onset of crack propagation was determined using a 0.1 Hz resonance frequency drop criterion. The resulting numbers of cycles were denoted as numbers of cycles to crack initiation ($N_i$).

### 2.4. Determination of Fatigue Crack Growth

Fatigue crack growth (FCG) measurements were conducted on a servohydraulic testing machine (MTS810, MTS System Corporation, Eden Prairie, MN, USA) using the SE(B)-specimen (single edge notched bending) geometry (see Figure 2a). In accordance with standards ASTM E1820 [54] and ISO 12135 [55], the following dimensions for the specimen and setup were satisfied in relation to the specimen width $W$: thickness $B = 0.5\ W$, span $S = 4\ W$, length $L > 4.5\ W$, roller pin diameter $0.5\ W > d > 0.25\ W$ and roller pin length

$b > 1.25\ W$. The stress intensity factor $K$ for the SE(B)-specimen geometry was calculated at a force $P$ according to ASTM 1820 [52] as follows:

$$K = \frac{P\,S}{B\,W^{3/2}}\,f\left(\frac{a}{W}\right), \tag{1}$$

with:

$$f\left(\frac{a}{W}\right) = \frac{3\left(\frac{a}{W}\right)^{1/2}\left[1.99 - \left(\frac{a}{W}\right)\left(1-\frac{a}{W}\right)\left(2.15 - 3.93\left(\frac{a}{W}\right) + 2.7\left(\frac{a}{W}\right)^2\right)\right]}{2\left(1+2\,\frac{a}{W}\right)\left(1-\frac{a}{W}\right)^{3/2}}, \tag{2}$$

where the stress intensity factor $K$ depends solely on the parameters describing the geometry of the complete sample as well as the crack length $a$. Thus, the crack driving force depending on the calculated stress intensity factor $K$ needs to be interpreted as a far-field crack driving force for the laminated specimens, as opposed to a local driving force for concepts taking into account the local variation of the stress intensity factor in a bimetal or laminate of dissimilar materials [56–58].

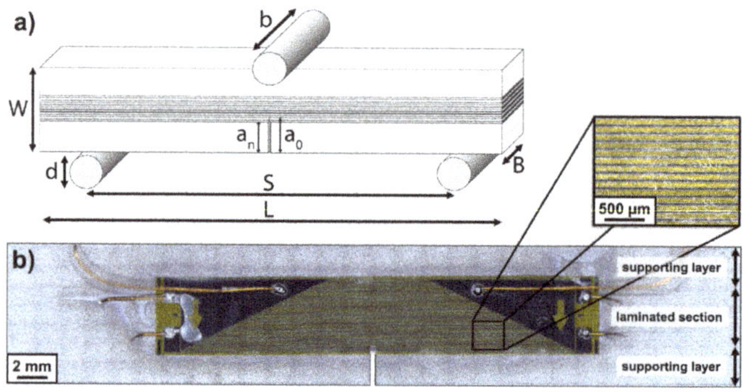


**Figure 2.** Single edge notched specimen geometry SE(B) for fatigue crack growth (FCG) measurements on laminated materials. (**a**) Notation of SE(B)-specimen dimensions; (**b**) Microscopical image of the specimen surface indicating the supporting layers and laminated structure (barely visible due to low contrast) as well as crack length measurement using crack propagation gauges.

The samples were machined from the rolled laminate and monolithic material sheets in T-S orientation (ASTM E1823 [59]) with following dimensions: $L$ = 50 mm (in traverse direction), $W$ = 8.75 mm (in normal direction) and $B$ = 4.4 mm (in rolling direction). Subsequently, a notch with notch length $a_n = 0.3\ W = 2.65$ mm was introduced using wire electrical discharge machining, resulting in a notch orientated perpendicular to the interfaces (crack arrester orientation) for laminate specimen. The specimen surfaces were prepared by grinding down to a grit size of 5 µm before testing. Crack propagation gauges (FAC-5, Tokyo Measuring Instruments Lab., Tokyo, Japan) with a resolution of 0.1 mm glued to the surface of each specimen, as can be seen in Figure 2b, allowed for automated and discrete measurement of crack extension during the test across the complete laminated section of the specimen. The difference between surface crack extension and crack propagation gauge measurement was detected to be below 0.2 mm. The FCG experiments were conducted under stress intensity $K$ control, an $R$-value of 0.1 and a testing frequency of 3 Hz. The $K$ controlled operation was realized using a closed loop feedback of crack extension measurement to automatically adjust the force loading to achieve the respective $K$ level (load shedding technique [60]). Prior to testing, the machined notch root was sharpened using razor blade grinding and subsequently pre-cracked at a constant stress intensity range of $\Delta K = 5\ \text{MPa}\sqrt{\text{m}}$ to a minimum crack extension of 0.3 mm to reach the initial crack length $a_0$ for the FCG tests.

The FCG tests on monolithic specimen were performed using $\Delta K$-increasing and $\Delta K$-decreasing procedures as well as $\Delta K$-constant tests additionally. As suggested by the ASTM E647 standard [60], for $\Delta K$-decreasing tests, a normalized $K$-gradient $C > -0.08$ mm$^{-1}$ was chosen to minimize effects of prior loading history on FCG rates below $10^{-5}$ mm/cycle. For $\Delta K$-increasing tests, the normalized $K$-gradient was set below $C < 0.08$ mm$^{-1}$ to minimize the influence of additional transient crack growth effects. $\Delta K$-constant tests were only performed on laminate specimen, as the external driving force for crack growth remains constant and changes in process zone size ahead of the crack tip only depend on local conditions associated with the laminated structure of materials with dissimilar mechanical properties. This aids in terms of a better understanding of effects on FCG properties associated with toughening mechanisms arising from interactions between the process zone ahead of the crack tip and the laminated structure. The FCG experiments were terminated as the cracks approached the end of the laminated structure. The crack growth rate *da/dN* was calculated as the average crack growth rate throughout the laminated section $0.3 < a/W < 0.7$ of each specimen (Figure 2) tested at a certain $\Delta K$ level.

### 2.5. Characterisation of Damage Mechanisms

In order to identify damage mechanisms in laminated structures post-mortem, the surface crack networks of the fatigued specimen were examined by scanning electron microscopy (Crossbeam 1540 EsB, Zeiss, Oberkochen, Germany) using secondary and backscattered electron contrast techniques as well as light microscopy (Axio Imager M1, Zeiss, Oberkochen, Germany) using brightfield and dark field contrast techniques.

Synchrotron X-ray computed microtomography (SXCT) experiments were performed at the BAMline [61] at the electron storage ring Bessy II (HZB, Berlin, Germany) on the crack network of a AA1050/AA5754 N3(Nxr) laminate specimen after the FCG experiment at constant $\Delta K$ = 17.5 MPa$\sqrt{m}$. A cuboidal sample with a cross section of 4 × 4 mm$^2$ (width $W$ × thickness $B$) and a length L of 20 mm containing the entire crack network was extracted metallographically from the laminated section of the SE(B)-specimen. Two scans were performed along the sample: One corresponding to the path of the primary crack network and another one adjacent to the prior scan to further enlarge the field of view to display the secondary crack network. The energy of the monochromatic and parallel beam was set to 30 keV using a double multilayer monochromator (DMM) with an energy resolution of ~3%. An effective pixel size of 2.2 μm was chosen using a corresponding microscope objective (2× magnification, Olympus, Hamburg, Germany) in an X-ray microscope setup (Optique-Peter, Lentilly, France) combined with a charge-coupled device (CCD)-based camera (4008 × 2672 pixels, PCO, Kelheim, Germany). Each tomographic scan comprised 2416 projections recorded within a total scan time of 2.5 h. The distance between the scintillator screen of the detector (~60 μm $CdWO_4$) and the investigated samples was 10 mm. An in-house routine developed on Python software [62] was used to reconstruct the projections. The reconstructed volume data was denoised using a non-local means filter, with Fiji software [63]. Subsequently, the cracks were identified by global threshold segmentation. Visualization of the segmented cracks was performed using the AvizoFire 9.4 software package [64].

### 2.6. In-Situ Characterisation of Damage Mechanisms

In order to gain additional information on crack propagation at the vicinity of interfaces and to investigate the prevalent damage mechanisms, specimen of the AA7075/DC05 N2 L1/4 laminate system were fatigued in-situ in a large chamber scanning electron microscope (LC-SEM, Visitec, Grevesmühlen, Germany), see [65–67] for details. Therefore, specimen with dimensions of 30 mm length (in rolling direction), 9.5 mm width (in traverse direction) and 3.0 mm height (in normal direction) were machined from the AA7075/DC05 N2 L1/4 laminate sheet. Prior to testing, the surfaces of the samples were prepared by grinding down to a grit size of 5 μm and mechanical polishing down to 1 μm grit size using diamond suspension. In-situ fatigue experiments in three-point bending mode were conducted in the LCF regime at a maximum stress amplitude $\Delta\sigma_{max}/2$ = 295 MPa as well as in the HCF regime at $\Delta\sigma_{max}/2$ = 210 MPa. In order to study crack propagation in the HCF regime in-situ, the specimen was pre-fatigued on a vibrophore testing machine (HFP 5100, Roell Amsler, Ulm, Germany) at $\Delta\sigma_{max}/2$ = 210 MPa. The pre-cycling test was terminated after 9.1 Mio cycles using a 0.1 Hz resonance frequency drop criterion, indicating the end of the crack nucleation phase.

The in-situ fatigue tests were conducted on a servohydraulic testing machine (MTS810, MTS System Corporation, Eden Prairie, MN, USA) which can be installed inside the vacuum chamber of the LC-SEM (Figure 3a). The electron gun was tilted at 90° and thus oriented perpendicular to the front surface of the specimen. This allowed for in-situ monitoring of crack nucleation and crack propagation at the front surface of the specimen during fatigue experiments (Figure 3b–d).

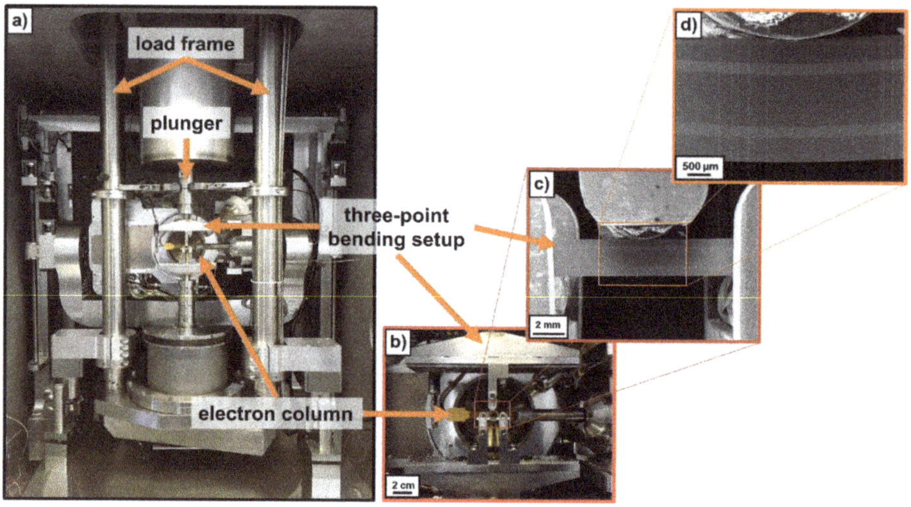


**Figure 3.** In-situ fatigue testing setup inside the large chamber scanning electron microscope (LC-SEM): (**a**) Electron gun oriented perpendicular to the front surface of specimen positioned inside the servohydraulic MTS810 testing machine; (**b**–**d**) Three-point bending setup and AA7075/DC05 N2 L1/4 laminated composite specimen at different magnifications, respectively.

The as prepared laminate sample for LCF fatigue testing and pre-fatigued laminate sample for HCF fatigue testing were positioned on the three-point bending fixture accordingly and the vacuum chamber was evacuated to a pressure below $5 \times 10^{-6}$ mbar for SEM operation. The fatigue tests were conducted in force control with an *R*-value of 0.1 and a testing frequency of 1 Hz. The crack propagation was monitored in-situ by scanning electron microscopy. At certain intervals, the fatigue test was interrupted to characterize the current state of crack propagation (interrupted monitoring) using the secondary electron contrast technique before continuing the fatigue experiment. Therefore, a constant loading of 0.2 $\sigma_{max}$ was applied on the one hand to open the crack wakes for SEM characterization as well as on the other hand to minimize possible transient crack growth effects during these periods of interrupted monitoring.

## 3. Results

### 3.1. Effects of a Hardness Gradient at Interfaces on Fatigue Life and Crack Propagation in LMCs

The AA1050/AA5754 N2 LMC systems were investigated in order to study the effects of integrating thin layers of a stronger material embedded in a softer matrix on the fatigue life and crack propagation. The local mechanical properties of the different layer materials after N2 processing can be seen in Table 2.

**Table 2.** Average hardness *H* of respective layers in the AA1050/AA5754 N2 LMC architectures as measured by nanoindentation measurements and resulting hardness gradient Δ*H* at LMC interfaces. Yield stress and ultimate tensile stress of the monolithic AA1050 N2 and AA5754 N2 materials as determined by uniaxial tensile tests.

| **Material** | **AA1050/AA5754 N2 LMC Systems** | | |
|---|---|---|---|
| | **AA1050 Layers** | **AA5754 Layers** | **Gradient (Δ*H*)** |
| Hardness *H*/GPa | 0.79 ± 0.07 | 1.37 ± 0.11 | 0.58 ± 0.13 |
| **Material** | **Monolithic Materials** | | |
| | **AA1050 N2 Mono** | **AA5754 N2 Mono** | **Difference** |
| Yield stress/MPa | 152 ± 5 | 332 ± 5 | 180 ± 7 |
| Ultimate tensile stress/MPa | 172 ± 8 | 361 ± 2 | 189 ± 8 |

The hardness was calculated as a mean value of the individual AA1050 and AA5754 layers, respectively, from both the L1/4 and L1/8 architecture. The hardness of the AA1050 layers is about 0.8 GPa and of the AA5754 layers about 1.4 GPa. Consequently, the hardness gradient at the laminate interfaces can be calculated to be about 0.6 GPa. However, it must be noted, that this estimation of hardness gradient at laminate interfaces does not take into account the formation of an interface affected zone at the immediate vicinity of an interface caused by different shearing behavior of dissimilar materials [14,68,69]. This must also be considered for the nanoindentation results presented in the following sections.

The fatigue life diagrams (*S-N* curves) of AA1050/AA5754 N2 laminate systems with two different architectures as well as of the constituent monolithic AA1050 N2 and AA5754 N2 materials are plotted in Figure 4a. For each specimen tested, the number of cycles to crack initiation ($N_i$) as well as the numbers of cycles to failure ($N_f$) are shown. The determination of the number of cycles to crack initiation using a 0.1 Hz drop criterion of resonance testing frequency can be seen in Figure 4c for a monolithic AA1050 sample and a AA1050/AA5754 laminated composite.

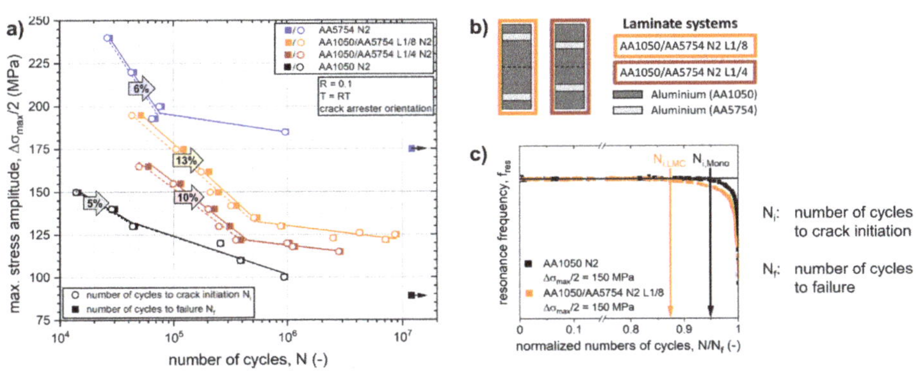


**Figure 4.** (**a**) Fatigue properties (*S-N* curves) of different AA1050/AA5754 N2 LMC architectures and constituent monolithic materials. Data partly adapted from [15,32,46]. The percentile of fatigue crack propagation on total fatigue life is indicated in the respective LCF regimes; (**b**) schematic illustration of the LMC architectures; (**c**) determination of the onset of fatigue crack propagation in the LCF regime for a monolithic and LMC sample using a 0.1 Hz resonance frequency drop criterion.

As can be seen in Figure 4a, the fatigue life of a soft AA1050 matrix material can be enhanced significantly by integrating thin and strong interlayers of AA5754 forming a laminated structure. This can be observed in both the LCF and the HCF regimes. The fatigue life of the L1/4 architecture is lower for both regimes compared to the L1/8 architecture, where the harder AA5754 layer is positioned more towards the surface (see Figure 4b). The increased fatigue life for the L1/8 architecture might be caused by effects of the interface affected zone [14,68,69], different shear strain distribution in the respective laminate surface layers during ARB processing [70] or a complex internal stress state upon cyclic loading due to the co-deformation of different aluminium layers with dissimilar hardness [71]. The transition between the LCF and HCF regime is correlated to a threshold value which depends on the material of the layer at the surface [16,32,46]. Below the threshold, in the HCF regime, the fatigue life of monolithic materials and laminated composites is primarily determined by the fatigue crack initiation stage: The numbers of cycles to crack initiation account for 95% to 99% of the numbers of cycles to failure for both monolithic materials and laminates. Above the threshold, in the LCF regime, the role of the crack propagation stage of fatigue life is significantly enhanced for the laminated composites compared to monolithic materials. The crack propagation stage accounts for 10% of numbers of cycles to failure for the AA1050/AA5754 N2 L1/4 composite and 13% for the AA1050/AA5754 N2 L1/8 composite compared to about 5 to 6% for the monolithic AA1050 N2 and AA5754 N2 materials, respectively.

Figure 4c shows a comparison of the drop of the resonance testing frequency between a laminated AA1050/AA5754 N2 L/8 specimen and a monolithic AA1050 N2 specimen tested in the LCF regime at 150 MPa (i.e., around 20 MPa above the threshold value). In both samples, the initiation of the macro-crack occurs at the respective surface layers consisting of the softer AA1050 material. As visible in the diagram, the resonance frequency of the laminated structure initially drops significantly slower with respect to the normalized numbers of cycles to failure $N/N_f$ as compared to the monolithic sample. This behavior can be attributed to a retardation of crack propagation at the vicinity of the interface to the harder AA5754 layer in the laminated composite [46]. Moreover, macro-crack propagation captures a significantly higher fraction of the total fatigue life in the LMCs than for the monolithic materials, as indicated by $N_{i,LMC}$ and $N_{i,Mono}$ in Figure 4c.

### 3.2. Effects of a Hardness Gradient at Interfaces on Fatigue Crack Growth (FCG) in LMCs

The effects of a periodic variation of strong and soft layers in a laminated composite and the associated result of having a periodic variation of the local driving forces at crack tips near the interfaces were studied using fatigue crack growth experiments on a AA1050/AA5754 N3(Nxr) laminate architecture. The local hardness of the different layers in the laminated section of the LMC architecture are listed in Table 3 as mean values of all AA1050 and AA5754 layers, respectively.

**Table 3.** Average hardness $H$ of constituent monolithic AA1050 N2(N1r) and AA5754 N2(N1r) materials and respective layers in the AA1050/AA5754 N3(Nxr) LMC architecture as well as the resulting hardness gradient $\Delta H$ at the LMC interfaces as measured by nanoindentation measurements.

| Material | AA1050 N2(N1r) Mono | AA5754 N2(N1r) Mono | AA1050/AA5754 N3(Nxr) LMC | | |
|---|---|---|---|---|---|
| | | | AA1050 Layers | AA5754 Layers | Gradient ($\Delta H$) |
| Hardness $H$/GPa | 0.71 ± 0.05 | 1.37 ± 0.10 | 0.73 ± 0.06 | 1.35 ± 0.12 | 0.62 ± 0.13 |

As shown in Table 3, no significant difference in hardness between AA1050 and AA5754 layers in the N3(Nxr) processed laminate and the hardness in the respective N2(N1r) processed monolithic materials can be observed. As the laminate and monolithic materials were recrystallized prior to each roll bonding process, this indicates that the local mechanical properties of the respective laminate layers and their constituent monolithic material can be considered comparable. The AA5754 layers in the LMC exhibit a hardness of 1.35 GPa and will be denoted as the stronger/harder layers in the following in opposition to the softer AA1050 layers with a hardness of 0.73 GPa. The resulting hardness gradient at the interfaces in the laminated composite can be calculated to be about 0.6 GPa.

Fatigue crack growth rates from increasing, decreasing, and constant stress intensity range tests for monolithic AA1050 and AA5754 as well as for constant stress intensity range tests for the AA1050/AA5754 laminated composite in crack arrester orientation are shown in Figure 5.

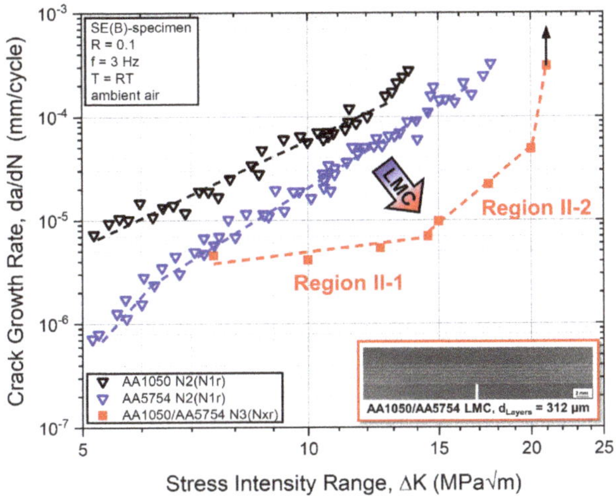


**Figure 5.** Fatigue crack growth rates in a AA1050/AA5754 N3(Nxr) laminated composite (crack arrester orientation) with a hardness gradient at interfaces and constituent monolithic materials AA1050 N2(N1r) and AA5754 N2(N1r) as a function of the applied stress intensity range $\Delta K$.

No significant differences between crack growth rates measured by increasing, decreasing, and constant stress intensity range tests were found for monolithic AA1050 and AA5754 specimen, respectively. This suggests that there are no effects on the crack growth rate from prior loading history for decreasing tests and from transient crack growth effects for increasing tests, since the normalized $K$-gradients for each test were chosen appropriately. The crack growth rate, da/dN, in monolithic materials between 5 and 12.5 MPa$\sqrt{m}$ for AA1050 and 6–16 MPa$\sqrt{m}$ for AA5754 can be described with respect to the applied stress intensity range, $\Delta K$, by the well-known Paris power-law relationship [72,73]:

$$\frac{da}{dN} = C\,\Delta K^{m}, \tag{3}$$

where $C$ and $m$ are the scaling constants, characterizing the crack growth behavior in Region II. The Paris equation exponents $m$ were found to be about $m = 3.4$ and 4.5 for the monolithic AA1050 and AA5754 materials, respectively (see Table 4). The crack growth rates of the monolithic AA5754 material are lower compared to the AA1050 alloy. At a stress intensity range below about 6 MPa$\sqrt{m}$, a deviation from Region II crack growth behavior can be observed for AA5754, indicating the transition towards near-threshold Region I fatigue crack growth. This suggests that the threshold $\Delta K_{th}$ of the technically pure aluminium AA1050 is lower than that of the AA5754 alloy. The near threshold fatigue crack growth behavior of the materials was not addressed experimentally, as measurement of crack growth rates below $5 \times 10^{-7}$ mm/cycle were associated with long measurement times at a testing frequency of 3 Hz on the servohydraulic testing system. Additionally, measurements of crack growth rates above $3 \times 10^{-4}$ mm/cycle were associated with increasing effects of plasticity resulting in mixed mode failure of specimen from plastic collapse and fatigue crack growth. The measurement data above this threshold value was discarded, as the crack length measured by the crack propagation gauges was considered inaccurate.

**Table 4.** Paris equation exponents $m$ of Region II fatigue crack growth in monolithic AA1050 and AA5754 materials and a laminated composite AA1050/AA5754 in crack arrester orientation.

| Material | AA1050 N2(N1r) Mono | AA5754 N2(N1r) Mono | AA1050/AA5754 N3(Nxr) LMC | |
|---|---|---|---|---|
| | | | Region II-1 | Region II-2 |
| Paris equation exponent $m$ | 3.37 | 4.51 | 0.89 | 5.72 |

Regarding the laminated composite, the crack growth behavior between 7.5 and 20 MPa$\sqrt{m}$ can be divided into two different zones, which are denoted as Region II-1 and Region II-2 in the following. The Paris equation exponents $m$ of the laminated composite were calculated to be to be about $m = 0.9$ and 5.7 for Region II-1 and Region II-2, respectively (Table 4).

As shown in Figure 5, the fatigue crack growth rate of the laminated AA1050/AA5754 composite is significantly reduced compared to crack growth rates in both constituent monolithic materials. In Region II-1, the fatigue crack growth rate in the laminated composite at a constant $\Delta K$ of 7.5 MPa$\sqrt{m}$ is about the same as in the monolithic AA5754 and about 20% of the crack growth rate in monolithic AA1050 material. For FCG tests at higher constant $\Delta K$ levels in Region II-1, further deviation of crack growth rates between the LMC and monolithic constituent material can be observed, as indicated by the lower exponent $m$ for the laminated composite. At $\Delta K = 14.5$ MPa$\sqrt{m}$, the fatigue crack growth rate behavior changes and can be described at subsequent higher stress intensity ranges using a different exponent $m$ for Region II-2. FCG tests at the transition between Regions II-1 and II-2 indicate the biggest decrease of fatigue crack growth rate in the laminated composite compared to the constituent materials (Figure 5). The FCG rate of the LMC at 14.5 MPa$\sqrt{m}$ was measured to be about 7% of the crack growth rate in monolithic AA5754 material.

A reduction of the fatigue crack growth rate in a laminated composite tested in crack arrester orientation of this magnitude compared to the constituent monolithic materials can only be explained by crack growth retardation effects associated with toughening mechanisms. In LMCs, these toughening mechanisms emerge from interactions between the process zone ahead of the crack tip and the periodic variation of mechanical properties at the interfaces due to the laminated architecture [27].

The surface crack networks of the LMC specimen were investigated after testing in order to assess the crack growth behavior perpendicular to the interfaces of the laminated composites. Figure 6 shows the fatigue crack growth paths in the laminated composite structure for experiments conducted at constant stress intensity ranges in Region II-1 (7.5 MPa$\sqrt{m}$ to 12.5 MPa$\sqrt{m}$) and Region II-2 (15 MPa$\sqrt{m}$ to 20 MPa$\sqrt{m}$), respectively.

At a low stress intensity range of 7.5 MPa$\sqrt{m}$, the crack path is orientated perpendicular to the layers and interfaces without being deflected significantly throughout the laminated structure. For increasing stress intensity ranges starting at 10 MPa$\sqrt{m}$, the crack path across the laminated section is increasingly impeded at the vicinity of interfaces. For experiments at constant stress intensity ranges of 12.5 MPa$\sqrt{m}$ and above, a clear distinction of different toughening mechanisms at the interfaces can be observed: (a) crack deflection (orange arrows, Figure 6) can be observed when the crack approaches interfaces from the softer (AA1050) layers towards the harder (AA5754) layers and (b) crack bifurcation (green arrows, Figure 6) can be found when the crack approaches the interfaces from the harder towards the softer layers.

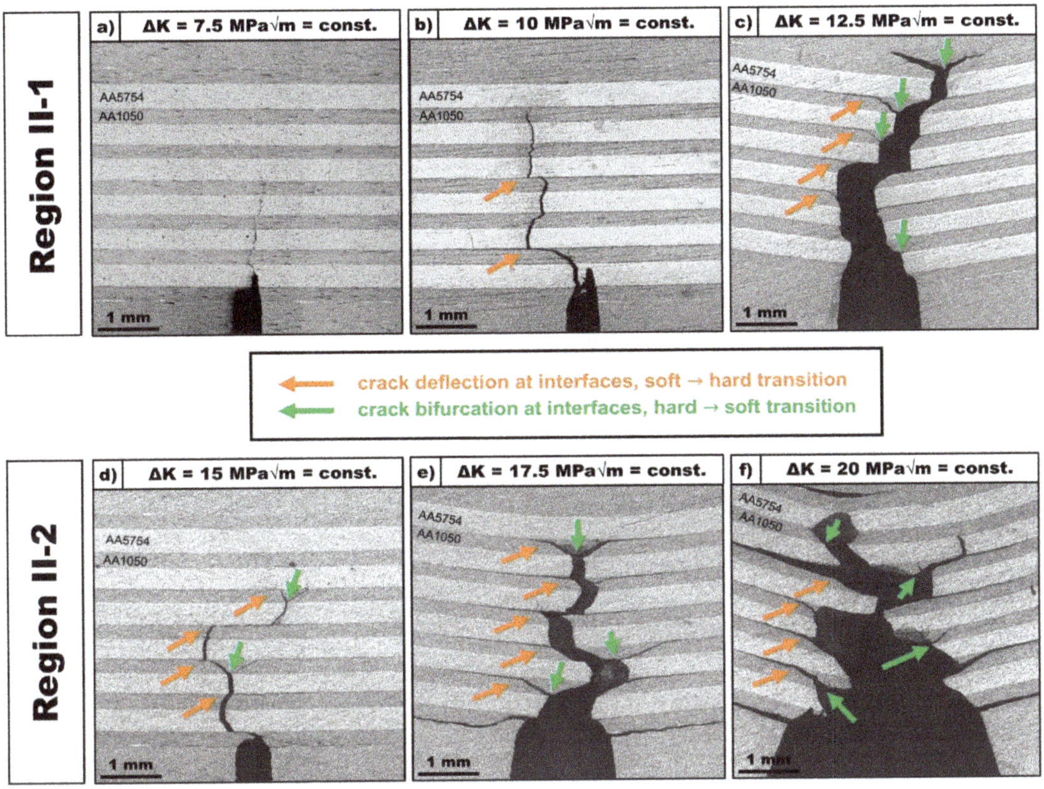


**Figure 6.** Light microscopical images of the surface crack networks in AA1050/AA5754 N3(Nxr) laminated composite structures consisting of soft AA1050 and hard/strong AA5754 layers. Identification of different toughening mechanisms at LMC interfaces resulting in impeded fatigue crack growth in crack arrester orientation for (**a–c**) Region II-1 FCG at constant $\Delta K = 7.5$ MPa$\sqrt{m}$, 10 MPa$\sqrt{m}$ and 12.5 MPa$\sqrt{m}$, respectively; (**d–f**) Region II-2 FCG at constant $\Delta K = 15$ MPa$\sqrt{m}$, 17.5 MPa$\sqrt{m}$ and 20 MPa$\sqrt{m}$, respectively.

The crack deflection mechanism at the interface results in crack growth along the interface. As the experiments were conducted using the SE(B)-specimen geometry, the maximum bending stress is located at the symmetry plane above the notch root and is reduced along the span $S$. This limits the crack growth along the interfaces, as the driving force is reduced gradually. The deflection of the crack path at the interfaces, when cracks approach interfaces from the soft layers towards the harder layers, and consequent crack growth along the interface imply, that further crack extension along the loading axis requires a new crack to be nucleated in the adjacent layer. As this crack re-nucleation phase depends on the local (micro-) structural characteristics and the tests were operated under a constant far-field stress intensity range, it is evident that the toughening mechanism of crack deflection strongly promotes the damage tolerant fatigue behavior of the LMC.

Bifurcation of the cracks can be observed for the opposite case, where a crack approaches the interface from the harder towards the softer layer. This mechanism leads to a branching of the original crack front into two new separate cracks. The bifurcation angles are found to be around 45° in relation to the symmetry plane above the notch root, where the bending stress is at a maximum. Bifurcation of the crack front leads to local redistribution of the far-field crack driving force that is remotely applied using a constant stress intensity range $\Delta K$, as it is reallocated across multiple crack tips. Crack growth across multiple crack fronts reduces the overall rate of fatigue crack growth and thus further enhances the damage tolerant fatigue properties of the laminated composite as seen in Figure 5.

Further investigation into these toughening mechanisms was done using synchrotron X-ray computed microtomography (SXCT), as these mechanisms described above were identified analyzing the crack networks at the surfaces of the laminated composites post-mortem. Results of the SXCT experiment on the AA1050/AA5754 N3(Nxr) laminated composite specimen fatigued at a constant stress intensity range of $\Delta K$ = 17.5 MPa$\sqrt{m}$, as can be seen in Figure 7.

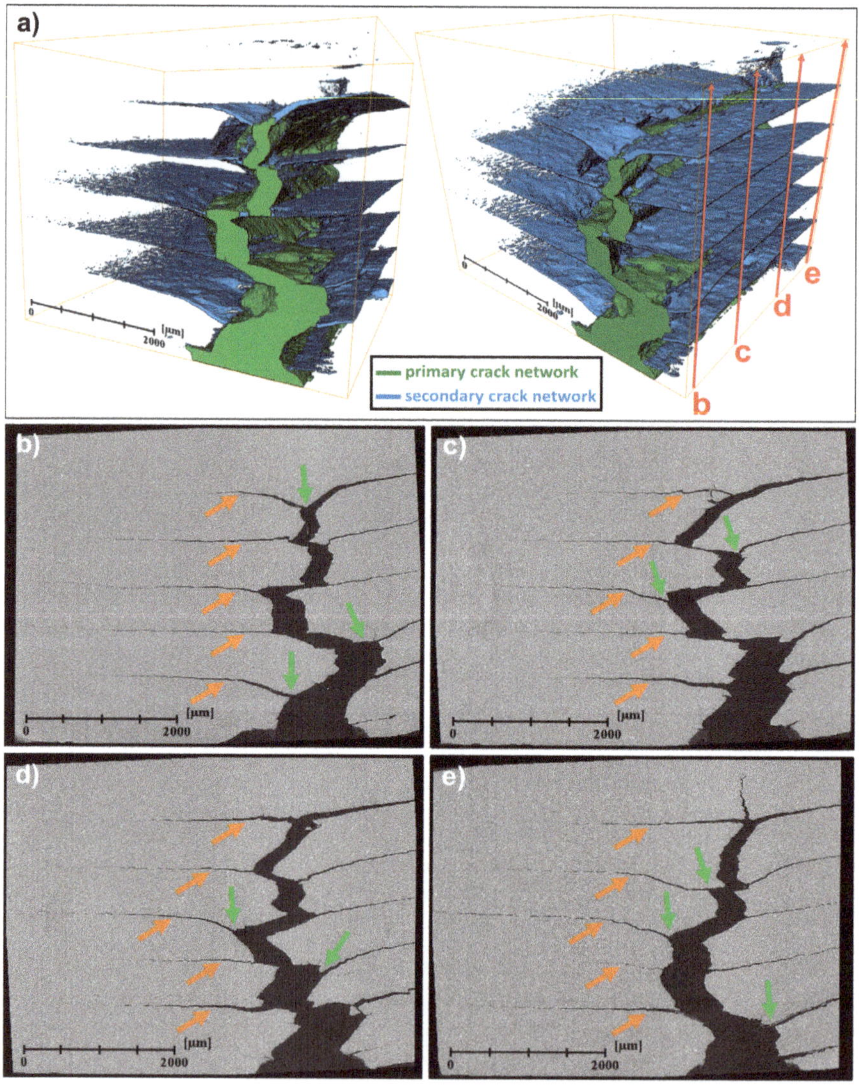


**Figure 7.** Synchrotron X-ray computed microtomography (SXCT) on a AA1050/AA5754 N3(Nxr) laminated composite specimen fatigued at a constant stress intensity range of $\Delta K$ = 17.5 MPa$\sqrt{m}$. (**a**) 3D-reconstruction of the crack network of the LMC specimen indicating the separation into a primary and secondary crack network; (**b–e**) cross sections extracted from the 3D-tomogram at different positions $z = 0.05\ B$, $z = 0.33\ B$, $z = 0.66\ B$ and $z = 0.95\ B$ (indicated by red arrows in (**a**)) in relation to the specimen thickness $B$, respectively. Identification of the toughening mechanisms (**a**) crack deflection (orange arrows) and (**b**) crack bifurcation (green arrows) throughout the crack networks in all cross sections.

The 3D-reconstruction of the crack network in the laminated specimen (Figure 7a) shows that the crack network can be divided into a primary (green) and a secondary (blue) crack network. The primary crack network starts at the notch root and subsequently grows along the symmetry plane of the laminated SE(B)-specimen. At this symmetry plane, the bending stress is at a maximum and promotes the highest tensile stresses perpendicular to the direction of the crack path and thus the highest driving forces at the crack tip. At the vicinity of the interfaces, the crack path is deflected due to interaction mechanisms between the process zone ahead of the crack tip and the variation of local mechanical properties at the interfaces. This results in the formation of the secondary crack network associated with the prevalent toughening mechanisms of crack deflection and crack bifurcation.

The crack networks in four cross sections of the 3D-tomogram at different positions $z = 0.05\ B$, $z = 0.33\ B$, $z = 0.66\ B$ and $z = 0.95\ B$ across the thickness $B$ of the LMC specimen were analyzed in Figure 7b–d, respectively. Crack deflection (orange arrows) and crack bifurcation mechanisms (green arrows) can be conclusively identified in cross sections of the crack network both near the surfaces (Figure 7b,e) as well as in the volume (Figure 7c,d) of the laminated composite sample.

These findings emphasize the magnitude of the effects on damage tolerant fatigue crack growth behavior in LMCs that toughening mechanisms at interfaces in laminated metallic composites with dissimilar hardness can produce.

### 3.3. *Effects of a Combined Gradient of Hardness and Elastic Properties at Interfaces on Fatigue Life and Crack Propagation in LMCs*

Investigations into the influence of a combined difference in strength and elastic properties at interfaces between layers on fatigue life and the role of toughening mechanisms at interfaces on crack propagation was studied on a AA7075/DC05 N2 L1/4 laminated composite architecture. The local mechanical properties in terms of hardness and Young's modulus of the different layer materials after N2 processing as well as the resulting gradients at interfaces can be seen in Table 5.

**Table 5.** Average hardness $H$ as measured by nanoindentation and Young's modulus $E$ of the respective AA7075 and DC05 layers in the AA7075/DC05 N2 L1/4 LMC architecture and resulting gradients $\Delta H$ and $\Delta E$ at interfaces. Yield stress and ultimate tensile stress of the monolithic AA7075 N2 and DC05 N2 materials as determined by uniaxial tensile tests.

| **Material** | **AA7075/DC05 N2 LMC System** | | |
|---|---|---|---|
| | **AA7075 Layers** | **DC05 Layers** | **Gradients ($\Delta H$, $\Delta E$)** |
| Hardness $H$/GPa | $1.55 \pm 0.08$ | $2.66 \pm 0.12$ | $1.11 \pm 0.14$ |
| Young's modulus $E$/GPa | 70 [1] | 210 [2] | 140 |
| **Material** | **Monolithic Materials** | | |
| | **AA7075 N2 Mono** | **DC05 N2 Mono** | **Difference** |
| Yield stress/MPa | $308 \pm 6$ | $575 \pm 8$ | $267 \pm 10$ |
| Ultimate tensile stress/MPa | $385 \pm 4$ | $620 \pm 7$ | $235 \pm 8$ |

[1] data based on [74]; [2] data based on [75].

The average hardness was measured to be 1.55 GPa in the AA7075 aluminium layers and 2.66 GPa in the DC05 steel layers of the laminated composite. Consequently, the hardness gradient $\Delta H$ at the laminate interfaces can be calculated to be about 1.1 GPa. The Young's moduli of the respective layers were assumed to be 70 GPa for AA7075 and 210 GPa for DC05 based on literature data [74,75], resulting in a gradient in elastic modulus $\Delta E$ of 140 GPa at the AA7075/DC05 interfaces.

Figure 8a plots the fatigue life diagrams ($S$-$N$ curves) of the AA7075/DC05 N2 L1/4 laminate architecture as well as of the constituent monolithic AA7075 N2 and DC05 N2 materials. Again, the number of cycles to crack initiation ($N_i$) as well as the numbers of cycles to failure ($N_f$) are shown for each fatigue test. The percentile of the fatigue crack propagation phase on the total fatigue life is specified for the LCF regimes of the laminate and monolithic materials, respectively. The S-N diagram reveals that the fatigue life of the AA7075 N2 matrix material can be enhanced in the LCF regime as well as in the HCF regime by integration of a thin DC05 steel layer.

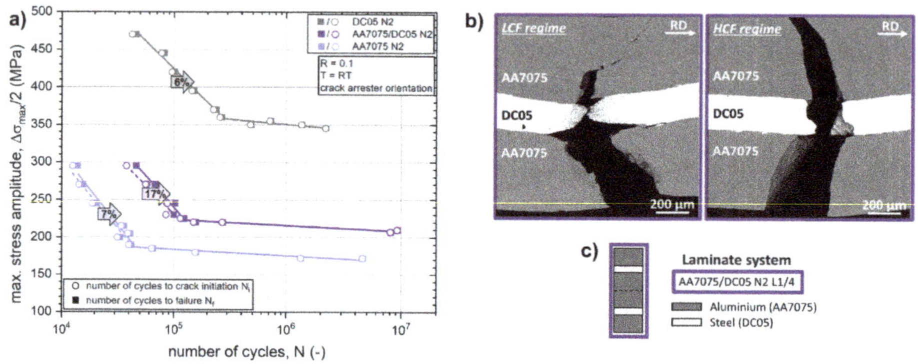


**Figure 8.** (**a**) Fatigue properties (*S-N* curves) of the AA7075/DC05 N2 L1/4 LMC architecture and constituent monolithic materials. The percentile of fatigue crack propagation on total fatigue life is indicated in the respective LCF regimes; (**b**) Fatigue crack propagation paths in AA7075/DC05 LMCs fatigued in the LCF and HCF regime, respectively; (**c**) Schematic illustration of the LMC architecture.

Previous studies by the authors [15,16] on different laminated Al/Steel composite systems with the same architecture as in Figure 8c revealed that the enhancement of the endurable maximum stress amplitudes $\Delta\sigma_{max}/2$ in the HCF regime correlates with the reduction of the maximum bending stress in the outer aluminium layer due to load transfer associated with the higher elastic modulus of the steel layer. FEM analysis of the stress distribution upon loading bending of Al/Steel laminated beams showed that the introduction of the steel layer near the surface of the laminate reduces the tensile stress at the aluminium surface layer. Using the FEM model presented by Kümmel et al. [15], the tensile stress at the outer AA7075 layer can be calculated to be reduced by 19% for the L1/4 architecture (Figure 8c) compared to monolithic AA7075 material. The fatigue limits of the AA7075/DC05 N2 L1/4 composite and the AA7075 N2 monolithic material can be estimated at 210 MPa and 170 MPa, respectively. Calculating an effective fatigue limit of the AA7075/DC05 composite and taking into account the FEM-calculated 19% reduction of the maximum tensile stress in the outer AA7075 layer resulting from the load transfer into the adjacent DC05 steel layer, the resulting effective endurable maximum stress amplitude $\Delta\sigma_{max,corr.}/2$ amounts to 170.1 MPa. This fits perfectly to the obtained fatigue limit of the monolithic AA7075 N2 material of 170 MPa. The numbers of cycles, which are consumed to initiate a crack in the HCF regime, amount to 96–99% of $N_f$ for both the laminated composite and constituent monolithic materials. Again, it becomes evident that in the HCF regime the fatigue life of the LMC is primarily determined by the fatigue crack initiation phase, as is the case for the monolithic AA7075 and DC05 materials and the overall role of crack propagation processes in the LMCs subjected to loadings in the HCF regime, which is relatively small.

This load transfer effect enhances the fatigue life of the LMC in the LCF regime as well, compared to the monolithic AA7075 material. Additionally, as reported in Section 3.1. for the Al/Al LMC systems, in the LCF regime, the role of fatigue crack propagation is promoted in the AA7075/DC05 LMC system compared to the constituent monolithic materials. As indicated in Figure 8a, the crack propagation phase accounts for 17% of the numbers of cycle to failure for the AA7075/DC05 composite compared to about 5 to 7% for the monolithic AA7075 N2 and DC05 N2 materials. This increased percentile of the crack propagation phase on the overall fatigue life in the laminated composite must be associated with toughening mechanisms impeding the propagating crack front at the vicinity of the interfaces.

Figure 8b shows SEM images of the fatigue crack propagation paths through the laminated structure of the AA7075/DC05 N2 L1/4 architecture of specimens fatigued in the LCF and HCF regime, respectively. A clear difference in the crack propagation behavior in the LMC architecture between the LCF regime compared to the HCF regime can be seen. In the HCF regime, the crack path propagates relatively straight through both the interfaces. A slight deviation in crack path trajectory away from the initial orientation perpendicular to the interfaces can be observed when the crack approaches the interface towards the harder and stiffer steel layer. Before penetrating the steel layer, the crack propagates about 100 µm along the first interface. In the LCF regime, a more distinct crack network can be observed. Crack propagation in 45° orientation towards the surface can be observed in the outer AA7075 layer. Some degree of necking occurring in the DC05 steel layer hints to promoted activities associated with plasticity involved in the crack propagation process at this stage. Applying only a post-mortem analysis of the crack networks in the laminated AA7075/DC05 composites does not deliver sufficient evidence to identify individual toughening mechanisms associated with the enhanced role of fatigue crack propagation in the LCF regime.

In order to gather a better understanding of the prevalent toughening mechanisms obstructing fatigue crack propagation at the vicinity of interfaces in AA7075/DC05 N2 L1/4 laminated composites, in-situ LCF and HCF fatigue experiments have been conducted inside the large chamber SEM.

The experiment in the LCF regime was conducted at a maximum bending stress amplitude $\Delta\sigma_{max}/2$ of 295 MPa as shown in Figure 9a. Figure 9b shows the crack network after 30.8k loading cycles. Multiple small cracks can be observed at the front surface. The formation of these small cracks in a AA7075 alloy is known to be associated with the brittle behavior of small intermetallic particles (e.g., $Mg_2Si$ and $Al_7Cu_2Fe$) [76]. Two larger cracks have been formed at the bottom surface of the specimen at this stage. The larger crack on the right side was nucleated at the surface aluminium layer below the loading anvil. At this position, the tensile stress at the surface is at a maximum, resulting from the cyclic loading in three-point bending mode. This crack has propagated through half of the outer AA7075 layer at an angle of about 45° with respect to the orientation of the interface. From this position on, the crack path trajectory changes and the crack propagates perpendicular to the interfaces (Figure 9c). As the crack approaches the immediate vicinity of the interface towards the steel layer after 31.5k loading cycles, the crack path gets deflected prior to the crack tip reaching the interface (Figure 9c). Upon reaching the aluminium/steel interface, crack propagation along the interface of about 350 µm was observed during the subsequent sequence of approximately 1k loading cycles (Figure 9d).

During this period, the formation of interface delamination cracks associated with stress redistribution effects at the opposing steel/aluminium interface was observed. After 32.8k loading cycles, further crack propagation into the inner AA7075 layer of the LMC architecture starting at one of the delamination cracks was observed (Figure 9e) before the original crack begins to penetrate the steel layer. This crack bridging mechanisms of the steel layer leads to the formation of a complex crack network with simultaneous crack propagation occurring in the aluminium and steel layers (Figure 9f, 33k loading cycles). The crack propagation in the DC05 steel layer is associated with visible plastic activity around the crack tip. Although no necking of the steel layer in the in-situ LCF experiment was found, as observed in Figure 8b, the plasticity is indicated by the formation of shear bands ahead of the crack tip at an angle of about 45° to 60° in relation to the direction of loading in Figure 9f. Upon coalition of the two individual cracks, the test was stopped, as the crack network reached the bonding plane of the final N2 ARB processing step (Figure 9a).

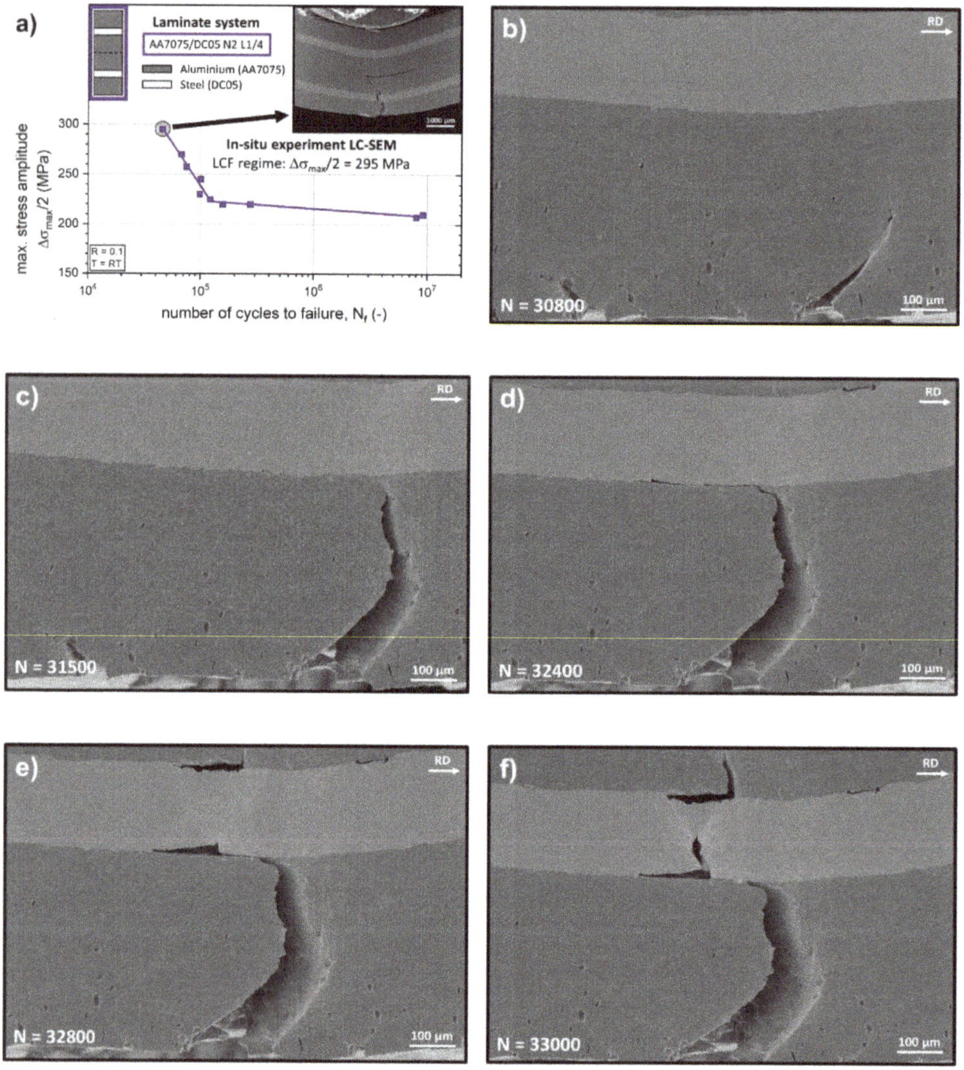


**Figure 9.** (**a**) Three-point bending fatigue experiment on a AA7075/DC05 laminated composite specimen fatigued in the LCF regime at a maximum stress amplitude of 295 MPa inside a large-chamber SEM; (**b**–**f**) In-situ observation of the fatigue crack propagation perpendicular to the interfaces (crack arrester orientation).

A second in-situ fatigue experiment was conducted in the HCF regime at a maximum bending stress amplitude $\Delta\sigma_{max}/2$ of 210 MPa, as shown in Figure 10a. Prior to the in-situ experiment in the large chamber SEM, the specimen was pre-fatigued externally on a vibrophore testing machine at the same maximum bending stress amplitude of 210 MPa for 9.1 Mio cycles. The initiated crack after 9.1 Mio loading cycles using the method described above can be seen in Figure 10b. The crack was nucleated at the position of the maximum tensile stress in the outer AA7075 layer resulting from the cyclic three-point bending loading and propagates at a perpendicular orientation towards the interfaces. As in the LCF experiment, multiple small cracks originating at intermetallic particles can be seen on the front surface. No major contributions of these cracks on the propagation of the major crack were observed during the in-situ experiments. After 20 k additional loading cycles, the crack reaches the aluminium/steel interface (Figure 10c). As in the LCF regime experiment, a deflection of the crack path can be observed before the crack tip reaches the interface. Consequently, the crack approaches the interface at a high deflection angle and propagates along the interface for about 40 k loading cycles, as can be seen in Figure 10d.

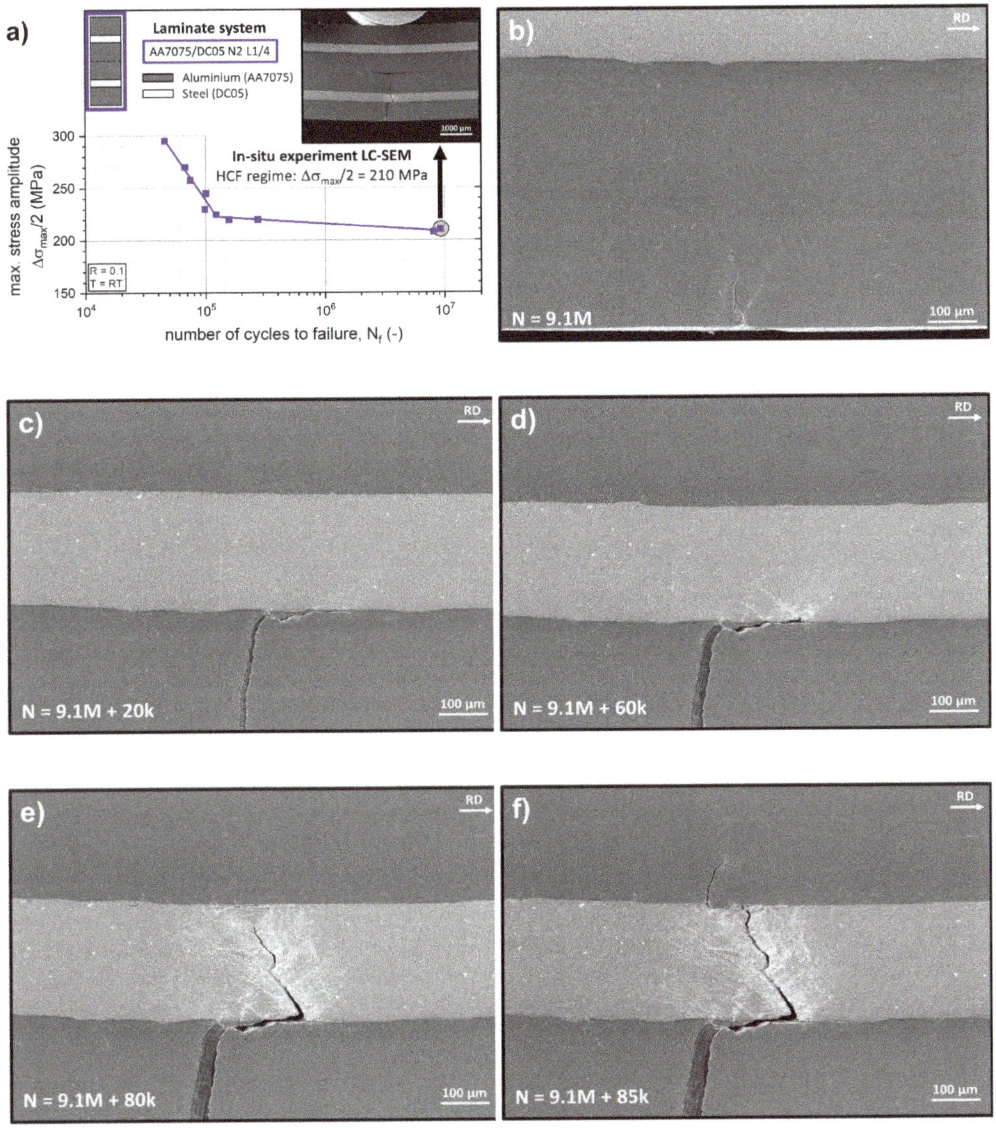


**Figure 10.** (**a**) Three-point bending fatigue experiment on a AA7075/DC05 laminated composite specimen fatigued in the HCF regime at a maximum stress amplitude of 210 MPa inside the large-chamber SEM. (**b**) Crack initiation occurring at 9.1 M cycles after external pre-fatigue on a vibrophore testing machine; (**c**–**f**) in-situ observation of the fatigue crack propagation perpendicular to the interfaces (crack arrester orientation).

At this stage, multiple events associated with local plasticity can be observed at different positions in the DC05 steel layer at the vicinity of the interface, indicated by secondary electrons contrast edge effects at roughened surfaces. Consequently, at one of these positions, a crack is nucleated into the adjacent steel layer. No formation of interface delamination cracks at the opposing steel/aluminium interface associated with stress redistribution effects was observed during this in-situ experiment. As can be seen in Figure 10e, after an additional 20 k cycles, the crack propagated through the steel layer at an angle, returning to the symmetry plane where the driving force on crack propagation is at a maximum due to the cyclic loading condition under three-point bending mode. After reaching the steel/aluminium layer, the crack deflects and propagates along the interface before penetrating the inner AA7075 layer (Figure 10f). This crack deflection at the steel/aluminium interface is much less pronounced when the length of crack that propagates along the interface (about 100 μm for the steel/aluminium interface vs. 250 μm for the aluminium/steel interface) is regarded. The less pronounced crack deflection is also visible by comparing the loading cycles for the crack propagation along the interfaces (below 5 k for the steel/aluminium interface vs. 60 k for the aluminium/steel interface). As for the loading cycle comparison, it must be stated again that the experiments were conducted under force control. The experiment was stopped once the crack reached the bonding plane of the final N2 ARB processing step. The crack network of this in-situ fatigued HCF specimen post-experiment, as can be seen in Figure 10a, resembles the crack network of the HCF fatigue-life specimen shown in Figure 8b, indicating an appropriate representation of HCF crack propagation using the described experimental procedure.

Using in-situ fatigue experiments, different toughening mechanisms associated with the interaction of the process zone ahead of the fatigue crack and the interfaces in laminated metallic composites with a combined gradient in hardness and elastic properties could be identified. The influence of these individual mechanisms was assessed in a semi-quantitative manner. The findings can explain the enhanced fatigue crack propagation behavior of the laminated composites in the LCF regime associated with the prevalent toughening mechanisms.

### *3.4. Effects of the LMC Architecture Combined with Gradients of Hardness and Elastic Properties at Interfaces on Fatigue Life and Crack Propagation*

Previous studies by the authors [15] revealed, that fatigue properties of LMCs can be specifically tailored by modifying the laminate architecture accordingly. In order to find optimized laminate architecture designs regarding fatigue life and crack propagation, laminates utilizing different combinations of material inhomogeneity effects in terms of gradients of hardness and elastic properties at interfaces were tested. The local mechanical properties in terms of hardness and Young's modulus of the different AA2024, Ti-G1 and DC05 layers as well as the resulting gradients at interfaces can be found in Table 6.

Based on the respective hardness and elastic properties of the individual layers (Table 6), the highest gradients $\Delta H$ and $\Delta E$ in the laminated composite systems can be found at the AA2024/DC05 interfaces and the smallest gradients $\Delta H$ and $\Delta E$ at the AA2024/Ti-G1 interfaces, with the respective gradients of hardness and elastic properties of the Ti-G1/DC05 interfaces ranging in between.

Figure 11a shows the architectures of the different tri-material laminated composites investigated, containing thin layers of AA2024, Ti-G1 and DC05 near the respective surfaces (shell structure) as well as a AA2024 material core structure. The near surface AA2024, Ti-G1, and DC05 layers in the shell structure of the laminate have about the same thickness in average in all three laminate architectures, leading to the same nominal density for all LMCs.

**Table 6.** Average hardness *H* as measured by nanoindentation and Young's modulus *E* of the respective AA2024, DC05, and Ti-G1 layers in AA2024/Ti-G1/DC05 N3 LMC systems and resulting gradients $\Delta H$ and $\Delta E$ at interfaces. Yield stress and ultimate tensile stress of the monolithic AA2024 N3, Ti-G1 N3, and DC05 N3 materials, as determined by uniaxial tensile tests.

| Material | AA2024/Ti-G1/DC05 N3 LMC Systems | | | |
|---|---|---|---|---|
| | **AA2024 Layers** | **Ti-G1 Layers** | **DC05 Layers** | **Gradients ($\Delta H$, $\Delta E$)** |
| Hardness $H$/GPa | 2.58 ± 0.05 | 2.78 ± 0.15 | 3.20 ± 0.13 | Ti-Al: 0.20 ± 0.16<br>Fe-Ti: 0.42 ± 0.20<br>Fe-Al: 0.62 ± 0.14 |
| Young's modulus $E$/GPa | 69 [1] | 103 [2] | 210 [3] | Ti-Al: 64<br>Fe-Ti: 107<br>Fe-Al: 141 |
| **Material** | **Monolithic Materials** | | | |
| | **AA2024 N3 Mono** | **Ti-G1 N3 Mono** | **DC05 N3 Mono** | **Difference** |
| Yield stress/MPa | 639 ± 10 | 538 ± 4 | 554 ± 1 | Al-Ti: 101 ± 11<br>Fe-Ti: 16 ± 5<br>Al-Fe: 85 ± 10 |
| Ultimate tensile stress/MPa | 654 ± 12 | 713 ± 2 | 704 ± 8 | Ti-Al: 59 ± 12<br>Ti-Fe: 9 ± 8<br>Fe-Al: 50 ± 14 |

[1] data based on [74]; [2] data based on [77]; [3] data based on [75].

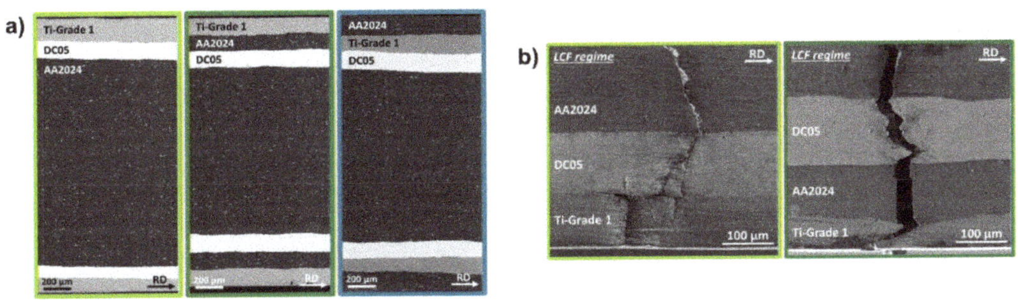


**Figure 11.** (**a**) SEM images of the tri-material laminated composite architectures Ti-G1/DC05/AA2024, Ti-G1/AA2024/DC05, and AA2024/Ti-G1/DC05, respectively; (**b**) Fatigue crack propagation paths in Ti-G1/DC05/AA2024 and Ti-G1/AA2024/DC05 LMCs fatigued in their respective LCF regimes.

The fatigue life diagrams (*S-N* curves) of the Ti-G1/DC05/AA2024, Ti-G1/AA2024/DC05, and AA2024/Ti-G1/DC05 N3 laminate architectures as well as of the constituent AA2024 N3 and Ti-G1 N3 materials are plotted in Figure 12a. In all three laminated composite systems, the fatigue life in both the LCF as well as HCF regime can be improved significantly by the integration of thin titanium and steel layers compared to the monolithic AA2024 N3 material. A comparison of the *S-N* curves at the respective LCF to HCF region transition areas reveals a gradual enhancement in terms of endurable maximum stress amplitude and numbers of cycles to failure, indicating increasing levels of optimized laminate architecture designs regarding fatigue life properties.

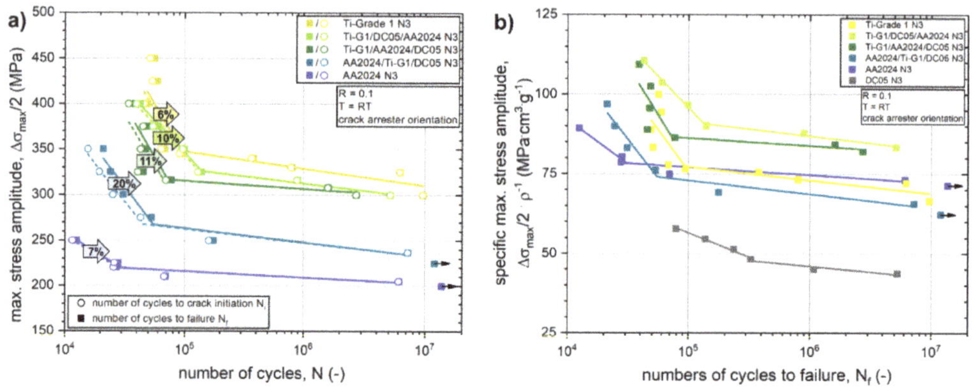


**Figure 12.** (**a**) Fatigue properties (*S-N* curves) of Al/Ti/Steel N3 LMC systems with different architectures (stacking sequences) and constituent monolithic Ti-G1 N3 and AA2024 N3 materials. The percentile of fatigue crack propagation on total fatigue life is indicated in the respective LCF regimes; (**b**) comparison of fatigue properties (*S-N* curves) of Al/Ti/Steel N3 LMC systems with different architectures and constituent monolithic Ti-G1 N3, AA2024 N3 and DC05 N3 materials in relation to their respective density $\rho$.

The best resistance against crack initiation for the different LMC systems could be achieved in laminates where the titanium layers are positioned at the respective surfaces: Ti-G1/DC05/AA2024 and Ti-G1/AA2024/DC05. Comparing these two laminate architectures, the resistance against crack initiation is enhanced for the Ti-G1/DC05/AA2024 laminate architecture compared to the Ti-G1/AA2024/DC05 architecture. This can be explained again by the load transfer from the titanium surface layer into the adjacent DC05 steel layer due to the gradient in elastic modulus, thereby reducing the effective tensile stresses at the bottom surface titanium layer.

The best resistance against crack propagation was observed for the AA2024/Ti-G1/DC05 laminate architecture, where the crack propagation phase accounted for 20% of the numbers of cycle to failure in the LCF regime compared to 11% for the Ti-G1/AA2024/DC05 architecture, 10% for the Ti-G1/DC05/AA2024 architecture and about 6 to 7% for the monolithic Ti-G1 N3 and AA2024 N3 materials. This coincides with the findings presented in Sections 3.1 and 3.3 and can be explained by a reduction of the local driving force at the crack tip, when the crack approaches the interfaces to the stronger and stiffer layers. For the AA2024/Ti-G1/DC05 architecture, this is the case for both the AA2024/Ti-G1 and the Ti-G1/DC05 interfaces (see Table 6), where the crack approaches the interface from the layer denoted first to the one denoted second, respectively. Figure 11b shows SEM images of the fatigue crack propagation paths through the laminated structure of the Ti-G1/DC05/AA2024 and Ti-G1/AA2024/DC05 LMCs fatigued in their respective LCF regimes.

In Figure 12b, the *S-N* curves are plotted in relation to the respective density ρ of the laminated composites and monolithic materials. The monolithic DC05 N3 steel exhibits by far the worst specific fatigue properties due to its high density. The specific fatigue life of the Ti-G1/DC05/AA2024 and Ti-G1/AA2024/DC05 LMC architectures are significantly improved compared to either of the constituent monolithic AA2024 N3, Ti-G1 N3, and DC05 N3 materials.

These findings clearly demonstrate that by an appropriate selection of the constituents, by adjusting their mechanical properties and by an intelligent design of the laminated metallic composite the fatigue life and cyclic crack propagation behavior of LMCs can be significantly improved.

## 4. Discussion

### *4.1. Effects of Toughening Mechanisms on Crack Propagation and Crack Growth in LMCs with a Hardness Gradient at Interfaces*

The material inhomogeneity effect at the interfaces associated with plasticity was characterized as a hardness gradient at the interfaces by means of nanoindentation, since this method is well suited to determine local mechanical properties in thin layers of a laminated metal composite. The hardness gradients $\Delta H$ at the interfaces in the AA1050/AA5754 N2 LMC systems and the AA1050/AA5754 N3(Nxr) LMC system were found to be very similar at about 0.6 GPa, respectively.

As seen in Figure 4a, the resistance against crack propagation was improved for the Al/Al-LMC systems by reinforcing the softer matrix material with thin layers of a stronger material (16.7 vol.%) utilizing the hardness gradient effect at the interfaces. A better resistance against crack propagation was achieved for the AA1050/AA5754 L1/8 architecture. As can be derived from Figure 4b, the stronger AA5754 layer is positioned more towards the sample surface in the AA1050/AA5754 L1/8 architecture than in the AA1050/AA5754 L1/4 architecture. Figure 4a shows that the crack propagation phase amounts for 13% of total numbers of cycle to failure for the L1/8 architecture compared to 10% for the L1/4 architecture. Since the fatigue experiments were conducted under force control, the remote driving force for crack propagation depends on the remaining load bearing cross section of the specimen. If a crack approaches the interface towards the harder AA5754 layer, the process zone ahead of the crack tip is affected by the variation of local mechanical properties at the interface and thus decreases the crack tip driving force locally (crack tip shielding effect [35,40,41]). This effect is superimposed by the remote (far-field) driving force on the crack. This leads to the following explanation: For the L1/4 architecture, the contribution of the remote driving force is much bigger than for the L1/8 architecture when the crack approaches the vicinity of the interface, since the remaining load bearing cross section of the L1/4 architecture is much smaller compared to the L1/8 architecture by then. This means, the contribution of local effects on the overall crack tip driving force of a crack at vicinity of the interface is smaller for the L1/4 architecture, which leads to a smaller percentile of the crack propagation phase on the overall fatigue life of this laminate architecture.

A significant improvement of the damage tolerant fatigue properties of dissimilar Al/Al-LMCs with an alternating layer structure and an inherent hardness gradient $\Delta H$ at the interfaces of 0.6 GPa was observed conducting FCG experiments (Figure 5). As highlighted by post-experiment analysis of the surface crack networks using light microscopy (Figure 6) and analysis of the 3D-crack network using synchrotron X-ray computed microtomography (Figure 7), this can be attributed to the presence of two different toughening mechanisms observed at the interfaces: crack deflection and crack bifurcation.

Crack deflection was observed at the interfaces when the crack approached the interfaces from the softer to the harder layers. This toughening mechanism is associated with a deflection of the crack path at the vicinity of the interface and subsequent crack growth along the interface. Kümmel et al. [32] proposed a schematic evolution of the crack deflection mechanism based on their findings regarding crack propagation in fatigue life experiments on Al/Al-LMCs with dissimilar hardness. Similar correlations can be observed for the crack networks of the FCG experiments in Figure 6.

At low constant far-field stress intensity ranges ($\Delta K = 7.5$ MPa$\sqrt{m}$), no crack deflection mechanism could be observed. Similar to the HCF experiments by Kümmel et al., no deflection of the crack growth path at the vicinity of interfaces was observed, as the crack grows perpendicular across the interfaces. Although no deflection in the crack path was observed at $\Delta K = 7.5$MPa$\sqrt{m}$, the crack growth rate in the AA1050/AA5754 LMC was determined to be about the same as for the monolithic AA5754 material and about 20% of the monolithic AA1050 material. This indicates that the periodic variation of the local mechanical properties and the associated variation of local crack driving forces in the laminated structures appears to have a retardation effect on the overall crack growth rate.

With an increase in far-field $\Delta K$ level in Region II-1, the appearance of the crack deflection mechanism can be observed more frequently, indicating a dependence of this mechanism on the remotely applied far-field driving force for crack growth.

At higher constant far-field stress intensity ranges ($\Delta K$ = 15 MPa$\sqrt{m}$ and above), the crack deflection mechanism can be observed at every interface for the transition from the softer to the harder layers. The crack deflection mechanism can be explained as follows: When the crack approaches the interface from the softer material AA1050 to the harder material AA5754, the process zone ahead of the crack tip reaches the interface and is affected by the change of the mechanical properties in the adjacent layer. As this layer has a higher flow stress (higher hardness), the process zone ahead of the crack tip will be smaller at the same remotely applied far-field stress intensity range. This leads to a change of size and shape of the process zone at the vicinity of the interface. As the crack growth path depends on local conditions around the crack tip, the change of the characteristics of the process zone leads to a deflection of the crack path. Subsequently, crack growth in both directions along the interface was observed. The driving force for crack growth along the interface can be attributed to the formation of a complex stress state of compressional stress in the harder AA5754 layer and tensile stress in the softer AA1050 layer due to the geometrically necessary co-deformation behavior at the interface during fatigue loading [32]. This leads to stress redistribution mechanisms at the respective interfaces [27]. Similar findings for the crack deflection mechanism were previously observed experimentally by Sugimura et al. [31] and Pippan et al. [30] in Steel/Steel bi-metal and interlayer systems at the vicinity of interfaces for the transition of the crack from the softer to the stronger material.

For the soft/hard transition at interfaces in Al/Al-LMCs with dissimilar hardness, the following effects contribute to the overall crack growth behavior at the vicinity of these interfaces. These effects either reduce or accelerate the overall crack growth rate: (a) Crack tip shielding effect, reduction; (b) Deflection of the crack path and crack growth along interfaces, reduction; (c) Crack growth on multiple crack fronts in both directions along the interface, reduction.

Crack bifurcation was observed at the interfaces when the crack approaches the interface from the harder to the softer layers. This toughening mechanism leads to simultaneous crack growth at two separate crack fronts into the adjacent softer AA1050 layers (Figure 6).

At lower constant far-field stress intensity ranges ($\Delta K$ = 7.5 MPa$\sqrt{m}$ and 10 MPa$\sqrt{m}$), no crack bifurcation mechanisms could be observed, whereas at higher stress intensity ranges in Region II-1 ($\Delta K$ = 12.5 MPa$\sqrt{m}$) and in Region II-2, this mechanism can be observed at multiple interfaces for the hard/soft transition in the respective crack networks.

For the time being, this mechanism is not fully understood. Two different considerations are suggested: The first consideration revolves around the anti-shielding effect [40,41]: When the crack approaches the interface from the harder to the softer material, the process zone ahead of the crack tip is affected by the change of the mechanical properties at the interface to the adjacent softer AA1050 layer. This leads to an enhancement of the process zone in size and a change in shape as can be seen in studies by Sistaninia et al. [39,78]. Consequently, the local crack tip driving force is enhanced significantly, as the crack reaches the interface to the softer material. This might lead to the crack bifurcation phenomenon and crack growth across multiple fronts inclined at 45–60° to the initial crack growth along regions of locally concentrated plastic deformation. Cyclic plastic deformation ahead of the crack tip is usually mainly concentrated in shear bands, inclined at an angle of 45–80° with regard to the direction of crack propagation [30]. The crack bifurcation mechanism was not observed experimentally at the respective interfaces in bi-metals [30,31]. Pippan et al. [30], however, observed a similar crack bifurcation phenomenon for crack growth into a softer $\alpha$-Fe interlayer embedded in a ferritic steel matrix. This leads to the second consideration, where the crack bifurcation phenomenon might be linked to interaction of the process zone ahead of the crack tip with the second to next interface, that is the next soft/hard transition interface, when the crack approaches the interface of the hard/soft transition. This has two implications: (1) The crack bifurcation phenomenon would not be observable in bi-metals, as there is no second interface present and (2) the crack bifurcation phenomenon is only present in interlayer and laminated structures, when the process zone ahead of the crack tip is significantly larger for both materials than the respective layer thickness (that is at high far-field stress intensity levels).

In the literature, the delamination and bifurcation of cracks at interfaces are investigated numerically using cohesive zone models [79] or virtual crack closure techniques [80]. As shown by a recent numerical study using a cohesive zone model [81], a weaker interface strength and increased yield strength mismatch leads to an additional shielding effect of the crack tip driving force and can cause the crack to bifurcate at the interface. However, the numerically described bifurcation phenomenon deviates to some extent from the appearance of the bifurcation mechanisms observed in this study.

Although the mechanisms behind crack bifurcation are not fully understood, we can summarize the following effects that contribute to the overall crack growth behavior at the vicinity of the interfaces at a hard/soft transition in Al/Al-LMCs with dissimilar hardness: (a) The crack tip anti-shielding effect leads to an acceleration of the overall crack growth rate, while (b) crack bifurcation and simultaneous growth at multiple crack fronts results in its reduction.

### *4.2. Effects of Toughening Mechanisms on Crack Propagation in LMCs with a Combined Gradient in Hardness and Elastic Properties at Interfaces*

The resistance against fatigue crack propagation in the LCF regime was improved for the Al/Steel-LMC system (Figure 8a) by reinforcing the softer and more compliant AA7075 matrix material with thin layers of a stronger and stiffer DC05 material (16.7 vol.%) utilizing gradient effect in hardness and elastic properties at the interfaces. In the HCF regime, the crack propagation phase plays a subsidiary role compared to the crack initiation phase. Consequently, the toughening mechanisms at the laminate interfaces do not contribute significantly to the overall fatigue life in the HCF regime.

Figure 13 summarizes the prevalent toughening mechanisms observed by means of in-situ fatigue experiments in three-point bending mode in the LCF and HCF regime, respectively, motivated on the distinction of previously identified toughening mechanisms in LMCs by Lesuer et al. [27].

The crack deflection mechanism at the vicinity of the Al/Steel layer follows the same evolution as described in Section 4.1, however the crack shielding effect associated with the gradient in hardness (yield stress inhomogeneity effect) is enhanced by an additional shielding effect based on the elastic inhomogeneity effect [34] as the AA7075 layers are both softer and more compliant than the DC05 steel layers.

The stress redistribution mechanism is associated with the formation of a complex stress state at the Al/Steel interfaces due to the geometrically necessary co-deformation behavior as a response to the cyclic loading. This results in compressional stress in the stronger and stiffer steel layer and tensile stress in the softer and more compliant aluminium layers. This complex stress state leads to interface delamination ahead of the crack tip at high remote loadings (LCF regime, Figure 9d,e) and consequently promotes crack propagation along the interface of the deflected crack in both the LCF and HCF regime (Figures 9d and 10d,f, respectively).

A crack bridging mechanism of the DC05 steel layer was identified at high remote loadings in the LCF regime (Figure 9e,f). On the macroscopic scale, this was observed in previous studies mainly for fatigue experiments on ductile/brittle laminate systems such as metal intermetal laminates (MILs), for example Nb/$Nb_3$Al-MILs [23] and Ti/$Ti_3$Al-MILs [24]. The crack bridging mechanism observed in this study is assumed to be closely linked to the stress redistribution mechanism at the Al/Steel interfaces, as interface delamination cracks are observed at the second interface (Steel/Al interface, Figure 9d,e). As explained above, inherent compressional stress in the steel layer and tensile stress in the aluminium layers are created due to the necessary co-deformation of layers with dissimilar (elastic) properties at the vicinity of the interfaces. This can explain the crack initiation into the inner AA7075 layer prior to the DC05 layer despite the remotely applied tensile stress level at the position of the steel layer being higher due to the loading in three-point bending mode.

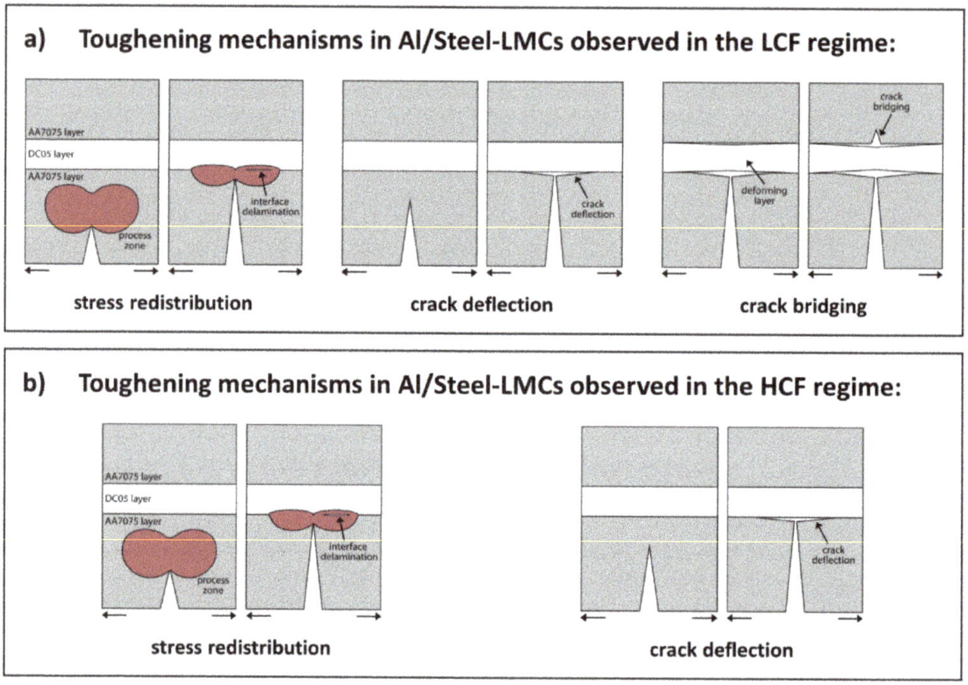


**Figure 13.** Toughening mechanisms in laminated aluminium/steel composites (AA7075/DC05) identified by means of in-situ fatigue experiments in three-point bending mode. Fatigue crack propagation perpendicular to the interfaces (crack arrester orientation). (**a**) LCF-regime: stress redistribution at interfaces leading to interface delamination, crack deflection at the interface towards the steel (DC05) layer, crack bridging of the deforming steel (DC05) layer; (**b**) HCF-regime: stress redistribution, crack deflection.

Regarding the overall contribution of the individual mechanisms on the retardation of crack propagation at the vicinity of interfaces of materials with dissimilar hardness and elastic properties from the in-situ experiments, the following conclusions can be drawn: The biggest effect on retardation of crack propagation in the LCF regime can be attributed to shielding effects at the Al/Steel interface due to the yield stress inhomogeneity and elastic inhomogeneity effects. These effects lead to the appearance of the crack deflection and stress redistribution mechanisms. The effect of the crack bridging mechanism was not pronounced as much in terms of loading cycles. However, it must be stated that the tests were conducted under force control and the crack bridging of the steel layer occurred at a state where the load bearing cross section of the specimen was already significantly reduced leading to high remote driving forces for crack propagation.

### 4.3. *Optimization of LMC Architectures for Light Weight Applications Subjected to Cyclic Loading Cases*

The *S-N* curves in Figure 12 highlight the lightweight potential of laminated metallic composites by utilizing interface effects associated with gradients in hardness and elastic properties and integrating these effects into the laminate architecture design of LMCs. As can be seen in Figure 12a, the Ti-G1/DC05/AA2024 laminate architecture can endure almost the same number of cycles to failure in the LCF regime as the monolithic Ti-G1 material at a nominal composition of 12.5 vol-% Ti-G1, 12.5 vol-% DC05 and 75 vol.% AA2024. This is mainly attributed to the good resistance against crack initiation in the Ti-G1 surface layer and is supported additionally by load transfer into the adjacent stiff DC05 layer in cyclic three-point bending mode loading. As an additional effect, the resistance against crack propagation is enhanced from 6% for the monolithic Ti-G1 material to 10% of the total number of cycles for the Ti-G1/DC05/AA2024 laminate. This increases the interval of total fatigue life, where the presence of propagating cracks can be detected prior to failure. Due to the lower density of the LMC structures, the specific fatigue properties of the Ti-G1/DC05/AA2024 and Ti-G1/AA2024/DC05 laminate systems exceed those determined in monolithic Ti-G1 material (Figure 12b).

The resistance against crack propagation in the AA2024/Ti-G1/DC05 laminate architecture was increased to 20% of the total fatigue life in the LCF regime for tests conducted under force control in three-point bending mode by shielding effects attributed to yield stress inhomogeneity and elastic inhomogeneity effects at the vicinity of the Al/Ti as well as Ti/Steel interfaces. This emphasizes the potential of utilizing inhomogeneity effects at interfaces in laminated metal composites and the role of an intelligently designed LMC architectures in promoting damage tolerant properties.

## 5. Conclusions

The influence of gradients in hardness and Young's modulus at interfaces in laminated metallic composites (LMCs) on fatigue crack propagation has been studied. The main conclusions are summarized in the following.

1. The resistance against crack propagation in the LCF regime is improved in Al/Al-LMC systems by reinforcing the softer AA1050 matrix with thin and harder AA5754 layers compared to monolithic AA1050 due to the hardness gradient at the interface.
2. The damage tolerant fatigue properties of dissimilar Al/Al-LMCs with an alternating layer structure are significantly enhanced in crack arrester orientation compared to both constituent monolithic AA1050 and AA5754 materials. Investigation of the surface and 3D-crack networks reveal the presence of two different toughening mechanisms at interfaces causing the reduction of crack growth in the LMCs. Crack deflection is observed when the crack approaches the interface from the softer towards the harder layers. Crack bifurcation happens in the opposite case. Both mechanisms appear to be more pronounced at higher remotely applied far-field stress intensity ranges. For the soft/hard transition at the interface in Al/Al-LMCs, a combination of the effects of crack tip shielding, deflection of the crack path and subsequent crack growth along the interfaces in both directions on multiple crack fronts contribute to the overall reduction of the local crack driving force at the vicinity of the interface. The local crack driving force at the vicinity of the hard/soft transition depends on the acceleration by anti-shielding effects and the superimposed reduction due to crack bifurcation and simultaneous crack growth along multiple crack fronts.
3. Resistance against crack propagation in the LCF regime is enhanced significantly in an Al/Steel-LMC system by reinforcing the softer and more compliant AA7075 matrix material with harder and stiffer DC05 steel layers, utilizing gradient effects in hardness and elastic properties at the interfaces. The prevalent toughening mechanisms observed in-situ are stress redistribution, crack deflection and crack bridging in the LCF regime and stress redistribution and crack deflection in the HCF regime. In the LCF regime, the enhanced crack propagation phase is attributed mainly to the crack tip shielding effects at the soft/hard and compliant/stiff transition at the Al/Steel interface, resulting in stress redistribution and crack deflection. The crack bridging of the steel layer appears to be promoted by interface delamination occurring at the inner Steel/Al interface due to stress redistribution and a complex stress state of compressional stress in the steel layer and tensile stress in the aluminium layers due to the geometrically necessary co-deformation of these layers at the interface upon cyclic loading.

4. The fatigue lives of Al/Ti/Steel tri-material laminated composites can be significantly improved compared to the monolithic AA2024 alloy by optimizing the laminate architecture, introducing thin Ti-G1 and DC05 layers. The specific fatigue properties of the LMC architectures containing Ti-G1 surface layers exceed those of the monolithic constituent materials significantly due to the lower density of the laminated composites. The effects of architecture optimization for laminated metal composites show promising potentials for LMCs in lightweight applications subjected to cyclic loading.

**Author Contributions:** Conceptualization, P.M.P. and H.W.H.; methodology, P.M.P., F.K., C.S., H.W.H., I.S.-M., H.M.; validation, H.W.H., M.G., H.M.; investigation, P.M.P., F.K., C.S., I.S.-M.; resources, M.G., H.M.; data curation, P.M.P.; writing—original draft preparation, P.M.P.; writing—review and editing, H.W.H., M.G., F.K., I.S.-M., H.M.; visualization, P.M.P., I.S.-M.; supervision, H.W.H., M.G.; project administration, H.W.H.; funding acquisition, H.W.H., M.G. All authors have read and agreed to the published version of the manuscript.

**Funding:** This research was partially funded by the German Research Council (DFG) which, within the framework of its 'Excellence Initiative' supports the Cluster of Excellence 'Engineering of Advanced Materials' at the University of Erlangen-Nürnberg, grant number EXC315.

**Institutional Review Board Statement:** Not applicable.

**Informed Consent Statement:** Not applicable.

**Data Availability Statement:** Not applicable.

**Acknowledgments:** The authors gratefully acknowledge the funding of the German Research Council (DFG) which, within the framework of its 'Excellence Initiative' supports the Cluster of Excellence 'Engineering of Advanced Materials' at the University of Erlangen-Nürnberg. We acknowledge the financial support from the German Research Council (DFG) via the research training school GRK1896: 'In-Situ Microscopy with Electrons, X-rays and Scanning Probes' and the 'Center for Nanoanalysis and Electron Microscopy' (CENEM) at FAU.

**Conflicts of Interest:** The authors declare no conflict of interest.

## References

1. Zhu, Y.; Ameyama, K.; Anderson, P.M.; Beyerlein, I.J.; Gao, H.; Kim, H.S.; Lavernia, E.; Mathaudhu, S.; Mughrabi, H.; Ritchie, R.O.; et al. Heterostructured materials: Superior properties from hetero-zone interaction. *Mater. Res. Lett.* **2021**, *9*, 1–31. [CrossRef]
2. Zhu, Y.; Wu, X. Perspective on hetero-deformation induced (HDI) hardening and back stress. *Mater. Res. Lett.* **2019**, *70*, 393–398. [CrossRef]
3. Wu, X.; Zhu, Y. Heterogeneous materials: A new class of materials with unprecedented mechanical properties. *Mater. Res. Lett.* **2017**, *5*, 527–532. [CrossRef]
4. Misra, A.; Göken, M.; Mara, N.A.; Beyerlein, I.J. Hierarchical and heterogeneous multiphase metallic nanomaterials and laminates. *MRS Bull.* **2021**, *46*, 1–8. [CrossRef]
5. Han, B.Q.; Huang, J.Y.; Zhu, Y.T.; Lavernia, E.J. Strain rate dependence of properties of cryomilled bimodal 5083 Al alloys. *Acta Mater.* **2006**, *54*, 3015–3024. [CrossRef]
6. Zhao, Y.; Topping, T.; Bingert, J.F.; Thornton, J.J.; Dangelewicz, A.M.; Li, Y.; Liu, W.; Zhu, Y.; Zhou, Y.; Lavernia, E.J. High Tensile Ductility and Strength in Bulk Nanostructured Nickel. *Adv. Mater.* **2008**, *20*, 3028–3033. [CrossRef]
7. Ruppert, M.; Schunk, C.; Hausmann, D.; Höppel, H.W.; Göken, M. Global and local strain rate sensitivity of bimodal Al-laminates produced by accumulative roll bonding. *Acta Mater.* **2016**, *103*, 643–650. [CrossRef]
8. Höppel, H.W.; Westermeyer, M.; Kümmel, F.; Göken, M. The Role of Interfaces on the Deformation Mechanisms in Bimodal Al Laminates Produced by Accumulative Roll Bonding. *Adv. Eng. Mater.* **2020**, *22*, 2000145. [CrossRef]
9. Fang, T.H.; Li, W.L.; Tao, N.R.; Lu, K. Revealing Extraordinary Intrinsic Tensile Plasticity in Gradient Nano-Grained Copper. *Science* **2011**, *331*, 1587–1590. [CrossRef]
10. Lu, K. Making strong nanomaterials ductile with gradients. *Science* **2014**, *345*, 1455–1456. [CrossRef]
11. Wu, X.; Jiang, P.; Chen, L.; Yuan, F.; Zhu, Y.T. Extraordinary strain hardening by gradient structure. *Proc. Natl. Acad. Sci. USA* **2014**, *111*, 7197–7201. [CrossRef]
12. Wu, X.; Jiang, P.; Chen, L.; Zhang, J.F.; Yuan, F.P.; Zhu, Y.T. Synergetic Strengthening by Gradient Structure. *Mater. Res. Lett.* **2014**, *2*, 185–191. [CrossRef]
13. Beyerlein, I.J.; Mara, N.A.; Carpenter, J.S.; Nizolek, T.; Mook, W.M.; Wynn, T.A.; McCabe, R.J.; Mayeur, J.R.; Kang, K.; Zheng, S.; et al. Interface-driven microstructure development and ultra high strength of bulk nanostructured Cu-Nb multilayers fabricated by severe plastic deformation. *J. Mater. Res.* **2013**, *28*, 1799–1812. [CrossRef]

14. Ma, X.; Huang, C.; Moering, J.; Ruppert, M.; Höppel, H.W.; Göken, M.; Narayan, J.; Zhu, Y. Mechanical properties of copper/bronze laminates: Role of interfaces. *Acta Mater.* **2016**, *116*, 43–52. [CrossRef]
15. Kümmel, F.; Diepold, B.; Sauer, K.F.; Schunk, C.; Prakash, A.; Höppel, H.W.; Göken, M. High Lightweight Potential of Ultrafine-Grained Aluminum/Steel Laminated Metal Composites Produced by Accumulative Roll Bonding. *Adv. Eng. Mater.* **2019**, *21*. [CrossRef]
16. Kümmel, F.; Diepold, B.; Prakash, A.; Höppel, H.W.; Göken, M. Enhanced monotonic and cyclic mechanical properties of ultrafine-grained laminated metal composites with strong and stiff interlayers. *Int. J. Fatigue* **2018**, *116*, 379–387. [CrossRef]
17. Misra, A.; Hirth, J.P.; Hoagland, R.G. Length-scale-dependent deformation mechanisms in incoherent metallic multilayered composites. *Acta Mater.* **2005**, *53*, 4817–4824. [CrossRef]
18. Misra, A.; Hoagland, R.G. Plastic flow stability of metallic nanolaminate composites. *J. Mater. Sci.* **2007**, *42*, 1765–1771. [CrossRef]
19. Bloyer, D.R.; Ritchie, R.O.; Rao, K.T. Fracture toughness and R-Curve behavior of laminated brittle-matrix composites. *Metall. Mater. Trans. A* **1998**, *29*, 2483–2496. [CrossRef]
20. Osman, T.M.; Hassan, H.A.; Lewandowski, J.J. Interface Effects on the Quasi-Static and Impact Toughness of Discontinuously Reinforced Aluminum Laminates. *Metall. Mater. Trans. A* **2008**, *39*, 1993–2006. [CrossRef]
21. Lesuer, D.R.; Syn, C.K.; Riddle, R.A.; Sherby, O.D. Extrinsic Fracture Mechanisms in Two Laminated Metal Composites. In *Intrinsic and Extrinsic Fracture Mechanisms in Inorganic Composite Systems*; Lewandowski, J.J., Hunt, W.H., Eds.; TMS: Warrendale, PA, USA, 1995; pp. 93–102.
22. Adharapurapu, R.R.; Vecchio, K.S.; Jiang, F.; Rohatgi, A. Fracture of Ti-Al3Ti metal-intermetallic laminate composites: Effects of lamination on resistance-curve behavior. *Metall. Mater. Trans. A* **2005**, *36*, 3217–3236. [CrossRef]
23. Bloyer, D.R.; Ritchie, R.O.; Rao, K.T. Fatigue-crack propagation behavior of ductile/brittle laminated composites. *Metall. Mater. Trans. A* **1999**, *30*, 633–642. [CrossRef]
24. Adharapurapu, R.R.; Vecchio, K.S.; Jiang, F.; Rohatgi, A. Effects of ductile laminate thickness, volume fraction, and orientation on fatigue-crack propagation in Ti-Al3Ti metal-intermetallic laminate composites. *Metall. Mater. Trans. A* **2005**, *36*, 1595–1608. [CrossRef]
25. Hassan, H.A.; Lewandowski, J.J.; El-Latif, M.H. Effects of lamination and changes in layer thickness on fatigue-crack propagation of lightweight laminated metal composites. *Metall. Mater. Trans. A* **2004**, *35*, 45–52. [CrossRef]
26. Hoffman, P.B.; Carpenter, R.D.; Gibeling, J.C. Fatigue Crack Growth in Aluminum Laminate Composites. *Mater. Res. Soc. Symp. Proc.* **1996**, *434*, 281–286. [CrossRef]
27. Lesuer, D.R.; Syn, C.K.; Sherby, O.D.; Wadsworth, J.; Lewandowski, J.J.; Hunt, W.H. Mechanical behaviour of laminated metal composites. *Int. Mater. Rev.* **1996**, *41*, 169–197. [CrossRef]
28. Ritchie, R.O. Mechanisms of fatigue crack propagation in metals, ceramics and composites: Role of crack tip shielding. *Mater. Sci. Eng. A* **1988**, *103*, 15–28. [CrossRef]
29. Ritchie, R.O.; Yu, W. Short crack effects in fatigue: A consequence of crack tip shielding. In *Small Fatigue Cracks*; Ritchie, R.O., Lankford, J., Eds.; TMS-AIME: Warrendale, PA, USA, 1986; pp. 167–189.
30. Pippan, R.; Flechsig, K.; Riemelmoser, F.O. Fatigue crack propagation behavior in the vicinity of an interface between materials with different yield stresses. *Mater. Sci. Eng. A* **2000**, *283*, 225–233. [CrossRef]
31. Sugimura, Y.; Lim, P.G.; Shih, C.F.; Suresh, S. Fracture normal to a bimaterial interface: Effects of plasticity on crack-tip shielding and amplification. *Acta Metall. Mater.* **1995**, *43*, 1157–1169. [CrossRef]
32. Kümmel, F.; Höppel, H.W.; Göken, M. Layer architecture and fatigue life of ultrafine-grained laminated metal composites consisting of different aluminum alloys. *Mater. Sci. Eng. A* **2017**, *702*, 406–413. [CrossRef]
33. Ohashi, Y.; Wolfenstine, J.; Koch, R.; Sherby, O.D. Fracture behavior of a laminated steel-brass composite in bend tests. *Mater. Sci. Eng. A* **1992**, *151*, 37–44. [CrossRef]
34. Kolednik, O.; Kasberger, R.; Sistaninia, M.; Predan, J.; Kegl, M. Development of Damage-Tolerant and Fracture-Resistant Materials by Utilizing the Material Inhomogeneity Effect. *J. Appl. Mech.* **2019**, *86*. [CrossRef]
35. Kolednik, O. The yield stress gradient effect in inhomogeneous materials. *Int. J. Solids Struct.* **2000**, *37*, 781–808. [CrossRef]
36. Simha, N.; Fischer, F.; Kolednik, O.; Chen, C. Inhomogeneity effects on the crack driving force in elastic and elastic–plastic materials. *J. Mech. Phys. Solids* **2003**, *51*, 209–240. [CrossRef]
37. Fratzl, P.; Gupta, H.S.; Fischer, F.D.; Kolednik, O. Hindered Crack Propagation in Materials with Periodically Varying Young's Modulus—Lessons from Biological Materials. *Adv. Mater.* **2007**, *19*, 2657–2661. [CrossRef]
38. Fan, K.; Wang, G.Z.; Tu, S.-T.; Xuan, F.Z. Effects of Toughness Mismatch on Failure Behavior of Bi-Material Interfaces. *Procedia Eng.* **2015**, *130*, 754–762. [CrossRef]
39. Sistaninia, M.; Kolednik, O. Improving strength and toughness of materials by utilizing spatial variations of the yield stress. *Acta Mater.* **2017**, *122*, 207–219. [CrossRef]
40. Sistaninia, M.; Kasberger, R.; Kolednik, O. To the design of highly fracture-resistant composites by the application of the yield stress inhomogeneity effect. *Compos. Struct.* **2018**, *185*, 113–122. [CrossRef]
41. Suresh, S.; Sugimura, Y.; Tschegg, E.K. The growth of a fatigue crack approaching a perpendicularly-oriented, bimaterial interface. *Scr. Metall. Mater.* **1992**, *27*, 1189–1194. [CrossRef]
42. Kolednik, O.; Predan, J.; Fischer, F.D.; Fratzl, P. Improvements of strength and fracture resistance by spatial material property variations. *Acta Mater.* **2014**, *68*, 279–294. [CrossRef]

43. Kolednik, O.; Predan, J.; Fischer, F.D.; Fratzl, P. Bioinspired Design Criteria for Damage-Resistant Materials with Periodically Varying Microstructure. *Adv. Funct. Mater.* **2011**, *21*, 3634–3641. [CrossRef]
44. Funari, M.F.; Lonetti, P. Initiation and evolution of debonding phenomena in layered structures. *Theor. Appl. Fract. Mech.* **2017**, *92*, 133–145. [CrossRef]
45. Martin, E.; Leguillon, D. A strain energy density criterion for the initiation of edge debonding. *Theor. Appl. Fract. Mech.* **2015**, *79*, 58–61. [CrossRef]
46. Kümmel, F.; Höppel, H.W.; Göken, M. Ultrafine-grained laminated metal composites: A new material class for tailoring cyclicially stressed components. *Adv. Eng. Mater.* **2021**, 2100070. [CrossRef]
47. Bryzgalin, G.I.; Tsvetkov, D.I.; Kartashov, G.G.; Nagibin, V.P.; Pisarev, A.V. Evaluation of the quality of laminated metallic materials on the basis of high-frequency endurance and damping capacity. *Strength Mater.* **1989**, *21*, 1652–1658. [CrossRef]
48. Saito, Y.; Tsuji, N.; Utsunomiya, H.; Sakai, T.; Hong, R.G. Ultra-fine grained bulk aluminum produced by accumulative roll-bonding (ARB) process. *Scr. Mater.* **1998**, *39*, 1221–1227. [CrossRef]
49. Rosochowski, A.; Olejnik, L. Severe plastic deformation for grain refinement and enhancement of properties. In *Microstructure Evolution in Metal Forming Processes*, 1st ed.; Woodhead Publishing: Cambridge, UK, 2012; pp. 114–141.
50. Hausöl, T.; Maier, V.; Schmidt, C.W.; Winkler, M.; Höppel, H.W.; Göken, M. Tailoring Materials Properties by Accumulative Roll Bonding. *Adv. Eng. Mater.* **2010**, *12*, 740–746. [CrossRef]
51. Ruppert, M.; Böhm, W.; Nguyen, H.; Höppel, H.W.; Merklein, M.; Göken, M. Influence of upscaling accumulative roll bonding on the homogeneity and mechanical properties of AA1050A. *J. Mater. Sci.* **2013**, *48*, 8377–8385. [CrossRef]
52. Oliver, W.C.; Pharr, G.M. An improved technique for determining hardness and elastic modulus using load and displacement sensing indentation experiments. *J. Mater. Res.* **1992**, *7*, 1564–1583. [CrossRef]
53. Hay, J.L.; Pharr, G.M. Instrumented Indentation Testing. In *ASM Handbook Volume 8-Mechanical Testing and Evaluation*; ASM International: Almere, The Netherlands, 2000; pp. 232–243.
54. ASTM International. ASTM E1820-11. In *Standard Test Method for Measuring of Fracture Toughness*; ASTM International: West Conshohocken, PA, USA, 2011. [CrossRef]
55. ISO. ISO 12135:2016. In *Metallic Materials-Unified Method of Test for the Determination of Quasistatic Fracture Toughness*; ISO: Geneva, Switzerland, 2016.
56. Barsoum, I.; Chandran, K.S. Stress intensity factor solutions for cracks in finite-width three layer laminates with and without residual stress effects. *Eng. Fract. Mech.* **2003**, *70*, 2015–2031. [CrossRef]
57. Yasniy, P.; Maruschak, P.; Lapusta, Y. Experimental study of crack growth in a bimetal under fatigue and fatigue-creep conditions. *Int. J. Fract.* **2006**, *139*, 545–552. [CrossRef]
58. Maruschak, P.O.; Konovalenko, I.V.; Maruschak, E.V.; Sorochak, A.P. Automated Method for Stereometric Study of Fatigue Failure Mechanisms. *Metallurgist* **2014**, *58*, 43–47. [CrossRef]
59. ASTM International. ASTM E1823-20b. In *Standard Terminology Relating to Fatigue and Fracture Testing*; ASTM International: West Conshohocken, PA, USA, 2020. [CrossRef]
60. ASTM International. ASTM E647-13a. In *Standard Test Method for Measurement of Fatigue Crack Growth Rates*; ASTM International: West Conshohocken, PA, USA, 2013. [CrossRef]
61. Görner, W.; Hentschel, M.P.; Müller, B.R.; Riesemeier, H.; Krumrey, M.; Ulm, G.; Diete, W.; Klein, U.; Frahm, R. BAMline: The first hard X-ray beamline at BESSY II. *Nucl. Instrum. Methods Phys. Res.* **2001**, *467–468*, 703–706. [CrossRef]
62. Van Rossum, G.; Drake, F.L. *Python 3 Reference Manual*; CreateSpace: Scotts Valley, CA, USA, 2009.
63. Schindelin, J.; Arganda-Carreras, I.; Frise, E.; Kaynig, V.; Longair, M.; Pietzsch, T.; Preibisch, S.; Rueden, C.; Saalfeld, S.; Schmid, B.; et al. Fiji: An open-source platform for biological-image analysis. *Nat. Methods* **2012**, *9*, 676–682. [CrossRef]
64. Avizo Software, Thermo Fisher Scientific. Available online: https://www.fei.com/software/avizo (accessed on 31 March 2021).
65. Höppel, H.W.; Göken, M.; Klein, M. Large Chamber Scanning Electron Microscopes: Challenges and Aspects for Materials Science. *Imaging Microsc.* **2004**, *3*, 2–3.
66. Nolte, R.; Höppel, H.W.; Göken, M. Großkammer-Rasterelektronenmikroskopie—Neue Dimensionen der Materialforschung und Schadensanalyse. In *Tagung Werkstoffprüfung—Fortschritte der Kennwertermittlung für Forschung und Praxis*; Stahleisen: Düsseldorf, Germany, 2009; pp. 195–202.
67. Nolte, R. Großkammer-Rasterelektronenmikroskopie: Inbetriebnahme, Optimierung und Experimentelle Methodik. Ph.D. Thesis, FAU Erlangen-Nürnberg, Erlangen/Nürnberg, Germany, 2011.
68. Schunk, C.; Nitschky, M.; Höppel, H.W.; Göken, M. Superior Mechanical Properties of Aluminum-Titanium Laminates in Terms of Local Hardness and Strength. *Adv. Eng. Mater.* **2019**, *21*, 1800546. [CrossRef]
69. Hausöl, T.; Höppel, H.W.; Göken, M. Tailoring materials properties of UFG aluminium alloys by accumulative roll bonded sandwich-like sheets. *J. Mater. Sci.* **2010**, *45*, 4733–4738. [CrossRef]
70. Lee, S.; Saito, Y.; Tsuji, N.; Utsunomiya, H.; Sakai, T. Role of shear strain in ultragrain refinement by accumulative roll-bonding (ARB) process. *Scr. Mater.* **2002**, *46*, 281–285. [CrossRef]
71. Kümmel, F.; Hausöl, T.; Höppel, H.W.; Göken, M. Enhanced fatigue lives in AA1050A/AA5005 laminated metal composites produced by accumulative roll bonding. *Acta Mater.* **2016**, *120*, 150–158. [CrossRef]
72. Paris, P.C.; Gomez, M.P.; Anderson, W.E. A Rational Analytic Theory of Fatigue. *Trend Eng.* **1961**, *13*, 9–14.
73. Paris, P.; Erdogan, F. A Critical Analysis of Crack Propagation Laws. *J. Basic Eng.* **1963**, *85*, 528–533. [CrossRef]

74. Villuendas, A.; Jorba, J.; Roca, A. The Role of Precipitates in the Behavior of Young's Modulus in Aluminum Alloys. *Metall. Mater. Trans. A* **2014**, *45*, 3857–3865. [CrossRef]
75. Luzin, V.; Banovic, S.; Gnäupel-Herold, T.; Prask, H.; Ricker, R.E. Measurement and Calculation of Elastic Properties in Low Carbon Steel Sheet. *Mater. Sci. Forum* **2005**, *495–497*, 1591–1596. [CrossRef]
76. Harlow, D.G.; Nardiello, J.; Payne, J. The effect of constituent particles in aluminum alloys on fatigue damage evolution: Statistical observations. *Int. J. Fatigue* **2010**, *32*, 505–511. [CrossRef]
77. Niinomi, M. Mechanical properties of biomedical titanium alloys. *Mater. Sci. Eng. A* **1998**, *243*, 231–236. [CrossRef]
78. Sistaninia, M.; Kolednik, O. Effect of a single soft interlayer on the crack driving force. *Eng. Fract. Mech.* **2014**, *130*, 21–41. [CrossRef]
79. Wang, B.; Siegmund, T. Simulation of fatigue crack growth at plastically mismatched bi-material interfaces. *Int. J. Plast.* **2006**, *22*, 1586–1609. [CrossRef]
80. Tabiei, A.; Zhang, W. Composite Laminate Delamination Simulation and Experiment: A Review of Recent Development. *Appl. Mech. Rev.* **2018**, *70*, 1–23. [CrossRef]
81. Pribe, J.D.; Siegmund, T.; Kruzic, J.J. The roles of yield strength mismatch, interface strength, and plastic strain gradients in fatigue crack growth across interfaces. *Eng. Fract. Mech.* **2020**, *235*, 107072. [CrossRef]

materials

MDPI

Article

# Temperature Dependence of Fracture Characteristics of Variously Heat-Treated Grades of Ultra-High-Strength Steel: Experimental and Modelling

Jaroslav Pokluda [1,2], Ivo Dlouhý [1,3], Marta Kianicová [2], Jan Čupera [1], Jana Horníková [1] and Pavel Šandera [1,*]

1 Faculty of Mechanical Engineering, Brno University of Technology, Technická 2, 616 69 Brno, Czech Republic; pokluda@fme.vutbr.cz (J.P.); idlouhy@ipm.cz (I.D.); cupera@fme.vutbr.cz (J.Č.); hornikova@fme.vutbr.cz (J.H.)
2 Faculty of Special Technology, Alexander Dubcek University of Trencin, Ku kyselke 469, 911 06 Trenčín, Slovakia; marta.kianicova@tnuni.sk
3 Institute of Physics of Materials AS CR, Žižkova 22, 61662 Brno, Czech Republic
* Correspondence: sandera@fme.vutbr.cz; Tel.: +420-541-142-827

**Abstract:** The temperature dependence of tensile characteristics and fracture toughness of the standardly heat-treated low-alloyed steel OCHN3MFA along with three additionally heat-treated grades was experimentally studied. In the temperature range of $\langle -196; 22 \rangle$ °C, all the additional heat treatments transferred the standard steel from a high- to ultra-high strength levels even with improved tensile ductility characteristics. This could be explained by a reduction of the inclusion content, refinement of the martensitic blocks, ductile retained austenite content, and homogenization of the shape ratio of martensitic laths as revealed by metallographic, X-ray, and EBSD techniques. On the other hand, the values of the fracture toughness of all grades were found to be comparable in the whole temperature range as the cause of a high stress triaxiality in the pre-cracked Charpy V-notch samples. The values of the fracture toughness of the standard steel grade could be predicted well using the fracture model proposed by Pokluda et al. based on the tensile characteristics. Such a prediction failed in the case of additionally heat-treated grades due to the different temperature dependence of the fracture mechanisms occurring in the tensile and fracture-toughness tests. While the tensile samples fractured in a ductile-dimple mode at all temperatures, the fracture-toughness specimens exhibited a transition from the ductile to quasi-brittle fracture mode with decreasing temperature. This transition could be interpreted in terms of a transfer from the model proposed by Rice and Johnson to the model of Tvergaard and Hutchinson.

**Keywords:** ultra-high steel grades; tensile characteristics; fracture toughness; temperature dependence; modelling fracture

**Citation:** Pokluda, J.; Dlouhý, I.; Kianicová, M.; Čupera, J.; Horníková, J.; Šandera, P. Temperature Dependence of Fracture Characteristics of Variously Heat-Treated Grades of Ultra-High-Strength Steel: Experimental and Modelling. *Materials* **2021**, *14*, 5875. https://doi.org/10.3390/ma14195875

Academic Editor: Andrea Di Schino

Received: 30 August 2021
Accepted: 30 September 2021
Published: 7 October 2021



## 1. Introduction

The global engineering approach to the assessment of the integrity of flawed mechanical structures directly results from linear or elastic-plastic fracture mechanics describing the fracture resistance in terms of the fracture toughness (FT) represented by the critical stress intensity factor $K_{Ic}$ or the critical J-integral $J_{Ic}$ [1,2]. The transfer of the data from laboratory specimens to engineering structures and components is not necessarily straightforward due to the constraint effects and, therefore, the two-parameter fracture mechanics is more relevant [3]. Nevertheless, the values of $K_{Ic}$ or $J_{Ic}$ are, besides the values of the Charpy-notch toughness (impact energy), crucial characteristics enabling a mutual comparison of engineering materials with respect to their fracture resistance. The FT tests remain the only relevant methods for determination of FT, but they are rather complicated and expensive. It is thus worthwhile to investigate the possibility of theoretical predictions of FT values using the data from much simpler and cheaper tests. Such predictions are usually based on models dealing with local approaches to fracture, both deterministic and probabilistic, which rely on the fact that it is possible to model macroscopic fracture behavior in terms of numerically calculated elastic-plastic stress-strain in the process zone combined with local fracture criteria.

Many advanced ceramics and composites as well as bcc metallic materials at lower temperatures exhibit brittle (cleavage) fracture. An important group of local cleavage-fracture models were proposed by the pioneering work of Beremin [4] and assume that fracture is initiated when a microcrack ahead of a crack front is subjected to a critical value of the Weibull stress and relies on a numerically calculated evolution of this local stress with the macroscopic applied stress (e.g., [5,6]). Kotrechko et al. [7] focused on the mechanisms of crack nucleation in ferritic steels by incorporating both the microscopic stress induced by elastic deformations and the stress caused by dislocation pile-ups into the probabilistic model. These statistical approaches predict the probability of brittle fracture for the given applied $K$-value along with the upper and lower bounds of the temperature dependence of FT. However, their application is rather cumbersome since it demands the determination of many material parameters from microstructure samples and numerous FT and tensile tests at multiple temperatures. Therefore, Yankova et al. [8] recently proposed a thinning function for the temperature dependence of cleavage initiators that promises a reduction of requested FT tests. Recently, the strain-gradient fracture mechanics coupled with atomistic approaches was utilized to eliminate the stress singularity at the crack tip and to predict FT for components sized in the entire range from macro to nano (e.g., [9]). This method is numerically complex and time consuming and, up to now, its verification was done only for single crystals of several pure metals and ceramics [10].

In high-strength steels with a basic martensitic structure and alloys with bcc matrix strengthened by fine particle dispersion, the specific transition behavior can be observed. In the upper part of the transition, plastic crack-tip blunting occurs before the unstable fracture associated with the microvoid coalescence mechanism and a ductile-dimple morphology of the fracture surfaces. With decreasing temperatures, fraction of the cleavage facets on the fracture surface dispersed between ductile-dimple areas increases and the microvoid coalescence starts to be controlled by decohesion of the particle/matrix interfaces and, due to a very low interparticle spacing, the stage of void growth is suppressed. However, shallow ductile dimples often remain observable even at very low temperatures. This was also the case of the high-strength steel OCHN3MFA (Russian GOST nomenclature) investigated in this work and, consequently, the local ductile-fracture models seemed to be more relevant for a description of its fracture process. Such models can be based either on the plastic work consumed in the plastic zone till the onset of fracture or on a separation energy related to the rupture of ligaments between the voids as reported in the classical works of Peel and Forsyth [11], Hahn and Rosenfield [12,13], McClintock [14], and Gurson [15]. Based on these assumptions, ductile fracture models were developed by Rice and Johnson [16], Pokluda et al. [17–19], and Tvergaard and Hutchinson [20]; see the Appendix A for more detail. These models are considered in this article since they enable a straightforward prediction of FT values using rather simple available tensile and microstructural material data. Some more recent works and reviews on ductile fracture [21–23], useful for comparative reasons and knowledge extension, are also worthwhile to mention here.

The fracture-toughness samples of the originally (standardly) heat-treated steel OCHN3MFA exhibited a ductile fracture morphology when tested at room temperature [24]. Some preliminary results indicated that, for this steel grade, the model of Pokluda and Šandera [18,19] could reasonably predict the $K_{Ic}$ (or $K_{Jc}$) values even at lower testing temperatures [25]. Industrial interest was then expressed for further verification of such a conclusion along with mechanical testing of additionally heat-treated steel grades. Therefore, the experimental research presented in this article was focused on the temperature dependence of the basic mechanical properties, FT, and temperature dependence of FT and the related fracture mechanisms as several OCHN3MFA steel grades. The capability of selected fracture models to predict the temperature dependence of FT and the related fracture mechanisms is also reported.

## 2. Material, Heat Treatment, and Microstructure

The high-strength low-alloy steel OCHN3MFA of the original (commonly applied) heat treatment consisting of quenching and stepwise tempering, along with another three grades of additional quenching and tempering at different temperatures were investigated. The chemical composition of the steel is displayed in Table 1. The heat treatment of original grade-conventional additionally heat-treated grades is shown in Table 2. The grades exhibited the following average values of Vickers hardness (5 indentations for each treatment): original treatment–502 HV, tempering 160 °C–725 HV, temp. 200 °C–706 HV, temp. 250 °C–696 HV.

**Table 1.** Chemical composition in wt %.

| | C | Mn | Si | Cr | Ni | Mo | V | P | S |
|---|---|---|---|---|---|---|---|---|---|
| Spectral analysis | 0.403 | 0.3 | 0.32 | 1.19 | 3.275 | 0.523 | 0.1363 | 0.01 | 0.01 |

**Table 2.** Heat treatment of OCHN3MFA grades: (a) the original grade-conventional treatment; (b) additionally heat-treated grades.

| | |
|---|---|
| (a) | *Quenching:* heating to 870 ± 10 °C: 2.5 h + 3 h dwell time. Cooling to 300 °C (water) in 3 min. + cooling to room temperature (oil) in 2 h.<br>*Tempering (stepwise):* 480 °C/5 h + 420 °C/5 h. |
| (b) | *Annealing:* 650 °C/5 h.<br>*Quenching:* 860 °C/1 h., cooling in oil.<br>*Tempering alternatives:* 160 °C/5 h, 200 °C/5 h, or 250 °C/5 h. |

The microstructures related to individual steel grades are documented in Figure 1 as obtained by a classical metallography technique. The microstructure of all grades consisted of fine tempered martensite laths containing low or no tetragonality martensite and carbide particles. It is well known that the population of carbides (not visible in Figure 1) depends on the tempering temperature. At the low tempering temperatures corresponding to the additionally treated grades, the martensitic matrix contains $Fe_{2,7}C$ carbides, which change to the cementite $Fe_3C$ particles at higher tempering temperatures applied to the originally treated samples (e.g., [26,27]). All originally and additionally heat-treated samples were machined from the same (single) forged structural component. Moreover, observable changes in the number and the size of inclusions in steels usually start only at annealing temperatures higher than 1250 K (e.g., [28]), which were not reached during the additional heat treatment. However, recent observations ([29]) indicated that repeated annealing, quenching, and tempering can significantly reduce both the size and the number of inclusions. The content of impurities in all grades is shown in Figure 2. One can see that the additional heat treatment really caused a reduction of the number of inclusions, thus confirming the conclusion of the paper [29].

The volume fraction of the retained austenite (RA) in steel grades was measured by the X-ray diffractometer system EMPYREA and the measurement program PANalytical with the following result: the additional treatment (160 °C) contained 11.2% RA; the additional treatment (200 °C) 9.4% RA, the additional treatment (250 °C) 3.2% RA, and the original treatment 0% RA. As expected, the percentage of RA decreased with the increasing final tempering temperature. With respect to [30], the retained austenite is expected to be thermo-mechanically stable in the elastic loading range and service temperatures between 0 and 250 °C.

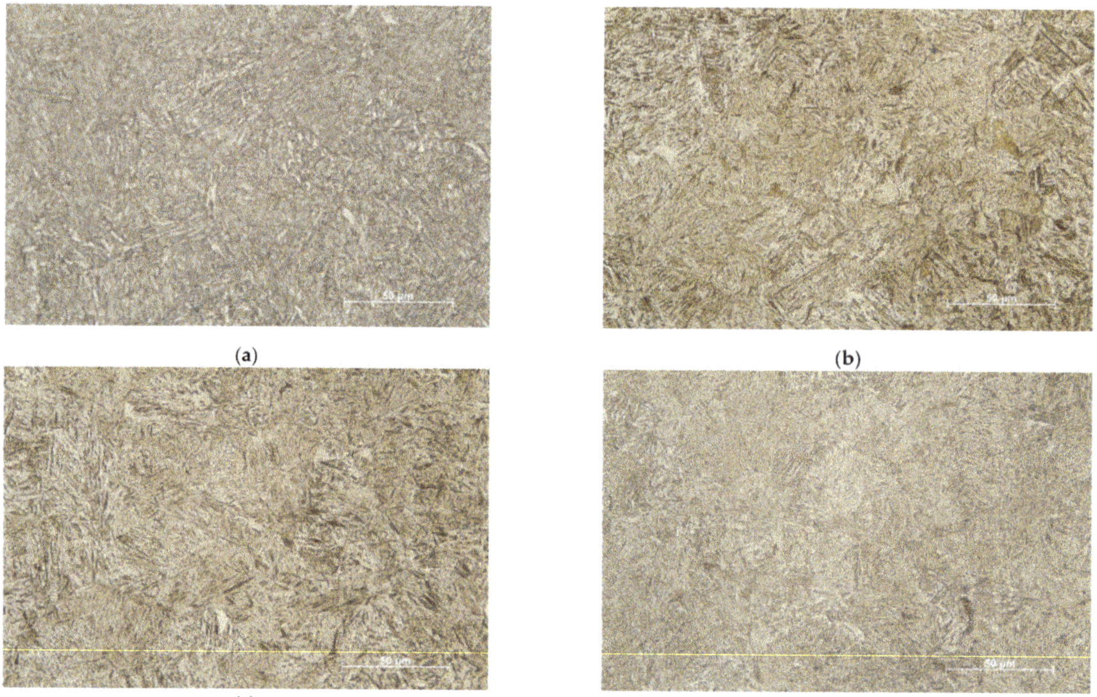


(a) (b) (c) (d)

**Figure 1.** Microstructures of the steel grades used in this study—(**a**) the original treatment, (**b**) the additional treatment (tempering at 160 °C), (**c**) the additional treatment (200 °C), and (**d**) the additional treatment (250 °C).

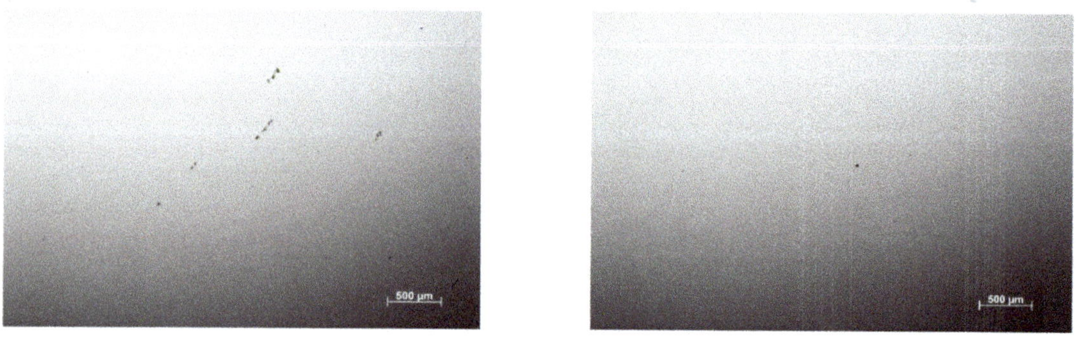


(a) (b)

**Figure 2.** *Cont.*

(c) (d)

**Figure 2.** Representative examples of the inclusion content in the OCHN3MFA steel: (**a**) original heat treatment, (**b**) additional treatment (160 °C), (**c**) additional treatment (200 °C), and (**d**) additional treatment (250 °C).

To find out possible differences in the structure of martensitic laths, the electron backscatter diffraction (EBSD) was employed on two originally treated (OT1, OT2) and two additionally treated (AT1:160 °C, AT2:250 °C) samples using FEG-SEM Carl Zeiss ULTRA PLUS (Carl Zeiss NTS GmbH, Oberkochen, Germany) equipped with an EBSD detector HKL Nordlys (Oxford Instruments, High Wycombe, UK). The EBSD data were collected by AZtech software and processed by Channel 5 software (both Oxford Instruments). A quantitative analysis of the shape and crystallographic orientation of nearly 3000 martensitic laths, digitally identified within the area defined by the matrix of 768 × 1024 pixels, was performed in each sample. The laths were approximated by the best-fit ellipses characterized by their area and aspect ratio (main/minor axes). All values of the aspect ratio were inside the range of (1; 15), which was divided into 100 equal segments. Inside each segment, the number of aspect ratios was weighted proportionally to the area of the related laths. Such obtained cumulative experimental dependences were normalized to obtain the cumulative distribution function ($CDF_{ex}$), with values lying in the range of $\langle 0, 1\rangle$, plotted for all samples in Figure 3. Observe that the $CDF_{ex}$ functions of the AT samples are smoother than those of the OT ones, exhibiting more wavy shapes. The values of the weighted arithmetic mean $A_{ex}$ and the median $M_{ex}$ obtained from $CDF_{ex}$ functions are shown in Table 3, showing that both the $A_{ex}$ and $M_{ex}$ values for the OT and AT samples are significantly different. The $CDF_{ex}$ functions were fitted by the three-parameter log-normal distribution $F(X)$:

$$\frac{dF(X)}{d(X)} = f(X) = \frac{1}{(X-\lambda)\sigma\sqrt{2\pi}} \exp \frac{[\ln(X-\lambda)-\mu]^2}{2\sigma^2}$$

where $f(X)$ is the probability density function. When selecting $\lambda = 1$ (the minimum of the aspect ratio), the parameters $\mu$ and $\sigma$ were determined (see Table 3) and the related $CDF_{LN}$ functions plotted in Figure 3 to see their high consistency with the experimental $CDF_{ex}$ functions. A difference between the values of both parameters $\mu$ and $\sigma$ for the OT and AT samples is also clearly visible in Table 3. In general, the analysis showed that the aspect-ratio values in the AT samples were rather uniformly distributed inside the range of $\langle 1, 15\rangle$, while high discontinuities of the package shapes existed in the microstructure of the OT specimens.

The maps of the crystallographic orientation of the laths depicted in Figure 4 show that the OT microstructure consisted of extended areas (martensitic blocks or even packets) with preferable [111] + [001] (blue + red) and [110] + [001] (green + red) orientations while the orientational distribution of the AT laths was more spatially refined and homogeneous, i.e., within smaller blocks. This means that, in the OT grades, more extended channels and corridors with a rather uniform crystallographic orientation were available for easy movement of dislocations.

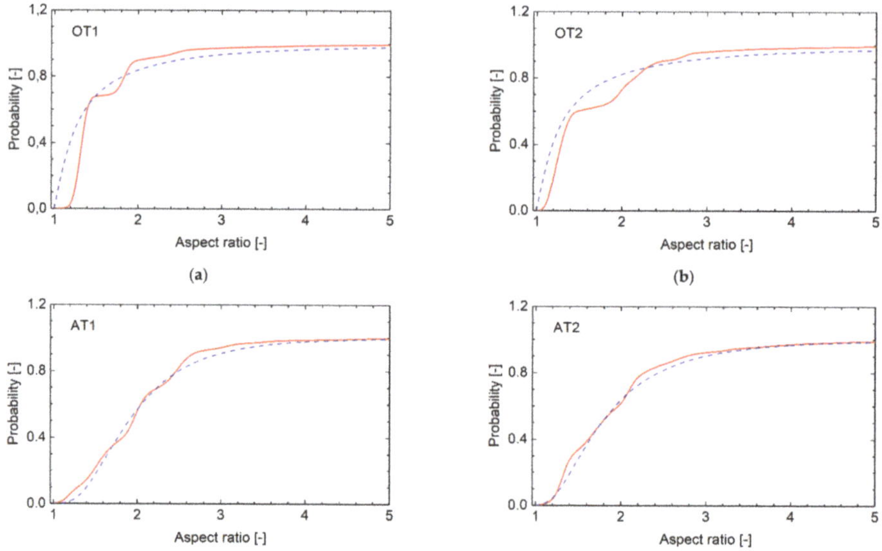


(a) (b) (c) (d)

**Figure 3.** The experimental cumulative distribution function ($CDF_{ex}$, full line) of the lath aspect ratio along with the log-normal approximation ($CDF_{LN}$, dashed line): (**a**,**b**) original samples OT1 and OT2; (**c**,**d**) additionally treated samples AT1 and AT2.

**Table 3.** The weighted arithmetic mean $A_{ex}$ and the median $M_{ex}$ obtained from experimental $CDF_{ex}$ functions and parameters $\mu$ and $\sigma$ of the related log-normal $CDF_{LN}$ functions.

| Sample | OT1 | OT2 | AT1 | AT2 |
|---|---|---|---|---|
| $A_{ex}$ | 0.66358 | 0.75314 | 1.07984 | 1.00900 |
| $M_{ex}$ | 0.25688 | 0.25688 | 0.89908 | 0.77064 |
| $\mu$ | −1.35914 | −1.35914 | −0.10638 | −0.26053 |
| $\sigma$ | 1.37770 | 1.46673 | 0.60530 | 0.73415 |

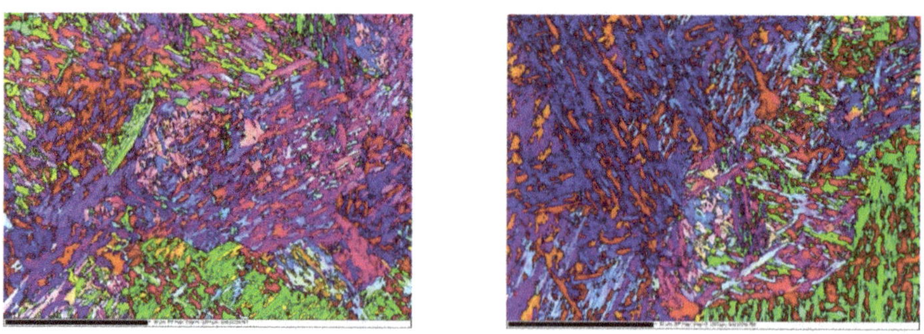

(a) (b)

**Figure 4.** *Cont.*

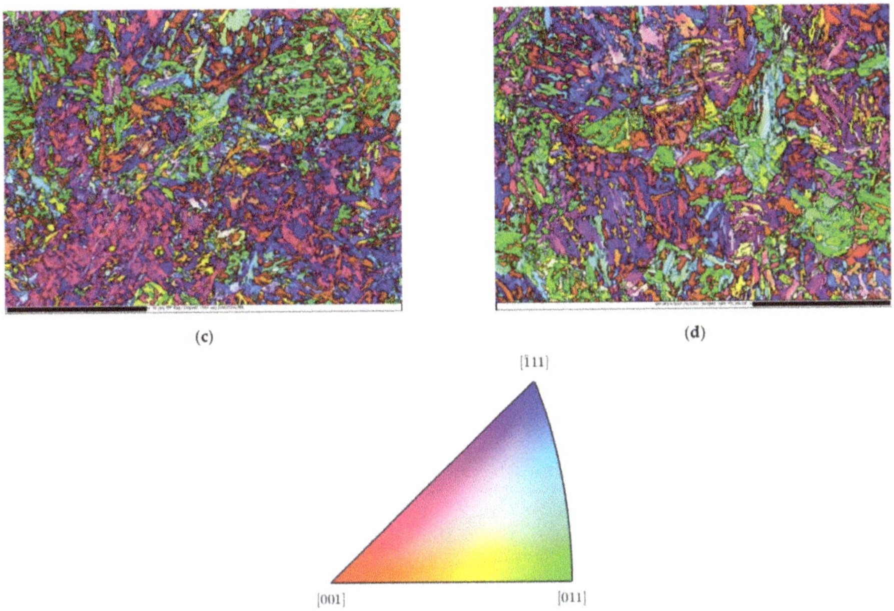


**Figure 4.** The EBSD crystallography maps: (**a**,**b**) the original heat treatment; (**c**,**d**) additional heat treatments (160 and 250 °C).

## 3. Tensile Characteristics of Steel Grades at Various Temperatures

Tensile tests were performed using the universal testing machine Zwick/Roell Z250 equipped with a cryogenic chamber. The testing temperatures were 22, 0, −20, −60, −80, −100, −120, and −196 °C, and the cross-head rate was 2 mm/min. The temperature was controlled by a thermocouple near the head of the tensile samples (bars), the scheme of which is shown in Figure 5. The samples were tempered at the requested temperature with an accuracy of ±1.5 °C for 20 min. The elongation was measured by the extensometer Multisens synchronized with the movement of the loading frame and the commercial software TestXpert was employed for a determination of basic tensile characteristics. The dimensions of fractured bars as the ultimate elongation and the smallest diameter of the neck were measured using the microscope Mitutoyo.

According to the ASTM E6-03 standard, the true stress $\sigma$ was calculated as the instantaneous normal stress, based on the instantaneous cross-sectional area, $A$. Since the strain data from the tensile test were obtained just using the extensometer, the values of $A$ were not directly measured. Up to the onset of necking, however, the true stress could be calculated as $\sigma = F/A_0\ (1 + \varepsilon)$, where $F$ is the load and $A_0$ is the original cross-section. Such a formula was also used after the onset of necking (after reaching the ultimate stress), but in that range, it did not give relevant true stress values and, obviously, the true strain data ($\varepsilon = \ln(1 + \varepsilon_{eng})$, where $\varepsilon_{eng}$ is the engineering strain) were not correctly calculated. Although this range of data was never used in the calculations, it is still plotted in Figures 6–9 and marked by dotted lines as invalid data. However, we could determine the true stress and true strain ($\sigma_f$, $\varepsilon_f$) at the point related to the final fracture of the specimen by measuring the diameter $d$ of the minimum cross-section area: $\sigma_f = \sigma_{eng}{\cdot}(d_0/d)^2$ and $\varepsilon_f = 2\ln(d_0/d)$, where $d_0$ is the original specimen diameter. Such calculated values of $\sigma_f$ and $\varepsilon_f$ correspond to the stars plotted in Figures 7–9.

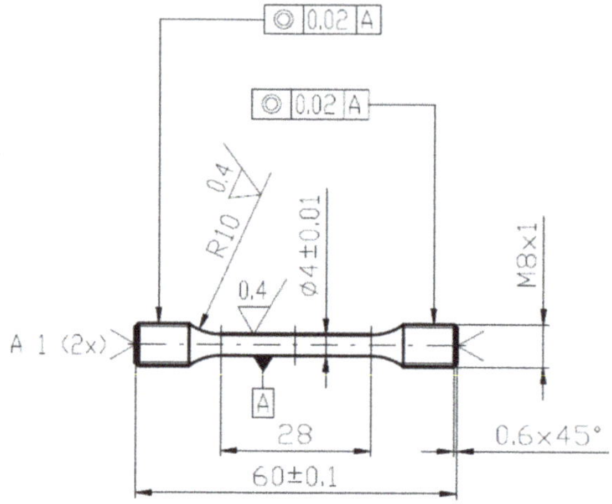


**Figure 5.** The scheme of tensile specimens.

Examples of the measured tensile true stress–strain curves for selected grades and testing temperatures are displayed in Figures 6–9. The tensile samples of the original grade (the current construction material) were denoted as OTx, where x = 1, 2, . . . , is the number of a sample. The tensile specimens of grades additionally treated by quenching and tempering at 160, 200, and 250 °C were denoted as AT(160)x, AT(200)x, and AT(250)x, respectively. The basic mechanical properties of the Young modulus $E$, yield strength $\sigma_y$, (0.2% proof stress), ultimate strength $\sigma_u$, uniform elongation $A_u$, elongation to fracture $A_f$, reduction in area $RA$, and fracture strain $\varepsilon_f$ are displayed for all steel grades and testing temperatures in Table 4, along with the values of $d_0$ and $d$. One can see that the originally heat-treated OCHN3MFA steel can be assigned to the category of high-strength steels, with $\sigma_u$ in the range of 1300–1800 MPa [31]. On the other hand, all the additionally heat-treated grades fall into the range of ultrahigh-strength steels, with $\sigma_u$ higher than 1900 MPa [32]. This strength improvement can be related to the refinement of martensitic blocks with a nearly uniform crystallographic orientation as identified by the EBSD measurement. In terms of the Hall–Petch concept, this means shorter segments, mean free paths, and pile-ups of dislocations and, consequently, a higher yield stress. A further reason was the reduction of both the number and the size of inclusions.

It should be emphasized that the additional heat treatment substantially improved not only the strength level but also dramatically raised the ductile characteristics $RA$ and $\varepsilon_f$ of the material at all testing temperatures. At the lowest testing temperature of −196 °C, indeed, the values of the ductile characteristics of all AT specimens became even three times higher than those of the OT specimens. This was due to the two-phase microstructure of the AT grades, particularly in the ductile phase of retained austenite (absent in the OT grades). Moreover, higher discontinuities identified in the lath shapes of the OT microstructure could produce some strain incompatibility during the tensile deformation, thus reducing the fracture strain of the OT samples compared to that of the AT specimens. A third reason could be seen in the reduction of the inclusion content.

**Table 4.** Basic mechanical properties of all steel grades at various testing temperatures.

| Sample | $t_{exp}$ | $E$ | $\sigma_y$ | $\sigma_u$ | $A_{gt}$ | $A_t$ | $RA$ | $d_0$ | $d$ | $\varepsilon_f$ |
|---|---|---|---|---|---|---|---|---|---|---|
| | °C | GPa | MPa | MPa | % | % | % | mm | mm | % |
| AT(160)1 | −196 | 236.8 | 1706 | 2644 | 7.3 | 10.3 | 31.1 | 3.914 | 3.250 | 37.2 |
| AT(160)2 | −120 | 233.7 | 1440 | 2375 | 6.6 | 10.3 | 46.1 | 3.949 | 2.900 | 61.8 |
| AT(160)3 | −60 | 232.5 | 1422 | 2287 | 5.6 | 8.2 | 44.3 | 3.954 | 2.950 | 58.6 |
| AT(160)4 | −20 | 228.5 | 1429 | 2273 | 6.0 | 8.8 | 40.5 | 3.914 | 3.020 | 51.9 |
| AT(160)5 | 22 | 212.4 | 1468 | 2250 | 6.6 | 9.8 | 40.9 | 3.955 | 3.040 | 52.6 |
| AT(200)1 | −196 | 239.6 | 1876 | 2456 | 5.9 | 9.1 | 37.3 | 3.929 | 3.110 | 46.8 |
| AT(200)2 | −120 | 230.9 | 1543 | 2253 | 6.1 | 10.1 | 46.2 | 3.941 | 2.890 | 62.0 |
| AT(200)3 | −60 | 220.6 | 1479 | 2127 | 4.6 | 6.2 | 44.0 | 3.927 | 2.940 | 57.9 |
| AT(200)4 | −20 | 211.3 | 1473 | 2058 | 4.1 | 4.6 | 45.8 | 3.938 | 2.900 | 61.2 |
| AT(200)5 | 22 | 208.6 | 1487 | 2111 | 6.6 | 10.3 | 40.3 | 3.936 | 3.040 | 51.7 |
| AT(250)1 | −196 | 241.4 | 1820 | 2317 | 5.7 | 10.1 | 36.6 | 3.944 | 3.140 | 45.6 |
| AT(250)2 | −20 | 232.7 | 1589 | 2068 | 4.9 | 8.7 | 43.2 | 3.942 | 2.970 | 56.6 |
| AT(250)3 | −60 | 223.7 | 1529 | 1998 | 4.8 | 8.9 | 45.7 | 3.949 | 2.910 | 61.1 |
| AT(250)4 | −20 | 222.0 | 1472 | 1945 | 4.4 | 9.0 | 50.4 | 3.919 | 2.760 | 70.1 |
| AT(250)5 | 22 | 211.3 | 1448 | 1902 | 4.8 | 9.4 | 46.8 | 3.950 | 2.880 | 63.2 |
| OT01 | −196 | 244.0 | 1833 | 2023 | 8.0 | 8.7 | 12.9 | 3.986 | 3.720 | 13.8 |
| OT03 | −100 | 238.2 | 1324 | 1580 | 5.7 | 6.3 | 31.9 | 4.011 | 3.309 | 38.5 |
| OT04 | −80 | 232.1 | 1384 | 1631 | 5.4 | 10.8 | 34.2 | 3.984 | 3.232 | 41.8 |
| OT05 | −60 | 226.6 | 1380 | 1609 | 5.6 | 10.8 | 32.0 | 3.995 | 3.295 | 38.5 |
| OT12 | 22 | 209.4 | 1309 | 1523 | 4.7 | 10.1 | 36.9 | 3.993 | 3.173 | 46.0 |

In general, the tensile characteristics of the additionally treated grades were comparable to high-end materials in the category of low-alloyed steels, such as AISI 4340 or 300 M, although the production technology of the latter steels includes, unlike that of the OCHN3MFA steel, purifying technologies, such as electro-slag refining or vacuum arc remelting [33,34]. Obviously, these technologies were, at least partially, substituted by the additional heat treatment.

The application of local approaches to the prediction of FT demands identification of the appropriate approximations of the true stress vs. true strain curve $\sigma$-$\varepsilon$ by the Hollomon function $\sigma = A\varepsilon_p{}^n$ ($\varepsilon_p$ is the true plastic strain) and by the Tvergaard and Hutchinson approximation in the form of Equation (A9b) in the Appendix A, hereafter called T-H approximation. The fitting parameters $A$, $E$, $n$, and $N$ were determined in the following way.

### 3.1. Approximation in the Elastic Region

The yield stress $\sigma_y$ corresponds to the plastic strain $\varepsilon_{p,y} = 0.002$ and the Young modulus can be obtained using two values $\sigma_y$ and $\varepsilon_{y,e} = \varepsilon(\sigma_y) - 0.002$ as $E = \sigma_y/\varepsilon_{y,e}$. In the whole stress range $0 \le \sigma \le \sigma_{max}$ then, the elastic part of strain corresponds to $\varepsilon_{el} = \sigma/E$ and the plastic part to $\varepsilon_p = \varepsilon - \varepsilon_{el}$.

### 3.2. Hollomon Approximation

For stress values $\sigma \ge \sigma_y$ and the plastic part of strain, the experimental dependence $\sigma$ vs. $\varepsilon$ was approximated by the function $\sigma = A\,\varepsilon_p{}^n$. The experimental data in the range $\sigma_{max} \ge \sigma \ge \sigma_y$ were then fitted by the linear function $\log(\sigma) = \log(A) + n\log(\varepsilon_p)$ using the least square method to obtain the values of $A$ and $n$.

### 3.3. T-H Approximation

The Young modulus in the elastic range $\sigma \leq \sigma_y$ was again obtained as $E = \sigma_y/\varepsilon_{y,e}$. The approximation $\varepsilon = \frac{\sigma_y}{E}\left(\frac{\sigma}{\sigma_y}\right)^{\frac{1}{N}}$ in the elastic-plastic range $\sigma_{max} \geq \sigma \geq \sigma_y$ (see Equation (A9) in the Appendix A) can then be written as a one-parametric function $\frac{\varepsilon}{\varepsilon_{y,e}} = \left(\frac{\sigma}{\sigma_y}\right)^{\frac{1}{N}}$. The experimental data were then fitted by the direct proportionality $N \log\left(\frac{\varepsilon}{\varepsilon_{y,e}}\right) = \log\left(\frac{\sigma}{\sigma_y}\right)$ using the least square method to obtain the value of $N$.

Such determined values of $A$, $n$, and $N$ for all grades and testing temperatures are summarized in Table 5. Examples of the regression curves for all grades and various testing temperatures are drawn in Figures 6–9. One can see a good matching of the experimental and approximated Hollomon and T-H curves up to the maximum uniform strain $\varepsilon_{max} = 0.05$ corresponding to the maximal loading force (ultimate strength) as shown in Figure 6. For higher strain values, the experimental dependence obtained by an extensometer naturally does not match the strain localization during the necking process. However, both approximations reproduce the localized strain well even up to the specimen fracture, which can be seen from the position of the point (star) corresponding to the fracture stress $\sigma_f$ and fracture strain $\varepsilon_f$ as obtained from the neck geometry of the fractured specimens in Figures 7–9. Indeed, these points lie close to both approximation curves, particularly to the Hollomon curve.

**Table 5.** Parameters $A$ and $n$ of the Hollomon approximation and the parameter $N$ of the T-H function for all steel grades and testing temperatures.

| Sample | $A$ | $n$ | $N$ |
|---|---|---|---|
| | MPa | - | - |
| AT(160)1 | 3733.3 | 0.124 | 0.19 |
| AT(160)2 | 3495.7 | 0.135 | 0.21 |
| AT(160)3 | 3417.8 | 0.131 | 0.21 |
| AT(160)4 | 3472.9 | 0.137 | 0.21 |
| AT(160)5 | 3621.6 | 0.150 | 0.20 |
| AT(200)1 | 3560.6 | 0.114 | 0.14 |
| AT(200)2 | 3145.0 | 0.115 | 0.16 |
| AT(200)3 | 3022.8 | 0.111 | 0.17 |
| AT(200)4 | 2931.7 | 0.105 | 0.17 |
| AT(200)5 | 2811.7 | 0.099 | 0.15 |
| AT(250)1 | 2939.1 | 0.077 | 0.12 |
| AT(250)2 | 2728.9 | 0.085 | 0.13 |
| AT(250)3 | 2625.7 | 0.083 | 0.13 |
| AT(250)4 | 2607.6 | 0.086 | 0.14 |
| AT(250)5 | 2532.8 | 0.085 | 0.14 |
| OT01 | 2284.5 | 0.027 | 0.09 |
| OT03 | 1492.7 | 0.040 | 0.05 |
| OT04 | 1829.1 | 0.038 | 0.08 |
| OT05 | 1839.9 | 0.042 | 0.07 |
| OT12 | 1951.7 | 0.064 | 0.09 |

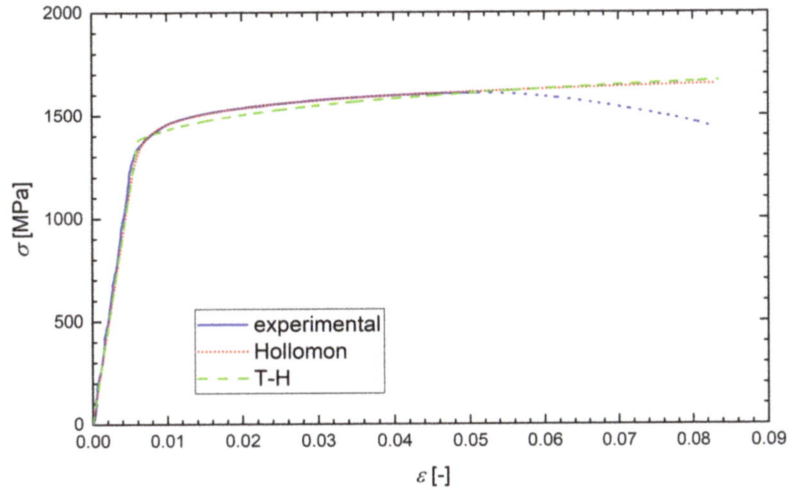


**Figure 6.** The experimental diagram of the true stress $\sigma$ vs. true strain $\varepsilon$ with its Hollomon and T-H approximations for a sample OT05 of the original grade tested at $-60$ °C. The dotted line corresponds to the invalid range of data.

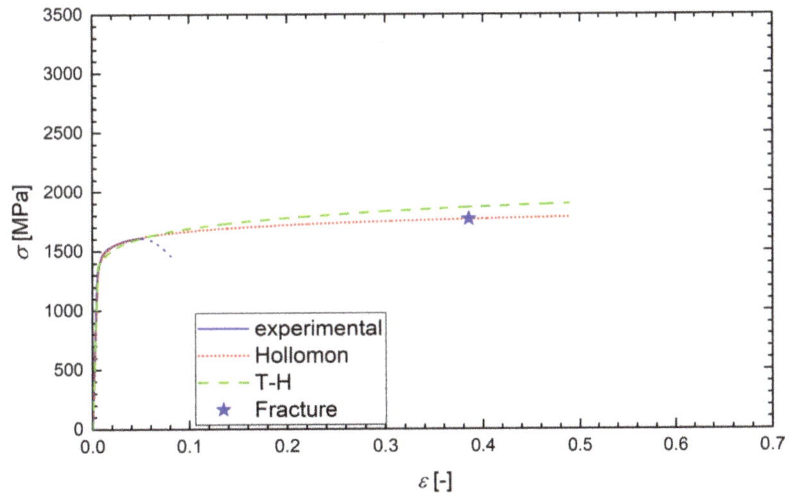


**Figure 7.** The experimental tensile diagram of the true stress $\sigma$ vs. true strain $\varepsilon$ for the sample OT05 (from Figure 6) with its Hollomon and T-H approximations plotted up to the fracture strain indicated by the star. The dotted line corresponds to the invalid range of data.

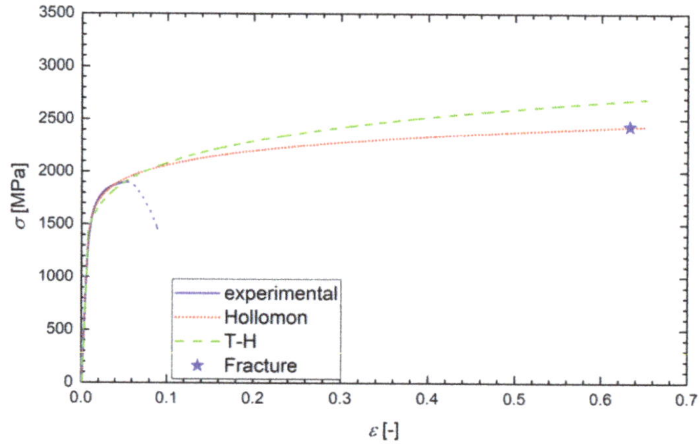


**Figure 8.** The experimental tensile diagram of the true stress $\sigma$ vs. true strain $\varepsilon$ for the sample AT(250)5 of the additionally treated grade tested at 22 °C. The Hollomon and T-H approximations are plotted up to the fracture strain indicated by the star. The dotted line corresponds to the invalid range of data.

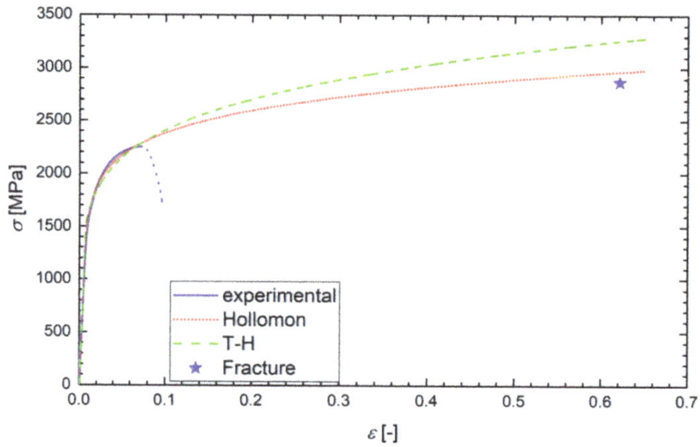


**Figure 9.** The experimental tensile diagram of the true stress $\sigma$ vs. true strain $\varepsilon$ for the sample AT(200)2 of the additionally treated grade tested at −120 °C. The Hollomon and T-H approximations are plotted up to the fracture strain indicated by the star. The dotted line corresponds to the invalid range of data.

## 4. Fracture Toughness of Steel Grades at Various Temperatures

The FT tests and their evaluation was performed according to the ISO standard [35]. The pre-crack generation was performed using the machine Zwick/Roell Z50 in four-point bending at room temperature. The chevron notch was cut by electric discharge machining. The FT of samples was then tested in three point bending with a 40 mm span of rollers. To reach the corresponding test temperature, the tests were carried out by applying a crosshead speed 1 mm/min in the cryogenic chamber of the Zwick/Roell Z50 (cooling by the liquid nitrogen vapors). In the load–displacement dependence, either the maximum load $F_C$ (brittle fracture) or the load $F_5$ (small preceding plasticity) were identified. The latter value was specified by the intersection of the linear-elastic part with the secant inclined by 5% tan $\alpha$. After identification of the initial crack length $a_0$ by means of the image analyses from optical macrographs of the fracture surfaces, the preliminary value $K_Q$ of FT was calculated using the values $F_C$ or $F_5$. If $K_Q$ fulfils the conditions of small-scale yielding and plane strain, then $K_Q = K_{Ic}$ as the valid linear-elastic FT. If these conditions were not fulfilled but the condition of a small subcritical crack advance $\Delta a$ (image correlation) was proved, the $K_{Jc}$-value was determined as:

$$K_{Ic} = \sqrt{\frac{E \cdot J_{Ic}}{1 - \nu^2}},$$

where $J_{Ic}$ is the critical value of the $J$-integral and ν is the Poisson ratio. If even the condition of a small subcritical crack advance failed, the calculated value was only considered a rough FT value and denoted as $K_{Ju}$. The geometry and dimensions of FT samples are shown in Figure 10. The specimens of the original material grade are denoted FT(O)x, x = 1, 2, ... , and the samples of additionally heat-treated grades, quenched and tempered to 160, 200, and 250 °C, are denoted as FT(160)x, FT(200)x, and FT(250)x, respectively.

The values of FT for the original FT(O) specimens at all testing temperatures are collected in Table 6 and plotted in Figure 11. All the $K_{Ic}$ values lying below the brittle-ductile transition temperature $t_{DBU} = -30$ °C are valid values of the linear-elastic plane-strain (LEPS) FT, except for the value denoted $K_{IJ}$ at −40 °C that was recalculated from the critical J-integral value. All the values lying above the $t_{DBU}$ temperature and denoted $K_{Ju}$ are invalid. The temperature dependence of valid $K_{Ic}$ data, corrected for the size effect, follows the concept of the Master curve well according to ASTM E1921-97 [36]. The Master curve is described by the formula:

$$20 + \{11 + 77\exp[0.019(T - T_0)]\}\left[\ln\left(\frac{1}{1-p}\right)\right]^{\frac{1}{4}},$$

where $T_0$ is the reference temperature and $p$ is the cumulative probability of fracture. In our case, $T_0 = -2.8$ °C, $p = 0.05$, and $p = 0.95$ stands for the lower and the upper bound curve, respectively, and $p = 0.5$ for the median curve.

On the other hand, all the data in the temperature range ⟨−196; 22⟩ °C displayed for the additionally treated grades in Table 6 are valid values of the linear-elastic FT. Unlike the tensile characteristics, the FT values of all grades were generally quite similar when compared at a given testing temperature. This was a consequence of a highly triaxial stress state at the crack front as discussed in the next section in more detail.

**Table 6.** Experimental values of the fracture toughness at all testing temperatures.

| | Original Steel Grade | | | | Additionally Treated 160 °C | | Additionally Treated 200 °C | | Additionally Treated 250 °C | |
|---|---|---|---|---|---|---|---|---|---|---|
| $t_{exp}$ °C | Sample | $K_{exp}$ MPa m$^{1/2}$ | Sample | $K_{exp}$ MPa m$^{1/2}$ | Sample | $K_{exp}$ MPa m$^{1/2}$ | Sample | $K_{exp}$ MPa m$^{1/2}$ | Sample | $K_{exp}$ MPa m$^{1/2}$ |
| −196 | FT(O)1 | 23.3 | FT(O)2 | 22.9 | FT(160)1 | 24.2 | FT(200)1 | 27.5 | FT(250)1 | 24.5 |
| −120 | | | | | FT(160)2 | 34.3 | FT(200)2 | 38.7 | FT(250)2 | 33.0 |
| −100 | FT(O)3 | 46.6 | | | | | | | | |
| −80 | FT(O)4 | 49.2 | | | | | | | | |
| −60 | FT(O)5 | 56.4 | FT(O)6 | 61.4 | FT(160)3 | 38.2 | FT(200)3 | 47.0 | FT(250)3 | 45.8 |
| −50 | FT(O)7 | 57.8 | | | | | | | | |
| −40 | FT(O)8 | 61.4 | FT(O)9 | 72.5 † | | | | | | |
| −20 | FT(O)10 | 72.0 * | FT(O)11 | 73.1 * | FT(160)4 | 44.8 | FT(200)4 | 52.4 | FT(250)4 | 52.5 |
| 22 | FT(O)12 | 69.1 * | | | FT(160)5 | 47.8 | FT(200)5 | 56.2 | FT(250)5 | 60.5 |

Note: * $K_{JU}$; † $K_{JC}$.

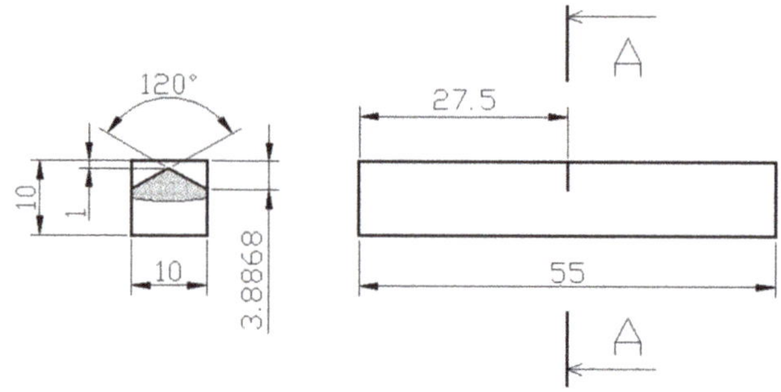


**Figure 10.** The specimen geometry used for the fracture toughness test.

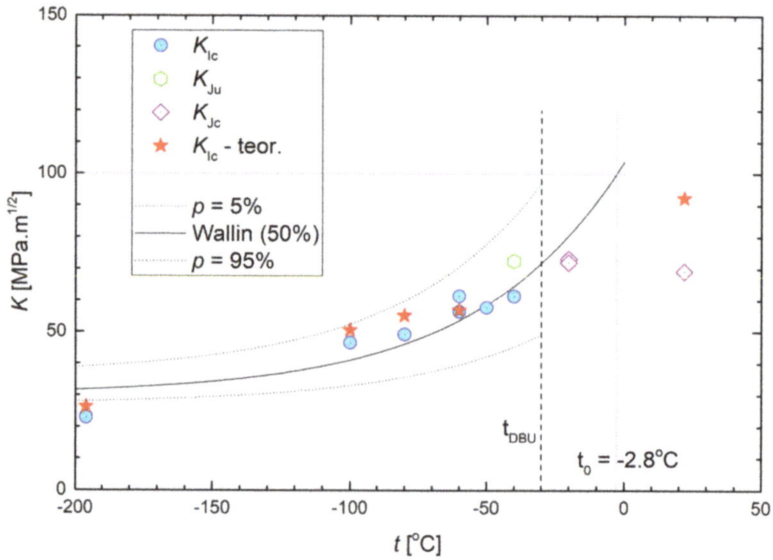


**Figure 11.** Temperature dependence of the measured and predicted fracture toughness of originally treated F(O) samples.

## 5. Morphology of Fracture Surfaces

The fracture morphology of all samples was studied in the scanning electron microscope (SEM) Zeiss-FEG SEM ULTRA PLUS. Macroscopic SEM pictures of typical fracture surfaces of the tensile and FT samples are depicted in Figures 12 and 13, respectively. The thickness reduction by necking of the tensile specimen and the circular shear ring surrounding the flat central fracture area are clearly visible in Figure 12. The morphology of the fatigue crack starting inside the chevron notch is depicted in Figure 13a, while the boundary (fatigue crack front) between the fatigue crack growth area (upper part) and the fast fracture area (bottom part) is clearly visible in Figure 13b.

The microscopical fracture surfaces of all tensile OT specimens in the temperature range $\langle -60; 22 \rangle$ °C were of a typical ductile dimple morphology as shown in Figure 14a,b. The fracture surfaces at −80 °C in Figure 14c exhibited features of microplasticity combined with some small cleavage facets, thus creating a mixed ductile/brittle morphology. At the testing temperature of −196 °C, a higher number of cleavage facets already occurred, and the morphology obtained a more brittle character (Figure 14d). As shown in Figure 15a,b, the morphology of FT(O) specimens in the temperature range $\langle -60; 22 \rangle$ °C was ductile dimple in correspondence with that of the OT samples. Note that the ductile-dimple area adjacent to the fatigue/final fracture boundary is the relevant morphology for the FT(O) specimen tested at −60 °C (Figure 15b) since it corresponds to the crack-tip process zone. In the FT(O) specimen tested at −80 °C, the mixed ductile-brittle morphology appeared near the fatigue-final fracture boundary as depicted in Figure 15c. At the lowest temperature of −196 °C, the fracture surface of the FT(O) sample again exhibited a more brittle character in correspondence with that of the tensile OT sample, as shown in Figures 14d and 15d.

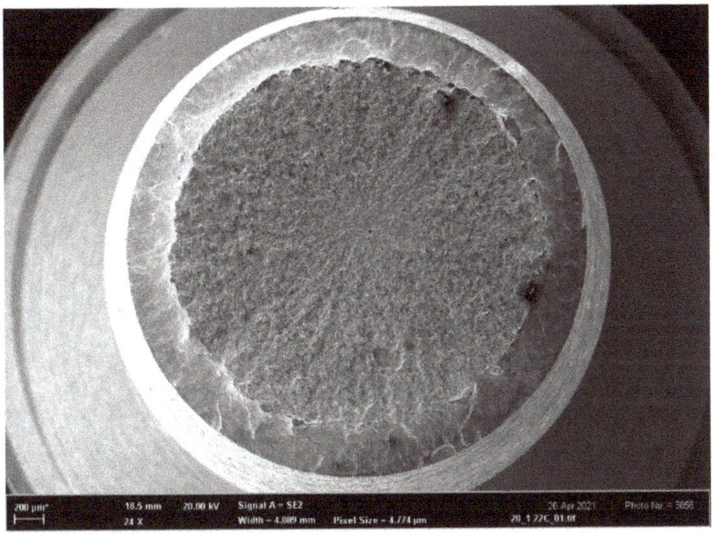


**Figure 12.** Fracture surface of the additionally treated tensile sample AT (200) tested at 22 °C. The reduction of the specimen thickness due to necking and the slant circular shear ring surrounding the flat central fracture area are clearly visible.

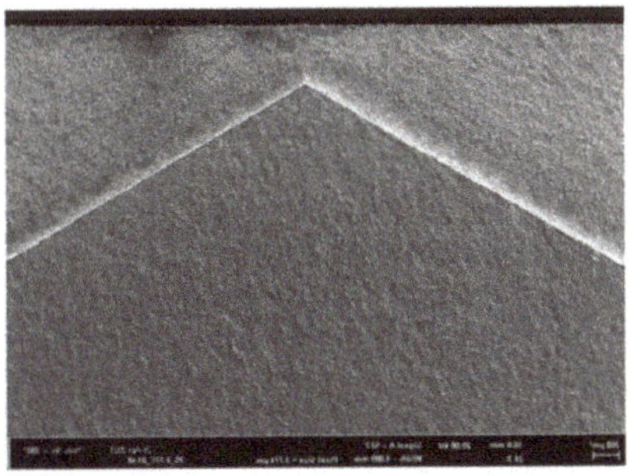

(a)

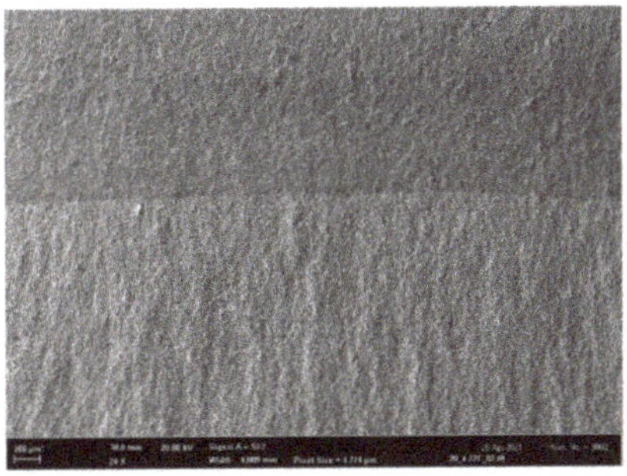

(b)

**Figure 13.** Examples of the macro-morphology of the fracture surfaces of samples FT (X): (**a**) the fatigue crack growth area inside the Chevron notch (middle and bottom part); (**b**) the boundary between the fatigue crack growth area (upper part) and the fast fracture area (bottom part).

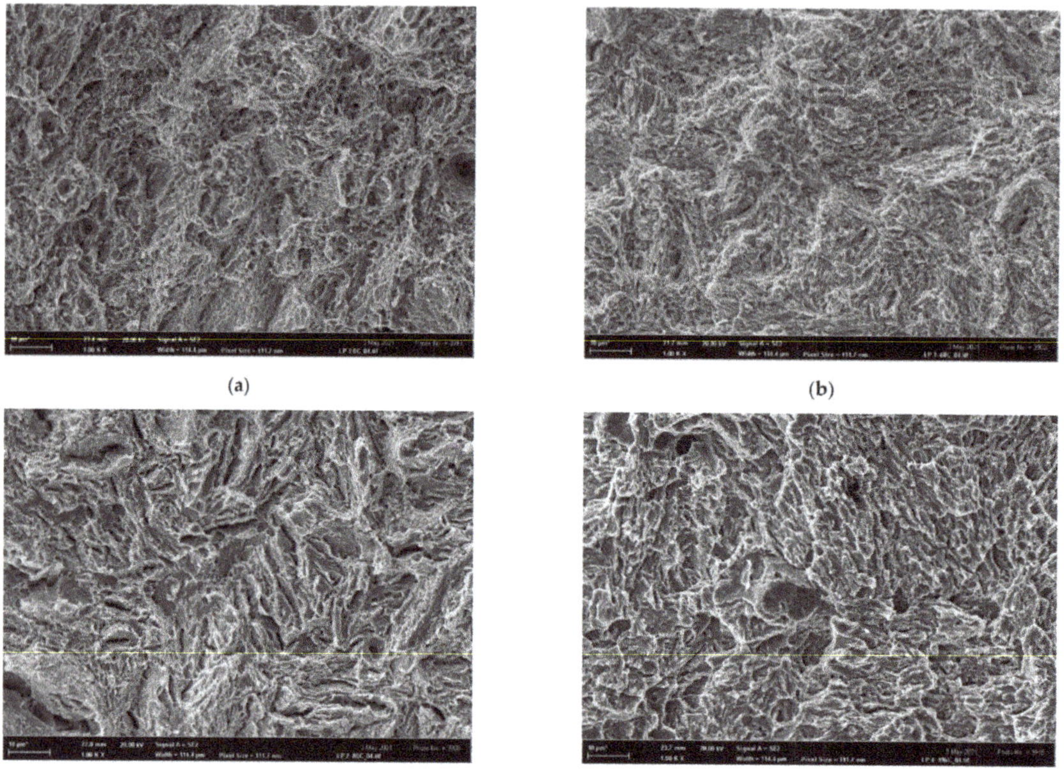

(a) (b) (c) (d)

**Figure 14.** The morphology of tensile OT samples: (**a**) ductile dimple, tested at 22 °C, (**b**) ductile dimple, tested at −60 °C, (**c**) mixed ductile-brittle, tested at −80 °C, (**d**) quasi-brittle, tested at −196 °C.

In general, there was good correspondence between the fracture morphology of the OT and FT(O) specimens at all testing temperatures, which documented a similarity in the fracture mechanisms operating in the tensile and FT tests in the whole temperature range.

On the other hand, all fracture surfaces of the tensile AT specimens were clearly of a ductile dimple morphology in the entire temperature range ⟨−196; 22⟩ °C, as shown in Figure 16. This documents an extended void-growth inside the well-developed necks, leading to superior tensile characteristics *RA* and $\varepsilon_f$ of all AT specimens. The fracture surfaces of all additionally treated fracture-toughness FT(X) samples in the temperature range ⟨−60; 22⟩ °C also exhibited a prevalent ductile dimple morphology (similar to the FT(O) samples) as documented in Figure 17a–c for the room temperature tests. At the temperature of −60 °C, however, some cleavage facets could already be observed as documented in Figure 17g for the FT (250) sample. The fracture morphology of FT(X) samples at the lowest temperature was of a mixed ductile/brittle or quasi-brittle character as depicted in Figure 17d–f. The much less ductility observed in the FT(X) samples compared to the AT(X) ones was a consequence of high tensile stress triaxiality at the pre-crack fronts in the FT(X) specimens. It accelerated the austenite/martensite strain-induced phase transformation, which, particularly during the low-temperature FT tests, quickly transferred the two-phase microstructure to a one-phase martensite structure similar to that of the FT(O) specimens. In the FT samples of all grades, moreover, the high stress triaxiality substantially reduced the values of the fracture strain compared to those in the tensile specimens, thus diminishing their influence on the values of the fracture toughness (see the $\varepsilon_f$-values of the OT samples in Table 4 and the $\varepsilon_{fc}$-values in Table 7. This corresponds to a higher initial. Both these effects led to a similarity in the fracture behavior of the FT(O) and FT(X) samples and caused a substantial difference in the morphology between AT specimens (ductile-dimple) and FT(X) specimens (quasi-brittle) as observed particularly in the low-temperature range. In contrast to the OT and FT(O) samples, therefore, there was a great difference in the fracture surface morphology of the AT and FT(X) specimens, which clearly indicated a dissimilarity of the fracture mechanisms in the tensile and FT tests of the additionally treated grades.

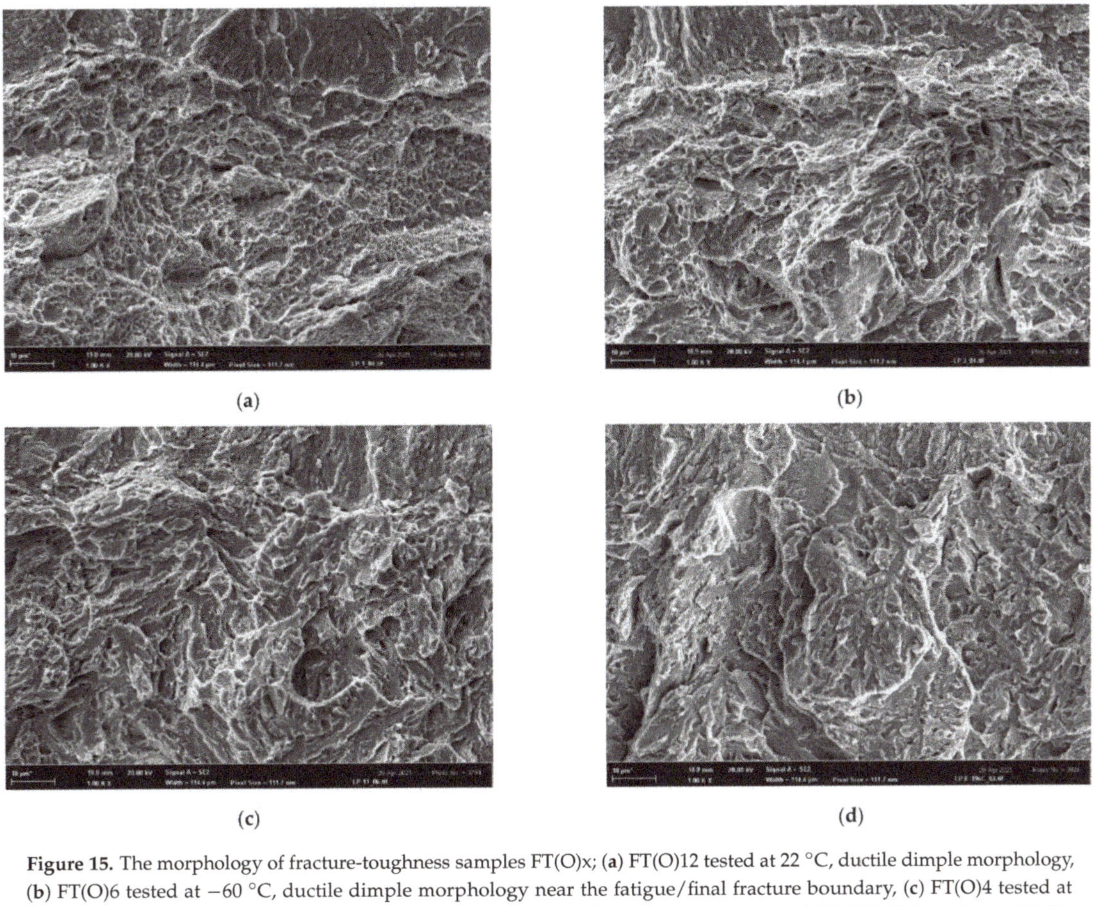

**Figure 15.** The morphology of fracture-toughness samples FT(O)x; (**a**) FT(O)12 tested at 22 °C, ductile dimple morphology, (**b**) FT(O)6 tested at −60 °C, ductile dimple morphology near the fatigue/final fracture boundary, (**c**) FT(O)4 tested at −80 °C, mixed ductile-brittle morphology near the fatigue/final fracture boundary, and (**d**) FT(O)1 tested at −196 °C, quasi-brittle morphology.

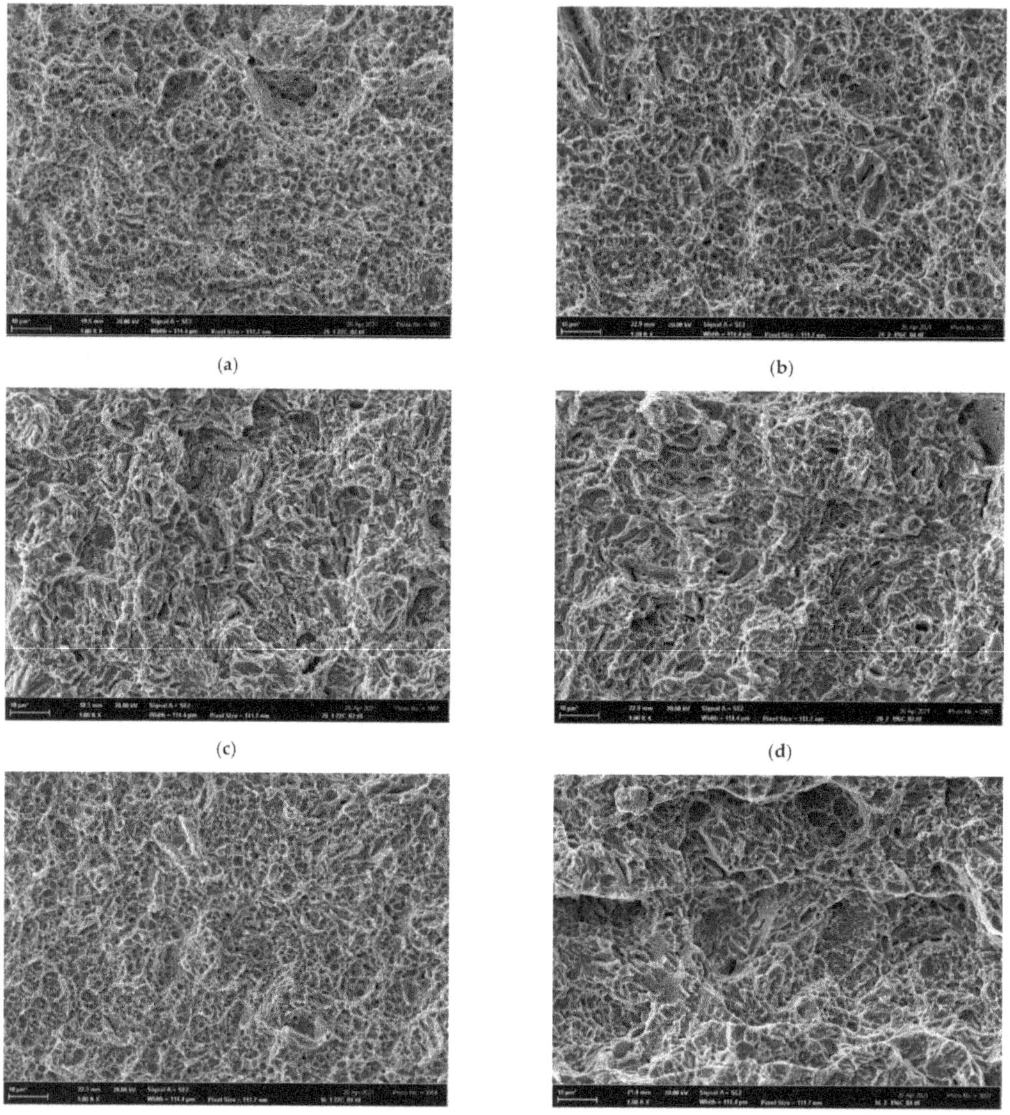

(a) (b) (c) (d) (e) (f)

**Figure 16.** Ductile dimple morphology of tensile AT(X)x specimens: (**a**) AT(250)5 tested at 22 °C, (**b**) AT(250)1 tested at −196 °C, (**c**) AT(200)5 tested at 22 °C, (**d**) AT(200)1 tested at −196 °C, (**e**) AT(160)5 tested at 22 °C and (**f**) AT(160)1 tested at −196 °C.

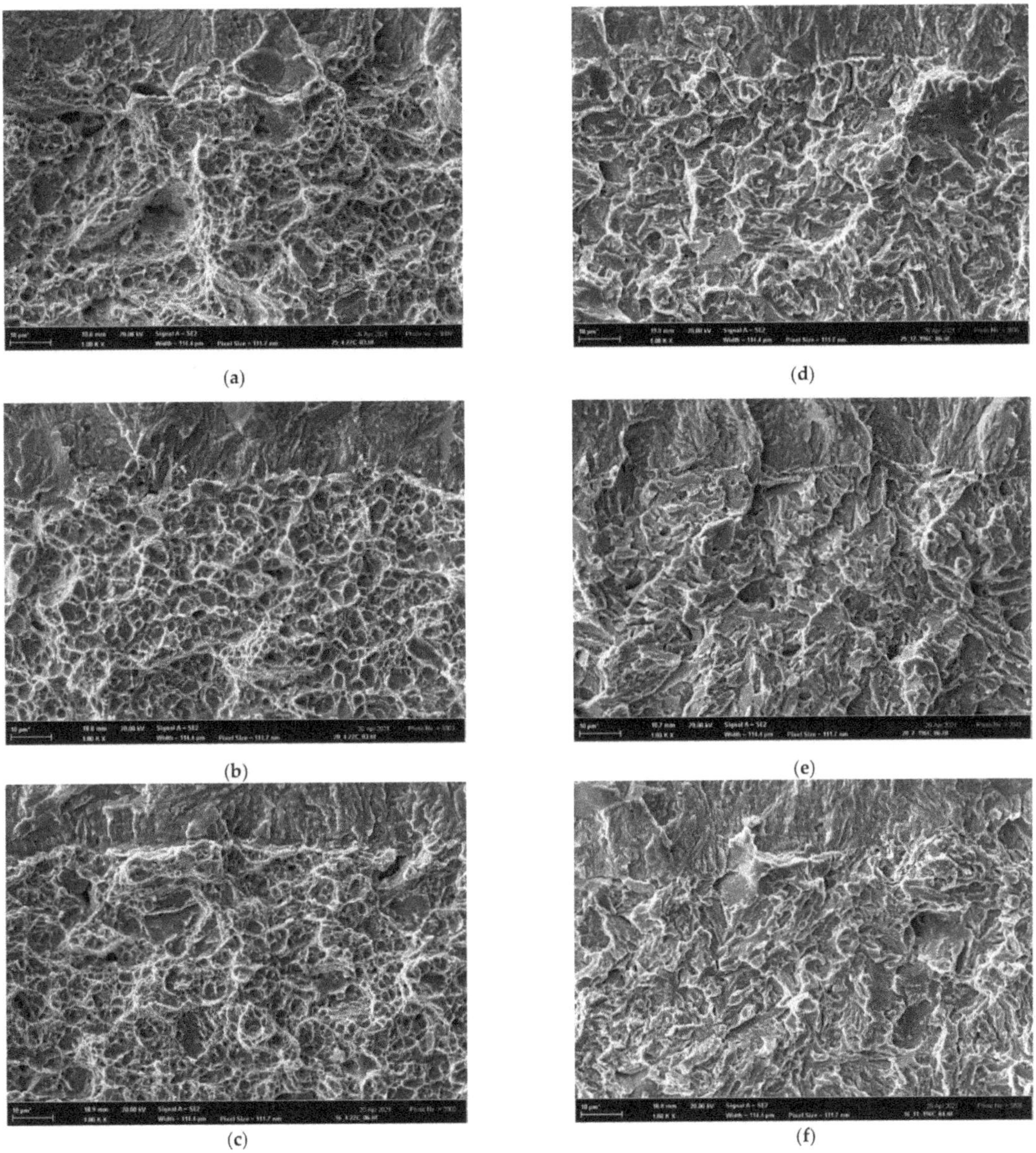

**Figure 17.** *Cont.*

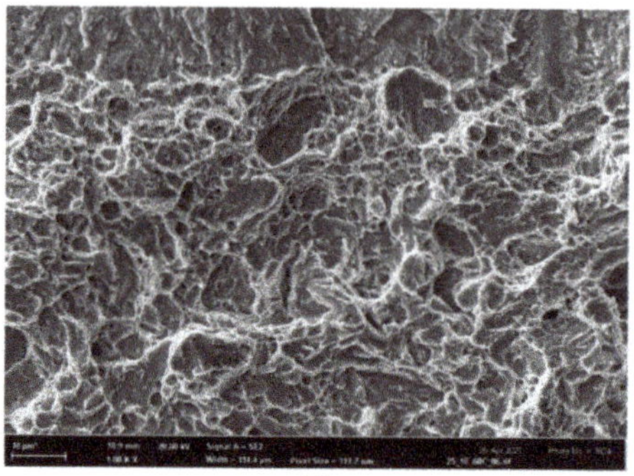

(g)

**Figure 17.** The morphology of FT(X)x samples: the ductile dimple morphology of (**a**) FT(250)5, (**b**) FT(200)5, (**c**) FT(160)5 tested at 22 °C; the ductile-brittle morphology of (**d**) FT(250)1, (**e**) FT(200)1 (**f**) FT(160)1 tested at −196 °C, and (**g**) the ductile-dimple morphology with some cleavage facets of FT(250)3 tested at −60 °C.

**Table 7.** Values of $\kappa_c$, $\varepsilon_{fc}$, and predicted $K_{Ic}{}^{pr}$ for original steel grades.

| Sample | $\kappa_c$ | $\varepsilon_{fc}$ | $K_{Ic}{}^{pr}$ |
|---|---|---|---|
| | - | % | MPa m$^{1/2}$ |
| FT(O)1 | 1.867 | 3.4 | 26.3 |
| FT(O)3 | 1.521 | 9.3 | 50.5 |
| FT(O)4 | 1.490 | 10.2 | 55.1 |
| FT(O)5 | 1.521 | 9.4 | 56.8 |
| FT(O)12 | 1.454 | 11.4 | 92.3 |

## 6. Theoretical Prediction and Interpretation of Fracture Toughness Values

Practically all the tested tensile and FT samples exhibited either a prevalent ductile-dimple morphology or at least some ductile markings. Therefore, models based on ductile fracture mechanisms were chosen to be more relevant for a prediction of FT and its temperature dependence. Among the ductile approaches mentioned in the introduction, only the models [16–20] provide theoretical formulae useful for prediction of FT in a straightforward manner. The principles of these models are described in Appendices A.1 and A.2 and their capability to predict experimental data and reflect the fracture mechanisms is presented hereafter.

The first model (Appendix A.1) enables a prediction of the temperature dependence of FT using the following data from tensile tests performed at corresponding temperatures (see Tables 4 and 5): $A$, $n$, $E$, $\nu$ = 0.3 (the Poisson ratio), $d_0$, and $d$. The solution of Equation (A7) yields the $\kappa_c$ value to obtain the fracture strain $\varepsilon_{fc}$ by introducing $\kappa = \kappa_c$ into Equation (A5) and, finally, to predict the $K_{Ic}$ value from Equation (A3). The values of $\kappa_c$, $\varepsilon_{fc}$ along with the predicted $K_{Ic}$ values (denoted $K_{Ic}{}^{pr}$) for the original steel grades are collected in Table 7 and a graphic comparison of the predicted and experimental values is presented in Figure 11. The agreement between the predicted and valid experimental values is acceptable, which implies that the fracture mechanisms in the tensile and FT tests were similar in the whole temperature range. Predicted and experimental data are covered by the 95% confidence band constructed according to the Master curve concept [34] as also shown in Figure 11.

On the other hand, the prediction of the temperature dependence of FT completely failed for the additionally treated grades. The predicted values were more than two times higher than the experimental data and they did not decrease with decreasing temperature. This is caused by the fact that the fracture mechanisms in the tensile and FT tests were different. Indeed, the ductile-dimple fracture mechanism in all tensile tests at all testing temperatures was associated with extremely high ultimate strength and fracture strain. It corresponded to very high values of the plastic work to fracture according to Equation (A2) while the more brittle fracture mechanism in the real FT tests demanded much less fracture energy. Consequently, the FT values predicted from Equation (A3) were within a highly overestimated range of (150; 300) MPa $m^{1/2}$ in the whole temperature range (−196; 22) °C.

One can see, however, that the experimental room temperature FT(X) values of the additionally treated grades were nearly two times greater than those pertaining to the lowest temperature. This indicates that the transition from the high-energy ductile fracture to the low-energy tearing with decreasing temperature and increasing yield stress could be assessed in terms of a transition from the single-void model of Rice and Johnson (R-J) [16] to the multiple-void model of Tvergaard and Hutchinson (T-H) [20]; see he Appendix A.2 in more detail. In the diagram $\sigma_y/E$ vs. $f_o$, where $f_o$ is the initial area void fraction, such a transfer is represented by a trajectory crossing the transition curve from its left- to the right-hand site. The transition curves in the diagram $\sigma_y/E$ vs. $f_o$ are related to steel grades by their $N$ values. The chart of transition curves for various $N$ depicted in Figure 18 was obtained by refining the originally published courser chart in Figure 12 [20] by employing an empirical nonlinear interpolation technique (Figure 12 contains only the transition curves for $N$ = 0, 0.1, and 0.2). When a combination of $\sigma_y$ and $f_o$ lies within the single-void regime to the left of the relevant transition curve in Figure 18, then the decreasing temperature (raising $\sigma_y$) derives the combination of $\sigma_y$ and $f_o$ across the transition into the low-energy fracture regime. Crossing the transition results in a drop in $K_{Ic}$ by a factor of two assuming that the yield stress $\sigma_y$ was increased by less than about a factor of 3. Let us now describe the construction of trajectories (straight lines) in the $\sigma_y/E$ vs. $f_o$ diagram in Figure 18 for additionally treated steel grades to see whether they cross the related transition curves.

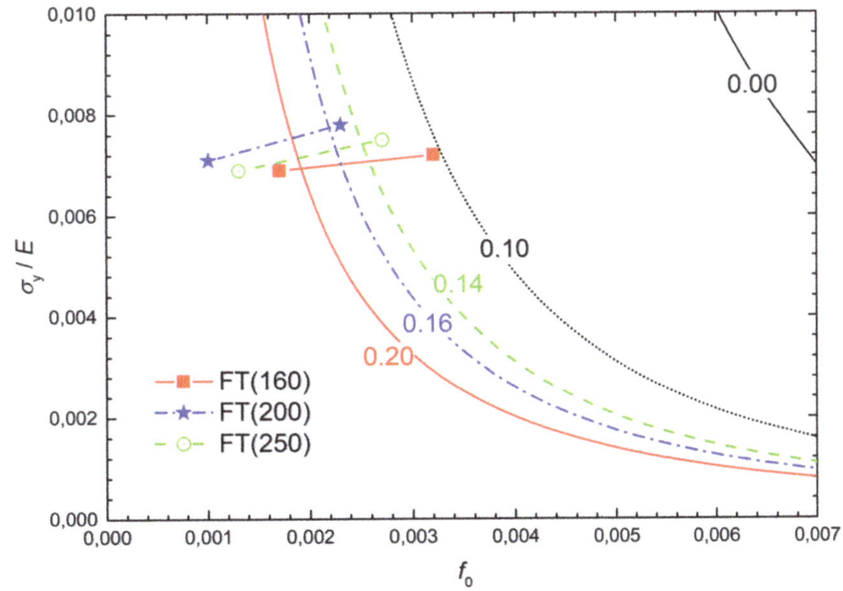


**Figure 18.** Linear trajectories FT(X) of additionally treated steel grades in the $\sigma_y/E$ vs. $f_o$ diagram. The corresponding transition curves are labelled by the values of the related hardening exponent $N$.

Given that all steel grades exhibit a single void fracture mechanism at room temperature, the corresponding points in the diagram $\sigma_y/E$ vs. $f_o$ must lie to the left of each related transition curve, i.e., the void fraction $f_o$ should be 0.001 in the order of magnitude and c = 2 in Equation (A13). The values of the averaged void spacing Λ for FT(X) samples for 22 °C calculated from Equation (A13) as:

$$\Lambda = \frac{1-\nu^2}{2Ec\sigma_y}K_{Ic}^2 \tag{1}$$

are collected in Table 8. Note that these values of 2–3 μm correspond well with the averaged distance between the void centers as documented by the SEM fractography in Figure 17a–c for samples fractured at room temperature. The area void fraction $f_o$ as a function of the ratio $\Lambda/r_o$, where Λ is the mean distance of voids and $r_o$ is the (initial) void radius, is plotted in Figure 19 for both triangle and square geometrical arrangements of the voids. From this graph, the values of $f_o$ for all grades can be obtained when considering the Λ values from Table 8 and selecting $r_o$ = 0.05 μm. These values correspond to the initial (room temperature) points ($f_o$, $\sigma_y/E$) of the trajectories in Figure 18 and ensure that these points lie in the R-J part of the diagram $\sigma_y/E$ vs. $f_o$, i.e., to the left of the relevant transition curves.

**Table 8.** Parameters Λ, λ, W, $f_o$, and $\sigma_y/E$ calculated for the initial and final points of the trajectories related to the additionally treated samples in Figure 18.

| Sample | $t_{exp}$ | Λ | λ | W | $f_o$ (Λ) | $f_o$ (λ) | $\sigma_y/E$ |
|---|---|---|---|---|---|---|---|
| | °C | μm | μm | - | ‰ | ‰ | - |
| FT(160)1 | −196 | | 1.63 | 0.81 | | 3.2 | 0.0072 |
| FT(160)5 | 22 | 2.23 | | | 1.7 | | 0.0069 |
| FT(200)1 | −196 | | 1.91 | 0.80 | | 2.3 | 0.0078 |
| FT(200)5 | 22 | 2.90 | | | 1.0 | | 0.0071 |
| FT(250)1 | −196 | | 1.78 | 0.70 | | 2.7 | 0.0075 |
| FT(250)5 | 22 | 2.59 | | | 1.3 | | 0.0069 |

The final points of the trajectories correspond to the testing temperature of −196 °C. Given that they should lie in the T-H part, the averaged void spacing should be determined from Equation (A11) as:

$$\lambda = \frac{1-\nu^2}{EW\sigma_y}K_{Ic}^2, \tag{2}$$

where $W$ is the work of separation per unit area divided by the product $\sigma_y\lambda$ (see Table 8). These values were obtained from Figure 10 in [20] with respect to the $N$ values for individual grades. The values of λ computed from Equation (2) and shown in Table 8 are slightly lower than those of Λ. This corresponds to a higher initial void fraction $f_o$ (associated with the higher yield stress $\sigma_y$ at −196 °C) and to higher values of $\sigma_y/E$ in Table 8. The final points for −196 °C determine the trajectories of additionally treated grades plotted in Figure 18. Note that the full-line (red) trajectory for the FT(160) grade crosses the associated (red and full-line) transition curve labelled by the value $N$ = 0.20, corresponding well with the N values for the AT(160)x samples in Table 5. This also holds for both the FT(200) and FT(250) grades related to the dashed-line (blue) and dashed-and-dot line (black) trajectories and transition curves for $N$ = 0.16 and $N$ = 0.14, respectively. Thus, all the trajectories for additionally treated grades cross the related transition curves from their left- to the right-hand side, which indicates a change of the fracture mechanisms in terms of the R-J and T-H models. The major part of the trajectory for the FT(160) grade lies in the low-energy fracture region, which corresponds to the occurrence of shallow dimples and cleavage facets on the fracture surfaces in the major part of the temperature range. The opposite is true for both the FT(200) and FT(250) grades, which matches the well-developed dimple fracture morphology observed even at low testing temperatures.

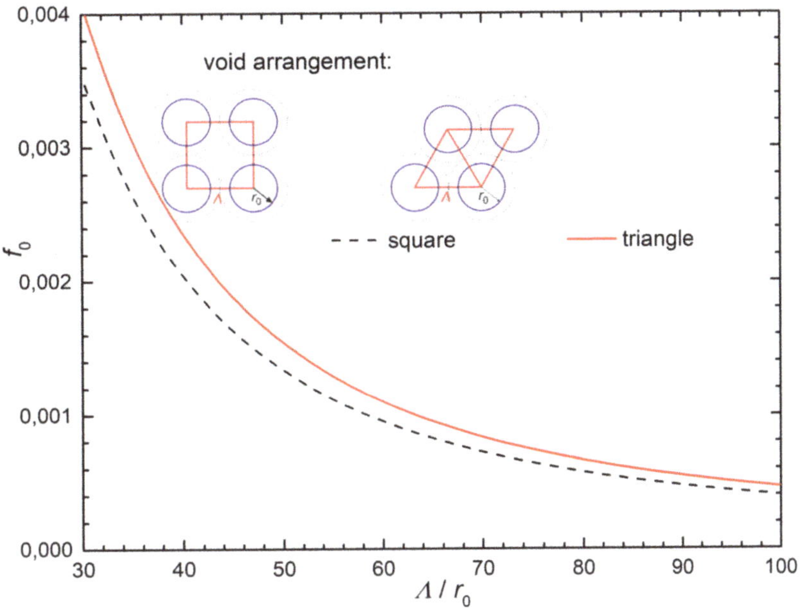


**Figure 19.** The area void fraction $f_0$ as a function of the ratio $\Lambda/r_0$, where $\Lambda$ is the mean distance of voids and $r_0$ is the (initial) void radius for the triangle and square geometrical arrangements of voids.

## 7. Conclusions

The results of the experimental investigation of the fracture characteristics of standard heat-treated low-alloyed steel OCHN3MFA along with its three additionally heat-treated grades were presented for the testing temperature range of $\langle -196; 22 \rangle$ °C. Modelling of the microstructure and fracture processes supported by X-ray and EBSD measurements enabled a physical interpretation of the obtained results, which can be summarized in the following points:

(i) All the additional heat treatments (annealing 650 °C, quenching and tempering to 160, 200, or 250 °C) transferred the standard steel from high- to ultrahigh strength levels even with an improved tensile ductility. The higher strength of the additionally treated grades could be understood in terms of the Hall–Petch relation since they exhibited finer regions of martensitic blocks with a similar crystallographic orientation. The higher ductility could be explained by their two-phase microstructure (martensite + retained austenite) and a more homogeneous distribution of the shape ratio of martensitic laths. A common reason for the increase in both the strength and the ductility was, most probably, the detected reduction of the inclusion content.

(ii) The values of the fracture toughness of all grades were found to be comparable in the whole temperature range due to a high tensile stress triaxiality localized at the process zone ahead of the pre-crack front. It accelerated the strain-induced transformation of austenite to martensite during fracture-toughness tests of the additionally treated grades. Moreover, the triaxiality three-times reduced the fracture strain (compared with the tensile test), thus diminishing its influence on the fracture toughness values.

(iii) The values of the fracture toughness of the standard steel grade could be predicted well using a fracture model proposed by Pokluda et al. based on the tensile characteristics. Such a prediction, however, failed in the case of the additionally heat-treated grades due to their extremely high fracture strains in the tensile tests and different temperature dependence of the fracture mechanisms in the tensile and fracture-toughness tests. While the tensile samples fractured in a ductile-dimple mode at all temperatures, the fracture-toughness specimens exhibited a transition from the ductile to a quasi-brittle fracture mode with a decreasing temperature. This transition was described in terms of a transfer from the void-crack interaction model of Rice and Johnson to the multi-void tearing model of Tvergaard and Hutchinson.

**Author Contributions:** Conceptualization, J.P.; methodology, J.P. and P.Š.; software, P.Š.; validation, J.P. and I.D.; formal analysis, J.H.; experimental investigation and sample preparation, M.K., I.D. and J.Č.; resources, J.P.; data curation, P.Š. and I.D.; writing—original draft preparation, J.P.; writing—review and editing, J.P., J.H. and P.Š.; visualization, P.Š.; supervision, J.P.; project administration, J.P. and M.K.; funding acquisition, J.P. All authors have read and agreed to the published version of the manuscript.

**Funding:** This research was funded by the Ministry of Education, Science, Research and Sport of the Slovak Republic, grant VEGA1/0346/19 and the Brno University of Technology, grant FSI-S-20-6266.

**Institutional Review Board Statement:** Not applicable.

**Informed Consent Statement:** Not applicable.

**Data Availability Statement:** Data supporting the findings of this study are available from the corresponding author upon request.

**Conflicts of Interest:** The authors declare no conflict of interest.

## Appendix A. Models of Fracture Toughness Based on the Localized Ductile Damage

### *Appendix A.1. The Fracture Energy Specified by the Critical Strain in the Process Zone*

The model introduced by Staněk and Pokluda [17] and further modified by Pokluda and Šandera [18,19] determines the critical fracture strain in the crack-tip process zone using the so-called *failure locus* as the dependence of the fracture strain on the stress triaxiality for a given material. The related FT value can then be predicted using the hypothesis of linear damage accumulation. This method was developed particularly for the assessment of $K_{\text{Ic}}$ values from the mechanical characteristics available from tensile tests of smooth and circumferentially notched cylindrical specimens. The method assumes that the crack tip blunting takes place during the initial phase of FT tests and the related damage process localized in the crack-tip plastic (process) zone precedes the unstable fracture. When performing the standard FT test of relatively small deformation rates under conditions of small-scale yielding and plane strain, the dissipation energy in the form of elastic waves can be neglected. Thus, the Griffith–Irwin–Orowan fracture criterion can be considered as:

$$\frac{1-\nu^2}{E}K_{Ic}^2 = 2\gamma + \omega_p\left(\varepsilon_{fc}\right), \tag{A1}$$

where $\gamma$ is the surface energy, $\omega_{\text{p}}$ $(K, \gamma)$ is the fracture energy, i.e., the work per unit area needed for building the plastic zone up to the moment of fracture, and $\varepsilon_{\text{fc}}$ is the critical (local) fracture strain in the process zone. Then, the following physically relevant presumptions are considered [19]:

(i) Nearly all energy supplied by external forces and/or released by elastic relaxation is consumed in the plastic zone during the ductile crack tip blunting process preceding the unstable crack advance. In other words, the second term in Equation (1) is much higher than the first one ($\omega_{\text{p}} \gg 2\gamma$).

(ii) The unstable fracture is controlled by reaching a critical value of the plastic strain $\varepsilon_{\text{fc}}$ in the process zone at the crack tip.

Given that the Hollomon approximation $\sigma = A\varepsilon_p{}^n$ describes the stress-plastic strain behavior during formation of the plastic zone sized $r_p = Bn^2$ ($B \approx 0.025$ m) [12]. Then, the plastic work consumed till the onset of unstable fracture can be expressed as:

$$\omega_p\left(\varepsilon_{fc}\right) = 2ABn^2 \int_0^{\varepsilon_{fc}} \varepsilon_p^n(\kappa)\mathrm{d}\varepsilon_p \tag{A2}$$

and Equation (A1) can be rearranged to:

$$K_{Ic} = \left[\frac{2ABE\varepsilon_{fc}^{n+1}}{(1-\nu^2)(n+1)}\right]^{1/2}. \tag{A3}$$

Owing to the highly triaxial tensile stress inside the crack-tip process zone and the void-coalescence fracture mechanism, the value of $\varepsilon_{fc}$ is much less than the fracture strain $\varepsilon_f$ of smooth tensile samples (e.g., [17,19]) and it must be determined by the following procedure. The dependence of the fracture strain on the stress triaxiality factor $\kappa$ (the ratio of the hydrostatic stress and the von Mises effective stress) can be expressed as:

$$\varepsilon_f(\kappa) = \frac{4}{5\kappa} \ln \frac{d_o}{d}. \tag{A4}$$

Equation (A4) is the simplest phenomenological approximation of the fracture locus. During the blunting process, the process zone elements experience a strain trajectory:

$$\varepsilon_p(\kappa) = 7.7 \exp(-2.9\kappa). \tag{A5}$$

starting at $\kappa = (n+1)/\sqrt{3}$ [17,37]. When adopting the hypothesis of the linear damage accumulation, the critical factor $\kappa_c$ associated with the onset of unstable fracture is determined by the integral equation:

$$\int_{\varepsilon_p((n+1)/\sqrt{3})}^{\varepsilon_p(\kappa_c)} \frac{\mathrm{d}\varepsilon_p(\kappa)}{\varepsilon_f(\kappa)} = \int_{(n+1)/\sqrt{3}}^{\kappa_c} -\frac{27.9\ \kappa\ \exp(-2.9\kappa)}{\ln(d_o/d)}\,\mathrm{d}\kappa = 1. \tag{A6}$$

When combining Equations (A4), (A6), and (A7), one obtains:

$$(7.7\kappa_c + 2.655)\ \exp\{-2.9\kappa_c\} = 0.8\ \ln\left(\frac{d_o}{d}\right) + 0.0205. \tag{A7}$$

The iterative solution of Equation (A7) yields the $\kappa_c$ value, and the fracture strain $\varepsilon_{fc}$ is then obtained from Equation (A5) by introducing $\kappa = \kappa_c$. Finally, the $K_{Ic}$ value is predicted from Equation (A3).

*Appendix A.2. The Separation Energy Related to Rupture of Void Ligaments*

The Tvergaard–Hutchinson (T-H) model [20] assumes a pre-existing population of roughly similar sized voids that give rise to an unstable fracture localized to a "void sheet" in the process zone. The void spacing is comparable to the initial thickness $\lambda$ of the void sheet, the length $l$ of which is much larger. The model constitutes a transition from the classical high-energy ductile fracture by a spatial coalescence of voids to the planar coalescence representing a sort of low-energy ductile tearing caused by many small voids interacting on a plane ahead of the crack tip.

The Gurson model [15] for an elastic-plastic solid containing voids is used to predict the traction–displacement law associated with the fracture process zone. The fracture localization occurs in a layer of initial thickness $\lambda$ identified with the average spacing of voids with the area fraction $f_o$. The yield condition of this model is:

$$\Phi(\sigma_c, \sigma_m, f) = \left(\frac{\sigma_c}{\sigma_{\text{eff}}}\right)^2 + 2q_1 f \cosh\left(\frac{3\sigma_m}{2\sigma_{\text{eff}}}\right) - \left[1 + \left(q_1 f^2\right)^2\right] = 0, \tag{A8}$$

where $\sigma_c$ is the effective stress, $\sigma_c = \sigma_{kk}/3$ is the mean stress, $\sigma_{\text{eff}}$ is the current effective stress associated with the matrix, $f$ is the current void volume fraction, and $q_1 = 3/2$ is the adjustment factor. The traction–displacement relation is computed assuming the failing layer undergoes uniaxial straining in the direction normal to the crack plane. The elastic-plastic solid has a tensile yield stress, $\sigma_y$, and a true stress logarithmic strain curve in uniaxial tension is specified by:

$$\begin{aligned} \varepsilon &= \tfrac{\sigma}{E} & \sigma \leq \sigma_y, \\ \varepsilon &= \tfrac{\sigma_y}{E}\left(\tfrac{\sigma}{\sigma_y}\right)^{\frac{1}{N}} & \sigma \geq \sigma_y \end{aligned} \tag{A9}$$

The work of separation per unit area is then computed as:

$$\Gamma_o = \int_0^{\delta_c} \sigma(\delta) \mathrm{d}\delta \tag{A10}$$

where $\sigma(\delta)$ is the traction–displacement law. The calculations revealed that the term $W = \Gamma_o/(\sigma_y\, \lambda)$ only slightly depends on $N$ but is insensitive to the area fraction $f_o$ and its value lies in the range of (0.7, 1). Assuming $\Gamma_o \equiv \omega_p$ in Equation (A1), the T-H model leads to the FT prediction:

$$K_{Ic} = \left[\frac{W}{(1-\nu^2)}\right]^{1/2} (E\sigma_y \lambda)^{1/2}. \tag{A11}$$

This formula is similar to that resulting from the model of Rice and Johnson (R-J) [16], which assumes the ductile fracture mechanism involving an interaction of just one void with the crack tip. The multiple-void T-H model transfers into the single-void R-J model when the geometrical condition $\lambda/l = 0.5$ is fulfilled. In the R-J model, the initiation of crack growth takes place when the void nearest to the tip begins to coalesce with the tip. This will happen when $\delta_t = c\Lambda$, where $\delta_t$ is the crack tip opening displacement, $\Lambda$ is the average void spacing, and $c$ is a number, which depends weakly on $f_o$ varying from about 2 for $f_o = 0.001$ to about 1 at $f_o = 0.05$. Using the small-scale yielding estimate:

$$\delta_t = 0.5\left(1 - v^2\right) K^2 / (E\sigma_y \Lambda)^{1/2}, \tag{A12}$$

one obtains the R-J prediction of FT as:

$$K_{Ic} = \left[\frac{2c}{(1-\nu^2)}\right]^{1/2} (E\sigma_y \Lambda)^{1/2}. \tag{A13}$$

If $\lambda$ in the T-H model is identified with $\Lambda$ in the R-J model, then (A11) and (A13) differ only in their numerical coefficients. The coefficient in $K_{Ic}$ for the R-J model is in the range of about two times that for the T-H model.

Note that both the T-H and R-J models predict an increase in $K_{Ic}$ with increasing $\sigma_y$ when lowering the testing temperature, but the opposite effect is commonly observed. Even if the initial void volume fraction $f_o$ is supposed to increase with $\sigma_y$ due to greater void nucleation at the higher stress, the dependence of each prediction on $f_o$ is too weak to result in a drop in $K_{Ic}$. A possible explanation for such a drop may be a transition from the single- to the multiple-void mechanism associated with the condition $\lambda/l = 0.5$. This condition corresponds to transition curves for different values of $N$ in the diagram $\sigma_y/E$ vs. $f_o$ [20]. The transfer from the single- to multiple-void mechanism then corresponds to crossing these curves from the left- to the right-hand site; see the Section 6 of this article for more detail.

## References


1. Standard test method for linear-elastic plane-strain fracture toughness KIc of metallic materials. In *Annual Book of ASTM-Standards*; ASTM E 399-20; ASTM International: Philadelphia, PA, USA, 2020.
2. Standard test method for JIC, a measure of fracture toughness. In *Annual Book of ASTM-Standards*; ASTM E 813-81; ASTM International: Philadelphia, PA, USA, 1986.
3. Joyce, J.A.; Link, R.E. Application of two parameter elastic-plastic fracture mechanics to analysis of structures. *Eng. Fract. Mech.* **1997**, *57*, 431–446. [CrossRef]
4. Beremin, F.M. A local criterion for cleavage fracture of a nuclear pressure vessel steel. *Metall. Trans. A* **1983**, *14*, 2277–2287. [CrossRef]
5. Pineau, A. Development of the local approach to fracture over the past 25 years: Theory and applications. *Int. J. Fract.* **2006**, *138*, 139–166. [CrossRef]
6. Andrieu, T.A.; Pineau, A.; Besson, J.; Ryckelynck, D.; Bouaziz, O. Beremin model: Methodology and application to the prediction of the Euro toughness data set. *Eng. Fract. Mech.* **2012**, *95*, 102–117. [CrossRef]
7. Kotrechko, S.; Strnadel, B.; Dlouhy, I. Fracture toughness of cast ferritic steel applying local approach. *Theor. Appl. Fract. Mech.* **2007**, *47*, 171–181. [CrossRef]
8. Yankova, M.S.; Jivkov, A.P.; Patel, R. Incorporation of obstacle hardening into local approach to cleavage fracture to predict temperature effects in the ductile to brittle transition regime. *Materials* **2021**, *13*, 1224. [CrossRef] [PubMed]
9. Profant, T.; Pokluda, J. The ab-initio aided strain gradient elasticity theory: A new concept for fracture nanomechanics. *Frat. Ed Integrità Strutt.* **2019**, *49*, 107–114. [CrossRef]
10. Kotoul, M.; Skalka, P.; Profant, T.; Řehák, P.; Šesták, P.; Černý, M.; Pokluda, J. A novel multiscale approach to brittle fracture of nano/micro-sized components. *Fat. Fract. Eng. Mater. Struct.* **2020**, *43*, 1630–1645. [CrossRef]
11. Peel, C.J.; Forsyth, P.J.E. The effect of composition changes on the fracture toughness of an Al-Zn-Mg-Cu-Mn forging alloy. *Metal Sci. J.* **1973**, *7*, 121–127. [CrossRef]
12. Hahn, G.T.; Rosenfield, A.R. Sources of fracture toughness: The relation between $K_{Ic}$ and the ordinary tensile properties of metals. In *Applications Related Phenomena in Titanium Alloys*; Conrad, H., Jaffee, R., Kessler, H., Minkler, W., Eds.; ASTM International: West Conshohocken, PA, USA, 1968; pp. 5–32.
13. Hahn, G.T.; Rosenfield, A.R. Metallurgical factors affecting fracture toughness of aluminum alloys. *Metall. Trans. A* **1975**, *6*, 653–668. [CrossRef]
14. McClintock, F.A. A criterion for ductile fracture by the growth of holes. *J. Appl. Mech.* **1968**, *35*, 363–371. [CrossRef]
15. Gurson, A.J. Continuum theory of ductile rupture by void nucleation and growth: Part i-yield criteria and flow rules for porous ductile media. *J. Eng. Mater. Technol.* **1977**, *99*, 2–15. [CrossRef]
16. Rice, J.R.; Johnson, M.A. The role of large crack tip geometry changes in plane strain fracture. In *Inelastic Behavior of Solids*; Kanninen, M.F., Adler, W.F., Rosenfield, A.R., Jaffee, R.I., Eds.; McGraw-Hill: New York, NY, USA, 1970; pp. 641–672.
17. Staněk, P.; Pokluda, J. On the theory of ductile fracture in the tensile test. *Metall. Mater.* **1984**, *22*, 710–719.
18. Pokluda, J.; Šandera, P. Simple method for calculation of $K_{Ic}$-value for materials with ductile fracture mechanism. In *Brittle Fracture of Materials and Structures*; ŠKODA Pilsen: Carlsbad, CA, USA, 1990; pp. 130–136. (In Czech)
19. Pokluda, J.; Šandera, P. *Micromechanisms of Fracture and Fatigue: In a Multiscale Context*; Springer Ltd.: London, UK, 2010.
20. Tvergaard, V.; Hutchinson, J.W. The relation between crack growth resistance and fracture process parameters in elastic-plastic solids. *J. Mech. Phys. Solids* **1992**, *40*, 1377–1397. [CrossRef]
21. Alexopoulos, N.D.; Titryakioglu, M. Relationship between fracture toughness and tensile properties of A357 cast aluminum alloy. *Metall. Mater. Trans. A* **2009**, *40*, 702–716. [CrossRef]
22. Tvergaard, V. Effect of stress-state and spacing on voids in a shear-field. *Int. J. Sol. Struct.* **2012**, *49*, 3047–3054. [CrossRef]
23. Needleman, A. Some issues in cohesive surface modeling. *Procedia IUTAM* **2014**, *10*, 221–246. [CrossRef]
24. Barényi, I.; Kianicová, M.; Majerík, J.; Pokluda, J. *Optimization of the Heat Treatment of the Pressure-Vessel Component of Military System Made of the High-Strength Stee*; Utility Model No. 9309; Industrial Property Office of the Slovak Republic: Bratislava, Slovakia, 2021.
25. Šandera, P.; Horníková, J.; Kianicová, M.; Dlouhý, I.; Pokluda, J. Determination of Ramberg-Osgood approximation for estimation of low-temperature fracture toughness. *AIP Conf. Proc.* **2020**, *2309*, 020024.
26. Wang, Y.; Denis, S.; Appolaire, B.; Archambault, P. Modelling of precipitation of carbides during tempering of martensite. *J. Phys. IV France* **2004**, *120*, 103–110. [CrossRef]
27. Krauss, G. Tempering of lath martensite in low and medium carbon steels: Assessment and challenges. *Steel Res. Int.* **2017**, *88*, 1700038. [CrossRef]
28. Cheng, G.; Li, W.; Zhang, X.; Zhang, L. Transformation of Inclusions in Solid GCr15 Bearing Steels During Heat Treatment. *Metals* **2019**, *9*, 642.
29. Villavicencio, J.; Ulloa, N.; Lozada, L.; Moreno, M.; Castro, L. The role of non-metallic Al2O3 inclusions, heat treatments and microstructure on the corrosion resistance of an API 5L X42 steel. *J. Mater. Res. Technol.* **2020**, *9*, 5894–5911. [CrossRef]

30. Hidalgo, J.; Findley, K.O.; Santofimia, M.J. Thermal and mechanical stability of retained austenite surrounded by martensite with different degrees of tempering. *Mater. Sci. Eng.* **2017**, *690*, 337–347. [CrossRef]
31. ASM Handbook Committee (Ed.) Properties and Selection: Irons, Steels, and High-Performance Alloys. In *ASM Handbook*; ASM International: Materials Park, OH, USA, 1990; Volume 1.
32. Malakondaiah, G.; Srinivas, M.; Rama Rao, P. Ultrahigh-strength low-alloy steels with enhanced fracture toughness. *Prog. Mater. Sci.* **1997**, *42*, 209–242. [CrossRef]
33. Wells, M.G.H.; Hauser, J.J. *Fracture Prevention and Control*; Metalworking Technology Series; ASM International: Los Angeles, CA, USA, 1972.
34. Banerji, S.K.; McMahon, C.J.; Feng, H.C. Intergranular fracture in 4340-type steels: Effects of impurities and hydrogen. *Metal. Trans. A* **1978**, *9*, 237–247. [CrossRef]
35. *ISO 12737:2010 Metallic Materials—Determination of Plane-Strain Fracture Toughness*; ISO Office: Geneva, Switzerland, 2010.
36. Standard test method for determination of reference temperature, T0, for ferritic steels in transition range. In *Annual Book of ASTM-Standards*; ASTM E1921-19be1; ASTM International: Philadelphia, PA, USA, 2019.
37. Hancock, J.W.; Cowling, M.J. Role of state of stress in crack-tip failure processes. *Metal Sci.* **1980**, *14*, 293–304. [CrossRef]

materials MDPI

*Article*

# Incorporation of Temperature and Plastic Strain Effects into Local Approach to Fracture

**Sergiy Kotrechko [1,2], Vladislav Kozák [3], Oleksandra Zatsarna [1], Galyna Zimina [1], Nataliya Stetsenko [1,*] and Ivo Dlouhý [3,4]**

1 G. V. Kurdyumov Institute for Metal Physics, National Academy of Sciences of Ukraine 36, Academician Vernadsky Blvd., UA-0380 Kyiv, Ukraine; serkotr@gmail.com (S.K.); alexsandra.ku@gmail.com (O.Z.); zimingal28@gmail.com (G.Z.)
2 E. O. Paton Institute for Materials Science and Welding, National Technical University of Ukraine "Igor Sikorsky Kyiv Polytechnic Institute" 37, Peremohy Ave., UA-03056 Kyiv, Ukraine
3 Institute of Physics of Materials, Academy of Sciences of the Czech Republic, Zizkova 22, 61669 Brno, Czech Republic; kozak@ipm.cz (V.K.); dlouhy@fme.vutbr.cz (I.D.)
4 Institute of Materials Science and Engineering, Brno University of Technology, Technicka 2, 61669 Brno, Czech Republic
* Correspondence: natusjaimp@gmail.com

**Abstract:** An unjustified simplification of the local quantitative criterion regarding cleavage nucleation is a key problem in the utilisation of the Local Approach to Fracture (LA), particularly to predict the fracture toughness within the ductile-to-brittle transition (DBT) region. The theoretical concept of the effect of both temperature and the plastic strain value on the crack nuclei (CN) generation rate in iron and ferritic steels is presented. It is shown how the plastic strain and temperature affect CN formation rate and, as a consequence, govern the shape of the temperature dependence of fracture toughness $K_{Jc}$ and its scatter limits. Within the framework of the microscopic model proposed, dependences of the CN bulk density on the plastic deformation value and temperature are predicted. Convenient approximation dependences for incorporating this effect into the LA are suggested. The experimental data of reactor pressure vessel steel and cast manganese steel demonstrate that the use of these dependences enables one to predict, with sufficient accuracy, the effect of temperature on the value of fracture toughness and its scatter limits over the DBT region. It is shown that accounting for both the temperature and strain dependence of CN bulk density gives rise to the invariance of parameters of the Weibull distribution to temperature.

**Keywords:** fracture toughness; ductile-to-brittle transition; local approach to fracture; ferritic steel

**Citation:** Kotrechko, S.; Kozák, V.; Zatsarna, O.; Zimina, G.; Stetsenko, N.; Dlouhý, I. Incorporation of Temperature and Plastic Strain Effects into Local Approach to Fracture. *Materials* **2021**, *14*, 6224. https://doi.org/10.3390/ma14206224

Academic Editor: Andrea Spagnoli

Received: 23 August 2021
Accepted: 13 October 2021
Published: 19 October 2021



## 1. Introduction

The Local Approach (LA) to fracture was introduced in the 1980s as a probabilistic method accounting for microstructural effects in the evaluation of global parameters. It aimed to solve key problems of fracture mechanics such as the prediction of the specimen geometry and statistical size effect on fracture toughness (transferability problem), as well as the effects of test temperature, operation temperature, and degradation of the material due to neutron irradiation, etc. The possibility of solving these complex problems was seen in the employment of statistical local criterion for initiating fracture in the vicinity of a crack, because this approach accounts for the key features of the mechanism initiating fracture at the micro-scale.

The Local Approach is based on the postulation that the probability of cleavage initiation within the vicinity of a crack/notch is described by the Weibull distribution. Accordingly, the parameters of this distribution are material constants that do not depend on its stress–strain state and temperature [1]. This approach describes the temperature dependence and the scatter limits of the fracture toughness $K_{IC}$. In this case, the temperature dependence of the value $K_{IC}$ is governed by the temperature dependence of the magnitude of yield strength. However, in the region of the ductile-to-brittle-transition, where there is a rapid increase in the fracture toughness against the background of a slight decrease in the yield strength, especially for high-strength steels, this approach underestimates the slope of the temperature dependence of the fracture toughness curve. In order to compensate for this shortcoming, one has to agree that shape parameter $m$ and scale factor $\sigma_u$ are not constants; their values depend on the temperature [2–8]. Moreover, the values of these parameters depend on specimen geometry [2,9,10] and local plastic strain value [6]. All this contradicts the initial statements of the Local Approach. A number of studies have attempted to solve this problem [11–16]. These works showed the need to address two key issues, namely: (i) the need to incorporate the effect of temperature and the magnitude of local plastic strain on the CN bulk density $\rho$; (ii) the need to account for the value of the threshold stress $\sigma_{th}$. As for the latter, in fact, this is a methodical problem. It consists of the development of a technique for the experimental determination of $\sigma_{th}$. Therefore, a simplified method for determining $\sigma_{th}$ for structural steels was proposed [17]. According to this technique, the value of $\sigma_{th}$ is determined based on the results of uniaxial tensile tests of 5–6 notched cylindrical specimens with special geometries (maximum diameter is 5 mm; minimum notched diameter is 5.2 mm; notch radius is 2 mm) at the boiling point of liquid nitrogen (−196 °C).

Ruggieri and Jivkov et al. [15,16] tried to incorporate the effect of plastic strain and test temperature on the number of CN forming in the local plastic zone ahead of a macrocrack tip. This was achieved by revising the expression for the Weibull stress. The nature of this revision was to consider the effect of both temperature and equivalent plastic strain on the CN ("cleavage initiators") number. This has been accomplished by introducing an empirical "thinning function" that makes it possible to predict "how many cleavage initiators must be generated relative to a reference case (for example all particles)" depending on temperature and the plastic strain value. The values of constants in this function were determined by a calibration procedure using experimental data. This enables one to predict, with high accuracy, the critical values of the $J_{IC}$ integral, both for low and high temperatures within the DBT region.

These studies have convincingly exhibited that the employment of LA within the DBT region requires the consideration of the temperature and deformation dependence of the CN bulk density. Here, in this contribution, the authors emphasised the need to explain the reasons for the significant effect of temperature on the number of formed "cleavage initiators". As attempts to answer this question, one considers the research results presented in [17–20], where a physical version of the LA to cleavage fracture was proposed and evidenced. This version of LA does not employ any empirical dependences for the probability of both CN formation and instability. These functions are derived from a detailed analysis of the processes of CN formation and unstable equilibrium in a polycrystalline aggregate. This approach enabled us to ascertain the influence of both the metal structure and the conditions of its loading on the probability of fracture. However, it turned out to be quite sophisticated for engineering calculations. At the same time, this approach can be used as a tool to analyse the key effects that control the initiation of cleavage in the vicinity of the macrocrack, and in particular to analyse the effect of temperature and the magnitude of plastic deformation on the CN formation rate inside of the "process zone".

The paper presents ideas about the physical nature of the effect of temperature and plastic strain value on the intensity of CN generation during the plastic deformation preceding cleavage. Physically substantiated dependences of CN density on temperature and the plastic strain value are obtained. To account for this effect within the LA, approximate dependences of CN density on temperature and the value of equivalent plastic strain are proposed. By the example of two structural steels, the suitability of using these dependences in predicting the value of fracture toughness within the DBT region is validated.

## 2. Theory

The inhomogeneity of microplastic deformation, which gives rise to plastic deformation incompatibility on grain and interphase boundaries, is a general reason for the CN formation in polycrystalline solids. A generalised model of CN formation in a polycrystalline aggregate was proposed [19,20].

The specific feature of this model is that it describes the formation and instability of the CN, accounting for both the main structural features of the polycrystal (inhomogeneous distribution of grain sizes and statistical distribution of their orientations) and for the action of microstresses generated by the incompatibility of microplastic deformations in polycrystal.

This model can be adopted for the prediction of the cleavage fracture in structural steels with ferritic basic microstructures arising from the expression for the bulk density ρ of CN formed at a given value of local plastic strain:

$$\rho = 2 \cdot \rho_{cb} \int_{t_C}^{t_r} g(t)\,dt \tag{1}$$

where

$$g(t) = \frac{1}{2\pi}\mathbf{exp}\left(\frac{t^2}{2}\right) \tag{2}$$

and $\rho_{cb}$ is the carbide particle density.

In this equation, $t$ is the magnitude of normalised shear microstresses $\xi_{ns}$ acting in slip systems:

$$t = \frac{\xi_{ns}}{\sqrt{D_{\xi_{ns}}}} \tag{3}$$

Here, $D_{\xi_{ns}}$ is the variance of shear microscopic stresses $\xi_{ns}$. Additionally $t_C$ is the critical value of normalised microstresses at which the CN are formed:

$$t_C = \frac{1}{k_\sigma} \cdot \left[M + \frac{1}{\bar{\sigma}} \cdot \sqrt{\frac{\xi_C}{Cd_{\max}}} - \beta \cdot \sqrt{\frac{\bar{e}}{d}}\right] \tag{4}$$

where $\bar{\sigma}$ and $\bar{e}$ are the equivalent macroscopic stresses and strains; $d$ and $d_{\max}$ are the average and maximum (with a given probability) ferritic grain sizes; $\xi_C$ is the critical stress of CN formation as a result of the carbide particle cleavage; $M$ is the orientation factor (for $\alpha - Fe\ M = 0.36$); $k_\sigma$, C, β are the coefficients (for ferritic steels $k_\sigma = 0.225$; C = 0.0336 N/m; $\beta \approx 2.57$ MPa $m^{0.5}$ [19]. Furthermore, $t_r$ is the critical value of normalised microstresses in which there is relaxation of incompatibilities in intergranular boundaries:

$$t_r = \frac{1}{k_\sigma} \cdot \left[M + \frac{1}{\bar{\sigma}}\left(\frac{\xi_Y}{m_b} \cdot \sqrt{\frac{r}{d_{\max}}} - \beta \cdot \sqrt{\frac{\bar{e}}{d}}\right)\right] \tag{5}$$

where $\xi_Y$ is the critical stress of the relaxation onset; $m_b$ is the orientation factor for relaxation slip systems in intergranular boundaries; $r$ is the distance from the grain boundary to the origin of microstress relaxation.

In Equation (5), the expression $\beta\sqrt{\bar{e}/d}$ specifies the value of shear microstresses caused by the interaction of a grain of average orientation $M$ with the surrounding matrix plastically deformed to strain $\bar{e}$. In polycrystals, these stresses arise as a result of the accumulation of microplastic strain incompatibilities at grain boundaries. The magnitude of the microplastic strain incompatibility changes nonmonotonically with an increase in macroplastic strain $\bar{e}$. At small macroplastic strains, it grows and then begins to decrease as a result of the predominance of relaxation processes in the near-boundary regions. A transition from growth to a decrease in these incompatibilities occurs at the critical strain $e_C$. At the macroscale, the value of $e_C$ can be estimated based on the dependence of the cleavage fracture stress of smooth (unnotched) specimens, $\sigma_f$, on the value of strain preceding fracture. When the strain $e_C$ is achieved, this stress $\sigma_f$ reaches its minimum value. For common structural steels with basic ferritic microstructures, $e_C \approx 0.02$ [10]. With an increase in ferrite grain size by annealing, the critical strain $e_C$ can increase to $\approx 0.05$ [10]. Thus, dependences (4) and (5) are valid for strains not exceeding the critical one $e_C$.

At strains $\bar{e}$, greater than $e_C$, the expressions for $t_C$ and $t_r$ may be presented as follows:

$$t_C = \frac{1}{k_\sigma} \cdot \left\{ M + \frac{1}{\bar{\sigma}} \left[ \sqrt{\frac{\xi_C}{Cd_{\max}}} - \beta \cdot \sqrt{\frac{e_C}{d}} + k_e \cdot \left( \frac{\bar{e}}{e_C} - 1 \right) \right] \right\} \tag{6}$$

where $k_e$ is the coefficient, characterising the intensity of relaxation process in the near-boundary regions. An estimate of this coefficient based on the [10] data (Figure 1) gives the value of $k_e \approx 1.52$ MPa m$^{0.5}$.

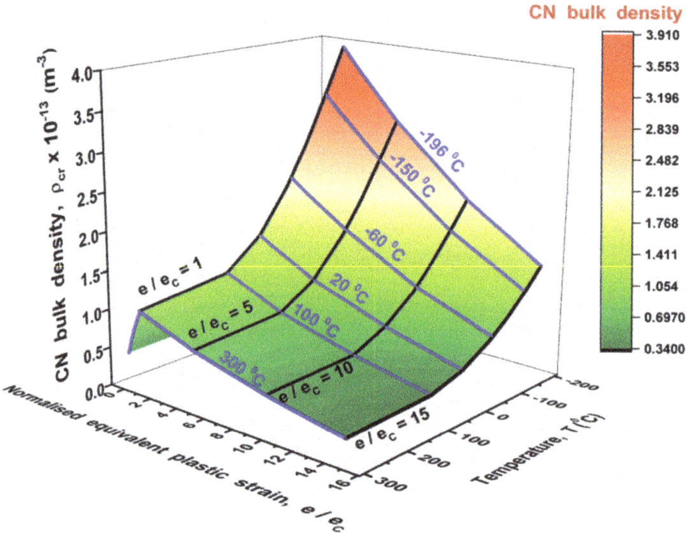


**Figure 1.** (Colour online). The effect of temperature $T$ and plastic strain value $\bar{e}$ on the CN density ρ for the RPV steel: $\bar{e}$ is the equivalent plastic strain; $e_C$ is the critical value of strain ($e_C = 0.02$).

Or, as the case may be, $t_r$:

$$t_r = \frac{1}{k_\sigma} \cdot \left\{ M + \frac{1}{\bar{\sigma}} \left[ \frac{\xi_Y}{m_b} \cdot \sqrt{\frac{r}{d_{\max}}} - \beta \cdot \sqrt{\frac{e_C}{d}} + k_e \cdot \left( \frac{\bar{e}}{e_C} - 1 \right) \right] \right\} \tag{7}$$

The theoretical dependences of CN bulk density ρ on the magnitude of plastic strain and temperature for reactor pressure vessel (RPV) RPV steel are shown in Figure 1.

The following values of microscopic parameters were used in their building: $r = 1$ µm (typical distance from grain boundary to stress relaxation source; this value is commensurate with the substructure parameters); $m_b = 0.1$ (the average value of orientation factor for the relaxation source); $d = 10$ µm and $d_{\max} = 30$ µm (mean and maximum ferrite grain values); $\xi_C = 7$ GPa ($\xi_C \approx 0.1G$ where $G \approx 70$ GPa is the iron carbide shear modulus) [21].

In the first approximation, the mean value of $\xi_Y$ may be estimated by the value of the thermally activated component of the yield strength $\xi_Y \approx \tau_Y$. The expression for $\tau_Y$ is the following [22]:

$$\tau_Y = 0.5 C_1 \mathbf{exp}\left[ -(C_2 - C_3 \mathbf{ln} \dot{e}) T \right] \tag{8}$$

where $\dot{e}$ is the plastic strain rate (for quasi-static tension $\dot{e} = 10^{-4} \cdot s^{-1}$); $C_1$, $C_2$ and $C_3$ are the constants, which values for typical ferritic steels are: $C_1$ = 1033 MPa, $C_2$ = 0.0068 $K^{-1}$, $C_3$ = 0.000415 $K^{-1}$ [22].

The equivalent stress value $\overline{\sigma}$ was calculated according to equation

$$\overline{\sigma} = \sigma_{0.2} \cdot \left( \frac{\overline{e}}{0.002} \right)^n \quad (9)$$

where $n$ is the work hardening exponent (value $n$ = 0.05 was used) and $\sigma_{0.2}$ is the yield strength:

$$\sigma_{0.2} = \sigma_a + C_1 \mathbf{exp}\left[ -\left( C_2 + C_3 \mathbf{ln} \dot{e} \right) T \right] \quad (10)$$

where $\sigma_a$ is the athermal component of $\sigma_{0.2}$.

For the investigated reactor pressure vessel (RPV) steel at room temperature ($T$ = 293 K), the yield strength was measured to be $\sigma_{0.2} = 610$ MPa. Accounting for Equation (10) and given the above $C_1$, $C_2$ and $C_3$ values, the athermal component of the yield strength is $\sigma_a$ = 564 MPa. The temperature dependences of both yield strength $\sigma_{0.2}$ and the work hardening exponent $n$ for RPV steel utilised in calculations are presented in Figure 2a.

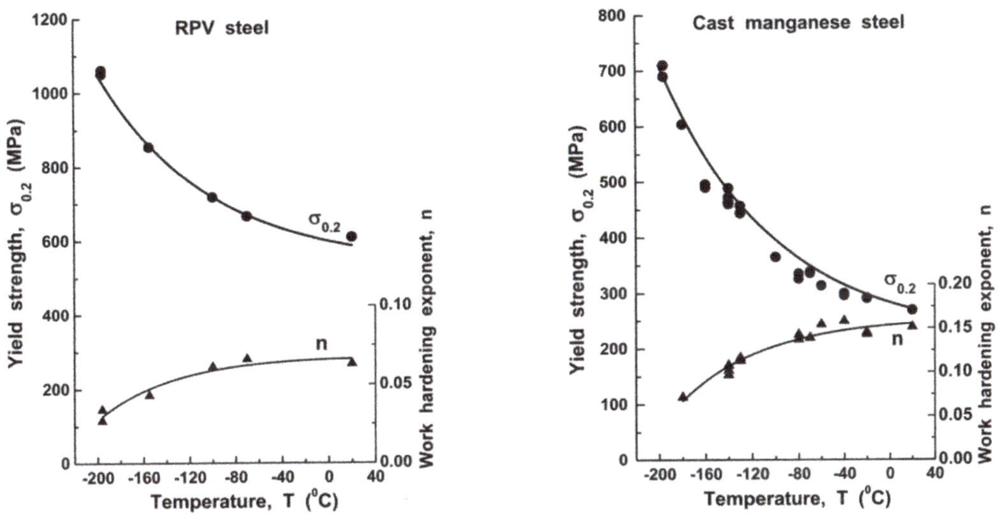


(a) (b)

**Figure 2.** The temperature dependences of yield strength $\sigma_{0.2}$ and work hardening exponent $n$ for RPV steel (**a**) and cast manganese steel (**b**).

According to the calculation results (Figure 1), the non-monotonic dependence of ρ on the plastic strain value is a characteristic feature of the CN formation in polycrystalline metals. As was shown above, this is due to non-monotonic change in shear microstresses, induced by microplastic strain incompatibility at the grain boundaries, with macroplastic strain $\overline{e}$ growth.

As the calculations show, the temperature increase should cause a monotonic decrease in the value of the CN density ρ. This is due to the thermally activated relaxation processes at the grain boundaries. In terms of the current model, this means a decrease in the value $t_r$ (dependences (7) and (5)) against the background of increasing the critical value of normalised stress of CN formation $t_C$ (dependences (6) and (4)).

Figure 1 shows that the effect of plastic strain magnitude and temperature on the CN bulk density ρ is a non-additive one. The maximum influence of plastic deformation on ρ is observed at strains close to the critical one $e_C$. With this deformation, the maximum sensitivity of ρ to temperature changes is also observed. At large strains, the sensitivity of ρ to changes in temperature decreases. In general, a comparison of the contribution of plastic strain and temperature to change in the intensity of the crack nuclei formation shows that the prevailing contribution is made by the change in temperature. An increase in temperature from +100 °C to −196 °C can lead to an almost fourfold increase in ρ. A decrease in temperature also increases the sensitivity of ρ to the plastic strain value. The dependences of CN density described above were obtained based on the analysis of the process of CN forming at microscale. Therefore, they enabled us to ascertain the fundamental regularities of the influence of temperature and plastic strain on the value of CN bulk density ρ, as well as determine the physical nature of this effect. However, these dependences *cannot be directly used* in LA since they contain microscopic parameters, the values of which can be estimated only with an accuracy of an order of magnitude. To incorporate these regularities into the LA, it is important to find the appropriate functions to approximate these dependences ρ on the temperature and the magnitude of plastic strain. The constants in these functions characterise the influence of microstructure parameters of specific steel on the sensitivity of ρ to both temperature and the magnitude of plastic strain. The values of these constants can be estimated from experimental evidence by means of calibration procedures.

In this case, as a first approximation, the following expressions can be used:

$$\rho = \rho_C - a \cdot \left(1 - \frac{\bar{e}}{e_C}\right) \text{ for } \bar{e} \le e_C \tag{11}$$

and

$$\rho = \rho_C - b \cdot \left(\frac{\bar{e}}{e_C} - 1\right) \text{ for } e_{\mathbf{max}} \ge \bar{e} \ge e_C \tag{12}$$

where

$$\rho_C = \rho_0 \cdot [1 - \mathbf{exp}(-\alpha\tau_Y)] \tag{13}$$

In the above equation, α and $\rho_0$ are coefficients, the values of which are calculated using experimental evidence by the calibration procedure, $e_C$ = 0.02 and $e_{\max} \le 0.3 \div 0.5$.

The dependences in Figure 1 give the following values: $a$ = 1.50 × $10^{13}$ $m^{-3}$; $b \approx 0.12 \times 10^{13}$ $m^{-3}$.

## 3. Experiment

Experimental investigations were performed on a reactor pressure vessel steel and on a low–alloyed manganese steel. To determine the value of fracture toughness, standard (ASTM E1921-05 [23]) specimens 1T CT (for the reactor pressure vessel steel) and 1T SENB (single edge notched bending) (for the low–alloyed manganese steel) were used. Chemical composition of the steels is presented in Table 1. More details on materials used can be found in our previous works, e.g., [17,24,25]. Both steels were investigated as received and had a ferritic structure. Mechanical tests of the pre-cracked specimens were executed in the DBT temperature range (Figure 3). The specimens were tested under crosshead speed of 1 mm/min in either environmental and/or cryogenic chamber. Experiments were carried out on a servo-hydraulic testing machine, Instron 8802. The value of $K_{Jc}$ was estimated according to the ASTM E1921-05 standard. According to this standard, the maximum limiting the valid values of elastic plastic fracture toughness, $K_{Jc(\text{limit})}$, was calculated according to the equation:

$$K_{Jc(\text{limit})} = \sqrt{\frac{Eb_0\sigma_{0.2}}{30(1-\nu^2)}} \tag{14}$$

where $E$ is the Young modulus, ($E$ = 210 GPa); $\nu$ is Poisson's ratio ($\nu$ = 0.27); $b_0$ is the unbroken ligament ($b_0$ = 22.4 mm); $\sigma_{0.2}$ is the yield strength (the temperature dependences for $\sigma_{0.2}$ of the investigated steels are presented in Figure 2). The temperature values of $K_{Jc(\mathrm{limit})}$ are presented in Figure 4. To determine the mechanical properties of the investigated steels and obtain true stress vs. true deformation curves, low-temperature tests for uniaxial tension of smooth (unnotched) specimens were performed. According to the results of these tests, the values of the yield stress and the strain hardening exponent were determined (Figure 2).

**Table 1.** Chemical composition of the investigated steels (wt %, remainder is *Fe*).

| C | Mn | Si | Cr | Ni | Mo | Cu | S | P | V | As | Co | Sn | Sb |
|---|---|---|---|---|---|---|---|---|---|---|---|---|---|
| | | | | | | | **RPV Steel** | | | | | | |
| 0.17 | 0.46 | 0.27 | 2.18 | 1.19 | 0.54 | 0.13 | <0.01 | 0.01 | 0.10 | <0.003 | <0.03 | 0.006 | <0.003 |
| | | | | | | | **Cast Ferritic Steel** | | | | | | |
| 0.09 | 1.18 | 0.37 | 0.12 | 0.29 | 0.03 | 0.29 | 0.025 | 0.01 | N/A | N/A | N/A | N/A | N/A |

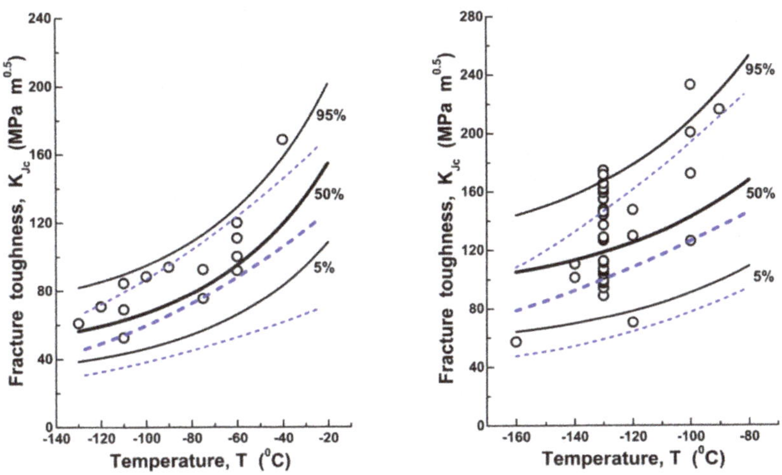


(a) (b)

**Figure 3.** (Colour online). The temperature dependences of fracture toughness $K_{Jc}$ for RPV steel (**a**) and cast low-alloyed manganese steel (**b**): circles represent the experimental values; lines are the calculation results (solid black lines are the results of $K_{Jc}$ calculation accounting for temperature dependence for the CN density $\rho$, dashed blue lines are the same at the constant values of $\rho$).

According to the results of these tests, the magnitudes of Weibull parameters were determined by calibration (shape parameter $m$ and scale parameter $\sigma_u$) as well as the values of $\rho_0$ and $\alpha$. To determine local parameters, a program was created that allows the calculation of local parameters using the maximum likelihood method for a three-parameter modification of the Beremin approach. This is a typical procedure for the Local Approach. The difference is only that $V_0$ is not constant but depends on $\rho$ ($V_0 = \sqrt{1/\rho}$),the value of which is determined by dependences (11) and (12).

The magnitude of threshold stress $\sigma_{th}$ was estimated by the technique described in [17] (Table 2).

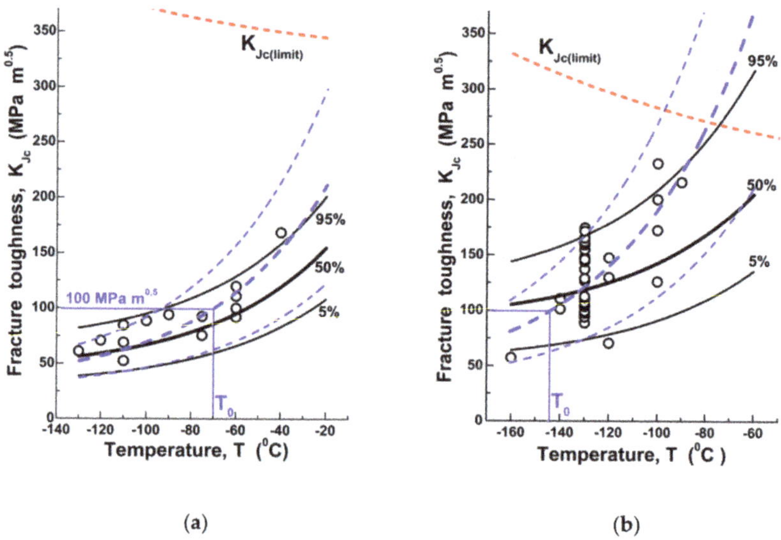


**Figure 4.** (Colour online). The temperature dependences of fracture toughness $K_{Jc}$ for RPV steel (**a**) and cast low-alloyed manganese steel (**b**): circles show the experimental evidence, lines are the calculation results (solid black lines are the results of $K_{Jc}$ calculation accounting for the CN density ρ and dashed blue lines are the results of $K_{Jc}$ calculation by the Master Curve method); $T_0$ is the reference temperature (short-dashed red lines are the temperature dependences of elastic plastic fracture toughness validity limit $K_{Jc(\text{limit})}$), calculated according to Equation (14).

**Table 2.** Mechanical properties of the tested steels ($\sigma_{0.2}$ is the yield strength determined as 0.2 pct proof stress; $\sigma_{ul}$ is the ultimate tensile strength; $e_u$ is the uniform elongation; ψ is the reduction in area—all at room temperature; $n$ is the work hardening exponent).

| Material | $\sigma_{0.2}$ (MPa) | $\sigma_{ul}$ (MPa) | $e_u$ (---) | ψ (---) | $n$ (---) |
|---|---|---|---|---|---|
| RPV steel | 610 | 714 | 0.07 | 0.75 | 0.06 |
| Cast steel | 319 | 481 | 0.26 | 0.74 | 0.15 |

## 4. Results and Discussion

Relatively small $m$ values ($m \leq 10$) are the consequence of accounting for the values of threshold in the calculation [26]. During the calibration procedure, it was possible to find such a set of Weibull parameters, at which the values of $\rho_0$ and $\alpha$ for both steels are almost the same (Table 3). This validates the ascertained values $\rho_0$ and $\alpha$ as those inherent to ferritic steels. In this case, the features of the microstructure of steels are accounted for by the values of the Weibull parameters. It should be emphasised that accounting for the ρ temperature dependence ρ (in terms of the Weibull distribution—for the parameter $V_0$ ($\rho = V_0^{-1}$) gives rise to the independence (invariance) of the Weibull parameters $\sigma_u$ and $m$ on temperature. This allows us to consider $\sigma_u$ and $m$ as material parameters. Contrarily, in the conventional versions of Local Approach, it is assumed that the quantity $V_0$ is constant. This leads to temperature dependences of parameters $\sigma_u$ and $m$, which contradicts the physical sense and complicates calculations.

**Table 3.** Calculation findings ($\sigma_{th}$ is the threshold stress; $m$ and $\sigma_u$ are the shape and scale parameters; $\rho_0$ and $\alpha$ are the coefficients; $T_0$ is the reference temperature).

| Material | $\sigma_{th}$ (MPa) | $\sigma_u$ (MPa) | $m$ | $\rho_0 \times 10^{13}$ (m$^{-3}$) | $\alpha$ (MPa$^{-1}$) | $T_0$ (°C) |
|---|---|---|---|---|---|---|
| RPV steel | 1100 | 6835 | 5.4 | 4 | 0.024 | −70 |
| Cast steel | 720 | 3700 | 8.0 | 4 | 0.020 | −145 |

Based on these data, temperature dependences of fracture toughness $K_{Jc}$ for the studied steels were plotted (Figure 3a,b). The algorithm of calculations of fracture toughness and fracture probabilities was as follows:

1. For each $j^{th}$ stage of loading $K_I^j$, the values of equivalent plastic strain $\bar{e}^i$ and normal tensile stresses $\sigma_{YY}^i$ in each $i^{th}$ finite element were calculated using the finite elements method (software ABAQUS, version 2018).

2. Further, the probability of the cleavage initiation in the $i^{th}$ element was calculated as

$$P_i = 1 - \mathbf{exp}\left[-\frac{V^i}{V_0^i} \times \left(\frac{\sigma_{YY}^i - \sigma_{th}}{\sigma_u}\right)^m\right] \tag{15}$$

where $V^i$ is the finite element volume and $V_0^i$ is the volume per one crack nucleus ($\rho = V_0^{-1}$). The value of $\rho^i$ was calculated according to Equations (11)–(13).

3. Then, the total probability of cleavage initiation in specimen is determined as

$$P_\Sigma = 1 - \prod_{n=1}^{n=N} [1 - P_i(n)] \tag{16}$$

where $N$ is the number of finite elements belonging to the yield region.

The step size and the number of loading stages were chosen in such a way as to obtain $K_I$ values when the probabilities of cleavage initiation $P_\Sigma$ are 5%, 50% and 95%.

In accordance with the calculation results, the theoretical curve describes the temperature dependence of fracture toughness $K_{Jc}$ with a 90% probability scatter band (Figure 3). It should be noted that such a coincidence is due to the fact that not only a decrease in yield strength, but also a reduction in the CN generation rate is the reason for $K_{Jc}$ growth with increasing temperature. If we do not take into account this fact, i.e., if we assume that $\rho = const$, as the case may be, $V_0 = const$ ($V_0 = 1/\rho$), then the slope of $K_{Jc}$ temperature dependence becomes unbelievably low (Figure 3, dashed lines). A particularly significant error occurs at low values of the failure probability, i.e., when determining $K_{Jc}$ threshold values). This is especially important for applications. It should be mentioned that accounting for $\rho$ temperature dependence provides invariance to temperature of the Weibull distribution parameters, which agrees well with the LA concept that considers $m$ and $\sigma_u$ as material parameters.

The proposed approach can also be used to analyse the Master Curve method. For this purpose, the values of reference temperatures $T_0$ were determined for the studied steels by the multitemperature method, and, according to the ASTM E1921-05 standard, the temperature dependences of fracture toughness were plotted (Figure 4).

According to these data, the Master Curve method overestimates the value $K_{Jc}$ at temperatures above the reference one $T_0$. This means the need to "correct" the width of the temperature range near $T_0$, within which the approximation of experimental evidence by the Master Curve method, is valid. The approach accounting for the temperature dependence of CN density formation may be used as a tool to fix this problem.

Fracture toughness is a stochastic characteristic. This is most pronounced within the DBT region. Therefore, it is necessary to pay attention to errors not only of $K_{Jc}$ absolute values, but also to errors of probabilities of their realisation. Figure 4 exhibits that the error in the $K_{Jc}$ value due to neglecting of $\rho$ temperature dependence is in the tens of percent.

At the same time, the error in probabilities of realisation of the given $K_{Jc}$ values can be quite high.

## 5. Conclusions

1. In a general case, the incompatibility of microplastic deformations at grain or interphase boundaries causes the crack nuclei (CN) formation. The value of this incompatibility depends on the magnitude of plastic strain and temperature. This is the reason for the influence of these factors on the intensity of the crack nuclei generation. The microscopic model presented in the work enables one to describe, quantitatively, the regularities of this effect.
2. The strain and temperature dependence of the CN bulk density results in the invariance of the Weibull distribution parameters both to the temperature and the plastic strain value, i.e., $\sigma_u$ and $m$, may be considered as material parameters. This is in line with the original concepts underlying the Local Approach.
3. Decreasing CN generation intensity with temperature growth is one of the key factors that governs the slope of $K_{Jc}$ temperature dependence and its scatter limits. Ignoring this effect underestimates the slope of $K_{Jc}$ temperature dependence and their absolute values. Especially large, up to 20–30%, is the error at low probabilities $P_\Sigma$ = 5% fracture (lower fracture toughness threshold).
4. The Master Curve method, vice versa, overestimates $K_{Jc}$ sensitivity to temperature. This may result in the overestimation of the fracture toughness value at temperatures higher than the reference one $T_0$.

**Author Contributions:** Conceptualization, S.K. and I.D.; methodology, S.K.; software, V.K.; validation, O.Z. and G.Z.; formal analysis, N.S.; investigation, O.Z.; data curation, G.Z.; writing—original draft preparation, S.K.; writing—review and editing, I.D.; visualization, N.S.; project administration, S.K.; funding acquisition, I.D. All authors have read and agreed to the published version of the manuscript.

**Funding:** This research was funded by the National Academy of Sciences of Ukraine, grant number 0121U107569; the APC is funded by Institute of Materials Science and Engineering, Brno University of Technology, Brno, Czech Republic.

**Institutional Review Board Statement:** Not applicable.

**Informed Consent Statement:** Not applicable.

**Data Availability Statement:** Data supporting the findings of this study are available from the corresponding author upon request.

**Conflicts of Interest:** The authors declare no conflict of interest.

## References

1. Beremin, F.M.; Pineau, A.; Mudry, F.; Devaux, J.-C.; D'Escatha, Y.; Ledermann, P. A local criterion for cleavage fracture of a nuclear pressure vessel steel. *Met. Trans. A* **1983**, *14*, 2277–2287. [CrossRef]
2. Wiesner, C.S.; Goldthorpe, M.R. The effect of temperature and specimen geometry on the parameters of the 'Local Approach' to cleavage fracture. *J. Phys. IV* **1996**, *6*, C6–295. [CrossRef]
3. Wasiliuk, B.; Petti, J.R.; Dodds, R.H. Temperature dependence of Weibull stress parameters: Studies using euro-material. *Eng. Fract. Mech.* **2006**, *73*, 1046–1059. [CrossRef]
4. Pineau, A. Development of the local approach to fracture over the past 25 years: Theory and applications. *Int. J. Fract.* **2006**, *138*, 139–166. [CrossRef]
5. Ruggieri, C.; Dodds, R.H. A local approach to cleavage fracture modelling: An overview of progress and challenge for engineering applications. *Eng. Fract. Mech.* **2018**, *187*, 381–403. [CrossRef]
6. Gao, X.; Dodds, R.H., Jr. Constraint effects on the ductile-to-brittle transition temperature of ferritic steels: A Weibull stress model. *Int. J. Fract.* **2000**, *102*, 43–69. [CrossRef]
7. Qian, G.; González-Albuixech, V.F.; Niffenegger, M. Calibration of Beremin model with the Master Curve. *Eng. Fract. Mech.* **2015**, *136*, 15–25. [CrossRef]
8. Gao, X.; Zhang GSrivatsan, T.S. A probabilistic model for prediction of cleavage fracture in the ductile-to-brittle transition region and the effect of temperature on model parameters. *Mat. Sci. Eng.* **2006**, *415*, 264–272. [CrossRef]

9. Cao, Y.; Hui, H.; Wang, G.; Xuan, F.-Z. Inferring the temperature dependence of Beremin cleavage model parameters from the Master Curve. *Nucl. Eng. Des.* **2011**, *241*, 39–45. [CrossRef]
10. Kotrechko, S.; Strnadel, B.; Dlouhý, I. Fracture toughness of cast ferritic steel applying local approach. *Appl. Fract. Mech.* **2007**, *47*, 171–181. [CrossRef]
11. Bordet, S.R.; Karstensen, A.D.; Knowles, D.M.; Wiesner, C.S. A new statistical local criterion for cleavage fracture in steel. Part I: Model presentation. *Eng. Fract. Mech.* **2005**, *72*, 435–452. [CrossRef]
12. Bordet, S.R.; Karstensen, A.D.; Knowles, D.M.; Wiesner, C.S. A new statistical local criterion for cleavage fracture in steel. Part II: Application to an offshore structural steel. *Eng. Fract. Mech.* **2005**, *72*, 453–474. [CrossRef]
13. Gao, X.; Zhang, G.; Srivatsan, T.S. Prediction of cleavage fracture in ferritic steel: A modified Weibull stress model. *Mat. Sci. Eng. A* **2005**, *394*, 210–219. [CrossRef]
14. Ruggieri, C.; Savioli, R.G.; Dodds, R.H., Jr. An engineering methodology for constraint corrections of elastic–plastic fracture toughness—Part II: Effects of specimen geometry and plastic strain on cleavage fracture predictions. *Eng. Fract. Mech.* **2015**, *146*, 185–209. [CrossRef]
15. Ruggieri, C.; Jivkov, A.P. A local approach incorporating the measured statistics of microcracks to assess the temperature dependence of cleavage fracture for a reactor pressure vessel steel. *Proc. Struct. Int.* **2019**, *18*, 28–35. [CrossRef]
16. Jivkov, A.P.; Burgos, D.S.; Ruggieri, C.; Beswick, J.; Savioli, R.G.; James, P.; Sherry, A. Use of local approaches to calculate changes in cleavage fracture toughness due to pre-straining and constraint effects. *Appl. Fract. Mech.* **2019**, *104*, 102380. [CrossRef]
17. Kotrechko, S.; Zatsarna, O.; Kozák, V.; Dlouhý, I. Threshold fracture stress: Theory and application. *Proc. Struct. Int.* **2019**, *23*, 413–418. [CrossRef]
18. Kotrechko, S. Physical Fundamentals of Local Approach to Analysis of Cleavage Fracture. In *Transferability of Fracture Mechanical Characteristics*; Dlouhý, I., Ed.; NATO Science Series (Series II: Mathematics, Physics and Chemistry); Springer: Dordrecht, The Netherlands, 2002; Volume 78, pp. 135–150. [CrossRef]
19. Kotrechko, S. The key problems of local approach to cleavage fracture. *J. Theor. Appl. Mech.* **2013**, *51*, 75–89.
20. Kotrechko, S.; Mamedov, S. Multi-scale local approach to cleavage fracture and its applications. In Proceedings of the 19th European Conference on Fracture (ECF19), Kazan, Russia, 26–31 August 2012; Curran Associates, Inc.: New York, NY, USA, 2016; pp. 971–982.
21. Mizubayashi, H.; Li, S.J.; Yumoto, H.; Shimotomai, M. Young's modulus of single phase cementite. *Scr. Mat.* **1999**, *40*, 773–777. [CrossRef]
22. Zerilli, F.J.; Armstrong, R.W. Dislocation-mechanics-based constitutive relations for material dynamics calculations. *J. Appl. Phys.* **1987**, *61*, 1816–1825. [CrossRef]
23. Standard Test Method for Determination of Reference Temperature, $T_0$, for Ferritic Steels in the Transition Range. Annual book of ASTM Standards 2005; Section 3: Metals Test Methods and Analytical Procedures; v.03.01: Metals–Mechanical Testing, Elevated and Low-Temperature Tests, Metallography. Available online: https://www.astm.org/Standards/E1921 (accessed on 18 October 2021).
24. Holzmann, M.; Jurášek, L.; Dlouhý, I. Master Curve Methodology and Data Transfer from Small on Standard Specimens. In *Transferability of Fracture Mechanical Characteristics*; Dlouhý, I., Ed.; NATO Science Series (Series II: Mathematics, Physics and Chemistry); Springer: Dordrecht, The Netherlands, 2002; Volume 78, pp. 225–242. [CrossRef]
25. Dlouhy, I.; Chlup, Z.; Kozak, V. Constraint effects at brittle fracture initiation in a cast ferritic steel. *Eng. Fract. Mech.* **2004**, *71*, 873–883. [CrossRef]
26. Ruggieri, C. Influence of threshold parameters on cleavage fracture predictions using the Weibull stress model. *Int. J. Fract.* **2001**, *110*, 281–304. [CrossRef]

*materials*

MDPI

*Article*

# Study on Microstructure and Fatigue Properties of FGH96 Nickel-Based Superalloy

**Yishan Bai [1], Shanglei Yang [1,2,*], Minqi Zhu [1] and Cong Fan [1]**

[1] School of Materials Engineering, Shanghai University of Engineering Science, Shanghai 201620, China; baiyishan1997@163.com (Y.B.); zhuminqi1997527@163.com (M.Z.); fc19960802@163.com (C.F.)
[2] Shanghai Collaborative Innovation Center for Laser Advanced Manufacturing Technology, Shanghai University of Engineering Science, Shanghai 201620, China
* Correspondence: yslei@sues.edu.cn

**Abstract:** In this study, using synchrotron radiation X-ray imaging, the microstructure, tensile properties, and fatigue properties of FGH96 nickel-based superalloy were tested, and the fatigue damage mechanism was analyzed. An analysis of the experimental results shows that the alloy structure is dense without voids or other defects. It was observed that the primary $\gamma'$ phase is distributed on the grain boundary in a chain shape, and the secondary $\gamma'$ phase is found inside the crystal grains. The X-ray diffraction (XRD) pattern indicates that no other phases were seen except for the $\gamma$ and $\gamma'$ phases. The tensile strength of the alloy is 1570 MPa and the elongation is 12.1%. Using data fitting and calculation, it was found that the fatigue strength of the alloy under the condition of $5 \times 10^6$ cycles is 620.33 MPa. A fatigue fracture has the characteristics of secondary crack, cleavage step, fatigue stripe, tire indentation, and dimple. The fracture is a mix of cleavage fracture and ductile fracture. Through a three-dimensional reconstruction of the alloy synchrotron radiation imaging area, it was found that the internal defects are small and mostly distributed at the edge of the sample. The dimple morphology is formed by cavity aggregation and cavity germination resulting from defects in the material itself, fracture of the second-phase particles, and separation of the second-phase particles from the matrix interface. By analyzing the damage mechanism of fatigue fractures, it is concluded that the cleavage step is formed by the intersection of cleavage planes formed by branch cracks, with the main crack of the confluence extending forward to form a cleavage fracture. The crack propagation path was also analyzed, and under the action of cyclic load and tip passivation, the crack shows Z-shaped propagation.

**Keywords:** synchrotron radiation X-ray imaging; fatigue performance; fatigue crack; FGH96 alloy; microstructure

**Citation:** Bai, Y.; Yang, S.; Zhu, M.; Fan, C. Study on Microstructure and Fatigue Properties of FGH96 Nickel-Based Superalloy. *Materials* **2021**, *14*, 6298. https://doi.org/10.3390/ma14216298

Academic Editors: Jaroslav Pokluda and Reinhard Pippan

Received: 23 September 2021
Accepted: 18 October 2021
Published: 22 October 2021



## 1. Introduction

The turbine disk is one of the most important components of the aero engine, and its working environment requires a high level of performance [1,2]. Powder metallurgy FGH96 nickel-based superalloy is used in the manufacture of turbine disks in the Chinese aviation industry due to its uniform structure, fine grain strengthening, high tensile strength, and high fatigue properties [3–6]. In the practical application of turbine disks, the fluctuation of air flow and the sharp changes in pressure produced by aircraft take-off and landing means there are higher requirements for the fatigue strength of turbine disks, and it is necessary to ensure that the turbine disk has sufficient anti-fatigue ability [7–9]. There is still a large gap between China and other countries in the manufacturing, microstructure and properties, service, and fatigue reliability of aviation turbine disk powder nickel-based superalloy, particularly with regard to the manufacturing process, impurity control, and fatigue properties [10–12].

Scholars at home and abroad have done much research on the fatigue properties of nickel-based superalloys. Hu et al. [13] studied the influence of the location of inclusions on the low-cycle fatigue life of the FGH96 turbine disk. The experimental results show that surface defects such as surface scratches and surface inclusions have a greater influence on the fatigue life, while the fatigue life of inclusions in the sample is usually longer. Jiménez et al. [14] studied the crack initiation and propagation of IN718 nickel-based superalloy under low-cycle fatigue by means of auxiliary in situ diffraction and phase contrast techniques. The results show that the propagation of short cracks in the microstructure is determined by the slip transfer path through the pre-existing grain boundary.

The synchrotron radiation X-ray imaging technology of the Shanghai light source can obtain the three-dimensional structure of defects without destroying the material structure [15,16]. The large advantage in the internal detection of materials has not been widely studied in the field of engineering application and applied basic research. There are few international and domestic researchers, and the test materials that have been studied are mostly aluminum alloy and titanium alloy [17–19].

In this paper, the size, shape, and distribution of the defects in FGH96 nickel-based superalloy are analyzed based on synchrotron radiation X-ray imaging. Testing the fatigue performance of the PM (powder metallurgy) FGH96 nickel-based superalloy turbine disk, determining the fatigue life limit of the material, and studying the fatigue damage behavior of the alloy have important basic theoretical significance for the development of new PM superalloy turbine disk materials and the safe operation of civil aircraft.

## 2. Experimental Materials and Methods

The experimental material is the FGH96 nickel-based superalloy aviation turbine disk, which was provided by Aviation Industry Corp (Beijing, China). The element content of the material is shown in Table 1.

**Table 1.** Chemical composition of FGH96 nickel-based superalloy (mass fraction, wt%).

| Material | Mass Fraction, % | | | | | | | | | | | |
|---|---|---|---|---|---|---|---|---|---|---|---|---|
| | C | Cr | Co | W | Mo | Ce | Al | Ti | Zr | Nb | B | Ni |
| FGH96 | 0.04 | 15.5 | 13.4 | 3.9 | 3.7 | 0.01 | 2.3 | 3.6 | 0.03 | 0.7 | 0.08 | Banlance |

The fatigue test dimensions were designed according to the national standard GB/t3075-2008, as shown in Figure 1a, the tensile test dimensions were designed according to GB/T228-2002, as shown in Figure 1b, and the synchrotron radiation X-ray computed tomography (SR-μCT) dimensions were designed according to the requirements of the Shanghai light source test samples, as shown in Figure 1c. The three samples were obtained by wire cutting. The microstructure observation, hardness test, fatigue test, fatigue, and tensile specimen fracture observation test were carried out on the material. The test methods and test equipment are shown in Table 2 below.

**Table 2.** Test, test equipment, and test methods.

| Test | Test Equipment | Test Method (Test Purpose) |
|---|---|---|
| Hardness testing | HV-1000 micro hardness meter (Shanghai Taiming Optical Instrument Company, Shanghai, China) | Hardness test every 0.25 mm |
| Microstructure observation | VHX-6000 super depth of field microscope (KEYENCE) S-3800 field emission scanning electron microscope (Japan Hitachi Co., Ltd., Tokyo, Japan) | The surface of the sample was polished and etched with corrosive agent (5 g $CuCl_2$ + 100 mL HCl + 100 mL $C_2H_5OH$) for 2 min |
| Fatigue test | Zwick Amsler 250 HFP 5100 testing machine (ZwickRoell) | The fatigue cycle mode is sine wave, the loading frequency is 20 Hz, and the stress ratio R = 0.1 |
| Tensile test | | The tensile load is uniform until the specimen is broken |
| Observation of fracture morphology | S-3800 field emission scanning electron microscope | The tensile and fatigue fracture surfaces were observed and the fatigue damage mechanism was analyzed |
| Phase analysis | S-3800 field emission scanning electron microscope | Phase composition of alloy surface was analyzed |

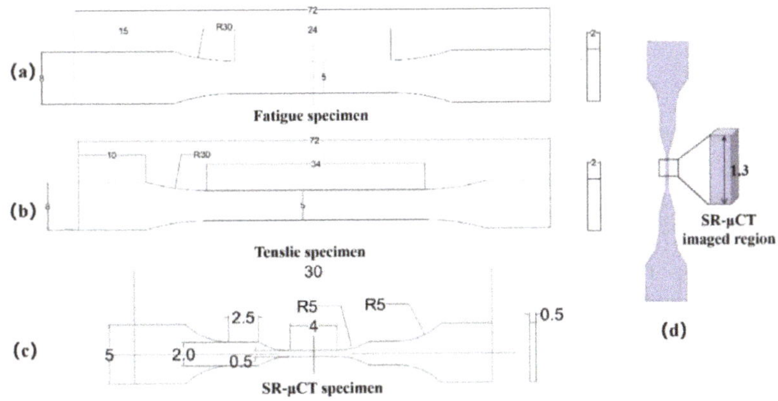


**Figure 1.** Specimen size: (**a**) fatigue specimen, (**b**) tensile specimen, (**c**) SR-μCT specimen, (**d**) synchrotron radiation imaging region (dimensions in mm).

The internal defects of the material were detected at BL13W1 of Shanghai synchrotron radiation source (see Figure 2). The SR-μCT sample was placed 30 cm away from the 4.0 SCMOS detector of a Hamamatsu flash lamp. To ensure penetration, the photon energy was 42 keV. The space pixel size was 0.65 μm. The exposure time was 4 s, and the imaging area is shown in Figure 1d. In the process of X-ray imaging, the sample is usually rotated 180° around the vertical axis, the rotation angle increment is 0.25°, and 720 projections can be collected at one time. Pitre and P3B software (PITRE V. 3.0, 2013, Shanghai Institute of Applied Physics, Chinese Academy of Sciences) were used to convert the data into 8-bit slices, and Avizo software (Avizo 9.5, 2019, Thermo Fisher Scientific, Waltham, MA, USA) was used to reconstruct the internal defects.

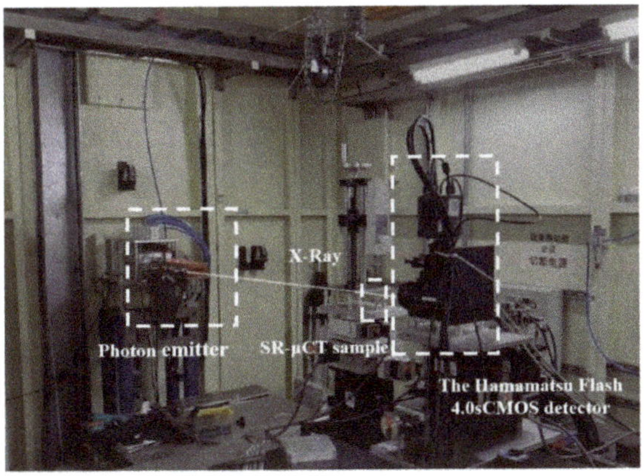


**Figure 2.** Image of the experimental device for detecting internal defects in the alloy at BL13W1.

## 3. Result Analysis and Discussion

### *3.1. Microstructure Analysis*

Figure 3 shows the microstructure of FGH96 nickel-based superalloy. Figure 3a shows the microstructure of 500 folds of the super depth of field. It can be seen from the figure that the grains are evenly distributed, the structure is dense, and there are no voids or cavities. According to the data, the primary $\gamma'$ phase diameter is approximately 200 nm, the secondary $\gamma'$ phase diameter is 50–200 nm, and the three times $\gamma'$ phase diameter is less than 50 nm [20]. Figure 3b is a 15,000 folds scanning electron microscopy (SEM) image. From Figure 3b, it can be seen that the primary $\gamma'$ phase is distributed on the grain boundary in a chain shape, and the secondary $\gamma'$ phase is uniformly distributed inside the grain boundary. Figure 3c is a 20,000 folds SEM image, in which the three times $\gamma'$ phase cannot be found. An energy dispersive spectrometer (EDS) analysis of the precipitates on the surface of microstructure was carried out, and the analysis results are shown in Figure 3d. The carbon content cannot be measured in the EDS analysis of the alloy. It can be judged that the material is $M_{23}C_6$ [16] by the presence of Cr and Mo.

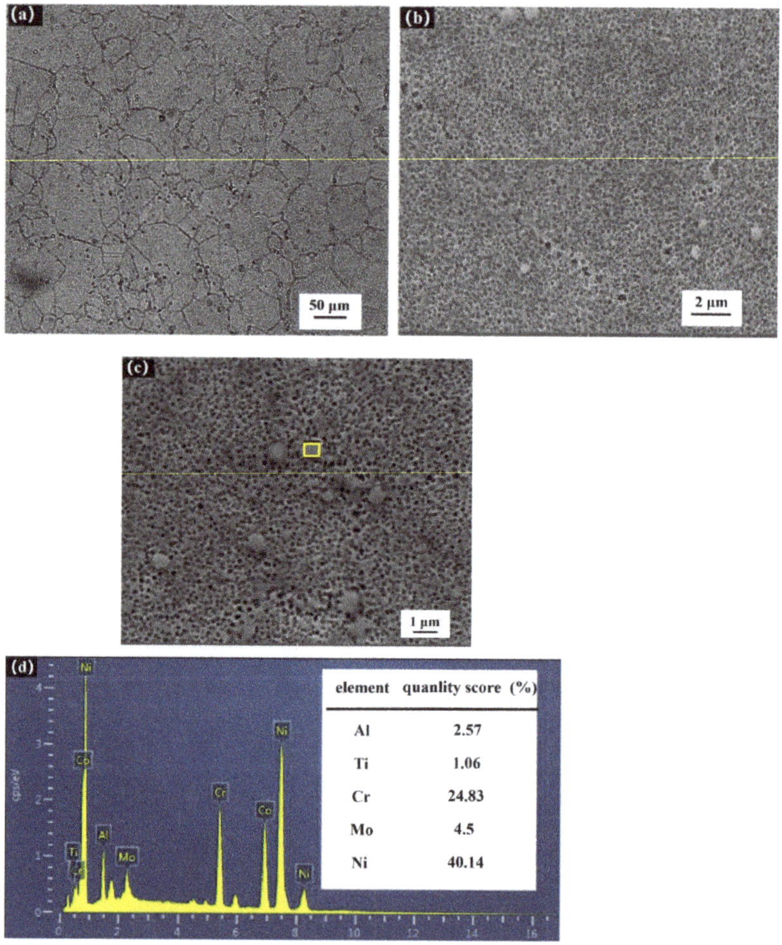


**Figure 3.** Microstructure of FGH96 alloy: (**a**) light microscope structure, (**b**) primary $\gamma'$ phase and secondary $\gamma'$ phase, (**c**) carbide $M_{23}C_6$, (**d**) EDS analysis results.

An XRD (X-ray diffraction) analysis of the surface structure of FGH96 nickel-based superalloy shows that it is mainly composed of $\gamma$ and $\gamma'$. The phase composition is shown in Figure 4, and the diffraction pattern shows that there are no other phases in the alloy.

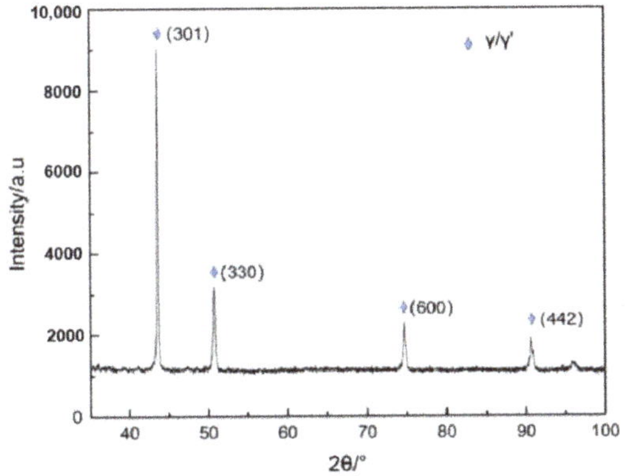


**Figure 4.** XRD diffraction pattern.

### 3.2. Mechanical Property Analysis

The room temperature hardness of the FGH96 nickel-based superalloy sample was measured. To reduce the hardness measurement error and avoid accidental data, it was measured every 0.25 mm from the first point. A total of 20 points were tested, as shown in Figure 5a. It can be seen from the figure that the hardness value does not change much, indicating that the $\gamma'$ phase distribution on the alloy surface is relatively uniform. After calculating the hardness data of multiple measurements, the average room temperature hardness of the FGH96 nickel-based superalloy is 470.6 HV.

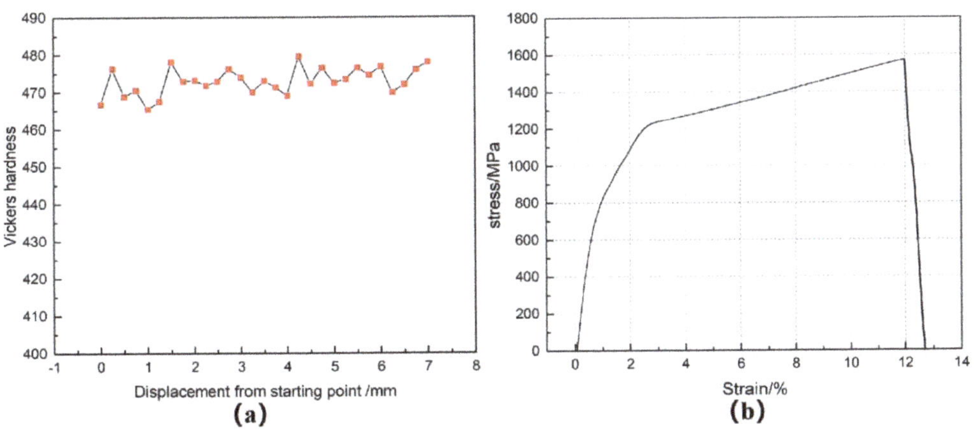


**Figure 5.** (**a**) Microhardness curve, (**b**) tensile stress–strain curve.

The tensile sample was tested. The relationship between material stress and strain is shown in Figure 5b. It is concluded that the tensile strength of FGH96 nickel-based superalloy is 1570 MPa and the elongation is δ 12.1%. FGH96 nickel-based superalloy relies on Ni to dissolve Cr, Co, W, Mo, and other alloy elements to strengthen the solid solution, weaken the element diffusion, form high-temperature resistant atomic groups, and reduce the stacking fault energy. At the same time, a large number of solute atoms play a pinning role in the dislocation aggregation [21] to reduce the dislocation activity. When the dislocation cuts through $\gamma'$, a higher external force is required to produce obvious second-phase strengthening. Through these two strengthening methods, the material has high tensile strength. In the plastic deformation process of FGH96 nickel-based superalloy, it mainly depends on the $\gamma'$ precipitate phase to hinder the dislocation movement. The tensile fracture morphology of FGH96 nickel-based superalloy is shown in Figure 6. In Figure 6a, the secondary crack (in the yellow ellipse) can be observed. The secondary crack grows inward perpendicular to the tensile fracture surface, which shows that the material has a certain brittleness in the tensile process. In Figure 6b, there are many equiaxed dimples, and the large number of dimples shows that the plasticity of the material is high.

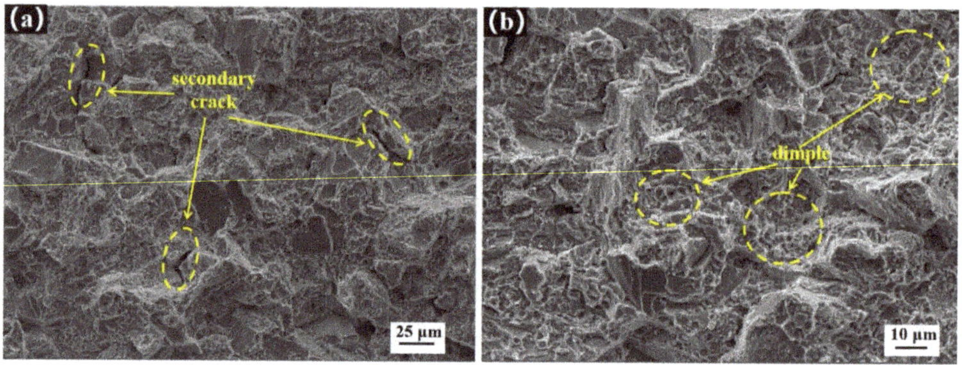


**Figure 6.** Tensile fracture morphology: (**a**) secondary cracks, (**b**) equiaxed dimples.

### 3.3. Fatigue Performance Analysis

According to the national standard GB/T15248-2008 "metallic materials axial constant-amplitude low-cycle fatigue test method", the fatigue test was carried out using the lifting method, and the data in Table 3 were obtained.

**Table 3.** Fatigue test data.

| Sample No. | Maximum Stress ($S_{max}$/MPa) | Stress Amplitude ($S_a$/MPa) | Life Cycle | Logarithm of $S_a$ | Logarithm of Life Cycle |
|---|---|---|---|---|---|
| 1 | 860 | 387 | 103,477 | 2.5877 | 5.0148 |
| 2 | 810 | 364.5 | 291,971 | 2.5616 | 5.4653 |
| 3 | 770 | 346.5 | 319,143 | 2.5397 | 5.5039 |
| 4 | 730 | 328.5 | 429,394 | 2.5165 | 5.6328 |
| 5 | 690 | 310.5 | 661,253 | 2.4921 | 5.8203 |
| 6 | 650 | 292.5 | 1,731,470 | 2.4661 | 6.2384 |
| 7 | 610 | 274.5 | 5,000,000 | 2.4385 | 6.699 |
| 8 | 570 | 256.5 | 5,000,000 | - | - |

The data in Figure 7a were obtained by fitting the life cycle and the maximum stress from Table 3. It can be seen from the figure that most points are near the fitting curve, indicating that the test results have a good correlation. As shown in Figure 7b, the fatigue strength of $5 \times 10^6$ cycles is 625.33 MPa by the linear fitting of $\log N$ and $\log S_a$.

Figure 8 shows the fatigue fracture morphology at 810 MPa, and the fatigue source area is shown in Figure 8a. It can be seen from the figure that the crack source is formed in the middle of the sample surface, and the crack diffuses around in a fan-shaped manner. At this time, the crack propagation rate is the slowest. Figure 8b shows the cleavage step in the fatigue source area. Figure 8c shows the stable stage of fatigue crack. There are many secondary cracks in the figure. The secondary crack shown in the yellow circle extends along the tear ridge, and the crack as a whole is Z-shaped. The secondary crack shown in the red circle passes through the tear ridge and consumes a large amount of energy, resulting in the stop of crack propagation. The fatigue stripe is a typical feature of stable crack growth. The fatigue stripe shown in Figure 8c is relatively smooth, so it is a plastic fatigue stripe, and the direction of the fatigue stripe is parallel to the direction of the secondary crack. The fatigue stripes shown in Figure 8d are brittle fatigue stripes with an uneven surface morphology, and there are cleavage steps formed by five cleavage surfaces above the stripes. Figure 8e,f show the morphology of the instantaneous fracture zone. Under a high stress cyclic load, tire indentation appears, as can be seen in Figure 8e,f, dimple morphology appears. In this stage, the crack is in the high-speed propagation phase. Under the action of fatigue cyclic load, dimples of different sizes gradually gather and connect with each other, and finally form a dimple fracture. Combined with the above characteristics, the fracture is a mix of cleavage fracture and ductile fracture.

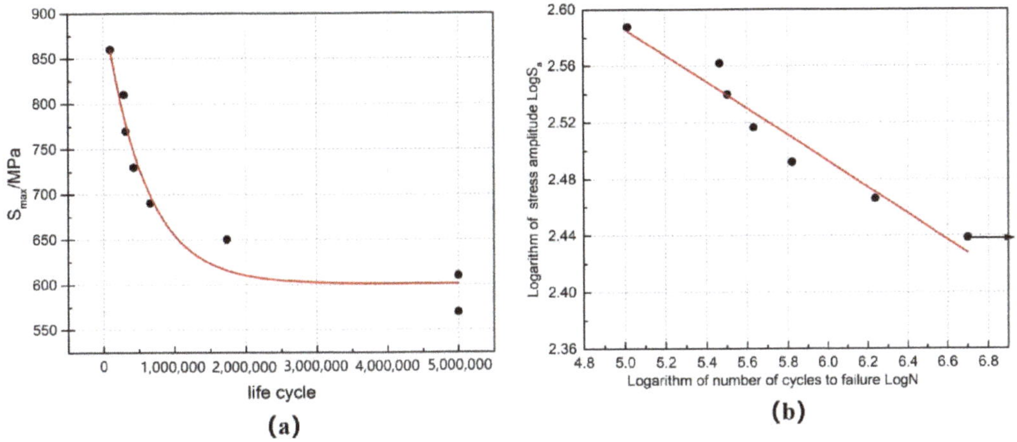


**Figure 7.** (**a**) $S_{max}$ and life cycle curves, (**b**) LogN and $LogS_a$ curves.

## *3.4. Analysis of Synchrotron Radiation X-Imaging*

Figure 9 shows the imaging of the SR-μCT sample with 3000 cycles under 950 MPa stress in the observation area of synchrotron radiation. The experimental data were sliced using Pitre software, and the three-dimensional image was then reconstructed using Avizo software. The volume of all defects in the sample was counted and analyzed, and the results shown in Figure 10b were obtained. There are 50 defects in the imaging area: where the volume is less than 50 μm$^3$, there are 40 defects; for 50–200 μm$^3$, there are six; and for more than 200 μm$^3$, there are four. The sphericity of several larger defects in Figure 9 was analyzed, and the sphericity of the three defects is greater than 0.9. From the sphericity, it can be judged that the three defects are cavities. The whole defect is concentrated in the middle of the sample due to the high energy required for imaging, which reduces the total number of photons passing through the sample, resulting in the imaging results being concentrated in the middle of the sample. Figure 10a is a top view of the imaging area, and a large number of defects are distributed at the edge of the specimen. During the fatigue loading cycle, the fatigue source area is easily formed in the area where these defects are concentrated.

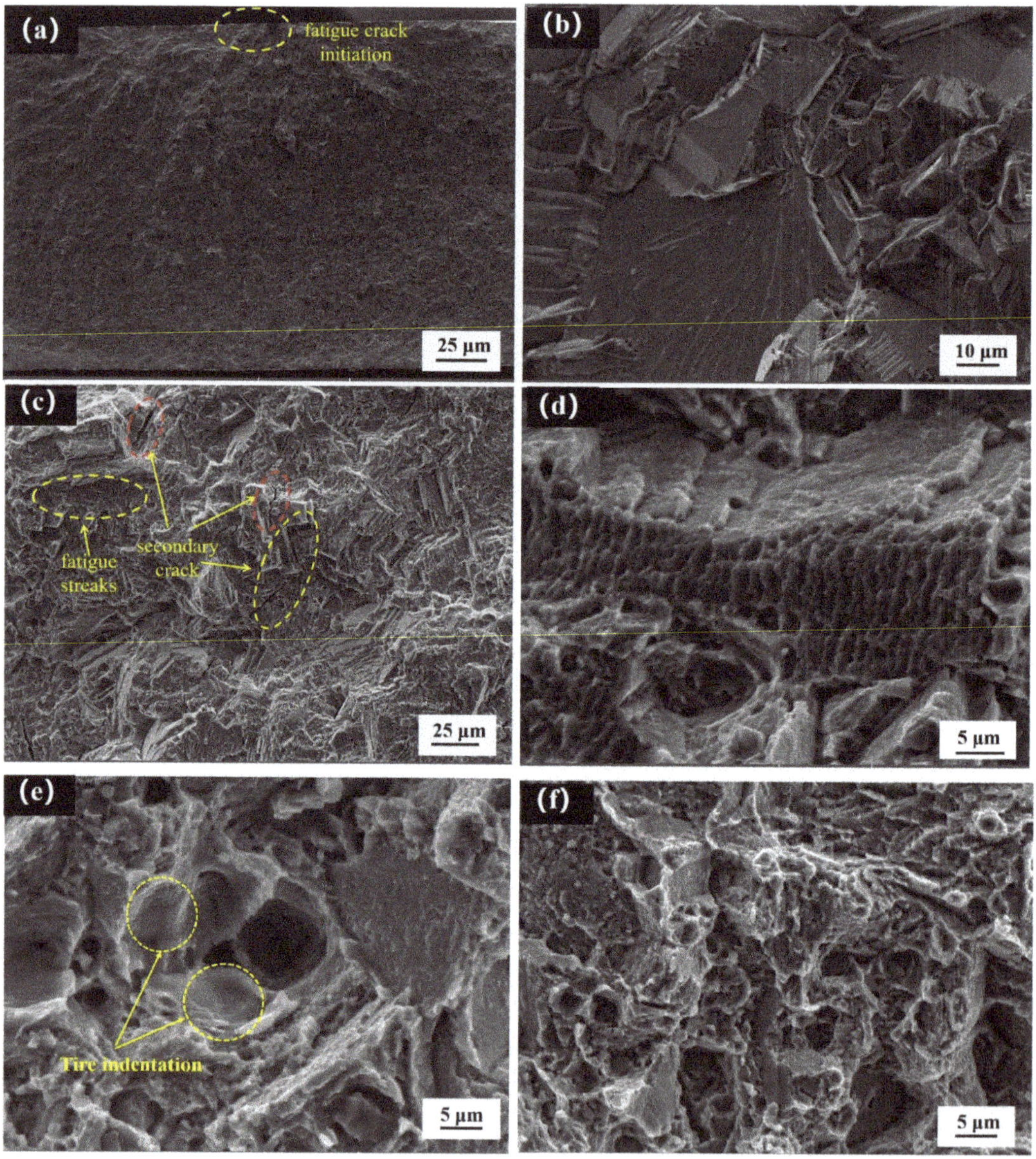


**Figure 8.** Fatigue fracture topography: (**a**) fatigue crack initiation, (**b**) cleavage step, (**c**) secondary cracks and fatigue streaks, (**d**) brittle fatigue streaks, (**e**) tire indentation, (**f**) dimples.

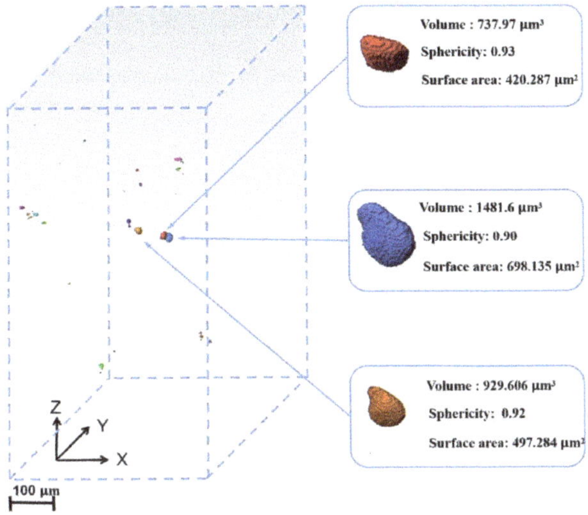


**Figure 9.** SR-μCT imaging area reconstruction showing representative defects in the micro sample.

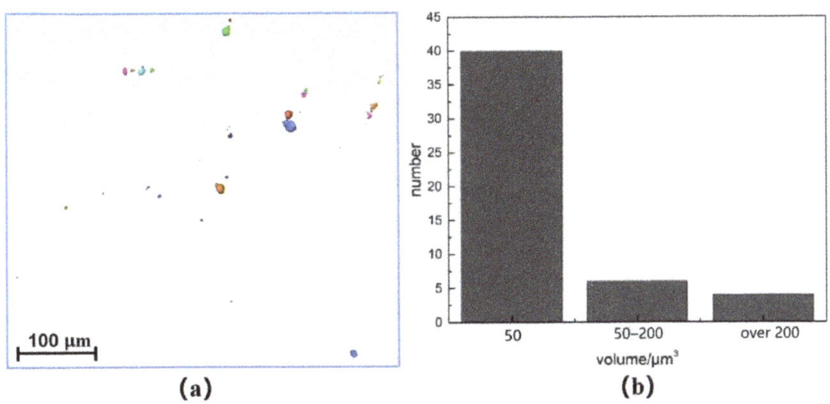


**Figure 10.** (**a**) Top view of imaging area, (**b**) statistical diagram of defect volume size.

The fatigue test of the sample was continued, and the sample fractured when the cumulative number of cycles reached 5625. The fatigue fracture was imaged by synchrotron radiation X-ray. Figure 11b is the side view of the fracture 3D reconstruction morphology. It can be seen from Figure 11b that most of the defects in Figure 11a are surface defects. Figure 12 is a SEM diagram of the fatigue fracture of the synchrotron radiation specimen, Figure 12a is the macro morphology of the fatigue fracture, and Figure 12b is the top view of the three-dimensional morphology. It can be seen that the defects shown in the blue box of Figure 11a are in the fatigue source area of Figure 12a. As the number of fatigue cycles is less, the characteristics of the fatigue source region are not obvious, and the whole fracture morphology is similar to that of a tensile fracture. Figure 12c shows the crack growth zone, in which secondary cracks appear, and Figure 12d shows the morphology of the instantaneous fracture zone, with a large number of equiaxed dimples. Through an enlarged observation of this area, it can be seen in Figure 12e that there are complete and broken particles at the lower part of the dimple. An EDS analysis of the particles was carried out. The analysis results are shown in Figure 12f. Through its element content ratio, it can be judged that the particles are $Ni_3Al$, which is an important strengthening phase of nickel-based superalloy.

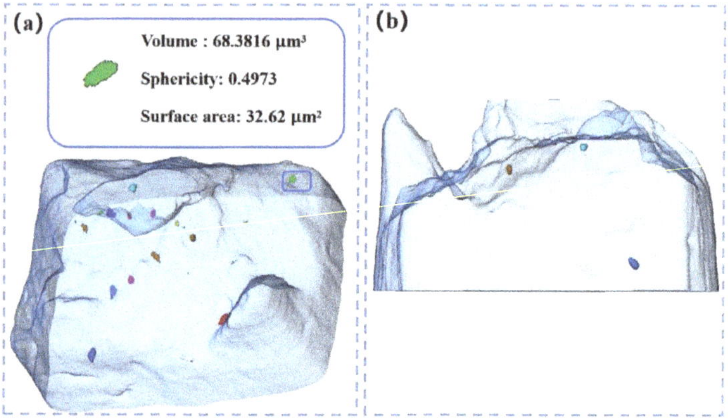


**Figure 11.** (**a**) Top view of fracture imaging, (**b**) side view of fracture imaging.

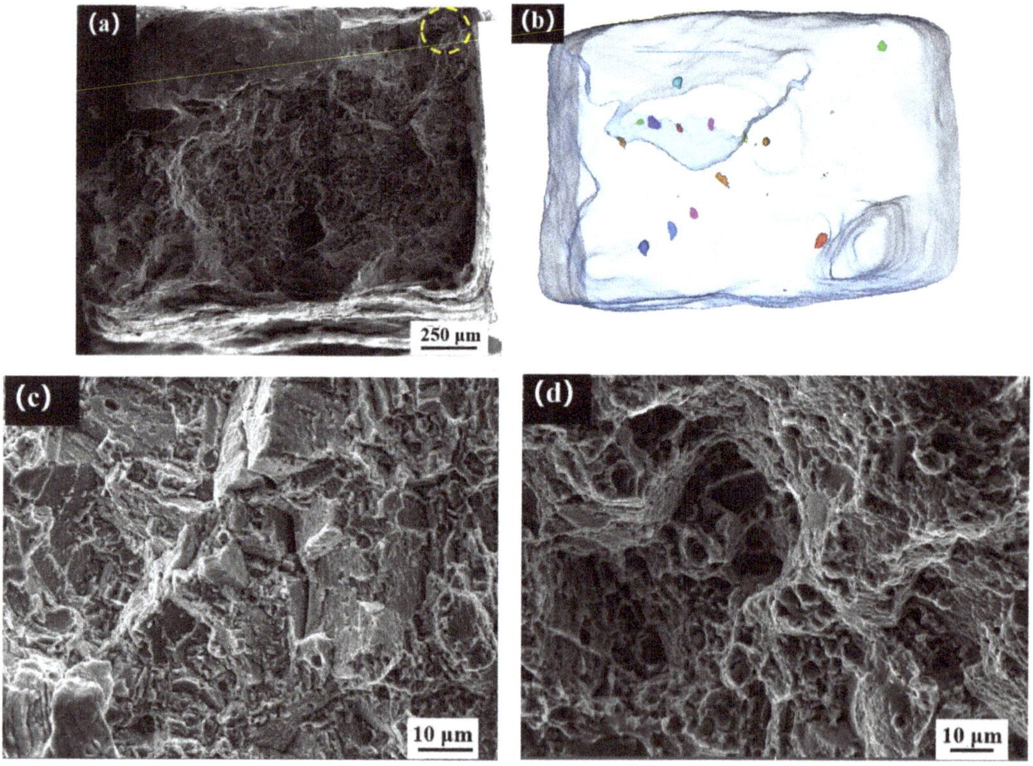


**Figure 12.** *Cont.*

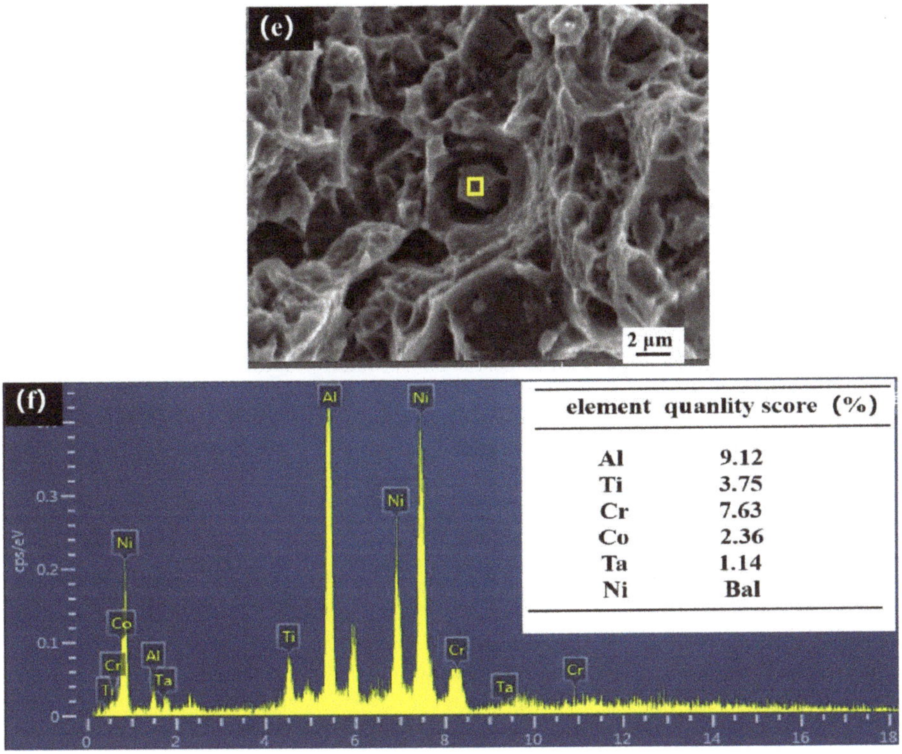


**Figure 12.** (**a**) Fatigue fracture morphology, (**b**) top view of fracture imaging, (**c**) second crack, (**d**) dimple morphology, (**e**) second-phase particles, (**f**) EDS analysis results.

To analyze this phenomenon, due to the face-centered cubic structure of Ni-based superalloy with good slip properties, the cavities grow up gradually under cyclic loading and connect with other cavities to form dimple fractures, as shown in Figure 13a. Figure 13b shows that the second-phase particles are separated from the matrix to form dimples. According to the dislocation theory, the dislocation loops are stacked around the second-phase particles under cyclic loading. When the external force is small, the dislocation loops keep equilibrium under the interaction of dislocation stacking stress and the repulsive force of the second-phase particles. When the external force gradually increases, the dislocation loops move towards the second-phase particles. When the accumulated elastic strain can overcome the interfacial bonding force between the second-phase particles or inclusions and the matrix to form a new surface, new micro voids can be formed. At the same time, the repulsive force of the second-phase particles on the dislocation ring is weakened, which makes the dislocation move to a new micro cavity under the action of external force and makes the cavity grow. With the disappearance of stacking dislocation constraints, new dislocations are generated from the dislocation sources, and the cavities expand and aggregate rapidly. Under the action of tensile stress, a large number of cavities grow up, and the cross-section of the matrix between the adjacent cavities gradually becomes smaller. When they connect with each other, fracture occurs, forming dimple fractures. The dimple formation mechanism shown in Figure 13c is similar to that in Figure 13b, except that Figure 13b shows the separation of the second-phase particles from the matrix, and Figure 13c shows the fracture of the second-phase particles. After the fracture of the second-phase particles, micro cavities are formed, which grow and gather to form dimple fractures.

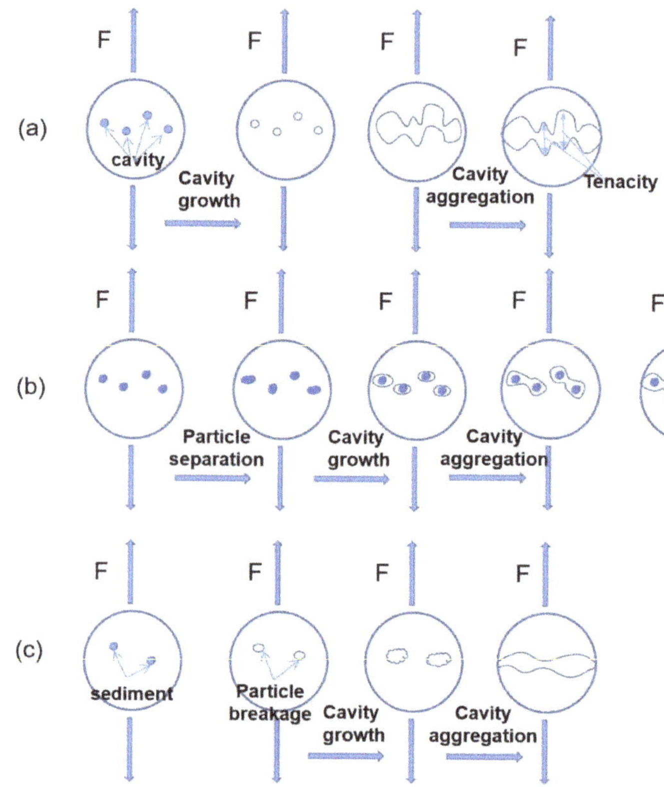


**Figure 13.** Formation mechanism of fatigue dimple: (**a**) cavity growth, (**b**) separation of second-phase particles, (**c**) fragmentation of second-phase particles.

Figure 14 shows two kinds of second-phase particles at the bottom of a fatigue dimple. Figure 14a shows the complete second-phase particles at the bottom of the dimple. Figure 14b shows the broken second-phase particles at the bottom of the dimple. The experimental results are consistent with the phenomenon of mechanism formation shown in Figure 13.

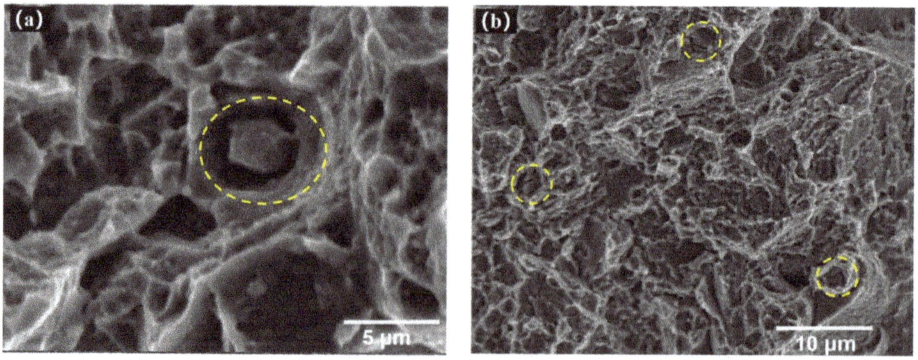


**Figure 14.** (**a**) Complete second-phase particle, (**b**) broken second-phase particle.

### 3.5. Fatigue Fracture Damage Mechanism

Figure 15a shows the cleavage planes and cleavage steps found in the fatigue test fracture. There are three cleavage planes in total. There are a large number of river patterns on one of the cleavage planes. The cleavage steps are generated by the propagation along different planes after the fatigue crack initiation, and are typical cleavage fractures.

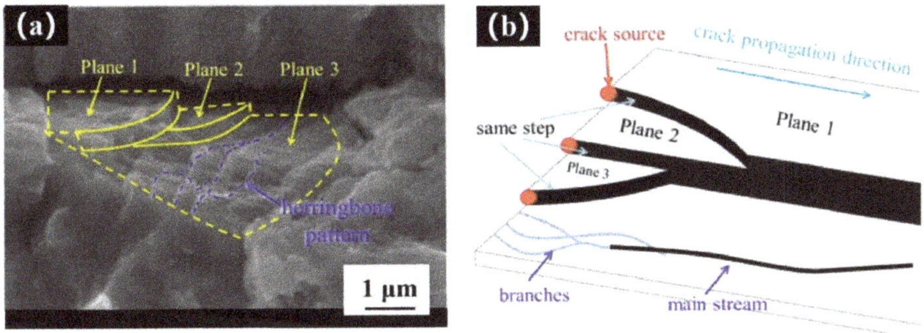


**Figure 15.** (**a**) SEM image of cleavage step, (**b**) schematic diagram of the formation mechanism of cleavage steps.

The cleavage plane and river pattern principle is shown in Figure 15b. The fatigue crack enters the propagation stage after cracking, but it does not propagate along one plane in the propagation process, instead it will propagate along several planes at the same time. The volume of the parallel end of the fatigue test sample at the same height is 40 folds that of the SR-μCT sample. From the imaging results, it can be judged that there are a large number of small defects in the fatigue sample. When a crack encounters internal defects, grains, and the second equal interface, it may deflect and expand in different planes. The cracks in different planes extend the same distance in the same cycle, and finally form a river pattern. Therefore, a river pattern also contains steps, which are generally shallow and difficult to distinguish from the figure. When steps are formed in the process of fatigue crack propagation, some energy will be lost due to the propagation in other directions. Therefore, a large number of cracks will converge into the main cracks in the process of propagation, and the tributaries will converge into a main stream. Therefore, according to the flow direction, we can know how a crack is propagated. The tributaries of the river pattern continue to gather to form the main stream, which will eventually form cleavage steps, and the fatigue cracks will continue to expand along the cleavage plane, eventually leading to a fracture. This is the formation mechanism of the cleavage plane and cleavage steps.

In the fracture morphology, many secondary cracks are found to expand as "Z". A model was established, as shown in Figure 16, to analyze the crack propagation mechanism. The cyclic stress of a fatigue test is represented as a sine wave, and Figure 16a shows the cyclic stress diagram of one cycle. The whole cycle was divided into five stages to analyze the Z-word expansion mechanism, as shown in Figure 16b. In the first stage, the crack is treated and closed. When the external force gradually increases, the crack begins to expand obliquely above, and then enters the second stage. In the third stage, when the external force reaches the apex, the crack tip is passivated, and the crack cannot continue to expand in the original direction due to a gradual reduction in the external force. In the fourth stage, the stress changes from tensile stress to compressive stress, the crack expands obliquely downward, and the crack tip changes into a folded ear shape under the increasing compressive stress. In the fifth stage, the compressive stress reaches maximum and the crack tip closes again until the end of this cycle. After a cycle, the crack can only advance a short distance, and this process is repeated until the sample breaks.

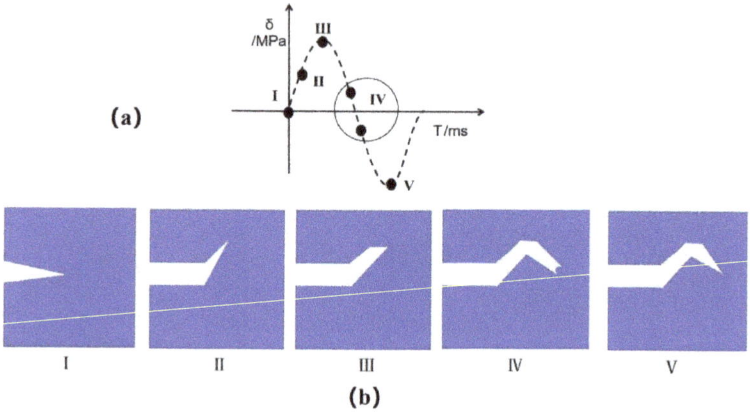


**Figure 16.** (**a**) Schematic diagram of cyclic stress, (**b**) Z-shaped expansion mechanism.

## 4. Conclusions

(1) The primary $\gamma'$ phase is distributed on the grain boundary in a chain shape, and the secondary $\gamma'$ phase is uniformly distributed inside the grain boundary. The average hardness of the alloy is 470.6 Hv, the tensile strength is 1570 MPa, and the elongation is 12.1%.

(2) Under the condition of $5 \times 10^6$ cycles, the fatigue strength of the alloy is 625.33 MPa, and the fatigue fracture is a mix of cleavage fracture and ductile fracture. In the crack growth zone, the secondary crack propagating along the tear ridge is the longest. Under the action of cyclic load and tip passivation, the crack propagates as a "Z" shape.

(3) Synchrotron radiation X-ray imaging shows that a defect volume less than 50 $\mu m^3$ accounts for 80% of the total. The dimple morphology is formed by the growth and aggregation of cavities in the material. There are three cases of cavity germination: defects of the material itself, fracture of the second-phase particles, and separation of the interface between the second-phase particles and the matrix.

(4) Cleavage fracture is caused by the development of multiple cleavage planes formed by branch cracks that converge to form a main crack, the intersection of different cleavage planes to form cleavage steps, and the main crack continuing to expand forward until fracture. Cleavage fracture is the main factor in fatigue damage.

(5) It is found that, compared with inclusions, voids are the main cause of fatigue fracture in FGH96 nickel-based superalloy. In future material manufacturing processes, while avoiding inclusions, the compactness of the material should be improved.

**Author Contributions:** Conceptualization, Y.B. and S.Y.; methodology, Y.B.; software, Y.B.; validation, Y.B., M.Z., C.F.; formal analysis, Y.B.; investigation, Y.B. and C.F.; resources, Y.B. and M.Z.; data curation, Y.B.; writing—original draft preparation, Y.B.; writing—review and editing, Y.B.; visualization, Y.B.; supervision, Y.B.; project administration, Y.B.; funding acquisition, Y.B. and S.Y. All authors have read and agreed to the published version of the manuscript.

**Funding:** This research was supported by the National Natural Science Foundation of China (51971129), Natural Science Foundation of Shanghai (19ZR1421200).

**Institutional Review Board Statement:** Not applicable.

**Informed Consent Statement:** Not applicable.

**Data Availability Statement:** The data used to support the findings of this study are included within the article.

**Acknowledgments:** Thank Shanglei Yang, Minqi Zhu and Cong Fan for their help.

**Conflicts of Interest:** The authors declare that they have no conflict of interest.

## References


1. Peng, Z.; Zou, J.; Wang, X. Microstructural characterization of dislocation movement during creep in powder metallurgy FGH96 superalloy. *Mater. Today Commun.* **2020**, *25*, 101361. [CrossRef]
2. Wang, Y.; Wang, X.; Zhong, B.; Wei, D.; Jiang, X. Estimation of fatigue parameters in total strain life equation for powder metallurgy superalloy FGH96 and other metallic materials. *Int. J. Fatigue* **2019**, *122*, 116–124. [CrossRef]
3. Chin, K.S.; Idapalapati, S.; Ardi, D.T. Fatigue of Surface-Treated Nickel-Based Superalloy at High Temperature. *J. Mater. Eng. Perform.* **2019**, *28*, 7181–7187. [CrossRef]
4. Wu, K.; Tan, L.; He, Y.; Deng, R.; Liu, F. Hot working behavior of powder superalloy U720Li. *Chin. J. Nonferrous Met.* **2019**, *29*, 1676–1683.
5. Wu, H.; Liu, M.; Wang, Y.; Huang, Z.; Tan, G.; Yang, L. Experimental study and numerical simulation of dynamic recrystallization for a FGH96 superalloy during isothermal compression. *J. Mater. Res. Technol.* **2020**, *9*, 5090–5104. [CrossRef]
6. Liu, Y.; Kang, M.; Wu, Y.; Wang, M.; Li, M.; Yu, J.; Gao, H.; Wang, J. Crack formation and microstructure-sensitive propagation in low cycle fatigue of a polycrystalline nickel-based superalloy with different heat treatments. *Int. J. Fatigue* **2018**, *108*, 79–89. [CrossRef]
7. Jiang, R.; Bull, D.; Proprentner, D.; Shollock, B.; Reed, P. Effects of oxygen-related damage on dwell-fatigue crack propagation in a P/M Ni-based superalloy: From 2D to 3D assessment. *Int. J. Fatigue* **2017**, *99*, 175–186. [CrossRef]
8. Wang, Y.; Wang, J.; Dong, J.; Li, A.; Li, Z.; Xie, G.; Lou, L. Hot deformation characteristics and hot working window of as-cast large-tonnage GH3535 superalloy ingot. *J. Mater. Sci. Technol.* **2018**, *34*, 2439–2446. [CrossRef]
9. Cantó, J.S.; Winwood, S.; Rhodes, K.; Birosca, S. A study of low cycle fatigue life and its correlation with microstructural parameters in IN713C nickel based superalloy. *Mater. Sci. Eng. A* **2018**, *718*, 114–125.
10. Hu, D.; Wang, T.; Ma, Q.; Liu, X.; Shang, L.; Li, D.; Pan, J.; Wang, R. Effect of inclusions on low cycle fatigue lifetime in a powder metallurgy nickel-based superalloy FGH96. *Int. J. Fatigue* **2018**, *118*, 237–248. [CrossRef]
11. Wang, J.J.; Wen, Z.X.; Zhang, X.H.; Zhao, Y.C.; Yue, Z.F. Effect mechanism and equivalent model of surface roughness on fatigue behavior of nickel-based single crystal superalloy. *Int. J. Fatigue* **2019**, *125*, 101–111. [CrossRef]
12. Chavoshi, S.Z.; Jiang, J.; Wang, Y.; Fang, S.; Wang, S.; Shi, Z.; Lin, J. Density-based constitutive modelling of P/M FGH96 for powder forging—ScienceDirect. *Int. J. Mech. Sci.* **2018**, *138*, 110–121. [CrossRef]
13. Luo, Y.; Wu, S.C.; Hu, Y.N.; Fu, Y.N. Cracking evolution behaviors of lightweight materials based on in situ synchrotron X-ray tomography: A review. *Front. Mech. Eng.* **2018**, *13*, 461–481. [CrossRef]
14. Jiménez, M.; Ludwig, W.; Gonzalez, D.; Molina-Aldareguia, J. The role of slip transfer at grain boundaries in the propagation of microstructurally short fatigue cracks in Ni-based superalloys. *Scr. Mater.* **2019**, *162*, 261–265. [CrossRef]
15. Hu, D.; Pan, J.; Mao, J.; Hu, S.; Liu, X.; Fu, Y.; Wang, R. Mechanical behavior prediction of additively manufactured components based on defect evolution observation by synchrotron radiation X-ray tomography. *Mater. Des.* **2020**, *198*, 109353. [CrossRef]
16. Callegari, B.; Campo, L.; Aristizabal, K.; Guitar, M.A.; Warchomicka, F.; Coelho, R.S.; Brito, P.P.; Soldera, F.A.; Mücklich, F.; Pinto, H.C. In situ assessment of isochronal phase transformations in a lamellar Ti–5Al–5Mo–5V–3Cr–1Zr alloy using synchrotron X-ray diffraction. *J. Alloy. Compd.* **2020**, *853*, 157105. [CrossRef]
17. Hu, Y.; Wu, S.; Withers, P.; Zhang, J.; Bao, H.; Fu, Y.; Kang, G. The effect of manufacturing defects on the fatigue life of selective laser melted Ti-6Al-4V structures. *Mater. Des.* **2020**, *192*, 108708. [CrossRef]
18. Xie, C.; Wu, S.; Yu, Y.; Zhang, H.; Hu, Y.; Zhang, M.; Wang, G. Defect-correlated fatigue resistance of additively manufactured Al-Mg4.5Mn alloy with in situ micro-rolling—ScienceDirect. *J. Mater. Process. Technol.* **2021**, *291*, 117039. [CrossRef]
19. Serrano-Munoz, I.; Shiozawa, D.; Dancette, S.; Verdu, C.; Buffiere, J.-Y. Torsional fatigue mechanisms of an A357-T6 cast aluminium alloy. *Acta Mater.* **2020**, *201*, 435–447. [CrossRef]
20. Hou, J.; Dong, J.X.; Yao, Z.H.; Jiang, H.; Zhang, M.C. Influences of PPB, PPB Affect Zone, Grain Boundary and Phase Boundary on Crack Propagation Path for a P/M Superalloy FGH4096. *Mater. Sci. Eng. A* **2018**, *724*, 17–28. [CrossRef]
21. Pollock, T.M.; Tin, S. Nickel-Based Superalloys for Advanced Turbine Engines: Chemistry, Microstructure and Properties. *J. Propuls. Power* **2006**, *22*, 361–368. [CrossRef]

materials

MDPI

*Article*

# Determination of Mechanical and Fracture Properties of Silicon Single Crystal from Indentation Experiments and Finite Element Modelling

**Petr Skalka [1] and Michal Kotoul [1,2,*]**

1 Institute of Solid Mechanics, Mechatronics and Biomechanics, Brno University of Technology, Technická 2896/2, 616 69 Brno, Czech Republic; skalka@fme.vutbr.cz

2 Faculty of Special Technology, Alexander Dubček University of Trenčín in Trenčín, Studentska 2, 911 50 Trenčín, Slovakia

* Correspondence: kotoul@fme.vutbr.cz

**Abstract:** It is well-known that cracks are observed around the impression during indentation of brittle materials. The cracks inception depends on load conditions, material and indenter geometry. The paper aims to use experimental micro-indentation data, FE simulations with cohesive zone modelling, and an optimisation procedure to determine the cohesive energy density of silicon single crystals. While previous studies available in the literature, which use cohesive zone finite element techniques for simulation of indentation cracks in brittle solids, tried to improve methods for the evaluation of material toughness from the indentation load, crack size, hardness, elastic constants, and indenter geometry, this study focuses on the evaluation of the cohesive energy density 2Γ from which the material toughness can be easily determined using the well-known Griffith-Irwin formula. There is no need to control the premise of the linear fracture mechanics that the cohesive zone is much shorter than the crack length. Hence, the developed approach is suitable also for short cracks for which the linear fracture mechanics premise is violated.

**Keywords:** micro-indentation; mechanical and fracture properties identification; finite element analysis; optimisation analysis

**Citation:** Skalka, P.; Kotoul, M. Determination of Mechanical and Fracture Properties of Silicon Single Crystal from Indentation Experiments and Finite Element Modelling. *Materials* **2021**, *14*, 6864. https://doi.org/10.3390/ma14226864

Academic Editors: Christian Motz and José Xavier

Received: 17 August 2021
Accepted: 8 November 2021
Published: 14 November 2021



## 1. Introduction

In recent decades, a number of studies devoted to identification of material properties such as Young's modulus, yield stress, and work hardening modulus by using experimental indentation data, finite element (FE) simulations, and optimisation procedures for solving inverse problems have occurred in the literature. Various optimisation techniques have been used by researchers, see, e.g., [1–4] to determine material properties from indentation load–displacement curves tests. Identification of elastic and/or elasto-visco-plastic constitutive laws from indentation tests in terms of general theoretical framework of inverse problems solution has been described in [5,6]. With respect to brittle materials, cohesive zone FE simulations of indentation cracking have been performed e.g., in [7–10] to investigate the crack morphology, the change of crack length with indenter shape, a quantitative evaluation of the threshold load for indentation fracture, and to explore a limitation of analytical models such as Lawn-Evans-Marshall model [11]. For indentation crack initiation and propagation modelling a cohesive interface consisting of cohesive elements is placed in the plane of potential cracking and only mode I type crack is considered. The behaviour of the cohesive elements in this interface is governed by a traction-separation law which mostly has the bilinear form characterised by three parameters- peak cohesive traction $\sigma_{\max}$, corresponding damage-initiating displacement $\Delta_c$ and failure displacement $\Delta_{sep}$. It was shown in [12] that the cohesive energy density and the peak cohesive traction play a far more important role than the shape of the cohesive traction—separation curve in predicting the final fracture behaviour. In case of the bilinear form of the cohesive traction—separation law the cohesive energy density (critical fracture energy) $2\Gamma$ can be calculated by $2\Gamma = \frac{1}{2}\sigma_{\max}\Delta_c$. Critical review of various cohesive zone models is given in [13]. Contrary to analytical approaches, cohesive interface FE simulations exhibit a natural advantage consisting in no need to specify the crack front a priori. Namely, the crack front is found as a result of the solution of the boundary value problem. Moreover, the influence of residual stresses developing under the indent due to inelastic compressive behaviour of brittle materials [10,14–17] is more reliably captured. Some care is needed with respect to the elasticity of the cohesive interface, specifically one should avoid double-counting the elasticity—once in the cohesive law and a second time as part of the bulk behaviour. Nevertheless, the effect of this issue is negligible when cohesive surfaces are only specified along a potential single crack path such as in the case of indentation cracking, or if the stiffness of cohesive surfaces is infinite [18]. Currently, there is an increasing effort to combine cohesive zone models with extended finite element method (XFEM) to model crack growth [19–21]. XFEM can avoid remeshing near the crack tip as the crack grows and all other difficulties connected with it. With respect to indentation crack modelling, remeshing is not needed as the indentation cracks extend only over short distances without kinking and a zone of the potential crack formation is covered with cohesive elements. Thus, the application of XFEM does not seem to bring any other benefits in this context. To interpret the results of simulation of the growth of indentation cracks in terms of the linear fracture mechanics, the cohesive (bridging) zone must be significantly smaller than the crack. Hence, great care is needed in applying the simulation results to short crack problems under indentation tests [22,23].

A direct application of the former macroscopic cohesive laws to cleavage fracture, which entails a simple separation of the atomic planes, is not easily workable. Consider (110) cleavage planes in Si crystal. Their interplanar spacing $d$ is 1.92 Å. The (110) cracks with $[\bar{1}10]$ crack front in Si crystal were analysed using ab initio and gradient elasticity theory in our study [24]. It was shown that the critical crack opening $\delta_c$ (interplanar separation) leading to the loss of the crystal bearing capacity is 0.2 nm, the corresponding peak stress is of the order of theoretical strength, and the cohesive energy density $2\Gamma \cong 5.2\ \mathrm{J/m^2}$. Moreover, the length of the cohesive zone is very small, approximately 0.6 nm. It means that macroscopic FE simulation would require extremely fine mesh, which is often unfeasible. Nguyen and Ortiz [25] suggested a way to the macroscopic form of the cohesive law by considering the cooperative behaviour of a large number $N$ of interatomic planes forming a cohesive layer. The thickness of the cohesive layer in FE simulations is given by the local element size $D$. Thus, the number of atomic planes in the cohesive layer is $N = 2D/d$, where the factor 2 was added due to symmetry. Nguyen and Ortiz showed that for sufficiently large $N$ the macroscopic critical opening displacement $\Delta_c$ and the corresponding macroscopic cohesive stress $\sigma_{\max}$ for the separation of a single atomic plane asymptotically scale as

$$\Delta_c = 2\sqrt{\frac{\Gamma N}{C}},\ \ \sigma_{\max} = 2\sqrt{\frac{\Gamma C}{N}}, \tag{1}$$

where the interplanar modulus $C$ depends on a specific material. For the interplanar cohesive potential suggested in [26]

$$\phi(\delta) = 2\Gamma - C\delta_c(\delta + \delta_c)e^{-\delta/\delta_c}, \tag{2}$$

the interplanar modulus $C$ for (110) planes in Si crystal is

$$C = \frac{2\Gamma}{\delta_c^2} \cong 1.3 \times 10^{20}\ \mathrm{J/m^4}. \tag{3}$$

For the element size $D$ = 0.25 μm, Equation (1) provides $\Delta_c \cong 14.4$ nm. While the critical opening displacement $\Delta_c$ and the corresponding macroscopic cohesive stress $\sigma_{\max}$ do depend on the element size, the cohesive energy density $2\Gamma$ is independent of the size element. The aim of this study is to use experimental microindentation data, FE simulations with cohesive zone modelling, and an optimisation procedure to determine the cohesive energy density of single crystals without having to check whether the size of the cohesion zone is considerably less than the crack size and thus to analyse the problems of short cracks. Obviously, such a procedure is particularly suitable for determining fracture properties of MEMS/NEMS parts or thin films using micro/nano indentation tests.

## 2. Materials and Methods

As-received Si crystals (100) with dimensions 50 mm × 50 mm × 3 mm were covered by poly(methyl methacrylate) PMMA before cutting and were precut to 1 cm × 1 cm samples by laser dicer. The cover layer PMMA was removed by acetone, isopropanol and deionised water in an ultrasound cleaner. The final cleaning step was etching of organic residues by oxygen plasma in Diener Plasma cleaner (Diener electronic GmbH, Ebhausen, Germany). Substrates were covered by a double layer of optics resists by spin-coating process. This double layer of resists is important for the lift-off process. The bottom resist was AR-BR 5460 (Allresist GmbH, Strausberg, Germany) and the top resist was AR-P 3540 (Allresist GmbH, Strausberg, Germany). The bottom resist is more sensitive than top resist and this combination of resists creates an undercut in resist layer. The exposure of the resist was done by UV Direct Write Laser system 66+ from Heidelberg Instruments (Heidelberg, Germany). After exposure, samples were developed by AR 300-47 (mixed with deionised water in the ratio 1:1, Allresist GmbH, Strausberg) for 60 s. The residues after developing were removed by oxygen plasma by reactive ion etching in Oxford Instruments Plasma Technology PlasmaPro NGP 80 (Oxford, UK). Optionally, the native oxide layer could be removed by buffered hydrofluoric acid (BOE 7:1 – $HF{:}NH_4F$ = 12.5:87.5%).

The indentation tests were load-controlled and performed using Fischerscope H100 XYp equipment (Riley Industries Ltd., Aldridge, UK) with maximal applied force to load cell of 1000 mN acting on standard Vickers diamond indenter with the centreline-to-face angle $\psi = 68^\circ$ which was aligned along the cleavage plane {101} in the direction <100> of Si crystal. The minimal applied force, which the equipment can detect, is 0.4 mN with the force resolution 20 μN and depth resolution ±2nm. In this study, the test forces of 300 mN, 500 mN, 750 mN, and 1000 mN, respectively, were applied. Loading stage lasted 20 s followed by 5 s creep and with 20 s long unloading stage. The unloading stage was followed by 5-s long period of constant loading of 0.4 mN. For each of applied forces 25 indentation tests were performed to minimise the experimental error. The crack length was measured by confocal laser microscope Olympus LEXT4000 (Olympus Corporation) from the centre of indentation. Figure 1 shows details of indentation with radial cracks after indentation tests. During indentation the crack behaviour is impossible to optically track because cracks spread under the surface, and they are thus invisible. The first visible radial cracks on the crystal top surface occur during the unloading stage of indentation tests.

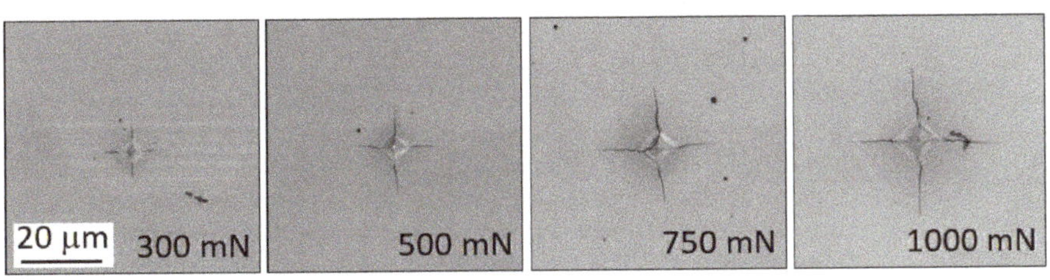


**Figure 1.** Photographs of indentation with radial cracks.

FE model for numerical simulation of the indentation test of Si crystal consisted of a cube with edge length of 200 μm (Si crystal), Vickers diamond indenter and the load cell, see Figure 2a.

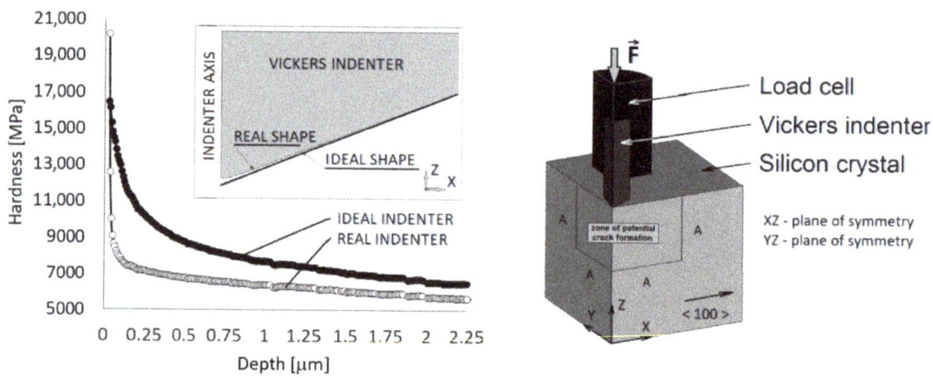


(**a**) (**b**)

**Figure 2.** (**a**) FE model of Vickers indentation test on silicon crystal. (**b**) Calibration of the tip shape based on the hardness.

The discretisation of the FE model was performed by linear solid elements (SOLID185 in ANSYS software (Release 19.2)) and the contact with friction coefficient $f$ = 0.05 between the Si crystal sample and Vickers indenter was defined. The sample was loaded through pushing the indenter into the sample and the indentation depth was gradually increased. As a result, the indenter was deformed, and the loading force was subsequently obtained as a reaction. Two planes of symmetry (XZ, YZ) and the plane XY with prescribed boundary condition ($U_Z$ = 0) representing the crystal storage were used in numerical simulations. Non-elastic response of the Si crystal (denoted as SC), that tends to accommodate the contact stresses under the indenter, was modelled in terms of ideally elastoplastic material defined by Young's modulus $E_{SC}$ =129.5 GPa, Poisson's ratio $\nu_{SC}$ = 0.278, the shear modulus $G_{SC}$ =79.6 GPa and the yield stress $\sigma_{y,SC}$ which is initially unknown. In this context it should be noted that the elastic-perfectly plastic material behaviour according to the von Mises yield condition accurately describes the compressive behaviour of many brittle materials [27,28]. Vickers diamond indenter (VDI) was considered as a linear isotropic body defined by Young´s modulus $E_{VDI}$ = 1220 GPa and Poisson´s ratio $\nu_{VDI}$ = 0.20. Elastic properties of individual components (crystal, indenter) were chosen on the basis of available literature data. Linear isotropic behaviour was also assumed for the load cell (LC) defined by Poisson´s ratio $\nu_{LC}$ = 0.3 and Young´s modulus $E_{LC}$ which takes the stiffness of the test equipment into account and is also initially unknown. Reduced Young´s modulus $E_r$ is then given by

$$\frac{1}{E_r} = \frac{1-\nu_{SC}^2}{E_{SC}} + \frac{1-\nu_{VDI}^2}{E_{VDI}} + \frac{1-\nu_{LC}^2}{E_{LC}}. \tag{4}$$

In the first step, the reduced Young´s modulus $E_r$ was searched together with the yield stress $\sigma_{k,SC}$ based on load-depth curves from indentation tests, see Figure 3a. The indenter tip shape deviation from the ideal shape, see Figure 2a, was also taken into account when searching for the yield strength. Crack initiation and growth was not considered in this stage. It should be pointed out that the effect of cracks on the force-depth curve is negligible for lower loading force values. The nonlinear least-squares routine to get the best fit between the given indentation data and the optimised indentation data, produced by FE analysis, was applied to determine the aforementioned parameters. The corresponding objective functional $\mathcal{F}(c)$ is given by, see [29–33]

$$\mathcal{F}(c) = \frac{1}{2}\sum_{i=1}^{N}\left[P_i^{comp}(c) - P_i^{exp}\right]^2 \min, \tag{5}$$

where $c$ is the optimisation variable set given above, $P_i^{comp}(c)$ and $P_i^{exp}$ are the predicted and experimental loading force, respectively and $i$ denotes a position along the force-depth curve. The yield stress $\sigma_{y,SC}$ was determined from unloading stage of each of the force-depth curves, where linear behaviour exists (approx. to 10% decrease from maximal value of the applied force) in accordance with Oliver and Pharr method [34]. The best fit was obtained with the yield stress $\sigma_{y,SC}$ = 6.4 GPa, the value which is close to the values applied for silicon in studies [7,8,10]. Further decrease in the applied force cannot be employed for a correct fitting because the FE model does not include pop-out effect which occurs approximately at 50% decrease of the applied force. The real shape of the indenter tip, which is used in numerical simulations, was found on the basis of a calibration curve of differential hardness, which is performed before the measurement itself. It is therefore a matter of finding a match between the calculated and measured dependence of load vs indentation depth. The shape of the indenter tip and at the same time the required yield strength are calibrated here. The calibration was performed using the universal hardness HU which takes elastic and plastic deformations into account and is defined by the following relation

$$\mathrm{HU} = \frac{P_{\max}}{S_c(h_{\max})}, \tag{6}$$

where $P_{\max}$ denotes the maximal force acting on the ideal Vickers indenter during a particular indentation test, $h_{\max}$ denotes the corresponding maximal depth of indentation into the Si crystal, and $S_c$ is the contact area between the indenter tip and the Si crystal. The calibration was solved as an inverse problem by using incremental iteration procedure where the universal hardness and the loading force are known, and the contact area is searched. When the contact area is found the shape of indenter tip is modified and the force is incrementally increased. This procedure runs until the maximal loading (here 1000 mN) is reached. Then the calibration procedure is finished. The ideal and real indenter tip shape of Vickers indenter are shown in Figure 2b. The difference between the ideal and the real shape of indenter tip is irrelevant in terms of the force-depth dependence but essential for the development of cracks in the near vicinity of the indenter tip.

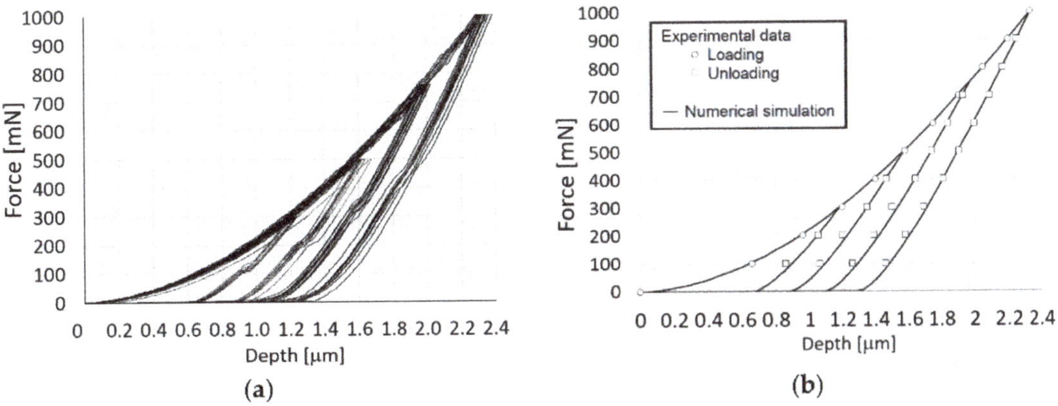


**Figure 3.** Representative force–depth curves from indentation tests: (**a**) experiment, (**b**) numerical simulation after calibration.

For the crack development modelling, a zone of the potential crack formation of the size $A \times A$ with $A$ = 50 μm was defined in both symmetry planes, see Figure 2a. The macroscopic cohesive potential can be obtained from Equations (2) and (3) in terms of the macroscopic opening displacement Δ as

$$\Phi(\Delta) = 2\Gamma - 2\Gamma\left(\frac{\Delta}{\Delta_c} + 1\right)e^{-\Delta/\Delta_c}, \tag{7}$$

or

$$\Phi(\Delta) = 2\Gamma - \sigma_{\max}(\Delta + \Delta_c)e^{1-\Delta/\Delta_c}. \tag{8}$$

The cohesive traction $T(\Delta)$ then follows from the derivative of the potential as

$$T(\Delta) = \Phi'(\Delta) = \frac{\sigma_{\max}}{\Delta_c}\Delta e^{1-\Delta/\Delta_c} = \frac{2\Gamma}{\Delta_c^2}\Delta e^{-\Delta/\Delta_c}. \tag{9}$$

Note that the tangential component was neglected due to the character of loading of potential cracks in Mode I and due to the mechanism of crack formation. It would be possible to extend the model with this feature, however, in our opinion it would not bring a desired benefit. The relationship between the normal traction $T$ and crack opening displacement is illustrated in Figure 4b. However, in Figure 4b due to symmetry, only half of the crack opening displacement is displayed. Hence, the area under the traction-half displacement curve is equal to the half of the cohesive energy density $2\Gamma$, that is to $\Gamma$. The cohesive crack zone is realised by means of nonlinear springs in tension. The nonlinear spring response in compression is considered as rigid and the tangential traction is ignored. The crack tip is defined as the point where the crack opening displacement is equal to $\Delta_c$ which corresponds to the maximal normal traction $\sigma_{\max}$. Numerical simulations were performed for radial cracks propagating along the (101) cleavage plane in the direction [100] and along the (011) cleavage plane in the direction [10], see Figure 2a. The complete elastic–plastic stress field during the unloading stage of the indentation is given by a superposition of the elastic contact stress field $\boldsymbol{\sigma}^m$ and a residual stress field $\boldsymbol{\sigma}^r$ generated due to the permanent deformation $\boldsymbol{\varepsilon}^p$ under the contact. While with decreasing contact force $P(t)$ the elastic contact stress field decreases, the residual stress field remains largely unchanged and promotes cracks extension. The boundary value problem to be solved during the unloading stage is to find the complete stress-strain field $\boldsymbol{\sigma} = \boldsymbol{\sigma}^m + \boldsymbol{\sigma}^r$, $\boldsymbol{\varepsilon} = \boldsymbol{\varepsilon}^m + \boldsymbol{\varepsilon}^r$:

$$\boldsymbol{\sigma}\cdot\boldsymbol{n} = \boldsymbol{p}(t) \text{ on } S_C(t), \;\; \boldsymbol{n}\cdot\boldsymbol{\sigma}\cdot\boldsymbol{n} = T(\Delta) \text{ on } S_{crack}(t), \tag{10}$$

$$\boldsymbol{u}^m + \boldsymbol{u}^r = \boldsymbol{u} = \boldsymbol{u}^0 \text{ on } S_u, \tag{11}$$

$$\boldsymbol{\sigma} = \boldsymbol{C} : (\boldsymbol{\varepsilon} - \boldsymbol{\varepsilon}^p), \;\; \boldsymbol{\varepsilon} = \boldsymbol{D} : \boldsymbol{\sigma} + \boldsymbol{\varepsilon}^p \text{ and } \nabla\cdot\boldsymbol{\sigma} = \boldsymbol{0} \text{ in } \Omega, \tag{12}$$

where $\boldsymbol{\varepsilon}$ satisfies compatibility, $S_C(t)$ denotes the actual contact surface with actual tractions $\boldsymbol{p}(t)$, $S_{crack}(t)$ is the actual crack surface, $S_u$ is the part of the boundary where displacements are prescribed, $\boldsymbol{C}$ and $\boldsymbol{D}$ are the stiffness and compliance tensors respectively, $\boldsymbol{n}$ is a unit normal to the surface. The actual tractions $\boldsymbol{p}(t)$ are related to the actual resultant contact force $P(t)$ by

$$P(t) = \int_{S_C(t)} \boldsymbol{n}\cdot\boldsymbol{p}(t)\mathrm{d}S. \tag{13}$$

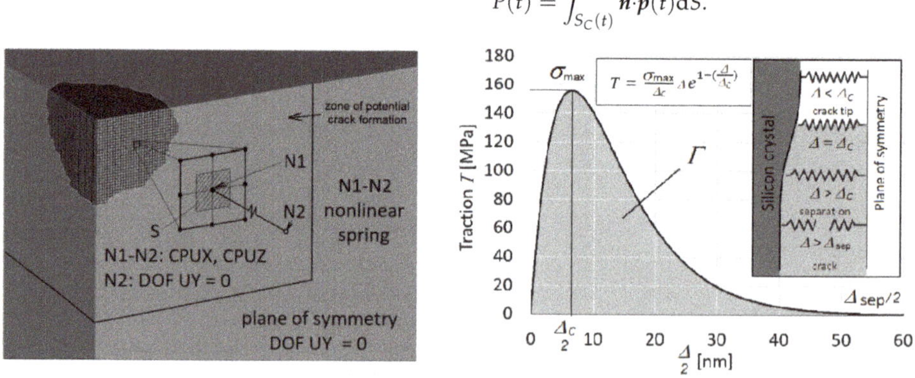


(a) (b)

**Figure 4.** Cohesive zone model (**a**) definition and boundary conditions, (**b**) the traction-crack opening length relationship.

The total energy $\in$ can be expressed in terms of mechanical and residual fields and the cohesive potential $\Phi(\Delta)$ as

$$\begin{array}{c}\in = \frac{1}{2}\int_{\Omega}\sigma : (\varepsilon - \varepsilon^p)d\Omega - \int_{S_C} p\cdot(u^m + u^r)dS + \int_{S_{crack}}\Phi(\Delta)dS = \\ \frac{1}{2}\int_{\Omega}(\varepsilon - \varepsilon^p) : C : (\varepsilon - \varepsilon^p)d\Omega - \int_{S_C} p\cdot(u^m + u^r)dS + \int_{S_{crack}}\Phi(\Delta)dS.\end{array} \tag{14}$$

Necessary first-order stationarity condition for the minimisation of the total energy reads

$$\int_{\Omega} C : (\varepsilon - \varepsilon^p) : \delta\varepsilon d\Omega - \int_{S_C} p\cdot\delta(u^m + u^r)dS + \int_{S_{crack}} \frac{2\Gamma}{\Delta_c^2} e^{-\Delta/\Delta_c}\delta\Delta dS + \Phi(\Delta_c)\delta S_{crack} = 0. \tag{15}$$

In case of full unloading $u^m = \varepsilon^m = p = 0$ Equation (15) reduces to:

$$\int_{\Omega} C : (\varepsilon^r - \varepsilon^p) : \delta\varepsilon^r d\Omega + \int_{S_{crack}} \frac{2\Gamma}{\Delta_c^2} e^{-\Delta/\Delta_c}\delta\Delta dS + \Phi(\Delta_c)\delta S_{crack} = 0, \tag{16}$$

where $\Phi(\Delta_c) = 2\Gamma(1 - 2e^{-1})$. Here, it should be emphasised again that the crack tip is defined as the point where the crack opening $\Delta$ is equal to $\Delta_c$ which corresponds to the maximal normal traction $\sigma_{\max}$. The virtual crack area increment $\delta S_{crack}$ is given by

$$\delta S_{crack} = \int_{\partial S_{crack}} \upsilon\cdot\delta L ds, \tag{17}$$

where $\partial S_{crack}$ is the crack front, $\upsilon$ is the local unit normal vector to the crack front and $\delta L$ denotes the local virtual crack extension. Observe that $\varepsilon^p$ and consequently also the residual strain field $\varepsilon^r$ depend on the maximal loading force $P_{\max}$.

The identification of the material parameters $\Gamma$ and $\Delta_c$ is based upon the best fit between the visible crack length on the top surface and its numerical prediction obtained by FE analysis under full unloading. The optimisation model is

$$J(\Gamma, \Delta_c) = \sum_{i=1}^{M}\left[L_{ij}^{exp} - L_j^{pred}(\Gamma, \Delta_c)\right]^2 \cdot \min, \tag{18}$$

where $L_{ij}^{exp}$ is a measured crack length on the top surface at i-th test, $L_j^{pred}(\Gamma, \Delta_c)$ is its theoretical counterpart, the subscript $j$ denotes j-th value of the maximal loading force $P_{\max,j}$ and $M$ = 25 is the number of performed indentation tests for each loading force. Simultaneously, the minimisation of the total energy is controlled. Observe, that as the independent cohesive material parameters also $\sigma_{\max}$ and $\Delta_c$ can be chosen, see Equation (9).

## 3. Results

This section is devoted to the evolution of crack front during indentation tests and determination of the cohesive material parameters using the FE analysis, experimental data and the optimisation model described above. Experimental values of the radial crack length measured on the top surface are shown in Figure 5 for several values of indentation depth and the corresponding maximal loading force. Moreover, linear regression of the experimental data is included revealing that within the applied loading range (up to 1 N) the crack length linearly depends on the indentation depth which corresponds to the maximal loading force $P_{\max}$. These data are used in the following subsection to find the cohesive energy density of the silicon crystal.

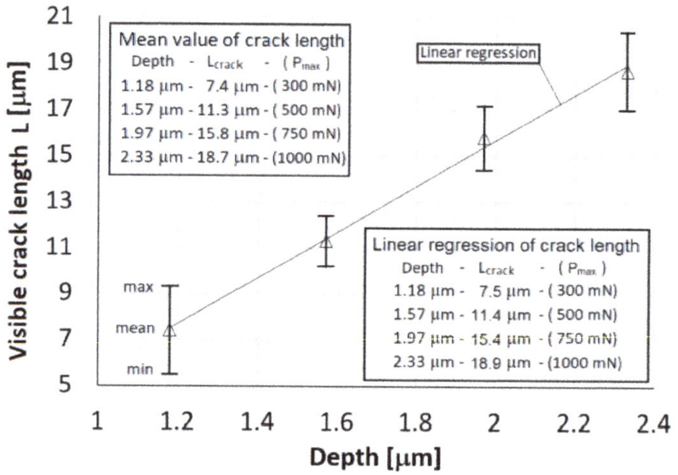


**Figure 5.** Experimental values of radial crack length.

### 3.1. *Cohesive Energy Density of Silicon Crystal*

The cohesive energy density was obtained from numerical simulations of the indentation cracking and the optimisation model. All the numerical simulations were treated as a direct problem and the radial crack length was one of output parameters. In all simulations the same element size of 0.25 µm was used. As already mentioned the crack length depends significantly on the cohesive energy density $2\Gamma$ which corresponds to the area under the curve representing the traction-displacement relationship and on the critical opening displacement $\Delta_c$. Both, the cohesive energy density $2\Gamma$ and the critical opening displacement $\Delta_c$ form the output of the inverse problem. The inverse analysis starts with an initial estimate of $\Gamma$. Subsequently $\Delta_c$ is sought so that the total energy $\in$ reaches a minimum. In the next step the cohesive energy density $2\Gamma$ is adjusted to minimise the discrepancy between the measured crack length on the top surface, $L_{ij}^{exp}$, and its theoretical prediction $L_j^{pred}(\Gamma, \Delta_c)$. With a new value of $\Gamma$ a corrected value of $\Delta_c$ is sought. This process is iteratively repeated until convergence criteria are met.

The above procedure was applied for all values of the maximal loading force $P_{\max}$ = 300 mN, 500 mN, 750 mN, and 1000 mN, and for each of 25 performed indentation tests corresponding to a particular value of $P_{\max}$. Subsequently, by averaging the iteratively received values of $\Gamma$, an estimate for the cohesive energy density of the analysed silicon crystal was obtained. The reliability of the used numerical model follows from the comparison of the determined values of $\Gamma$ with values reported in literature. Figure 6 shows the dependency of crack length on the indentation depth and the iteratively received values of $\Gamma$ for particular loading force. If a particular crack length for an appropriate indentation depth is selected in Figure 6, a corresponding value of the cohesive energy density $2\Gamma$ can be read off. For tested forces/depths, see Figure 5, we get $\Gamma = 3.06$ J/m$^2$ (crack length 7.5 µm and the indentation depth 1.18 µm), $\Gamma = 2.81$ J/m$^2$ (crack length 11.4 µm and indentation depth 1.57 µm), $\Gamma = 2.63$ J/m$^2$ (crack length 15.4 µm and indentation depth 1.97 µm) and $\Gamma = 2.70$ J/m$^2$ (crack length 18.9 µm and indentation depth 2.33 µm). It is seen that with increasing indentation depth the estimate of the cohesive energy density converges to the value $2\Gamma = 5.30$ J/m$^2$ which agrees well with the silicon cohesive energy density values reported in literature. This convergence is due to a decrease in the measurement error with increasing indentation depth. In general, measurements at a lower indentation depth (a lower applied force) are subject to a larger error. The critical crack opening displacement is $\Delta_c = 13$ nm. Let us however point out again that the critical opening displacement $\Delta_c$ does depend on the element size $D$, c.f. Equation (1), which, as already mentioned, is 0.25 µm.

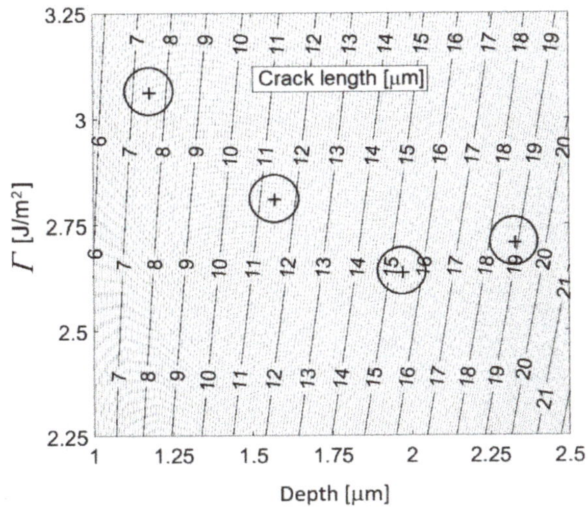


**Figure 6.** Crack length as function of the indentation depth and the cohesive energy density.

*3.2. Crack Extension during Indentation Test*

With determined parameters $\Gamma, \Delta_c$, crack initiation and growth modelling can be attempted during loading and unloading phases of the indentation test. It is well-known that cracks initiate during the loading phase below the plastic zone which develops under the contact. Further increase of indenter loading leads to crack extension. Figure 7 shows several stages of crack development during the indentation test modelled for maximal loading force of 300 mN.

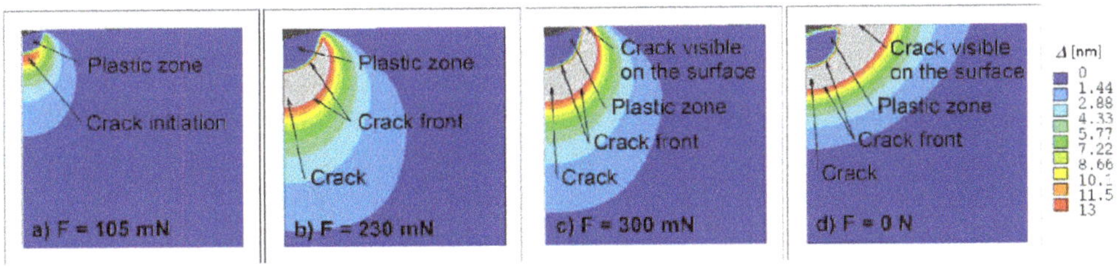


**Figure 7.** Crack extension during the indentation test for maximal loading force of 300 mN: loading force (**a**) F = 105 mN, (**b**) F = 230 mN, (**c**) F = 300 mN and (**d**) F = 0 N.

The instant when this crack occurs can be experimentally captured by monitoring the differential hardness dependence on indentation depth and/or applied force which provides an efficient tool to visualise the indentation induced changes in a tested material such as inception of cracks [35].

From measurements, the value of the acting force was approximately 90 mN, see Figure 8a, which is in a good accordance with numerical simulation results (F = 105 mN). Figure 8b presents the same measurements, however the differential hardness is plotted against the recorded indentation depth. One of the advantages of numerical simulations is the visualisation of the invisible crack inception and crack extension. During the unloading stage of indentation test, the originally invisible crack grows to the top surface of the specimen and becomes visible when the applied force decreases from 300 mN to 230 mN, see Figure 7. With further force decrease the crack grows in the radial direction and after complete indenter unloading one can observe cohesive behaviour near the crack tip manifested by cups-like closure, see Figure 9. It is a matter of interest to display numerically predicted distribution of the crack opening along the crack flanks. The results computed for the indentation test with $P_{max}$ = 750 mN are displayed in Figure 10. Figure 10 shows the distribution of crack opening along the crack flanks within the cohesive area which allows to identify the crack front. It is clearly seen that crack grows during the unloading stage of indentation test due to residual stress field.

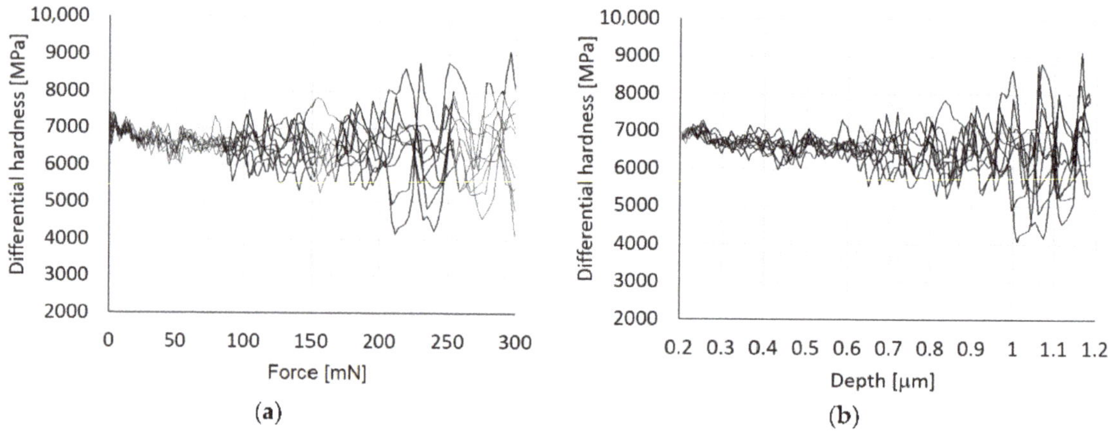


**Figure 8.** Dependence of differential hardness on (**a**) force, (**b**) depth.

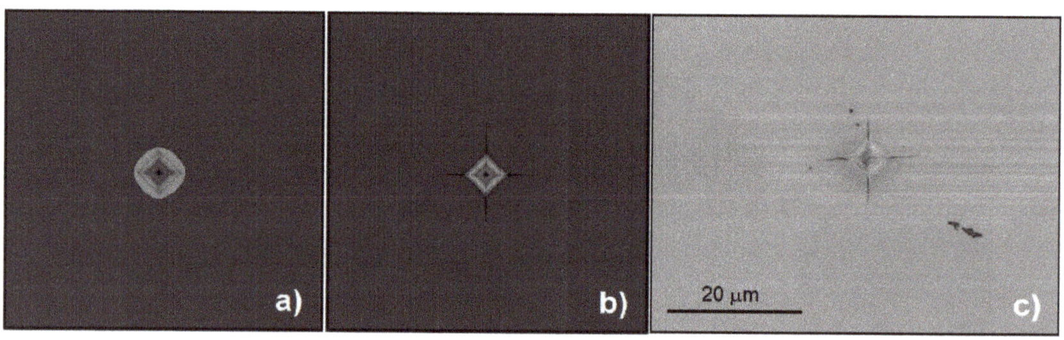


**Figure 9.** Indentation with the maximal loading force $P_{max}$ = 300 mN: (**a**) state at maximal load (numerical simulation), (**b**) state after complete unloading (numerical simulation), (**c**) state after complete unloading (experimental observation).

### 3.3. Mesh Density of Cohesive Zone Area

It was already pointed out that the critical opening displacement $\Delta_c$ depends on the element size $D$. As mentioned in Introduction, Nguyen and Ortiz [25] suggested a way to the macroscopic form of the cohesive law by considering the cooperative behaviour of a large number $N$ of interatomic planes forming a cohesive layer. This approach then shows that $\Delta_c$ scales with $\sqrt{D}$. A distinctively weaker dependency on the element size $D$ used for discretisation of the cohesive zone area, see Figure 2, was observed for the radial crack length and for the work of cohesive forces as well. Several loading forces were tested and for each the size of the cohesive zone area was adjusted, and thus the discretisation density with respect to the crack surface area. The greater the loading force and, as a result, the greater the crack surface area, the lower is the discretisation error due to the size of the element used. For that reason, the sensitivity to the size of the element (ESIZE) was performed, especially because the ideal size of the element would be at the atomic level—this would of course correspond to a different traction-separation $T(\Delta)$ dependence. Figure 11 shows a linear regression of the dependency of the crack length on the element size which was used for prediction of the crack length in case when the element size is approaching zero. These data were determined for each indentation loading force and then they were compared with experimental observation and measurement of the radial crack length, see Equation (18).

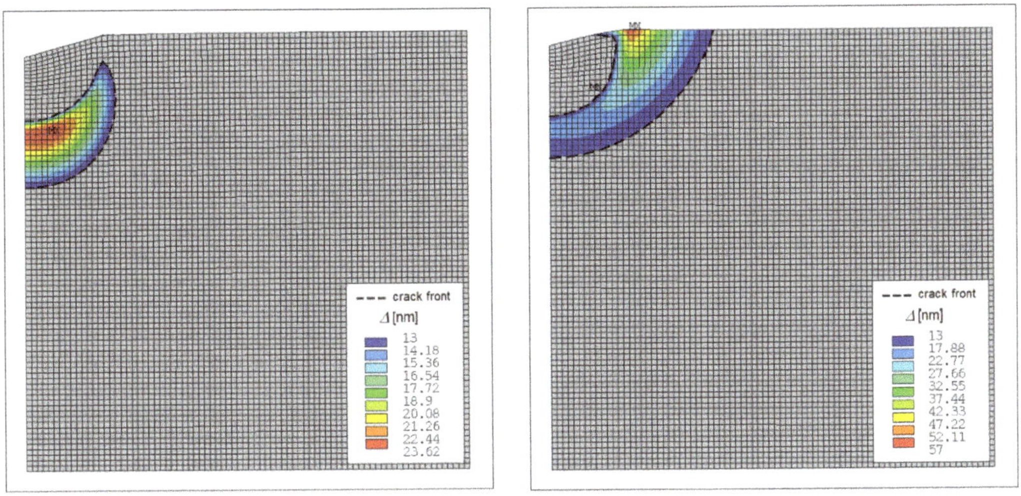


(a) (b)

**Figure 10.** Distribution of the crack opening Δ along the crack flanks for $P_{max}$ = 300 mN (**a**) maximal loading, (**b**) complete unloading.

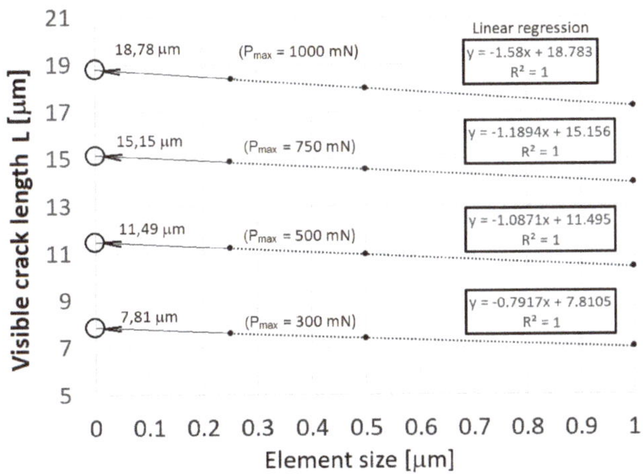


**Figure 11.** Size effect of mesh density on crack length.

## 4. Discussion and Conclusions

While previous studies available in the literature, e.g., [7–10], which use cohesive zone finite element techniques for simulation of indentation cracks in brittle solids, tried to improve methods for the evaluation of material toughness from the indentation load, crack size, hardness, elastic constants, and indenter geometry, this study focuses on the evaluation of the cohesive energy density 2Γ from which the material toughness can be easily determined using the well-known Griffith-Irwin formula

$$K_{IC} = \sqrt{\frac{2\Gamma E}{1-\upsilon^2}}. \quad (19)$$

With the cohesive energy density determined as 2Γ = 5.30 J/m$^2$ the Formula (19) gives $K_{IC}$ = 0.86 MPa·m$^{1/2}$. In contrast to the previous studies, there is no need to control the premise of the linear fracture mechanics that the cohesive zone is much shorter than the crack length. Hence, the developed approach is suitable also for short cracks for which the linear fracture mechanics premise is violated. Besides, in spite of the previous improvements of the indentation cracking formulas, they are still relatively inaccurate to predict the fracture toughness in comparison to the proposed approach based on evaluation of the cohesive energy density.

An integral part of the analysis is modelling of the permanent deformation under the contact since it gives rise to a residual stress field which is the primary driving force for cracks during the unloading process. A reliable model of the permanent deformation and the related residual stress/strain field requires knowledge of the yield stress, the reduced Young´s modulus of the whole system consisting of the load cell, Vickers indenter and the silicon crystal, and also the real shape of the indenter. All these parameters were found using an optimisation procedure which provides the best fit between the experimental indentation data and the optimised indentation data, produced by FE analysis. Specifically, optimal value of the yield strength $\sigma_{y,\,SC}$ was 6.4 GPa. The computed residual stress/strain field enters the analysis of the inverse problem for identification of the cohesive energy density 2Γ and the critical crack opening displacement $\Delta_c$. The inverse problem solution requires to find the best fit between the visible crack length on the top surface of the silicon crystal and its numerical prediction obtained by FE analysis under full unloading and simultaneously to ensure minimisation of the total energy. The solution results are presented in the form of a diagram which links together the cohesive energy density, the crack length, and the indentation depth, from which the cohesive energy density 2Γ can be easily read off for particular crack length and indentation depth. Nevertheless, in case of a lower indentation depth the measurements are subject to a larger error which is reflected in the estimation error of the cohesive energy density. As the indentation depth increases, the error decreases and the estimate of the cohesive energy density converges to the value 2Γ = 5.30 J/m$^2$.

There are several conflicts concerning the selection of the cohesive interface parameters $\sigma_{\max}$, $\Delta_c$ in FE modelling of cleavage fracture including the indentation cracks in brittle materials using the cohesive interface model. As already mentioned in Introduction, the ab initio calculations show that the crystal loses its bearing capacity after an interplanar separation of only a few Angstroms. Simultaneously, the peak stresses within the interplanar separation zone are of the order of theoretical strength of crystal. To reach such values in macroscopic FE simulation, extremely fine mesh would be required and full atomistic resolution in the vicinity of the crack would be necessitated, which is however unfeasible and impractical. Therefore, a suitable transformation of atomistic binding relation leading to macroscopic cohesive law is needed. A way to the macroscopic form of the cohesive law was suggested by Nguyen and Ortiz [25] as mentioned in Introduction. As a result, the cohesive interface parameters $\sigma_{\max}$, $\Delta_c$ in any macroscopic cohesive law for the cleavage fracture differ by orders from their physical atomistic counterparts, however with the cohesive energy density (critical fracture energy) 2Γ remaining unchanged. There are other limitations for the choice of the parameter $\sigma_{\max}$ which were thoroughly investigated and discussed in [7]. It was shown that the parameter $\sigma_{\max}$ should be chosen to be lower than $\approx$0.2 $\sigma_y$ to ensure initiation of crack in a linear fracture mechanics context. Moreover, care is needed when changing $\sigma_{\max}$ because then the crack bridging zone also changes which affects the choice of cohesive element size $D$. In the papers [7,10] $\sigma_{\max}$ was chosen from the range <0.5, 1> GPa, the typical value of the yield strength was $\sigma_y = 5$ GPa, and the typical value of the Young modulus was $E$ = 200 GPa. The fracture toughness $K_{IC}$ ranged from 0.7 to 1 MPa.m$^{1/2}$. All these papers used the bilinear form of the cohesive traction—separation law. In this paper, the parameter $\sigma_{\max}$, and/or the critical opening displacement $\Delta_c$ together with the cohesive energy density $2\Gamma$ were selected to provide best fit between the visible crack length on the top surface and its numerical prediction according to Equation (18). The optimal values were found as $\sigma_{\max} = 150$ MPa, $\Delta_c$ = 13 nm, and $2\Gamma$ = 5.30 J/m$^2$. It is interesting to notice that the asymptotic scaling rule in Equation $(1)_1$ predicts the macroscopic critical opening displacement as $\Delta_c \cong 14.4$ nm, see the text below Equation (3).

The reliability of the developed model is evidenced by the determined value of the cohesive energy density, which is in a good accordance with the values reported in the literature. Furthermore, the reliability of the model follows from the comparison of numerical simulation results with the measured differential hardness data which provide an estimate of the loading force for the indentation induced cracks inception during the loading stage. The proposed approach also seems appropriate for toughness evaluation of hard coatings bonded to a brittle substrate. Studies along this line will be left for a future work.

**Author Contributions:** P.S.—numerical model creation and computations, M.K.—methodology, writing—original draft preparation and editing. All authors have read and agreed to the published version of the manuscript.

**Funding:** The authors gratefully acknowledge a financial support of the Czech Science Foundation under the Project No.17-18566S.

**Institutional Review Board Statement:** Not applicable.

**Informed Consent Statement:** Not applicable.

**Data Availability Statement:** Data is contained within the article.

**Conflicts of Interest:** The authors declare no conflict of interest.

## Nomenclature

| | |
|---|---|
| $c$ | optimisation variable |
| $d$ | interplanar spacing |
| $h_{\max}$ | maximal depth of indentation |
| $f$ | friction coefficient |
| $i$ | position along the force-depth curve |
| $\mathbf{p}$ | actual tractions |
| $\mathbf{n}$ | unit normal to the surface |
| $A$ | size of cohesive zone |
| $C$ | interplanar modulus |
| $D$ | finite element size |
| $\mathbf{C}$ | stiffness tensor |
| $\mathbf{D}$ | compliance tensor |
| $\mathcal{F}$ | objective functional |
| $HU$ | universal hardness |
| $N$ | number of interatomic planes forming a cohesive layer |
| $E_{LC}$ | Young´s modulus of test equipment load cell |
| $E_r$ | Reduced Young´s modulus |
| $E_{SC}$ | Young´s modulus of the Si crystal |
| $E_{VDI}$ | Young´s modulus of Vickers diamond indenter |
| $G_{SC}$ | shear modulus of the Si crystal |
| $K_{IC}$ | material fracture toughness |
| $L_{ij}^{exp}$ | measured crack length on the top surface at i-th test |
| $L_j^{pred}$ | predicted crack length |

| | |
|---|---|
| $P$ | contact force |
| $P^{comp}$ | predicted loading force |
| $P^{exp}$ | experimental loading force |
| $P_{\max}$ | maximal force acting on the ideal Vickers indenter |
| $S_c$ | contact area between the indenter tip and the Si crystal |
| $S_{crack}$ | crack surface |
| $S_u$ | part of the boundary where displacement are prescribed |
| $T$ | cohesive traction |
| $\boldsymbol{\varepsilon}^p$ | permanent deformation |
| $\delta$ | microscopic crack opening |
| $\delta_c$ | critical microscopic crack opening (interplanar separation) |
| $\delta \boldsymbol{L}$ | local virtual crack extension |
| $\boldsymbol{\upsilon}$ | local unit normal vector to the crack front |
| $\boldsymbol{\nu}_{SC}$ | Poisson´s ratio of the Si crystal |
| $\nu_{VDI}$ | Poisson´s ratio of Vickers diamond indenter |
| $\nu_{LC}$ | Poisson´s ratio of test equipment load cell |
| $\psi$ | centreline-to-face angle |
| $\in$ | total energy |
| $\boldsymbol{\sigma}$ | complete stress-strain field |
| $\boldsymbol{\sigma}^m$ | elastic contact stress field |
| $\boldsymbol{\sigma}^r$ | residual stress field |
| $\sigma_{y,\,SC}$ | yield stress of the Si crystal |
| $\sigma_{max}$ | peak cohesive traction |
| $2\Gamma$ | cohesive energy density (critical fracture energy) |
| $\Delta$ | macroscopic opening displacement |
| $\Delta_c$ | corresponding damage-initiating displacement |
| $\Delta_{sep}$ | failure displacement |
| $\phi$ | interplanar cohesive potential |

## References


1. Luo, J.; Lin, J.; Dean, T.A. A study on the determination of mechanical properties of a power law material by its indentation force–depth curve. *Philos. Mag.* **2006**, *86*, 2881–2905. [CrossRef]
2. Luo, J.; Lin, J. A study on the determination of plastic properties of metals by instrumented indentation using two sharp indenters. *Int. J. Solids Struct.* **2007**, *44*, 5803–5817. [CrossRef]
3. Chaiwut, G.; Esteban, P.B. Characterization of elastoplastic properties based on inverse analysis and finite element modeling of two separate indenters. *J. Eng. Mater. Technol.* **2007**, *129*, 603–608.
4. Horníková, J.; Šandera, P.; Černý, M.; Pokluda, J. Multiscale Modelling of Nanoindentation Test in Coper Crystal. *Engng. Fract. Mech.* **2008**, *75*, 3755–3762. [CrossRef]
5. Bonnet, M.; Constantinescu, A. Inverse problems in elasticity. *Inverse Probl.* **2005**, *21*, R1–R50. [CrossRef]
6. Constantinescu, A.; Tardieu, N. On the identification of elastoviscoplastic constitutive laws from indentation tests. *Inverse Probl. Eng.* **2001**, *9*, 19–44. [CrossRef]
7. Lee, J.H.; Gao, Y.F.; Johanns, K.E.; Pharr, G.M. Cohesive interface simulations of indentation cracking as a fracture toughness measurement method for brittle materials. *Acta Mater.* **2012**, *60*, 5448–5467. [CrossRef]
8. Hyun, H.C.; Rickhey, F.; Lee, J.H.; Hahn, J.H.; Lee, H. Characteristics of indentation cracking using cohesive zone finite element techniques for pyramidal indenters. *Int. J. Solids Struct.* **2014**, *51*, 4327–4335. [CrossRef]
9. Rickhey, F.; Lee, J.H.; Lee, H. XFEM investigation on Knoop indentation cracking: Fracture toughness and aspect-ratio of radial-median cracks. *Mater. Des.* **2016**, *107*, 393–405. [CrossRef]
10. Rickhey, F.; Marimuthu, K.P.; Lee, H. Investigation on Indentation Cracking-Based Approaches for Residual Stress Evaluation. *Materials* **2017**, *10*, 404. [CrossRef]
11. Lawn, B.R.; Evans, A.G.; Marshall, D.B. Elastic/plastic indentation damage in ceramics: The median/radial crack system. *J. Am. Ceram. Soc.* **1980**, *63*, 574–581. [CrossRef]
12. Tvergaard, V.; Hutchinson, J.W. The relation between crack growth resistance and fracture process parameters in elastic-plastic solids. *J. Mech. Phys. Solids* **1992**, *40*, 1377–1392. [CrossRef]
13. Park, K.; Paulino, G.H. Cohesive zone models: A critical review of traction-separation relationships across fracture surfaces. *Appl. Mech. Rev.* **2011**, *64*, 060802. [CrossRef]
14. Zeng, K.; Rowcliffe, D. Experimental measurement of residual stress field around sharp indentation in glass. *J. Am. Ceram. Soc.* **1994**, *77*, 524–530. [CrossRef]
15. Kese, K.; Rowcliffe, D.J. Nanoindentation method for measuring residual stress in brittle materials. *J. Am. Ceram. Soc.* **2003**, *86*, 811–816. [CrossRef]

16. Zeng, K.; Giannakopoulos, A.E.; Rowcliffe, D.J. Vickers indentations in glass-II. Comparison of finite elementanalysis and experiments. *Acta Metall. Mater.* **1995**, *43*, 1945–1954. [CrossRef]
17. Peitl, O.; Serbena, F.C.; Mastelaro, V.R.; Zanotto, E.D. Internal residual stress measurements in a bioactive glass–ceramic using Vickers indentation. *J. Am. Ceram. Soc.* **2010**, *93*, 2359–2368. [CrossRef]
18. Xu, X.P.; Needleman, A.; Abraham, F.F. Effect of inhomogeneities on dynamic crack growth in an elastic solid. *Modeling Simul. Mater. Sci. Eng.* **1997**, *5*, 489–516. [CrossRef]
19. Zi, G.; Belytschko, T. New crack-tip elements for XFEM and applications to cohesive cracks. *Int. J. Numer. Methods Eng.* **2003**, *57*, 2221–2240. [CrossRef]
20. Mubashar, A.; Ashcroft, A.; Crocombe, A.D. Modelling damage and failure in adhesive joints using a combined XFEM-cohesive element methodology. *J. Adhes.* **2014**, *90*, 682–697. [CrossRef]
21. Stuparu, F.; Constantinescu, D.M.; Apostol, D.A.; Sandu, M. A Combined Cohesive Elements—XFEM Approachfor Analyzing Crack Propagation in Bonded Joints. *J. Adhes.* **2016**, *92*, 535–552. [CrossRef]
22. Bao, G.; Suo, Z. Remarks on crack-bridging concepts. *Appl. Mech. Rev.* **1992**, *45*, 355–366. [CrossRef]
23. Gao, Y.F.; Bower, A.F. A simple technique for avoiding convergence problems in finite element simulations of crack nucleation and growth on cohesive interfaces. *Modell. Simul. Mater. Sci. Eng.* **2004**, *12*, 453–463. [CrossRef]
24. Kotoul, M.; Skalka, P.; Profant, T.; Friák, M.; Řehák, P.; Šesták, P.; Černý, M.; Pokluda, J. Ab Initio Aided Strain Gradient Elasticity Theory in Prediction of Nanocomponent Fracture. *Mech. Mater.* **2019**, *136*, 103074. [CrossRef]
25. Nguyen, O.; Ortiz, M. Coarse-graining and renormalization of atomistic binding relations and universal macroscopic cohesive behaviour. *J. Mech. Phys. Solids* **2002**, *50*, 1727–1741. [CrossRef]
26. Rose, J.H.; Smith, J.R.; Ferrante, J. Universal features of bonding in metals. *Phys. Rev. B-Condens. Matter* **1983**, *28*, 1835–1845. [CrossRef]
27. Francois, P.; Lefebvre, A.; Vanderschaeve, G. Low temperature plasticity of brittle materials. A new device for compressive testing under confining pressure. *Phys. Status Solidi* **1988**, *109*, 187–192. [CrossRef]
28. Cook, R.F.; Pharr, G.M. Direct obseravation and analysis of indentation cracking in glasses and ceramics. *J. Am. Ceram. Soc.* **1990**, *73*, 787–817. [CrossRef]
29. Kang, J.J.; Becker, A.A.; Sun, W. Determining elastic–plastic properties from indentation data obtained from finite element simulations and experimental results. *Int. J. Mech. Sci.* **2012**, *62*, 34–46. [CrossRef]
30. Gu, Y.; Nakamura, T.; Prchlik, L.; Sampath, S.; Wallace, J. Micro-indentation and inverse analysis to characterize elastic-plastic graded materials. *Mater. Sci. Eng. A* **2003**, *345*, 223–233. [CrossRef]
31. Jiang, J.; Fasth, A.; Nylen, P.; Choi, W.B. Microindentation and inverse analysis to characterize elastic-plastic properties for thermal sprayed Ti2AlC and NiCoCrAlY. *Therm. Spray Technol.* **2009**, *18*, 194–200. [CrossRef]
32. Sun, G.; Xu, F.; Li, G.; Huang, X.; Li, Q. Determination of mechanical properties of the weld line by combining micro-indentation with inverse modelling. *Comput. Mater. Sci.* **2014**, *85*, 347–362. [CrossRef]
33. De Bono, D.M.; London, T.; Baker, M.; Whiting, M.J. A robust inverse analysis method to estimate the local tensile properties of heterogeneous materials from nano-indentation data. *Int. J. Mech. Sci.* **2017**, *123*, 162–176. [CrossRef]
34. Oliver, W.C.; Pharr, G.M. An improved technique for determining hardness and elastic modulus using load and displacement sensing indentation experiments. *J. Mater. Res.* **1992**, *7*, 1564–1583. [CrossRef]
35. Bursikova, V.; Blahova, O.; Karaskova, M.; Zajickova, L.; Jasek, O.; Franta, D.; Bursik, J. Mechanical properties of ultrananocrystalline thin films deposited using dual frequency discharges. *Chem. Listy* **2011**, *105*, 98–101.

*materials*

*Article*

# Microstructure and Fatigue Damage of 316L Stainless Steel Manufactured by Selective Laser Melting (SLM)

**Zhentao Wang [1], Shanglei Yang [1,2,*], Yubao Huang [1], Cong Fan [1], Zeng Peng [1] and Zihao Gao [1]**

1 School of Materials Engineering, Shanghai University of Engineering Science, Shanghai 201600, China; wangzhentao23@163.com (Z.W.); huangyb16@163.com (Y.H.); fc19960802@163.com (C.F.); pengzeng98@163.com (Z.P.); gzh_960303@163.com (Z.G.)

2 Shanghai Research and Development Center for Key Technologies of Ultra-Intense Laser Processing, Shanghai University of Engineering Science, Shanghai 201600, China

* Correspondence: yslei@sues.edu.cn

**Abstract:** In this paper, 316L stainless steel powder was processed and formed by selective laser melting (SLM). The microstructure of the sample was studied using an optical microscope, and the fatigue failure of the sample and the characteristics of crack initiation and propagation were analyzed, providing a research basis for the application of SLM-316L. Due to the influence of microstructure and SLM process defects, the fatigue cracks of SLM-316L mainly emerged due to defects such as lack of fusion and pores, while the cracks of rolled 316L initiated at the inclusions near the surface of the specimen. After fatigue microcrack initiation of the SLM-316L specimen, due to the existence of shear stress and tear stress, the crack tip was passivated and Z-shaped propagation was formed. The existence of internal defects in SLM-316L made the microcrack initiation random and diverse. At the same time, the existence of defects affected the crack propagation in the form of bending, bifurcation and bridge, which made the main crack propagation deviate from the maximum load direction.

**Keywords:** 316L stainless steel; selective laser melting; microstructure; fatigue damage

**Citation:** Wang, Z.; Yang, S.; Huang, Y.; Fan, C.; Peng, Z.; Gao, Z. Microstructure and Fatigue Damage of 316L Stainless Steel Manufactured by Selective Laser Melting (SLM). *Materials* **2021**, *14*, 7544. https://doi.org/10.3390/ma14247544

Academic Editors: Jaroslav Pokluda and Reinhard Pippan

Received: 29 September 2021
Accepted: 23 November 2021
Published: 8 December 2021



## 1. Introduction

Selective laser melting (SLM) technology is a kind of additive manufacturing technology with great development potential. Based on the discrete-stack principle, this technology uses the laser beam as a heat source under the control of a galvanometer, and is guided by computer-aided design data to melt the selective area of metal powder layer by layer, so that the metal powder accumulates into a solid [1,2]. Recent studies have shown that the SLM process exhibits good mechanical properties due to its rapid cooling and unique multi-scale (from nano to macro scale) heterogeneous structure [3–5]. SLM is not only a useful supplement to traditional processing methods such as casting, forging, welding and machining, but also a new manufacturing mode of metal parts. SLM has broad prospects for development in the biomedical, aerospace and energy industries [6,7].

316L stainless steel is a kind of ultra-low carbon austenitic stainless steel, which is widely used in the aerospace and automobile industries due to its good comprehensive mechanical properties [8,9]. In recent years, the precise fabrication of complex parts for the aerospace industry using 316L stainless steel by SLM has become a research hotspot [10]. Compared to traditional molding technology, SLM has the advantage of not requiring development of a mold and having a short manufacturing cycle. Despite these irreplaceable advantages, the complex thermal effect and the typical rapid melting and cooling involved [3,11] means that SLM austenitic stainless steel has inherent defects, such as internal defects (especially unmelted hole defects), surface defects and residual stress, which strongly affect the fatigue performance of the processed materials, thereby reducing the reliability of engineering applications [12,13]. At present, many scholars have conducted a lot of research on the powder characteristics, optimization of different process parameters, density and mechanical properties of 316L stainless steel [14–16]. However, there are few reports investigating the fatigue failure mechanism of SLM-316L, yet fatigue failure is one of the most common failure forms of structural components under alternating loads [17–20]. The research on fatigue failure of SLM-316L is of great significance to the development of SLM technology. Riemer et al. used the fatigue behavior of 316L manufactured by SLM to check the crack initiation and crack growth behavior [21]. The results obtained clearly show that 316L is a promising candidate for cyclically loaded parts manufactured by SLM. Primarily attributed to the high ductility that directly follows the SLM process, the 316L stainless steel shows fatigue properties similar to that of a conventionally processed material in its as-built model. Sarkar et al. studied the fatigue properties and fracture types of SLM stainless steel under different average stress modes [22]. The fatigue properties and fracture characteristics of the samples under different average stress modes were significantly different, were mainly manifested as the average stress in tensile mode was not conducive to the fatigue performance, while the average stress in compression mode could improve the fatigue performance. Pegues et al. studied the fatigue crack initiation characteristics and the mechanism of AM-304L by the in situ microstructure observation of an interrupted fatigue test [23], and considered that the microstructure refinement and the decrease in high angle grain boundary density (HAGBs) improved the fatigue properties of AM-304L. Although relevant studies have reported the fracture process of static tensile failure and the corresponding potential failure mechanism, there are few studies on the fatigue failure of SLM-316L, which is insufficient for such an important failure. In order to understand the fatigue failure mechanism of SLM-316L, a series of fatigue tests were carried out on SLM-316L and rolled 316L by using a HB250 electro-hydraulic servo dynamic fatigue testing machine. The fatigue fracture surfaces of the two were observed by scanning electron microscopy (SEM) and optical microscopy (OM), and then the fatigue failure mechanism and crack propagation of SLM-316L were analyzed. These results provide a new understanding for us to better understand the fatigue failure of SLM-316L stainless steel.

## 2. Test Materials and Methods

The composition of 316L powder and traditional 316L stain-less steel is shown in Table 1. Before laser additive manufacturing, the powder was heated in a drying furnace to remove the moisture on the surface of the metal powder and increase the fluidity in the powder to avoid defects such as pores inside the additive manufacturing sample.

**Table 1.** Composition of powder and rolled 316L.

| Composion | Cr | Mn | Ni | Mo | Cu | C | P | S | Fe |
|---|---|---|---|---|---|---|---|---|---|
| 316L powder | 16.97 | 0.8 | 12.2 | 2.89 | 0.01 | 0.003 | 0.002 | 0.006 | Bal. |
| Rolled 316L | 16–18 | ≤2.0 | 10–14 | 2.0–3.0 | — | ≤0.03 | ≤0.045 | ≤0.03 | Bal. |

The forming equipment used in this study was an HBD-100SLM, produced by Hanbon Technology Co., Ltd. (Shanghai, China), as shown in Figure 1d. The schematic diagram of the 316L laser selective forming process and the scanning strategy is shown in Figure 1a–c. The process parameters used in this study are shown in Table 2. Before printing, the oxygen in the molding chamber was reduced to less than 1000 ppm, 316L stainless steel was used as the substrate, and the shielding gas was argon. In the printing process, a rotation of 67°, layer by layer, was performed using the scanning strategy shown in Figure 1c. The forming samples were prepared according to the two-dimensional data of the software imported before the test and the process parameters in Table 2. After preparing SLM-316L sample, the sample was cut according to the size diagram in Figure 1e.

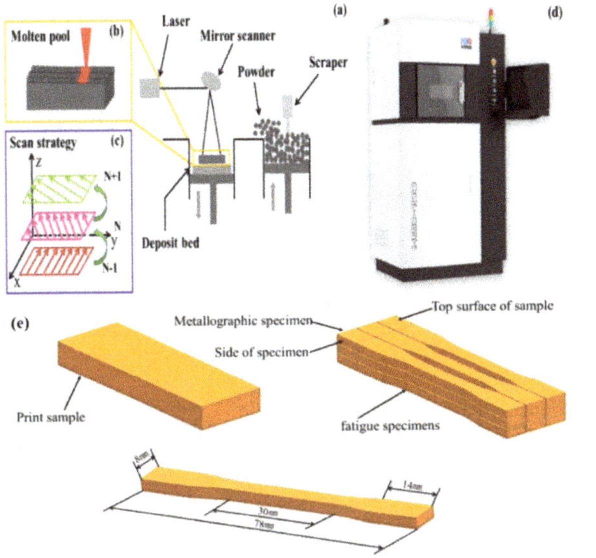


**Figure 1.** SLM forming equipment HBD-100D and cutting diagram of metallographic and fatigue samples. (**a**) Printing process, (**b**) molten pool, (**c**) Printing strategy, (**d**) Print instrument, (**e**) cutting diagram of metallographic and fatigue sam-ples.

**Table 2.** SLM process parameters.

| Materials | Scanning Distance (mm) | Laser Power (W) | Scanning Speed (mm/s) | Layer Thickness (μm) | Spot Distance (μm) |
|---|---|---|---|---|---|
| 316L | 0.07 | 160 | 1000 | 30 | 70 |

After cutting, the samples were sanded by sandpapers with different levels of roughness until the surface scratches were shallow. Next, the samples were polished until no obvious scratches were observed under the microscope. Once polished, the samples were etched with corrosives ($HCl:HNO_3$ = 3:1) for about 10 s.

According to GB/T 26077-2010 standard, and in a room temperature (25 ± 3 °C) environment with relative humidity (40–60%), the rolling 316L and SLM-316L were subjected to fatigue tests using the symmetric tension-tension cycle of sinusoidal loading waveforms until the sample reached $10^6$ cycles or fracture. The stress ratio (R = σmin/σmax) and load frequency were 0.1 and 15 Hz, respectively. All fatigue samples were cyclically loaded at different stress ranges (Δσ = σmax−σmin). The ultimate fatigue strength (Nf ≥ $10^6$) of rolled 316L and SLM-316L was analyzed by an SN curve. All fatigue tests were carried out on a Zwick/Roell Amsler HB250 hydraulic servo material testing machine, and the fatigue fracture was observed and analyzed by an S-3400N scanning electron microscope (SEM).

## 3. Results and Discussion

### *3.1. Microstructure and Defect Analysis*

The sample was based on the XOY surface as the bottom surface and the Z axis as the construction direction, as shown in Figures 1c and 2a,b, respectively, which illustrates the metallographic microstructure of samples in the vertical and parallel directions of construction. Once the additive manufacturing commenced, the laser was directed in a predetermined path, melting the metal powder to form a molten micro-molten pool, where the molten metals were bonded to each other in order to achieve the bonding of atoms. Figure 2a shows the microstructure of the sample perpendicular to the additive manufacturing direction, and the traces formed by the laser scanning powder can be seen. Figure 2b shows the microstructure of the sample parallel to the additive manufacturing direction. Visibly, a lot of fish-scale-shaped channels are stacked together. As the heat source of nearly Gaussian heat, the laser beam had a higher energy density in the center and a lower energy density in the edge. During the process of melting, the penetration of powder scanned by the center of laser beam was greater than that scanned by the edge of laser beam. After a lot of pre-laid powders were scanned one-dimensionally, two-dimensionally and then three-dimensionally by laser beam, a lot of overlapping fish scale interfaces were formed. The liquid molten pool nucleated directly on the contact surface between the molten metal and the semi molten grains and grew epitaxially. In addition, at the bottom of the micro molten pool, the temperature gradient was high, and the nucleation grains grew along the direction perpendicular to the temperature gradient, forming a series of long columnar crystals that grew through multiple powder layers [24,25].

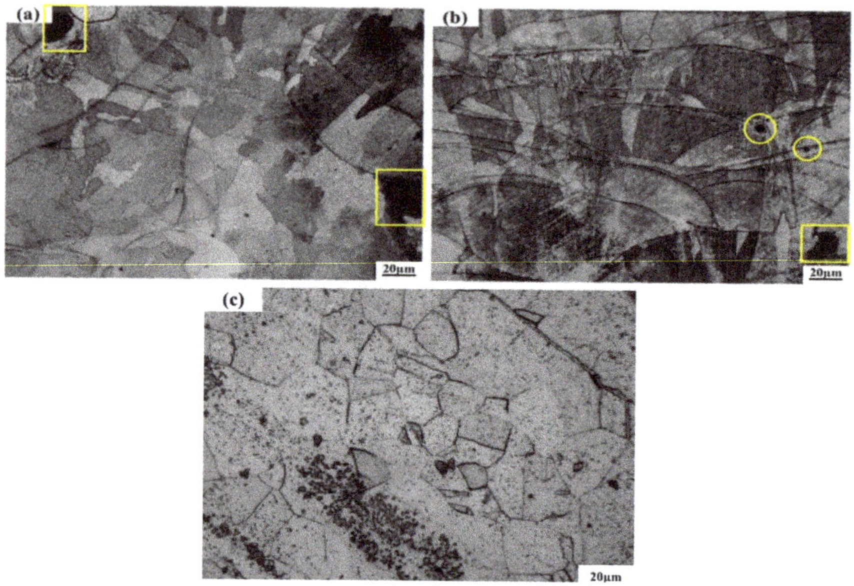


**Figure 2.** Microstructure of SLM-316L stainless steel samples in different building directions and rolling 316L: (**a**) microstructure of SLM-316L stainless steel sample perpendicular to the building direction, (**b**) microstructure of SLM-316L stainless steel sample parallel to the building direction, (**c**) microstructure of rolling 316L.

Figure 2a,b show there were some defects such as pores (in yellow circles) and unfused porosity (in yellow boxes), and the gaps appeared where the melting channels made contact. Weingarten et al. studied the pore composition of laser selective melting samples. Their research found that the $H_2O$ attached to the powder surface decomposed and saturated in the molten metal to form hydrogen when the laser interacted with the metal powder. In the subsequent solidification process of the metal, the hydrogen overflowed to form pores. Before the metal powder was melted, there were some voids between the powders. During the solidification of molten metal, the gas in the voids will also produce pores without overflow. Figure 2c shows the microstructure of rolled 316L. Many rectangular grains can be seen, and the grain size distribution is uniform. The formation of splash and unfused porosity is shown in Figure 3. In the process of laser selective additive manufacturing, laser irradiation melts the metal powder to form a liquid molten pool. With the accumulation of heat, and after reaching the boiling point of the molten pool, the metal vapor is formed. Under the action of recoil pressure, the molten pool splashes. The existence of spatter can be seen in Figure 3b. In addition, as a near Gaussian heat source, the laser has high central energy density and small edges. During the laser scanning process, the molten pool is unstable, which can easily to lead to incomplete fusion at the overlap between the weld bead and the weld bead [26]. The presence of non-fusion can be seen in Figure 3c, and there is still unmelted metal powder inside the non-fusion in Figure 3d.

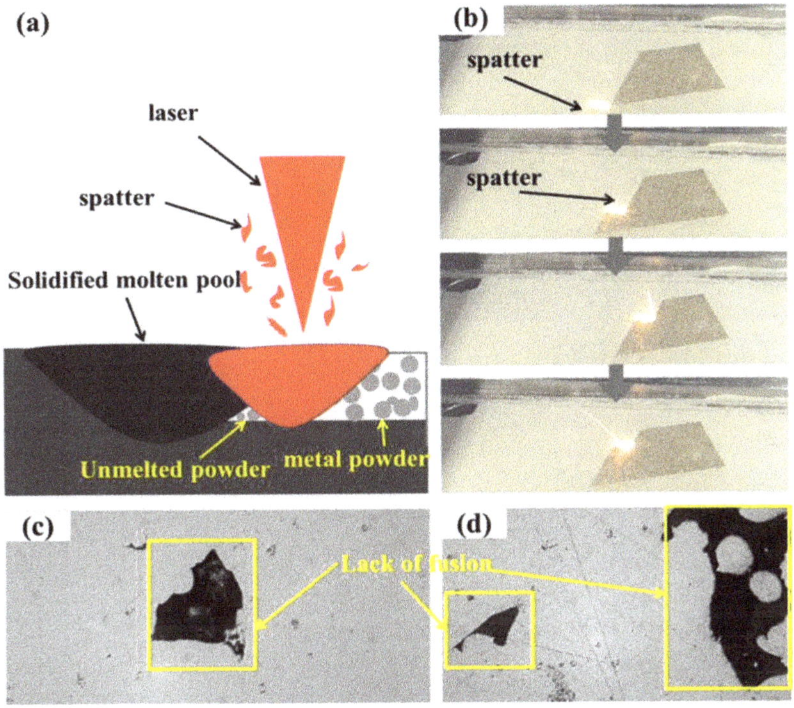


**Figure 3.** Mechanism of spatter and incomplete fusion in SLM forming process: (**a**) laser scanning single channel metal powder model, (**b**) forming process, (**c**) lack of fusion, (**d**) lack of fusion with unmelted metal powder.

### 3.2. Fatigue Test Analysis

During the course of this study, a laboratory hydraulic servo low-frequency fatigue testing machine was used to apply a load to simulate the actual service process of the sample. The national standard GB/T 3075-2008 "Metal Material Fatigue Test Axial Force Control Method" was followed, using the lifting method to perform the fatigue test with $10^6$ cycles for rolling 316L and laser selective additive manufacturing 316L. Fatigue data results are shown in Table 3. According to the data in Table 3, the fatigue limits of rolling 316L and SLM-316L specimens were 370 MPa and 200 Mpa, respectively, under the condition that no fracture occurred after $10^6$ times. Without breaking after $10^6$ times, the stress level of SLM-316L only reached 54% of that of rolled 316L. As the stress level increased, the fatigue life of rolled 316L decreased rapidly. The fatigue life of laser-selected area additive manufacturing 316L decreased slowly with the increase in stress level. Sometimes with the increase in stress level, life expectancy will increase. Figure 4a–c show the fracture morphologies of rolled 316 B5, and Figure 4a shows the macroscopic morphology of sample B5. Visibly, there are fewer internal defects, an obvious plastic deformation in the fracture, and obvious inclusion defects in the lower left corner of the fracture. The inclusion defects are located on the surface of the sample and spread radially around. Figure 4b shows an enlarged view of the fatigue source area of sample B5, which is also a further enlargement of Figure 4a. Noticeably, fatigue cracks originate from inclusion defects on the surface of the sample. After the crack is formed, it starts from the defect and spreads radially around. Defects under cyclic loading will cause the stress concentration to be higher than the surroundings, resulting in microcracks in the sample without a large external load. Figure 4c is an enlarged view of the defect of the fracture surface of sample B5, which is also a further enlargement of Figure 4b. One can see the existence of the tear ridge (inside the yellow ellipse), which is left when the cracks in the expansion process meet once they have expanded in all directions.

**Table 3.** Fatigue test data of rolled 316L and SLM-316L.

| Type | Sample No. | Maximum Stress (Smax/Mpa) | Stress Amplitude (Sa/Mpa) | Life Cycle |
|---|---|---|---|---|
| Rolled | B-1 | 460 | 207 | 241,870 |
| Rolled | B-2 | 440 | 198 | 262,350 |
| Rolled | B-3 | 420 | 189 | 354,397 |
| Rolled | B-4 | 400 | 180 | 461,752 |
| Rolled | B-5 | 380 | 171 | 600,691 |
| Rolled | B-6 | 370 | 166.5 | 1,000,000 |
| Rolled | B-7 | 360 | 162 | 1,000,000 |
| SLM | S-1 | 300 | 135 | 421,906 |
| SLM | S-2 | 280 | 126 | 399,303 |
| SLM | S-3 | 260 | 117 | 465,199 |
| SLM | S-4 | 240 | 108 | 454,712 |
| SLM | S-5 | 220 | 99 | 530,287 |
| SLM | S-6 | 210 | 94.5 | 580,493 |
| SLM | S-7 | 200 | 90 | 1,000,000 |

Figure 4d–f show the fracture morphology of SLM-316L sample S6. Figure 4d shows the macro fracture morphology of sample S6. The figure illustrates that there are defects such as sintered powder (in the yellow frame) and unfused pores (in the yellow circle) inside the sample. Figure 4e–h are an enlarged view of the fracture morphology of sample S6, which is also a further enlargement of Figure 4d. As shown in Figure 4e, you can see the tearing ridges produced by the cracks on the left that meet the defects caused by unfusion. The cracks will expand further after encountering the defects during the propagation process. Furthermore, as shown in Figure 4f, around the sintered powder, the cracks generated will expand to the surroundings. As shown in Figure 4g, cracks are generated on the surface of the sample, and then expand forward, and tear ridges will also be generated when intersecting with other cracks. This differs from rolled 316L in that, depending on the type of cracks, microcracks may occur on the surface of defects such as sintered powder, unfused, and pores. According to the location of the crack, the fatigue source of laser-selected area additive manufacturing 316L not only exists on the surface of the sample, but also exists inside the sample, and may also exist near the surface. When cyclic loading, the stress concentration is prone to occur near the defect, especially on the surface of some sharp notches, as the stress concentration factor is much higher than the surroundings. There are microcracks near defects such as sintered powder, LOF, pores, etc., indicating that these process defects promote the formation and propagation of cracks during the actual loading process of the sample. The materials used this time are added with the same parameters. There are defects in samples S1–S7, however some samples have many defects and some samples have few defects. As a result, the fatigue life of laser selective additive manufacturing of 316L is not obvious, or the fatigue life is increased by the increase in stress level. If other parameters of SLM-316L are selected, the internal defects of 316L may be reduced and the fatigue performance may be better, however the defect cannot be completely eliminated. Defects such as sintered powder, LOF, and pores are common in laser additive manufacturing. These defects can be divided into two categories according to the causes. The first type is pores, which are caused by unstable small holes in the sidewalls that are prone to local severe necking and collapse during

laser additive manufacturing. The necking and collapse of the liquid metal on the sidewall of the small hole wraps the gas in the small hole to form pores and cavities. The second category is non-fusion defects. There are two reasons for the occurrence of non-fusion defects: The first aspect is that the instability of laser additive manufacturing causes the hole wall to collapse and the powder particles in the bubbles of the liquid molten pool are not discharged in time. After the solidification of the molten pool, the powder particles are solidified in the weld to form particle slag. As shown in Figure 4h, on the other hand, the particle size is different. Large particles need more heat to melt. The laser is characterized by fast heating speed and fast cooling speed. The laser cannot melt particles with a larger diameter at one time, and the particles with a smaller diameter have a large specific surface area, a large heating area per unit, and high heat absorption. During laser additive process, the small particles will first melt and stick around the large particles to further block the transmission of laser heat. Finally, the large particles and the small particles below the large particles cannot be completely melted to form defects, as shown in Figure 4f.

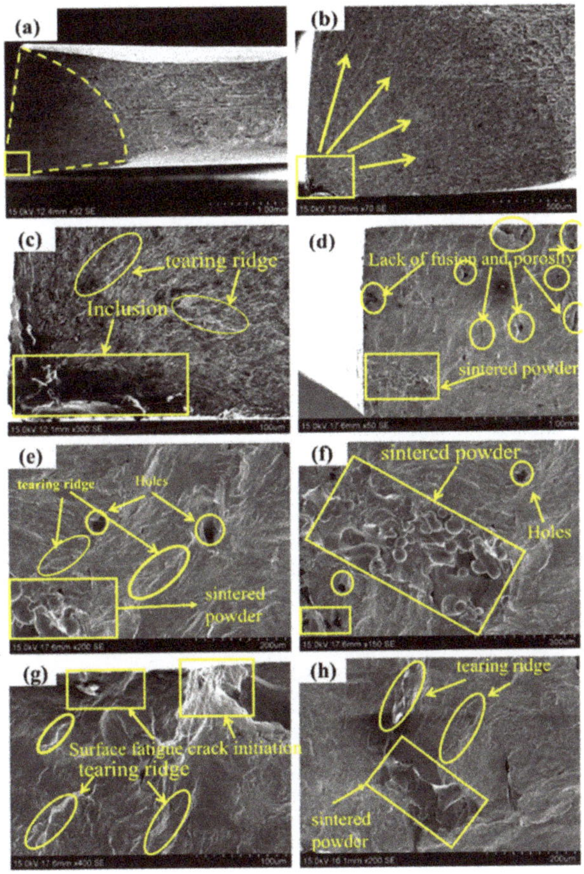


**Figure 4.** Fracture morphology of rolzSLM-316L fatigue initiation: (**a–c**) rolling 316L, (**d–h**) SLM-316L.

### 3.3. *Surface Fatigue Damage Analysis*

Figure 5 shows the crack propagation morphology on the surface of SLM-316L specimens under different stress conditions. When the stress is 250 MPa, the main crack begins on the surface of the sample, and then expands in a Z-shape way under cyclic loading. Holes are found on the surface of the sample. These holes are the weak areas in the process of crack growth, where the tip of the crack tends to propagate. Through the observation of the overall propagation path of cracks in Figure 5a, we found that multiple deflections had occurred during the propagation, and the secondary microcracks that were propagating along the surface defects of the specimen are differentiated at I, II and III in Figure 5a. Among them, the microcracks at I and II no longer propagated forward after extending a certain distance, while the main cracks at III were divided into microcracks that propagated in two different directions. Careful observation shows that when crack 1 did not extend to the microcracks at position III, secondary microcracks had been generated around the holes. These microcracks shared the crack driving force at the tip of crack 1 and reduced the driving force at crack 1, making this insufficient for crack 1 to continue to expand and therefore rendering crack 1 blunt. At this time, crack 1 and the defects in the upper left corner of the specimen had formed a tip plastic zone in a certain area. With the further expansion of the crack, crack 1 may have expanded to the defects in the upper left corner. Further amplification of the crack origin is shown in Figure 5d. Visibly, there are pore defects holes on the surface of the specimen. During the fatigue test, the stress concentration occurs at the hole, resulting in the initiation of microcracks. Obvious slip can be found at the edge of fracture. From Figure 5e,i, it can be seen that when the cyclic stress is 270 MPa and 290 MPa, respectively, the main cracks of SLM-316L samples originated from the surface of the samples and the number of deflections during the process of crack propagation was more than that of 250 MPa. In addition, slip can be found near the crack of all the fracture samples.

Figure 6 shows the fatigue damage of the sample surface after the fatigue test of SLM-316L was interrupted under different stresses. Figure 6a shows the plastic damage surface of the sample when the maximum cyclic stress was 250 Mpa. Visibly, there is a plastic zone at the crack tip, and a certain part of the crack tip has been enlarged and is visible in Figure 6b. From Figure 6b, we can see a lot of dislocation slip bands with a slip distance of 0.5–1 μm, and different types of slip are marked. Type I slip is distributed regularly in a cluster, and this slip originates from the grain boundaries. Type II slip exists in isolation, and this slip is often located close to the grain boundary. Type III slip is the formation of very small slip marker clusters inside the grains and near the grain boundaries. Two sliding mark directions are usually observed. The first one is more significant and occupies most of the surface of the cluster. This is probably related to the slip zone of the main slip system. The second one is less obvious, and is called the secondary slip zone [27].As shown in Figure 6a,b, we can see that a large amount of type I slip exists, and this slip originates from the grain boundary. In Figure 6a, we can see the existence of type II slip. Type II slip exists in isolation and is not distributed in a certain regular cluster, like type I slip. This type of slip is often located near the grain boundary. No obvious type III slip was found. Figure 6c shows the plastic damage surface of the sample when the maximum cyclic stress is 270 MPa, and the crack tip is enlarged to obtain Figure 6d. From Figure 6d, a large number of dislocation slip bands can also be seen. Compared with the stress of 250 MPa, only I-type slip is found, and the distance is similar to the stress of 250 MPa. Figure 6e shows the plastic damage surface of the sample when the maximum cyclic stress is 290 MPa. Figure 6f is obtained by enlarging a certain part of the crack tip. In Figure 6f, a large number of dislocation slip bands can be seen. Compared with 250 MPa stress, the distance is similar to 250 MPa stress. From Figure 6e, we can also find Type I slip, however the amount of Type I slip is reduced. In addition to Type I slip, there is also Type II slip. The slip pattern is affected by the material properties, stress and frequency during fatigue. The slip form is influenced by material properties, fatigue stress and frequency. Different types of slip are only affected by the stress amplitude. Under higher stress amplitude, the number of type I slip decreases and the number of type II slip increases with the decrease in stress amplitude. At very low stress amplitudes, type III slip dominates, while type I slip rarely occurs. When the early type I slip occurs on the surface of the sample, more parallel slip bands are formed inside the grain with the increase in the number of cycles. For type II slips, most are formed at the twin boundary. For type III slip, the amount of slippage in the early stage is very small, and during the cyclic loading process, the slip also changes to type I and type II. The existence of slip is caused by the plastic deformation of the sample during the fatigue test. In the loading process of the sample subjected to cyclic stress, a lot of dislocation movements occur inside the sample, a large number of dislocation movements produce plastic deformation, and the dislocation movement is along a certain direction and that of the slip plane.

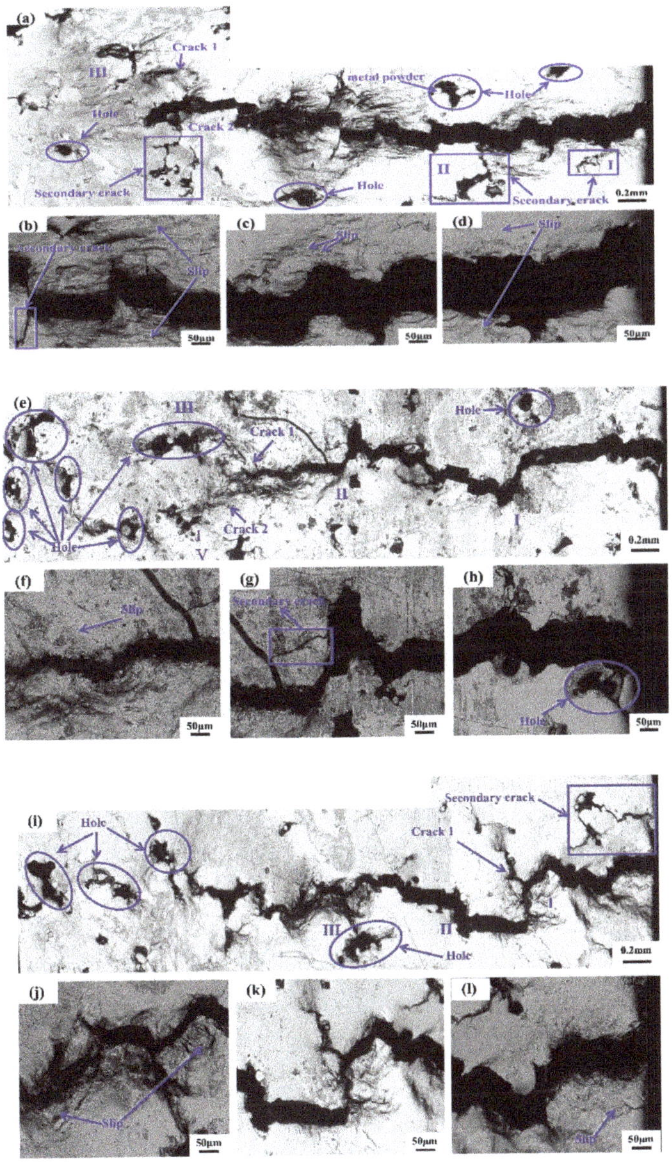


**Figure 5.** Surface crack growth of SLM-316L under different stresses: (**a**–**d**) 250 MPa, (**e**–**h**) 270 MPa, (**i**–**l**) 290 MPa.

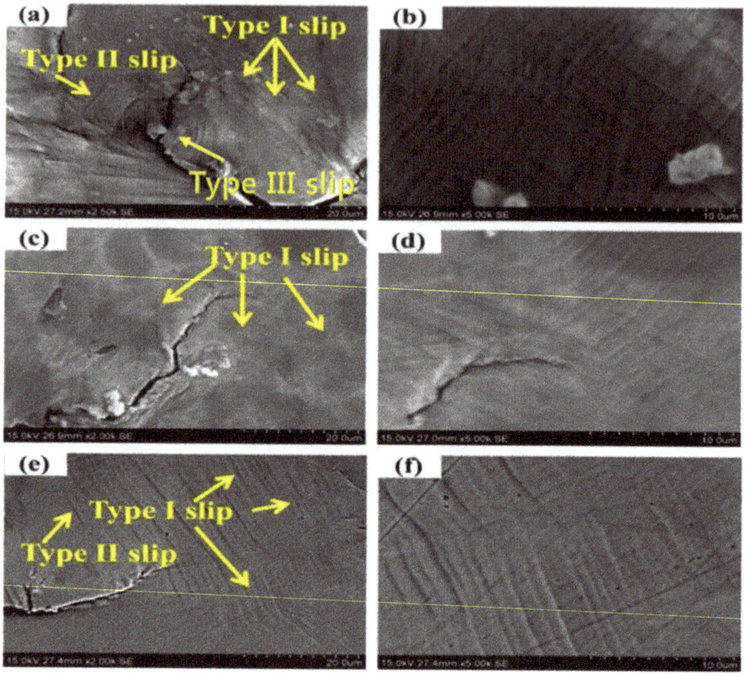


**Figure 6.** Surface fatigue damage morphology of SLM-316L under different stresses: (**a**,**b**) 250 MPa, (**c**,**d**) 270 MPa, (**e**,**f**) 290 MPa.

## 4. Discuss

### *4.1. Fatigue Crack Initiation Mechanism*

We can see from Figure 5 that the crack originates from the sample surface, and defects such as holes are found at the crack source of the sample. Figure 6 shows that a large number of slip bands exist at the crack tip, and the fracture mechanism of the sample can be reflected through the analysis of the pores and slip bands. In this paper, the corresponding model diagram was established to analyze the crack formation of the sample, as shown in Figure 7. Prior to the fatigue test, the surface of the sample was smooth, without obvious unevenness. When the fatigue loading started, a lot of slip bands shown in Figure 6 indicated that some areas on the sample surface would slip during cyclic loading, resulting in the resident slip band, which is a dislocation wall formed by dislocation. With the continuous fatigue loading, a lot of non-uniform regions formed by intrusive extrusion are generated on the surface of the sample. As shown in Figure 7a, with the continuous cyclic loading, slip occurs inside the sample. In the process of actual fatigue loading and reverse loading, material properties change, and this results in the slip that does occur not being completely reversible. Intrusive extrusion steps are generated at some positions on the surface of the SLM-316L sample, which makes the sample surface no longer smooth and forms an uneven area. With further fatigue loading, as shown in Figure 7b, this uneven plastic deformation region is prone to causing stress concentration during cyclic loading, which promotes the initiation of microcracks. During the SLM process, the molten pool exists for a short time and the fluidity of the molten metal decreases. Warpage deformation, tensile residual stress, porosity, unmelted powder, and unfused (LOF) defects are inevitable [28]. The formed 316L sample may have defects such as holes inside or on the surface. According to Griffith's theory, the stress value in the absence of holes is one third of that in the presence of circular holes. Under a certain stress, with the continuous fatigue loading, microcracks will be easily generated around the holes and other defects due to stress concentration, and the existence of stress concentration will also affect the propagation of cracks. In the actual fatigue loading process, as shown in Figure 5, the defects on the surface and inside of the specimen are mostly irregular shapes, which make the stress concentration coefficient at these defects larger, and microcracks tend to initiate at these locations. As shown in Figure 7c, the microcracks initiate in multiple defects and continue to expand under the action of fatigue load. The expanded microcracks will connect with each other to form the main crack, and ultimately make the specimen fracture in advance.

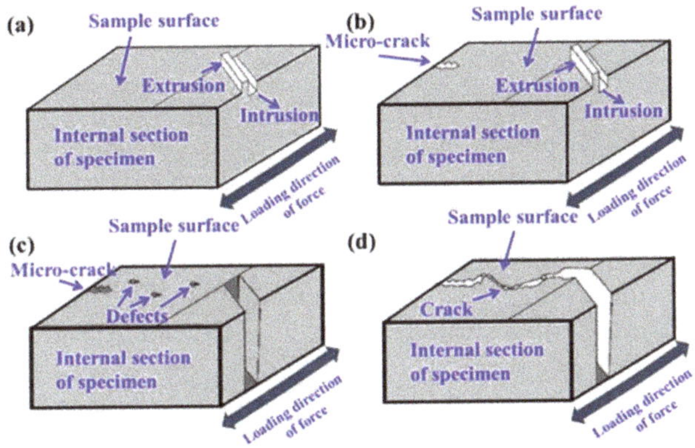


**Figure 7.** Schematic diagram of fatigue crack formation mechanism; (**a**) Schematic diagram of slip generation, (**b**) Schematic diagram of microcrack generation, (**c**) Schematic diagram of crack growth, (**d**) Schematic diagram of crack propagation path.

### *4.2. Characteristics of Fatigue Crack Growth*

In general, under certain stress, the main crack generated by fatigue loading extends along the direction perpendicular to the load. However, due to the influence of impurities, holes and other defects, the crack can propagate in different directions. The research of Beachem and Yoder studied crack propagation [29]. After the slip system at a certain position of the specimen is activated to produce microcracks, the tip of the microcrack will continue to undergo passivation, making its propagation path appear Z-shaped. In this paper, the crack propagation characteristics of the sample with defects are analyzed through the model. As shown in Figure 8a, when the stress is low, there is little possibility of microcracks occurring at the defect near the initiation of the main crack. When the stress is high, the defects near the main crack initiation zone will also produce microcracks, and the microcracks will expand a certain distance under the action of tensile stress, and some cracks may bridge with the main cracks. During the forward propagation of the main crack, a certain angle deflection always occurs to the area where the defect exists, as shown in Figure 8c,d. Many bridging areas are formed at the tip of the main crack and at the defect near the tip, which is more conducive to the accelerated propagation of fatigue crack.

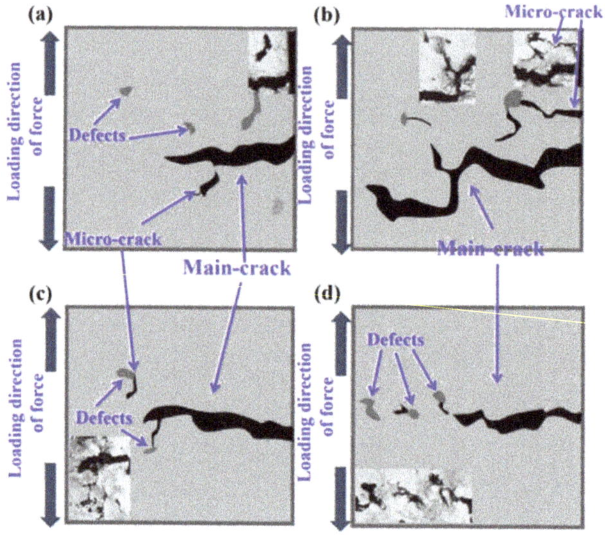


**Figure 8.** Sketch of fatigue crack propagation characteristics in the presence of defects; (**a**,**b**) Schematic diagram of main defect expansion, (**c**,**d**) Schematic diagram of the effect of defects on crack propagation.

## 5. Conclusions

(1) The surface morphologies of SLM-316L stainless steel samples with different forming directions are different. In the direction perpendicular to the laser additive manufacturing, the melting tracks of the sample are well bonded; In the sample parallel to the additive manufacturing direction, a lot of fish scale melting tracks can be found, which are superimposed together. The surface of these tracks adheres to a lot of unmelted metal powder, and there are small holes in the overlap between the tracks.

(2) Fatigue tests of SLM-316L and rolled 316L were carried out $1 \times 10^6$ times under different stresses. Due to the existence of defects, the fatigue performance of SLM-316L specimen is inferior to that of rolled 316L specimen. The fatigue fracture surface of SLM-316L sample was smooth, and no obvious plastic deformation traces were found. The fracture mode was brittle fracture. However, there are radiation patterns at the fatigue source of rolled 316L, and clear dimples can be observed on the fracture surface. The fracture mode is ductile fracture.

(3) In the process of cyclic loading, some areas on the surface of SLM-316L sample slipped to produce resident slip bands, and a lot of non-uniform areas formed by intrusive extrusion were generated on the surface of the sample. The existence of hole defects promoted the initiation of microcracks, and notably, they also accelerated the propagation of cracks.

(4) After the fatigue microcracks of SLM-316L sample were generated, the crack tip expanded along the front defect due to the influence of external load, which was generally in a Z-shaped way. The inevitable defects during the SLM process affected the propagation of fatigue within the main crack in the form of bending, bifurcation and bridge, which always deviates from the maximum load direction in the process of main crack propagation.

**Author Contributions:** Conceptualization, Z.W. and S.Y.; methodology, Z.W., Z.G.; validation, Z.W., Y.H., C.F.; formal analysis, Z.W., Y.H.; investigation, Z.W., Y.H., S.Y.; and C.F.; resources, Z.W. and Z.P.; data cura-tion, Z.W., Y.H.; writing—original draft preparation, Z.W., Y.H.; writing—review and editing, Z.W., C.F.; visual-ization, Z.W.; supervision, Z.W.; project administration, Z.W.; funding acquisition, Z.W. and S.Y. All authors have read and agreed to the published version of the manuscript.

**Funding:** This research was supported by the National Natural Science Foundation of China (51971129), Natural Science Foundation of Shanghai (19ZR1421200).

**Institutional Review Board Statement:** Not applicable.

**Informed Consent Statement:** Not applicable.

**Data Availability Statement:** The data used to support the findings of this study are included within the article.

**Acknowledgments:** Thank Shanglei Yang, Yubao Hang, Cong Fan, Zeng Peng and Zihao Gao for their help.

**Conflicts of Interest:** The authors declare that they have no conflict of interest.

## References


1. Song, Y.; Sun, Q.; Guo, K.; Wang, X.; Sun, J. Effect of scanning strategies on the microstructure and mechanical behavior of 316L stainless steel fabricated by selective laser melting. *Mater. Sci. Eng. A* **2020**, *793*, 139879. [CrossRef]
2. Zhao, X.; Wei, Q.S.; Gao, N.; Zheng, E.L.; Shi, Y.S.; Yang, S.F. Rapid fabrication of TiN/AISI 420 stainless steel composite by selective laser melting additive manufacturing. *J. Mater. Process. Technol.* **2019**, *270*, 8–19. [CrossRef]
3. Jeon, J.M.; Park, J.M.; Yu, J.H.; Kim, J.G.; Seong, Y.; Park, S.H.; Kim, H.S. Effects of microstructure and internal defects on mechanical anisotropy and asymmetry of selective laser-melted 316L austenitic stainless steel. *Mater. Sci. Eng. A* **2019**, *763*, 138152. [CrossRef]
4. Wang, Y.M.; Voisin, T.; McKeown, J.T.; Ye, J.; Calta, N.P.; Li, Z.; Zeng, Z.; Zhang, Y.; Chen, W.; Roehling, T.T.; et al. Additively manufactured hierarchical stainless steels with high strength and ductility. *Nat. Mater.* **2018**, *17*, 63–71. [CrossRef]
5. Suryawanshi, J.; Prashanth, K.G.; Ramamurty, U. Mechanical behavior of selective laser melted 316L stainless steel. *Mater. Sci. Eng. A* **2017**, *696*, 113–121. [CrossRef]
6. Huang, Y.B.; Yang, S.L.; Gu, J.X.; Xiong, Q.; Duan, C.; Meng, X.; Fang, Y. Microstructure and wear properties of selective laser melting 316L. *Mater. Chem. Phys.* **2020**, *254*, 123487. [CrossRef]
7. Casati, R.; Lemke, J.; Vedani, M. Microstructure and Fracture Behavior of 316L Austenitic Stainless Steel Produced by Selective Laser Melting. *J. Mater. Sci. Technol.* **2016**, *32*, 738–744. [CrossRef]
8. Shang, Y.T.; Yuan, Y.P.; Li, D.F.; Li, Y.; Chen, J. Effects of scanning speed on in vitro biocompatibility of 316L stainless steel parts elaborated by selective laser melting. *Int. J. Adv. Manuf. Technol.* **2017**, *92*, 4379–4385. [CrossRef]
9. Antony, K.; Arivazhagan, N.; Senthilkumaran, K. Numerical and experimental investigations on laser melting of stainless steel 316L metal powders. *J. Manuf. Process.* **2014**, *16*, 345–355. [CrossRef]
10. Santos, E.C.; Shiomi, M.; Osakada, K.; Laoui, T. Rapid manufacturing of metal components by laser forming. *Int. J. Mach. Tools Manuf.* **2006**, *46*, 1459–1468. [CrossRef]
11. Xia, M.J.; Gu, D.D.; Yu, G.Q.; Dai, D.; Chen, H.; Shi, Q. Porosity evolution and its thermodynamic mechanism of randomly packed powder-bed during selective laser melting of Inconel 718 alloy. *Int. J. Mach. Tools Manuf.* **2017**, *116*, 96–106. [CrossRef]
12. Zhang, H.Z.; Xu, M.T.; Liu, Z.D.; Li, C.; Kumar, P.; Liu, Z.; Zhang, Y. Microstructure, surface quality, residual stress, fatigue behavior and damage mechanisms of selective laser melted 304L stainless steel considering building direction. *Addit. Manuf.* **2021**, *46*, 102147. [CrossRef]
13. Qiu, C.L.; Panwisawas, C.; Ward, M.; Basoalto, H.C.; Brooks, J.W.; Attallah, M.M. On the role of melt flow into the surface structure and porosity development during selective laser melting. *Acta Mater.* **2015**, *96*, 72–79. [CrossRef]
14. Kamath, C.; El-Dasher, B.; Gallegos, G.F.; King, W.E.; Sisto, A. Density of additively-manufactured, 316L SS parts using laser powder-bed fusion at powers up to 400W. *Int. J. Adv. Manuf. Technol.* **2014**, *74*, 65–78. [CrossRef]
15. Lima, M.; Sankare, S. Microstructure and mechanical behavior of laser additive manufactured AISI 316 stainless steel stringers. *Mater. Des.* **2014**, *55*, 526–532. [CrossRef]
16. Hanzl, P.; Zetek, M.; Baksa, T.; Kroupa, T. The Influence of Processing Parameters on the Mechanical Properties of SLM Parts. *Procedia Eng.* **2015**, *100*, 1405–1413. [CrossRef]
17. Uhlmann, E.; Fleck, C.; Gerlitzky, G.; Faltin, F. Dynamical Fatigue Behavior of Additive Manufactured Products for a Fundamental Life Cycle Approach. *Procedia CIRP* **2017**, *61*, 588–593. [CrossRef]
18. Zhang, C.; Cao, D.; You, W.; Wei, D.; Wu, C.; Yang, G. Fatigue failure of welded details in steel bridge pylons. *Eng. Fail. Anal.* **2021**, *127*, 105530. [CrossRef]
19. Sajith, S.; Shukla, S.S.; Murthy, K.; Robi, P.S. Mixed mode fatigue crack growth studies in AISI 316 stainless steel-ScienceDirect. *Eur. J. Mech. A/Solids* **2020**, *80*, 103898. [CrossRef]
20. He, L.; Akebono, H.; Sugeta, A. Effect of high-amplitude loading on accumulated fatigue damage under variable-amplitude loading in 316 stainless steel. *Int. J. Fatigue* **2018**, *116*, 388–395. [CrossRef]
21. Riemer, A.; Leuders, S.; Thöne, M.; Richard, A.; Tröster, T.; Niendorf, T. On the fatigue crack growth behavior in 316L stainless steel manufactured by selective laser melting. *Eng. Fract. Mech.* **2014**, *120*, 15–25. [CrossRef]

22. Sarkar, S.; Kumar, C.S.; Nath, A.K. Effect of mean stresses on mode of failures and fatigue life of selective laser melted stainless steel. *Mater. Sci. Eng. A* **2017**, *700*, 92–106. [CrossRef]
23. Pegues, J.W.; Roach, M.D.; Shamsaei, N. Additive Manufacturing of Fatigue Resistant Austenitic Stainless Steels by Establishing the Process-Structure-Property Relationships. *Mater. Res. Lett.* **2020**, *8*, 8–15. [CrossRef]
24. Yadroitsev, I.; Krakhmalev, P.; Yadroitsava, I.; Johansson, S.; Smurov, I. Energy input effect on morphology and microstructure of selective laser melting single track from metallic powder. *J. Mater. Process. Technol.* **2013**, *213*, 606–613. [CrossRef]
25. Das, S. Physical Aspects of Process Control in Selective Laser Sintering of Metals. *Adv. Eng. Mater.* **2003**, *5*, 701–711. [CrossRef]
26. Zhou, X.; Wang, D.; Liu, X.; Zhang, D.D.; Qu, S.; Ma, J.; London, G.; Shen, Z.; Liu, W. 3D-imaging of selective laser melting defects in a Co-Cr-Mo alloy by synchrotron radiation micro-CT. *Acta Mater.* **2015**, *98*, 1–16. [CrossRef]
27. Phung, N.L.; Favier, V.; Ranc, N.; Vales, F.; Mughrabi, H. Very high cycle fatigue of copper: Evolution, morphology and locations of surface slip markings. *Int. J. Fatigue* **2014**, *63*, 68–77. [CrossRef]
28. Meng, L.X.; Ben, D.D.; Yang, H.J.; Ji, H.B.; Lian, D.L.; Zhu, Y.K.; Chen, J.; Yi, J.L.; Wang, L.; Yang, J.B.; et al. Effects of embedded spherical pore on the tensile properties of a selective laser melted Ti6Al4V alloy. *Mater. Sci. Eng. A* **2021**, *815*, 141254. [CrossRef]
29. Beachem, C.D.; Yoder, G.R. Elastic-plastic fracture by homogeneous microvoid coalescence tearing along alternating shear planes. *Metall. Trans.* **1973**, *4*, 1145–1153. [CrossRef]

MDPI
St. Alban-Anlage 66
4052 Basel
Switzerland
Tel. +41 61 683 77 34
Fax +41 61 302 89 18
www.mdpi.com

*Materials* Editorial Office
E-mail: materials@mdpi.com
www.mdpi.com/journal/materials

www.ingramcontent.com/pod-product-compliance
Lightning Source LLC
Chambersburg PA
CBHW061614220926
51850CB00005B/505